ABOUT THE AUTHORS

James W. Dally obtained a bachelor of science degree and a master of science degree, both in mechanical engineering, from the Carnegie Institute of Technology. He obtained a doctoral degree in mechanics from the Illinois Institute of Technology. He has taught at Cornell University, the Illinois Institute of Technology, and the University of Rhode Island. He is currently a professor in the Department of Mechanical Engineering at the University of Maryland.

Professor Dally has also worked at the Mesta Machine Company, Armour Research Foundation, and IBM, Federal Systems Division. He is a fellow of the American Society of Mechanical Engineers, Society for Experimental Mechanics, and American Academy of Mechanics. He was elected to the National Academy of Engineering in 1984 and was appointed as an honorary member of the Society for Experimental Mechanics in 1983.

Professor Dally has coauthored five books: *Experimental Stress Analysis*, *Photoelastic Coatings*, *Instrumentation for Engineering Measurement*, *Static and Dynamic Photoelasticity and Caustics*, and *Packaging of Electronic Systems*. He has written over 100 scientific papers and holds five patents.

William F. Riley holds the rank of Distinguished Professor Emeritus in the College of Engineering at Iowa State University. He received a bachelor of science degree in mechanical engineering from the Carnegie Institute of Technology in 1951 and a master of science degree in mechanics from the Illinois Institute of Technology in 1985.

He worked from 1951 to 1954 as a mechanical engineer for the Mesta Machine Company in the design and development of steel-mill equipment and heavy machine tools. From 1954 to 1966 he engaged in experimental stress analysis activities at IIT Research Institute as a research engineer, manager of the experimental stress analysis group, and finally science adviser to the mechanics research division. During the summers of 1966 and 1970 he served as a consultant for USAID/NSF in India in their summer institute program for college teachers. Since 1966 he has been an engineering educator at Iowa State University.

Professor Riley was elected a fellow of the Society for Experimental Mechanics in 1977 and an honorary member in 1984. In 1977 he received the M. M. Frocht Award of the Society for Experimental Mechanics for excellence in engineering education, and in 1990 was the recipient of the Mechanics Division Distinguished Educator Award of the American Society for Engineering Education.

Professor Riley has published approximately 50 technical papers dealing with stress analysis topics in national and international journals and has co-authored six books: *Experimental Stress Analysis, Introduction to Photomechanics, Essentials of Mechanics, Mechanics of Materials, Instrumentation for Engineering Measurements*, and *Introduction to Mechanics of Materials.*

**Books are to be returned on or before
the last date below.**

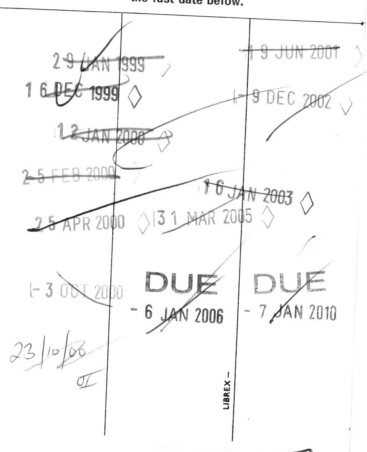

2 9 JAN 1999

1 6 DEC 1999

1 2 JAN 2000

2 5 FEB 2000

2 5 APR 2000

- 3 OCT 2000

23/10/00

1 9 JUN 2001

1 9 DEC 2002

1 6 JAN 2003

3 1 MAR 2005

DUE
- 6 JAN 2006

DUE
- 7 JAN 2010

LIBREX —

EXPERIMENTAL STRESS ANALYSIS

Also Available from McGraw-Hill

Schaum's Outline Series in Mechanical Engineering

Most outlines include basic theory, definitions, and hundreds of solved problems and supplementary problems with answers.

Titles on the Current List Include:

Acoustics
Basic Equations of Engineering
Continuum Mechanics
Engineering Economics
Engineering Mechanics, 4th edition
Fluid Dynamics
Fluid Mechanics & Hydraulics, 2d edition
Heat Transfer
Introduction to Engineering Calculations

Lagrangian Dynamics
Machine Design
Mathematical Handbook of Formulas
  & Tables
Mechanical Vibrations
Operations Research
Strength of Materials, 2d edition
Theoretical Mechanics
Thermodynamics, 2d edition

Schaum's Solved Problems Books

Each title in this series is a complete and expert source of solved problems containing thousands of problems with worked out solutions.

Related Titles on the Current List Include:

3000 Solved Problems in Calculus
2500 Solved Problems in Differential Equations
2500 Solved Problems in Fluid Mechanics and Hydraulics
1000 Solved Problems in Heat Transfer
3000 Solved Problems in Linear Algebra
2000 Solved Problems in Mechanical Engineering Thermodynamics
2000 Solved Problems in Numerical Analysis
700 Solved Problems in Vector Mechanics for Engineers: Dynamics
800 Solved Problems in Vector Mechanics for Engineers: Statics

Available at your College Bookstore. A complete list of Schaum titles may be obtained by writing to: Schaum Division
             McGraw-Hill, Inc.
             Princeton Road, S-1
             Hightstown, NJ 08520

EXPERIMENTAL STRESS ANALYSIS

Third Edition

James W. Dally

Professor of Mechanical Engineering
University of Maryland, College Park

William F. Riley

Distinguished Professor Emeritus of Engineering Science and Mechanics
College of Engineering
Iowa State University

McGraw-Hill, Inc.

New York   St. Louis   San Francisco   Auckland   Bogotá   Caracas
Hamburg   Lisbon   London   Madrid   Mexico   Milan   Montreal
New Delhi   Paris   San Juan   São Paulo   Singapore   Sydney   Tokyo   Toronto

This book was set in Times Roman by TechSet.
The editors were Lyn Beamesderfer, John Corrigan, and John M. Morriss.
The cover was designed by Rafael Hernandez.
Project supervision was done by The Total Book.
R. R. Donnelley & Sons Company was printer and binder.

EXPERIMENTAL STRESS ANALYSIS

1 2 3 4 5 6 7 8 9 0 DOC DOC 9 0 9 8 7 6 5 4 3 2 1

ISBN 0-07-015218-7

Library of Congress Cataloging-in-Publication Data

Dally, James W.
    Experimental stress analysis/James W. Dally, William F. Riley.
    3rd ed.
       p.   cm.
    Includes index.
    ISBN 0-07-015218-7
    1. Strains and stresses.   2. Photoelasticity.   3. Strain gages.
I. Riley, William F. (William Franklin), (date).   II. Title.
TA407.D32   1991
620.1'123—dc20

CONTENTS

Part III   Optical Methods of Stress Analysis

Part IV   Coating Methods

PREFACE

In this revised third edition we have retained the essential features of the previous two editions of *Experimental Stress Analysis*, which is a text intended for upper-division undergraduate students or graduate students beginning to study experimental methods of mechanics. We have made extensive revisions which reflect the changes in experimental mechanics that have occurred during the past 12 years and have added a significant amount of new material on fracture mechanics, optical methods, and statistical evaluation of experimental data. The book's material has been organized into five parts:

Part I, "Elementary Elasticity and Fracture Mechanics," contains three chapters on elasticity and a new introductory chapter on elementary fracture mechanics.

Part II, "Strain-Measurement Methods and Related Instrumentation," contains five chapters, beginning with an introductory chapter describing the general aspects of strain measurement. One chapter covers electrical-resistance strain gages, and another covers strain-gage-signal conditioning circuits. The chapter on recording methods has been changed significantly to reflect the many new developments in recording instruments and to introduce digital conversion and digital data storage and processing. The chapter on analysis methods has been expanded to illustrate techniques for determining stress intensity factors, crack initiation toughness, and residual stresses.

Part III, "Optical Methods of Stress Analysis," contains five chapters, beginning with a chapter on basic optics which includes a description of the method of caustics. The treatment of moiré has been expanded to a complete chapter, which covers both the classical moiré methods and the more modern procedures of moiré interferometry. The treatment of photoelasticity has been reduced to two chapters to provide space for the more modern experimental methods included in this revision. A new chapter describing optical methods for determining fracture parameters has been added to extend the coverage to include both stress and fracture analysis.

Part IV, "Coating Methods," contains a single chapter which combines the experimental approaches of birefringent coatings and brittle coatings. The development of the stresses in both of these coatings is identical although the response of each coating to these stresses is different. Application of a coating to a prototype is another common feature that is emphasized in this chapter.

Part V, "Application of Statistics," contains a single chapter added in response to the suggestions of several reviewers. The coverage is specifically focused to show the application of statistics in enhancing the accuracy of experiments and in improving the method of reporting experimental results which show variation.

Each part of the book is essentially independent so that the instructor can be quite flexible in selection of the course content. For instance, a two- or three-credit course on strain gages can be offered by using two chapters of Part I and all of Part II. Parts I and III can be combined to provide a thorough three- or four-credit course on photoelasticity and moiré. Selected chapters from the first four parts can be organized to introduce the broader field of experimental stress analysis. A complete, detailed treatment of the subject matter covered in the text and supplemented with laboratory expercises on brittle coatings, strain gages, photoelasticity, and moiré will require six to eight credit hours.

The essential feature of the text is its completeness in introducing the entire range of experimental methods to the student. A reasonably deep coverage is presented of the theory required to understand experimental stress analysis and of the four primary methods employed: brittle coatings, strain gages, photo-elasticity, and moiré. While the primary emphasis is placed on the theory of experimental stress analysis, the important experimental techniques associated with each of the four major methods are covered in sufficient detail to permit the student to begin laboratory work with a firm understanding of experimental procedures. Exercises designed to support and extend the treatment and to show the application of the theory have been placed at the end of each chapter.

Laboratory exercises have not been included, since the laboratory work will depend strongly on local conditions such as the equipment and supplies available, the interests of the instructor and students, and the local industrial problems of current interest. It is believed that the instructor is best qualified to specify the associated laboratory exercises on the basis of interest, equipment, supplies, and time available for this important supplement to the course.

A significant amount of new material has been added to this third edition of the book; however, space limitations and publishing costs did not permit coverage of several important topics such as holography, failure theories, and nondestructive testing. It is anticipated that the instructor will, in certain instances, treat these topics by using his or her own lecture notes or by using recent papers published in the technical journals. The authors hope that most instructors will find most of the fundamental material required to present a complete and practical course on the theory of experimental stress analysis in the text.

The material presented here has been assembled by both authors over a period of 35 years. Courses have been developed on experimental stress analysis, photoelasticity, and photomechanics at the Illinois Institute of Technology, Cornell University, Iowa State University, and the University of Maryland. The material has been shown to be interesting and comprehensible by the students participating in these courses. The mathematics employed in the treatment can easily be understood by senior undergraduates. Cartesian notation and/or vector notation has been used to enhance understanding of the field equations. A great deal of effort was devoted to the selection and preparation of the illustrations employed. These illustrations complement the text and should aid appreciably in presenting the material to the student.

The contributions of many investigators working in experimental mechanics should be acknowledged. This edition represents a summary of many of the more mature methods which have been developed by the combined efforts of many excellent researchers. In particular, we wish to thank Drs. A. J. Durelli and G. R. Irwin, our mentors in experimental stress analysis and fracture mechanics for many years. Also, J. W. Dally would like to acknowledge the many valuable conversations he has had with Drs. R. J. Sanford and J. Sirkis, colleagues at the University of Maryland, which provided feedback on the new material included in this revision. Next, we acknowledge the excellent illustrations provided by Fu-pen Chiang, R. Chona, A. J. Durelli, J. F. Kalthoff, W. P. T. North, D. Post, R. J. Sanford, W. N. Sharpe, Jr., C. W. Smith, C. E. Taylor, and the many suppliers of commercial equipment. Thanks are also due to Robert Davis, U.S. Military Academy, West Point; Gaza Kardos, Carleton University; Barney Klamecki, University of Minnesota; R. E. Rowlands, University of Wisconsin, Madison; Harvey Sharfstein, San Jose State University; C. W. Smith, Virginia Polytechnic University; and A. A. Sukere, University of Missouri, Columbia for their careful review of the manuscript.

James W. Dally
William F. Riley

LIST OF SYMBOLS

a amplitude of a light wave

A area

b aperture width

B strength of a magnetic field

B_L lateral boundary

c relative stress optical coefficient

c velocity of light in a vacuum

c_1, c_2 stress optical coefficients

C capacitance

C count

C galvanometer constant

C_c coating coefficient of sensitivity

C_s specimen coefficient of sensitivity

C_v coefficient of variation

C_v Poisson's ratio mismatch correction factor

d degree of damping

d_x mean deviation

D damping coefficient

D diameter

D volume dilatation $(\varepsilon_1 + \varepsilon_2 + \varepsilon_3)$

D_0 fluid damping constant

D_0 fog density of a photographic film

e electron charge

E electromotive force or voltage

E exposure

E magnitude of a light vector

E modulus of elasticity

E	potential gradient
E_b	bias voltage
E^c	modulus of elasticity of a coating
E_m	back electromotive force
E_0	exposure inertia of a photographic film
E_R	reference voltage
E^s	modulus of elasticity of a specimen
E^*	modulus of elasticity of a calibrating beam
f	focal length of a lens
f	frequency
f_a	aliasing frequency
f_{bw}	bandwidth
f_s	sampling rate
f_ε	material strain-fringe value
f_σ	material stress-fringe value
F	force
F_{CB}	bending correction factor
F_{CR}	reinforcing correction factor
$\mathbf{F}_n$	resultant force
F_x, F_y, F_z	cartesian components of the body-force intensity
F_r, F_θ, F_z	polar components of the body-force intensity
g	gravitational constant
$\mathscr{G}$	energy release rate or crack extension force
G	shear modulus of elasticity
G	torsional spring constant
G_D	gray level difference
h	thickness
i	current density
I	current
I	intensity of light
I	moment of inertia
I_e	intensity of emerging light
I_g	gage current
I_G	galvanometer current
I_i	intensity of incident light
I_r	intensity of reflected light
I_1, I_2, I_3	first, second, and third invariants of stress
J	polar moment of inertia
J_1, J_2, J_3	first, second, and third invariants of strain
k	dielectric constant
K	bulk modulus
K	coefficient of elasticity

K	optical strain coefficient
K	strength of a light source
K_I	stress intensity factor (opening mode)
K_Ic	crack initiation toughness
K_II	stress intensity factor (shear mode)
K_III	stress intensity factor (tearing mode)
K_t	transverse-sensitivity factor
K_T	compressibility constant
l	length
l_g	gage length
l_0	gage length
L	length
$\mathscr{L}$	loss factor
M	magnification factor
n	index of refraction
n	integer
n_0	index of refraction in an unstressed medium
n_1, n_2, n_3	index of refraction along the principal directions
n_1, n_2, n_3	principal directions
N	cycles of relative retardation
N	fringe order
N	number of calibration values
N	number of charge carriers
N	number of cycles
p	pitch of a moiré grating
p	pressure
P	force, applied load
P	power
P_D	power density
P_g	power dissipated by a gage
P_s	load shedding due to yielding
q	resistance ratio
Q	figure of merit
r	resistance ratio
r_p^*	apparent distance
$\mathscr{R}$	crack growth resistance
R	radius
R	range
R	reflection coefficient
R	resistance
R	resolution
R_b	ballast resistor

$\mathscr{R}_c$ capacitive reactance

R_e equivalent resistance

R_g gage resistance

R_G galvanometer resistance

R_M measuring-circuit resistance

R_p parallel resistor

R_s series resistor

R_x external resistance

s distance

s_1, s_2 curvilinear coordinates along an isostatic

S sensitivity index

S_a axial strain sensitivity of a gage

S_A strain sensitivity of a material

S_c circuit sensitivity

S_{CG} galvanometer-circuit sensitivity

S_g strain sensitivity of a gage, gage factor

S_{sc} strain sensitivity of a semiconductor material

S_t gage sensitivity to time

S_t transverse strain sensitivity of a gage

S_T gage sensitivity to temperature

S_x standard deviation

$S_{\bar{x}}$ standard error

S_x^2 variance

S_ε strain sensitivity

S_θ galvanometer sensitivity

S_σ stress sensitivity

t time

t_r risetime

T period

T temperature

T torque

T transmission coefficient

$\mathbf{T}_n$ resultant stress

T_{nx}, T_{ny}, T_{nz} cartesian components of the resultant stress

T_x, T_y, T_z cartesian components of the surface tractions

u, v, w cartesian components of displacement

u_r, u_θ, u_z polar components of displacement

U strain energy

V voltage

V volume

V_B bridge excitation voltage

w width

w_0	gage width
W	strain energy density
x, y, z	cartesian coordinates
$\bar{x}$	sample mean
Z	Westergaard complex stress function
α	angle of incidence
$\alpha, \beta, \theta, \phi$	angles
α	coefficient of thermal expansion
α_p	polarizing angle
β	angle of reflection
β	coefficient of thermal expansion
γ	angle of refraction
γ	shear-strain component
γ	surface energy
γ	temperature coefficient of resistivity
$\gamma_{r\theta}$	shear-strain component in polar coordinates

$$\left.\begin{array}{l} \gamma_{xy} = \gamma_{yx} \\ \gamma_{yz} = \gamma_{zy} \\ \gamma_{zx} = \gamma_{xz} \end{array}\right\} \text{cartesian shear-strain components}$$

δ	displacement
δ	linear phase difference
Δ	relative phase difference
Δ	relative retardation
ΔR_T	resistance change due to temperature
ΔR_ϵ	resistance change due to strain
ϵ	normal strain
ϵ_a	axial strain
ϵ_c	calibration strain
ϵ_n	normal-strain component
$\epsilon_r, \epsilon_\theta, \epsilon_z$	normal-strain components in polar coordinates
ϵ_t	transverse strain
ϵ_{t*}	threshold strain for a brittle coating under a uniaxial state of stress
$\epsilon_{xx}, \epsilon_{yy}, \epsilon_{zz}$	normal-strain components in cartesian coordinates
$\epsilon_1, \epsilon_2, \epsilon_3$	principal normal strains
$\epsilon_1^c, \epsilon_2^c$	principal normal strains in a coating
$\epsilon_1^s, \epsilon_2^s$	principal normal strains in a specimen
η	nonlinear term
θ	angular deflection
θ_s	steady-state deflection of a galvanometer
λ	Lame's constant
λ	wavelength
μ	mobility of charge carriers

μ shear modulus

μ true arithmetic mean of a population

v Poisson's ratio

v^c Poisson's ratio of a coating

v^s Poisson's ratio of a specimen

v^* Poisson's ratio of a calibrating beam

ζ wave number

ζ, η elliptic coordinates

π piezoresistive proportionality constant

ρ radius of curvature

ρ resistivity coefficient

ρ specific resistance

σ normal-stress component

σ true standard deviation of a population

σ_n normal component of the resultant stress

$\sigma_{rr}, \sigma_{\theta\theta}, \sigma_{zz}$ normal-stress components in polar coordinates

σ_{uc}^c ultimate compressive strength of a brittle coating

σ_{ut}^c ultimate tensile strength of a brittle coating

$\sigma_{xx}, \sigma_{yy}, \sigma_{zz}$ normal-stress components in cartesian coordinates

σ_{ys} yield strength

σ_0 applied stress

$\sigma_1, \sigma_2, \sigma_3$ principal normal stresses

σ_1^c, σ_2^c principal normal stresses in a coating

σ_1^s, σ_2^s principal normal stresses in a specimen

σ_1', σ_2' secondary principal stresses

σ_η normal-stress component in elliptic coordinates

σ^* normal stress in a calibrating beam

τ shear-stress component

τ_n shear-stress component of the resultant stress

$\tau_{r\theta}$ shear-stress component in polar coordinates

$$\left.\begin{array}{l} \tau_{xy} = \tau_{yx} \\ \tau_{yz} = \tau_{zy} \\ \tau_{zx} = \tau_{xz} \end{array}\right\}$$ shear-stress components in cartesian coordinates

ϕ Airy's stress function

ω angular frequency

Ω body-force function

EXPERIMENTAL
STRESS
ANALYSIS

PART
I

ELEMENTARY ELASTICITY AND FRACTURE MECHANICS

CHAPTER
1

STRESS

1.1 INTRODUCTION

An experimental stress analyst must have a thorough understanding of stress, strain, and the laws relating stress to strain. For this reason, the first three chapters of this text have been devoted to the elementary concepts of the theory of elasticity. The first chapter deals with stresses produced in a body due to external and body-force loadings. The second chapter deals with deformations and strains produced by the loadings and with relations between the stresses and strains. The third chapter covers plane problems in the theory of elasticity, important since a large part of a first course in experimental stress analysis deals with two-dimensional problems. Also treated is the stress-function approach to the solution of plane problems. Upon completing the subject matter of the first three chapters of the text, the student should have a firm understanding of stress and strain and should be able to solve some of the more elementary two-dimensional problems in the theory of elasticity by using the Airy's-stress-function approach.

Failures under loading conditions well below the yield stress often occur in structures with small crack or cracklike flaws. Such failures show that conventional strength studies, no matter how accurately conducted, are not always sufficient to guarantee structural integrity under operational conditions. The field of study which considers crack-extension behavior as a function of applied loads is known as *fracture mechanics*. As a result of research during the past decade, fracture mechanics is now used to solve many practical engineering problems in failure analysis, material selection, and structural-life prediction. For this reason, Chap. 4 has been introduced to provide the student with a brief treatment of some of the fundamental concepts of fracture mechanics.

1.2 DEFINITIONS

Two basic types of force act on a body to produce stresses. Forces of the first type are called *surface forces* for the simple reason that they act on the surfaces of the body. Surface forces are generally exerted when one body comes in contact with another. Forces of the second type are called *body forces* since they act on each element of the body. Body forces are commonly produced by centrifugal, gravitational, or other force fields. The most common body forces are gravitational, being present to some degree in almost all cases. For many practical applications, however, they are so small compared with the surface forces present that they can be neglected without introducing serious error. Body forces are included in the following analysis for the sake of completeness.

Consider an arbitrary internal or external surface, which may be plane or curvilinear, as shown in Fig. 1.1. Over a small area ΔA of this surface in the neighborhood of an arbitrary point P, a system of forces acts which has a resultant represented by the vector $\Delta \mathbf{F}_n$ in the figure. It should be noted that the line of action of the resultant force vector $\Delta \mathbf{F}_n$ does not necessarily coincide with the outer normal n associated with the element of area ΔA. If the resultant force $\Delta \mathbf{F}_n$ is divided by the increment of area ΔA, the average stress which acts over the area is obtained. In the limit as ΔA approaches zero, a quantity defined as the resultant stress $\mathbf{T}_n$ acting at the point P is obtained. This limiting process is illustrated in equation form below.

$$\mathbf{T}_n = \lim_{\Delta A \to 0} \frac{\Delta \mathbf{F}_n}{\Delta A} \qquad (1.1)$$

The line of action of this resultant stress $\mathbf{T}_n$ coincides with the line of action of the resultant force $\Delta \mathbf{F}_n$, as illustrated in Fig. 1.2. It is important to note at this point that the resultant stress $\mathbf{T}_n$ is a function of both the position of the point P in the body and the orientation of the plane which is passed through the point and identified by its outer normal n. In a body subjected to an arbitrary system of loads, both the magnitude and the direction of the resultant stress $\mathbf{T}_n$ at any point P change as the orientation of the plane under consideration is changed.

As illustrated in Fig. 1.2, it is possible to resolve $\mathbf{T}_n$ into two components: one σ_n normal to the surface is known as the resultant normal stress, while the component τ_n is known as the resultant shearing stress.

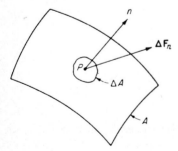

FIGURE 1.1
Arbitrary surface (either internal or external) showing the resultant of all forces acting over the element of area ΔA.

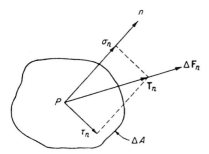

FIGURE 1.2
Resolution of the resultant stress $\mathbf{T}_n$ into its normal and tangential components σ_n and τ_n.

Cartesian components of stress for any coordinate system can also be obtained from the resultant stress. Consider first a surface whose outer normal is the positive z direction, as shown in Fig. 1.3. If the resultant stress $\mathbf{T}_n$ associated with this particular surface is resolved into components along the x, y, and z axes, the cartesian stress components τ_{zx}, τ_{zy}, and σ_{zz} are obtained. The components τ_{zx} and τ_{zy} are shearing stresses since they act tangent to the surface under consideration. The component σ_{zz} is a normal stress since it acts normal to the surface.

If the same procedure is followed using surfaces whose outer normals are in the positive x and y directions, two more sets of cartesian components, τ_{xy}, τ_{xz}, σ_{xx}, and τ_{yx}, τ_{yz}, σ_{yy}, respectively, can be obtained. The three different sets of three cartesian components for the three selections of the outer normal are summarized in the array below:

$$\left\|\begin{array}{ccc} \sigma_{xx} & \tau_{xy} & \tau_{xz} \\ \tau_{yx} & \sigma_{yy} & \tau_{yz} \\ \tau_{zx} & \tau_{zy} & \sigma_{zz} \end{array}\right\| \quad \begin{array}{l} \text{outer normal parallel to the } x \text{ axis} \\ \text{outer normal parallel to the } y \text{ axis} \\ \text{outer normal parallel to the } z \text{ axis} \end{array}$$

From this array, it is clear that nine cartesian components of stress exist. These components can be arranged on the faces of a small cubic element, as shown in Fig. 1.4. The sign convention employed in placing the cartesian stress components on the faces of this cube is as follows: if the outer normal defining the cube face

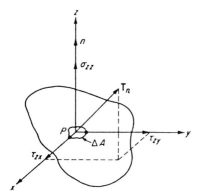

FIGURE 1.3
Resolution of the resultant stress $\mathbf{T}_n$ into its three cartesian components τ_{zx}, τ_{zy}, and σ_{zz}.

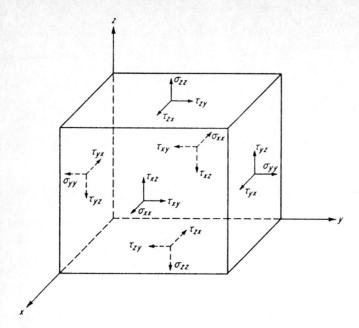

FIGURE 1.4
Cartesian components of stress acting on the faces of a small cubic element.

is in the direction of increasing x, y, or z, then the associated normal and shear stress components are also in the direction of positive x, y, or z. If the outer normal is in the direction of negative x, y, or z, then the normal and shear stress components are also in the direction of negative x, y, or z. As for subscript convention, the first subscript refers to the outer normal and defines the plane upon which the stress component acts, whereas the second subscript gives the direction in which the stress acts. Finally, for normal stresses, positive signs indicate tension and negative signs indicate compression.

1.3 STRESS AT A POINT

At a given point of interest within a body, the magnitude and direction of the resultant stress $\mathbf{T}_n$ depend upon the orientation of the plane passed through the point. Thus an infinite number of resultant-stress vectors can be used to represent the resultant stress at each point since an infinite number of planes can be passed through each point. It is easy to show, however, that the magnitude and direction of each of these resultant-stress vectors can be specified in terms of the nine cartesian components of stress acting at the point. This can be seen by considering equilibrium of the elemental tetrahedron shown in Fig. 1.5. In this figure the stresses acting over the four faces of the tetrahedron are represented by their average values. The average value is denoted by placing a $\sim$ sign over the stress

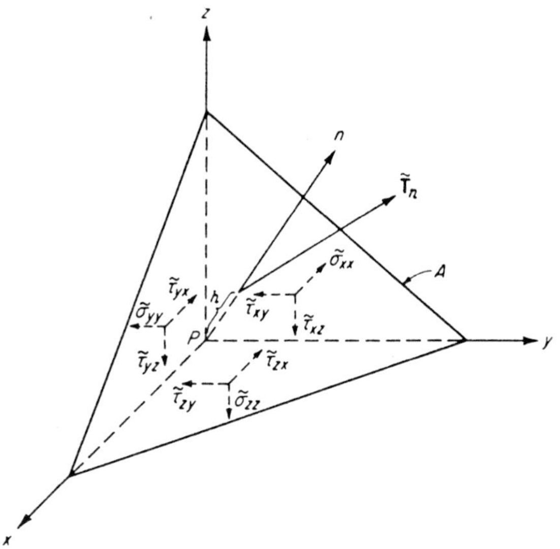

FIGURE 1.5
Elemental tetrahedron at point P showing the average stresses which act over its four faces.

symbol. In order for the tetrahedron to be in equilibrium, the following condition must be satisfied. First consider equilibrium in the x direction:

$$\tilde{T}_{nx} A - \tilde{\sigma}_{xx} A \cos (n, x) - \tilde{\tau}_{yx} A \cos (n, y) - \tilde{\tau}_{zx} A \cos (n, z) + \tilde{F}_x \tfrac{1}{3} hA = 0$$

where h = altitude of tetrahedron
  A = area of base of tetrahedron
  $\tilde{F}_x$ = average body-force intensity in x direction
  $\tilde{T}_{nx}$ = component of resultant stress in x direction

and $A \cos (n, x)$, $A \cos (n, y)$, and $A \cos (n, z)$ are the projections of the area A on the yz, xz, and xy planes, respectively.

By letting the altitude $h \to 0$, after eliminating the common factor A from each term of the expression, it can be seen that the body-force term vanishes, the average stresses become exact stresses at the point P, and the previous expression becomes

$$T_{nx} = \sigma_{xx} \cos (n, x) + \tau_{yx} \cos (n, y) + \tau_{zx} \cos (n, z) \qquad (1.2a)$$

Two similar expressions are obtained by considering equilibrium in the y and z directions:

$$T_{ny} = \tau_{xy} \cos (n, x) + \sigma_{yy} \cos (n, y) + \tau_{zy} \cos (n, z) \qquad (1.2b)$$

$$T_{nz} = \tau_{xz} \cos (n, x) + \tau_{yz} \cos (n, y) + \sigma_{zz} \cos (n, z) \qquad (1.2c)$$

Once the three cartesian components of the resultant stress for a particular plane have been determined by employing Eqs. (1.2), the magnitude of the resultant stress $\mathbf{T}_n$ can be determined by using the expression

$$|\mathbf{T}_n| = \sqrt{T_{nx}^2 + T_{ny}^2 + T_{nz}^2}$$

The three direction cosines which define the line of action of the resultant stress $\mathbf{T}_n$ are

$$\cos(T_n, x) = \frac{T_{nx}}{|\mathbf{T}_n|} \qquad \cos(T_n, y) = \frac{T_{ny}}{|\mathbf{T}_n|} \qquad \cos(T_n, z) = \frac{T_{nz}}{|\mathbf{T}_n|}$$

The normal stress σ_n and the shearing stress τ_n which act on the plane under consideration can be obtained from the expressions

$$\sigma_n = |\mathbf{T}_n| \cos(T_n, n) \qquad \text{and} \qquad \tau_n = |\mathbf{T}_n| \sin(T_n, n)$$

The angle between the resultant-stress vector $\mathbf{T}_n$ and the normal to the plane n can be determined by using the well-known relationship

$$\cos(T_n, n) = \cos(T_n, x) \cos(n, x) + \cos(T_n, y) \cos(n, y)$$
$$+ \cos(T_n, z) \cos(n, z)$$

It should also be noted that the normal stress σ_n can be determined by considering the projections of T_{nx}, T_{ny}, and T_{nz} onto the normal to the plane under consideration. Thus

$$\sigma_n = T_{nx} \cos(n, x) + T_{ny} \cos(n, y) + T_{nz} \cos(n, z)$$

Once σ_n has been determined, τ_n can easily be found since

$$\tau_n = \sqrt{T_n^2 - \sigma_n^2}$$

1.4   STRESS EQUATIONS OF EQUILIBRIUM

In a body subjected to a general system of body and surface forces, stresses of variable magnitude and direction are produced throughout the body. The distribution of these stresses must be such that the overall equilibrium of the body is maintained; furthermore, equilibrium of each element in the body must be maintained. This section deals with the equilibrium of the individual elements of the body. On the element shown in Fig. 1.6, only the stress and body-force components which act in the x direction are shown. Similar components exist and act in the y and z directions. The stress values shown are average stresses over the faces of an element which is assumed to be very small. A summation of forces in the x direction gives

$$\left(\sigma_{xx} + \frac{\partial \sigma_{xx}}{\partial x} dx - \sigma_{xx}\right) dy\, dz + \left(\tau_{yx} + \frac{\partial \tau_{yx}}{\partial y} dy - \tau_{yx}\right) dx\, dz$$

$$+ \left(\tau_{zx} + \frac{\partial \tau_{zx}}{\partial z} dz - \tau_{zx}\right) dx\, dy + F_x\, dx\, dy\, dz = 0$$

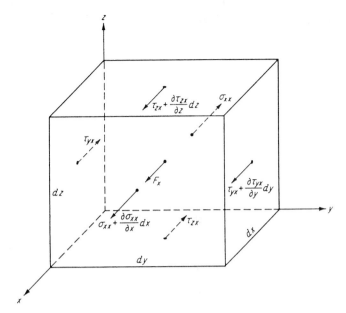

FIGURE 1.6
Small element removed from a body, showing the stresses acting in the x direction only.

Dividing through by $dx\, dy\, dz$ gives

$$\frac{\partial \sigma_{xx}}{\partial x} + \frac{\partial \tau_{yx}}{\partial y} + \frac{\partial \tau_{zx}}{\partial z} + F_x = 0 \qquad (1.3a)$$

By considering the force and stress components in the y and z directions, it can be established in a similar fashion that

$$\frac{\partial \tau_{xy}}{\partial x} + \frac{\partial \sigma_{yy}}{\partial y} + \frac{\partial \tau_{zy}}{\partial z} + F_y = 0 \qquad (1.3b)$$

$$\frac{\partial \tau_{xz}}{\partial x} + \frac{\partial \tau_{yz}}{\partial y} + \frac{\partial \sigma_{zz}}{\partial z} + F_z = 0 \qquad (1.3c)$$

where F_x, F_y, F_z are body-force intensities (in lb/in^3 or N/m^3) in the x, y, and z directions, respectively.

Equations (1.3) are the well-known stress equations of equilibrium which any theoretically or experimentally obtained stress distribution must satisfy. In obtaining these equations, three of the six equilibrium conditions have been employed. The three remaining conditions can be utilized to establish additional relationships between the stresses.

Consider the element shown in Fig. 1.7. Only those stress components which will produce a moment about the y axis are shown. Since the coordinate system

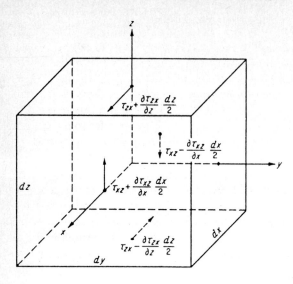

FIGURE 1.7
Small element removed from a body, showing the stresses which produce a moment about the y axis.

has been selected with its origin at the centroid of the element, the normal stress components and the body forces do not produce any moments.

A summation of moments about the y axis gives the following expression:

$$\left(\tau_{zx} + \frac{\partial \tau_{zx}}{\partial z}\frac{dz}{2}\right) dx\, dy\, \frac{dz}{2} + \left(\tau_{zx} - \frac{\partial \tau_{zx}}{\partial z}\frac{dz}{2}\right) dx\, dy\, \frac{dz}{2}$$

$$- \left(\tau_{xz} + \frac{\partial \tau_{xz}}{\partial x}\frac{dx}{2}\right) dy\, dz\, \frac{dx}{2} - \left(\tau_{xz} - \frac{\partial \tau_{xz}}{\partial x}\frac{dx}{2}\right) dy\, dz\, \frac{dx}{2} = 0$$

which reduces to

$$\tau_{zx}\, dx\, dy\, dz - \tau_{xz}\, dx\, dy\, dz = 0$$

Therefore,
$$\tau_{zx} = \tau_{xz} \tag{1.4a}$$

The remaining two equilibrium conditions can be used in a similar manner to establish that

$$\tau_{xy} = \tau_{yx} \tag{1.4b}$$

$$\tau_{yz} = \tau_{zy} \tag{1.4c}$$

The equalities given in Eqs. (1.4) reduce the nine cartesian components of stress to six independent components, which may be expressed in the following array:

$$
\begin{array}{ccc}
\sigma_{xx} & \tau_{xy} & \tau_{zx} \\
\tau_{xy} & \sigma_{yy} & \tau_{yz} \\
\tau_{zx} & \tau_{yz} & \sigma_{zz}
\end{array}
$$

1.5 LAWS OF STRESS TRANSFORMATION

It has previously been shown that the resultant-stress vector $\mathbf{T}_n$ acting on an arbitrary plane defined by the outer normal n can be determined by substituting the six independent cartesian components of stress into Eqs. (1.2). However, it is often desirable to make another transformation, namely, that from the stress components σ_{xx}, σ_{yy}, σ_{zz}, τ_{xy}, τ_{yz}, τ_{zx}, which refer to an $Oxyz$ coordinate system, to the stress components $\sigma_{x'x'}$, $\sigma_{y'y'}$, $\sigma_{z'z'}$, $\tau_{x'y'}$, $\tau_{y'z'}$, $\tau_{z'x'}$, which refer to an $Ox'y'z'$ coordinate system. The transformation equations commonly used to perform this operation will be developed in this section.

Consider an element similar to Fig. 1.5 with an inclined face having outer normal n. Two mutually perpendicular directions n' and n'' can then be denoted in the plane of the inclined face, as shown in Fig. 1.8. The resultant stress $\mathbf{T}_n$ acting on the inclined face can be resolved into components along the directions n, n', and n'' to yield the stresses σ_{nn}, $\tau_{nn'}$, and $\tau_{nn''}$. This resolution of the resultant stress into components can be accomplished most easily by utilizing the cartesian components T_{nx}, T_{ny}, and T_{nz}. Thus

$$\sigma_{nn} = T_{nx} \cos(n, x) + T_{ny} \cos(n, y) + T_{nz} \cos(n, z)$$

$$\tau_{nn'} = T_{nx} \cos(n', x) + T_{ny} \cos(n', y) + T_{nz} \cos(n', z)$$

$$\tau_{nn''} = T_{nx} \cos(n'', x) + T_{ny} \cos(n'', y) + T_{nz} \cos(n'', z)$$

If the results from Eqs. (1.2) and (1.4) are substituted into these expressions, the following important equations are obtained:

$$\begin{aligned} \sigma_{nn} = {} & \sigma_{xx} \cos^2(n, x) + \sigma_{yy} \cos^2(n, y) + \sigma_{zz} \cos^2(n, z) \\ & + 2\tau_{xy} \cos(n, x) \cos(n, y) + 2\tau_{yz} \cos(n, y) \cos(n, z) \\ & + 2\tau_{zx} \cos(n, z) \cos(n, x) \end{aligned} \qquad (1.5a)$$

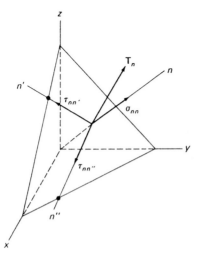

FIGURE 1.8
Resolution of $\mathbf{T}_n$ into three cartesian components σ_{nn}, $\tau_{nn'}$, and $\tau_{nn''}$.

$$\tau_{nn'} = \sigma_{xx} \cos(n, x) \cos(n', x) + \sigma_{yy} \cos(n, y) \cos(n', y)$$
$$+ \sigma_{zz} \cos(n, z) \cos(n', z)$$
$$+ \tau_{xy}[\cos(n, x) \cos(n', y) + \cos(n, y) \cos(n', x)]$$
$$+ \tau_{yz}[\cos(n, y) \cos(n', z) + \cos(n, z) \cos(n', y)]$$
$$+ \tau_{zx}[\cos(n, z) \cos(n', x) + \cos(n, x) \cos(n', z)] \qquad (1.5b)$$

$$\tau_{nn''} = \sigma_{xx} \cos(n, x) \cos(n'', x) + \sigma_{yy} \cos(n, y) \cos(n'', y)$$
$$+ \sigma_{zz} \cos(n, z) \cos(n'', z)$$
$$+ \tau_{xy}[\cos(n, x) \cos(n'', y) + \cos(n, y) \cos(n'', x)]$$
$$+ \tau_{yz}[\cos(n, y) \cos(n'', z) + \cos(n, z) \cos(n'', y)]$$
$$+ \tau_{zx}[\cos(n, z) \cos(n'', x) + \cos(n, x) \cos(n'', z)] \qquad (1.5c)$$

Equations (1.5) provide the means for determining normal- and shear-stress components at a point associated with any set of cartesian reference axes provided the stresses associated with one set of axes are known.

Expressions for the stress components $\sigma_{x'x'}$, $\sigma_{y'y'}$, $\sigma_{z'z'}$, $\tau_{x'y'}$, $\tau_{y'z'}$, $\tau_{z'x'}$ can be obtained directly from Eq. (1.5a) or Eq. (1.5b) by employing the following procedure.

In order to determine $\sigma_{x'x'}$, select a plane having an outer normal n coincident with x'. A resultant stress $T_n = T_{x'}$ is associated with this plane. The normal stress $\sigma_{x'x'}$ associated with this plane is obtained directly from Eq. (1.5a) by substituting x' for n. Thus

$$\sigma_{x'x'} = \sigma_{xx} \cos^2(x', x) + \sigma_{yy} \cos^2(x', y)$$
$$+ \sigma_{zz} \cos^2(x', z) + 2\tau_{xy} \cos(x', x) \cos(x', y)$$
$$+ 2\tau_{yz} \cos(x', y) \cos(x', z) + 2\tau_{zx} \cos(x', z) \cos(x', x) \qquad (1.6a)$$

By selecting n coincident with the y' and z' axes and following the same procedure, expressions for $\sigma_{y'y'}$ and $\sigma_{z'z'}$ can be obtained as follows:

$$\sigma_{y'y'} = \sigma_{yy} \cos^2(y', y) + \sigma_{zz} \cos^2(y', z)$$
$$+ \sigma_{xx} \cos^2(y', x) + 2\tau_{yz} \cos(y', y) \cos(y', z)$$
$$+ 2\tau_{zx} \cos(y', z) \cos(y', x) + 2\tau_{xy} \cos(y', x) \cos(y', y) \qquad (1.6b)$$

$$\sigma_{z'z'} = \sigma_{zz} \cos^2(z', z) + \sigma_{xx} \cos^2(z', x)$$
$$+ \sigma_{yy} \cos^2(z', y) + 2\tau_{zx} \cos(z', z) \cos(z', x)$$
$$+ 2\tau_{xy} \cos(z', x) \cos(z', y) + 2\tau_{yz} \cos(z', y) \cos(z', z) \qquad (1.6c)$$

The shear-stress component $\tau_{x'y'}$ is obtained by selecting a plane having outer normal n coincident with x' and the in-plane direction n' coincident with y', as shown in Fig. 1.9. The shear stress $\tau_{x'y'}$ is then obtained from Eq. (1.5b) by

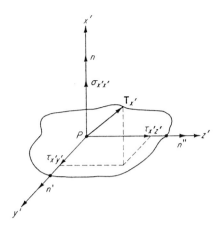

FIGURE 1.9
Resolution of $\mathbf{T}_{x'}$ into three cartesian stress components $\sigma_{x'x'}$, $\tau_{x'y'}$, and $\tau_{x'z'}$.

substituting x' for n and y' for n'. Thus

$$
\begin{aligned}
\tau_{x'y'} = {}& \sigma_{xx} \cos (x', x) \cos (y', x) \\
& + \sigma_{yy} \cos (x', y) \cos (y', y) + \sigma_{zz} \cos (x', z) \cos (y', z) \\
& + \tau_{xy}[\cos (x', x) \cos (y', y) + \cos (x', y) \cos (y', x)] \\
& + \tau_{yz}[\cos (x', y) \cos (y', z) + \cos (x', z) \cos (y', y)] \\
& + \tau_{zx}[\cos (x', z) \cos (y', x) + \cos (x', x) \cos (y', z)] \qquad (1.6d)
\end{aligned}
$$

By selecting n and n' coincident with the y' and z', and z' and x' axes, additional expressions can be developed for $\tau_{y'z'}$ and $\tau_{z'x'}$, respectively, as follows:

$$
\begin{aligned}
\tau_{y'z'} = {}& \sigma_{yy} \cos (y', y) \cos (z', y) \\
& + \sigma_{zz} \cos (y', z) \cos (z', z) + \sigma_{xx} \cos (y', x) \cos (z', x) \\
& + \tau_{yz}[\cos (y', y) \cos (z', z) + \cos (y', z) \cos (z', y)] \\
& + \tau_{zx}[\cos (y', z) \cos (z', x) + \cos (y', x) \cos (z', z)] \\
& + \tau_{xy}[\cos (y', x) \cos (z', y) + \cos (y', y) \cos (z', x)] \qquad (1.6e)
\end{aligned}
$$

$$
\begin{aligned}
\tau_{z'x'} = {}& \sigma_{zz} \cos (z', z) \cos (x', z) \\
& + \sigma_{xx} \cos (z', x) \cos (x', x) + \sigma_{yy} \cos (z', y) \cos (x', y) \\
& + \tau_{zx}[\cos (z', z) \cos (x' \; x) + \cos (z', x) \cos (x', z)] \\
& + \tau_{xy}[\cos (z', x) \cos (x', y) + \cos (z', y) \cos (x', x)] \\
& + \tau_{yz}[\cos (z', y) \cos (x', z) + \cos (z', z) \cos (x', y)] \qquad (1.6f)
\end{aligned}
$$

These six equations permit the six cartesian components of stress relative to the $Oxyz$ coordinate system to be transformed into a different set of six cartesian components of stress relative to an $Ox'y'z'$ coordinate system.

1.6 PRINCIPAL STRESSES

In Sec. 1.2 it was noted that the resultant-stress vector $\mathbf{T}_n$ at a given point P depended upon the choice of the plane upon which the stress acted. If a plane is selected such that $\mathbf{T}_n$ coincides with the outer normal n, as shown in Fig. 1.10, it is clear that the shear stress τ_n vanishes and that $\mathbf{T}_n$, σ_n, and n are coincident.

If n is selected so that it coincides with $\mathbf{T}_n$, then the plane defined by n is known as a principal plane. The direction given by n is a principal direction, and the normal stress acting on this particular plane is a principal stress. In every state of stress there exist at least three principal planes, which are mutually perpendicular, and associated with these principal planes there are at most three distinct principal stresses. These statements can be established by referring to Fig. 1.10 and noting that

$$T_{nx} = \sigma_n \cos (n, x) \qquad T_{ny} = \sigma_n \cos (n, y) \qquad T_{nz} = \sigma_n \cos (n, z) \qquad (a)$$

If Eqs. (1.2) are substituted into Eqs. (a), the following expressions are obtained:

$$\sigma_{xx} \cos (n, x) + \tau_{yx} \cos (n, y) + \tau_{zx} \cos (n, z) = \sigma_n \cos (n, x)$$
$$\tau_{xy} \cos (n, x) + \sigma_{yy} \cos (n, y) + \tau_{zy} \cos (n, z) = \sigma_n \cos (n, y) \qquad (b)$$
$$\tau_{xz} \cos (n, x) + \tau_{yz} \cos (n, y) + \sigma_{zz} \cos (n, z) = \sigma_n \cos (n, z)$$

Rearranging Eqs. (b) gives

$$(\sigma_{xx} - \sigma_n) \cos (n, x) + \tau_{yx} \cos (n, y) + \tau_{zx} \cos (n, z) = 0$$
$$\tau_{xy} \cos (n, x) + (\sigma_{yy} - \sigma_n) \cos (n, y) + \tau_{zy} \cos (n, z) = 0 \qquad (c)$$
$$\tau_{xz} \cos (n, x) + \tau_{yz} \cos (n, y) + (\sigma_{zz} - \sigma_n) \cos (n, z) = 0$$

Solving for any of the direction cosines, say $\cos (n, x)$, by determinants gives

$$\cos (n, x) = \frac{\begin{vmatrix} 0 & \tau_{yx} & \tau_{zx} \\ 0 & \sigma_{yy} - \sigma_n & \tau_{zy} \\ 0 & \tau_{yz} & \sigma_{zz} - \sigma_n \end{vmatrix}}{\begin{vmatrix} \sigma_{xx} - \sigma_n & \tau_{yx} & \tau_{zx} \\ \tau_{xy} & \sigma_{yy} - \sigma_n & \tau_{zy} \\ \tau_{xz} & \tau_{yz} & \sigma_{zz} - \sigma_n \end{vmatrix}} \qquad (d)$$

It is clear that nontrivial solutions for the direction cosines of the principal plane will exist only if the determinant in the denominator is zero. Thus

$$\begin{vmatrix} \sigma_{xx} - \sigma_n & \tau_{yx} & \tau_{zx} \\ \tau_{xy} & \sigma_{yy} - \sigma_n & \tau_{zy} \\ \tau_{xz} & \tau_{yz} & \sigma_{zz} - \sigma_n \end{vmatrix} = 0 \qquad (e)$$

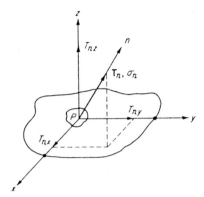

FIGURE 1.10
Coincidence of $\mathbf{T}_n$ with the outer normal n indicates that the shear stresses vanish and that σ_n becomes equal in magnitude to $\mathbf{T}_n$.

Expanding the determinant after substituting Eqs. (1.4) gives the following important cubic equation

$$
\begin{aligned}
\sigma_n^3 &- (\sigma_{xx} + \sigma_{yy} + \sigma_{zz})\sigma_n^2 \\
&+ (\sigma_{xx}\sigma_{yy} + \sigma_{yy}\sigma_{zz} + \sigma_{zz}\sigma_{xx} - \tau_{xy}^2 - \tau_{yz}^2 - \tau_{zx}^2)\sigma_n \\
&- (\sigma_{xx}\sigma_{yy}\sigma_{zz} - \sigma_{xx}\tau_{yz}^2 - \sigma_{yy}\tau_{zx}^2 - \sigma_{zz}\tau_{xy}^2 + 2\tau_{xy}\tau_{yz}\tau_{zx}) = 0 \qquad (1.7)
\end{aligned}
$$

The roots of this cubic equation are the three principal stresses. By substituting the six cartesian components of stress into this equation, one can solve for σ_n and obtain three real roots. Three possible solutions exist.

1. If σ_1, σ_2, and σ_3 are distinct, then n_1, n_2, and n_3 are unique and mutually perpendicular.
2. If $\sigma_1 = \sigma_2 \neq \sigma_3$, then n_3 is unique and every direction perpendicular to n_3 is a principal direction associated with $\sigma_1 = \sigma_2$.
3. If $\sigma_1 = \sigma_2 = \sigma_3$, then a hydrostatic state of stress exists and every direction is a principal direction.

Once the three principal stresses have been established, they can be substituted individually into Eqs. (c) to give three sets of simultaneous equations which together with the relation

$$
\cos^2(n, x) + \cos^2(n, y) + \cos^2(n, z) = 1
$$

can be solved to give the three sets of direction cosines defining the principal planes. A numerical example of the procedure used in computing principal stresses and directions is given in the exercises at the end of the chapter.

In treating principal stresses it is often useful to order them so that $\sigma_1 > \sigma_2 > \sigma_3$. When the stresses are ordered in this fashion, σ_1 is the normal stress having the largest algebraic value at a given point and σ_3 is the normal stress having the smallest algebraic value. It is important to recall in this ordering process that tensile stresses are considered positive and compressive stresses are considered negative.

Another important concept is that of stress invariants. It was noted in Sec. 1.5 that a state of stress could be described by its six cartesian stress components with respect to either the $Oxyz$ coordinate system or the $Ox'y'z'$ coordinate system. Furthermore, Eqs. (1.6) were established to give the relationship between these two systems. In addition to Eqs. (1.6), three other relations exist which are called the three invariants of stress. To establish these invariants, refer to Eq. (1.7), which is the cubic equation in terms of the principal stresses σ_1, σ_2, and σ_3. By recalling that σ_1, σ_2, and σ_3 are independent of the cartesian coordinate system employed, it is clear that the coefficients of Eq. (1.7) which contain cartesian components of the stresses must also be independent or invariant of the coordinate system. Thus, from Eq. (1.7) it is clear that

$$
\begin{aligned}
I_1 &= \sigma_{xx} + \sigma_{yy} + \sigma_{zz} = \sigma_{x'x'} + \sigma_{y'y'} + \sigma_{z'z'} \\
I_2 &= \sigma_{xx}\sigma_{yy} + \sigma_{yy}\sigma_{zz} + \sigma_{zz}\sigma_{xx} - \tau_{xy}^2 - \tau_{yz}^2 - \tau_{zx}^2 \\
&= \sigma_{x'x'}\sigma_{y'y'} + \sigma_{y'y'}\sigma_{z'z'} + \sigma_{z'z'}\sigma_{x'x'} - \tau_{x'y'}^2 - \tau_{y'z'}^2 - \tau_{z'x'}^2 \\
I_3 &= \sigma_{xx}\sigma_{yy}\sigma_{zz} - \sigma_{xx}\tau_{yz}^2 - \sigma_{yy}\tau_{zx}^2 - \sigma_{zz}\tau_{xy}^2 + 2\tau_{xy}\tau_{yz}\tau_{zx} \\
&= \sigma_{x'x'}\sigma_{y'y'}\sigma_{z'z'} - \sigma_{x'x'}\tau_{y'z'}^2 - \sigma_{y'y'}\tau_{z'x'}^2 - \sigma_{z'z'}\tau_{x'y'}^2 + 2\tau_{x'y'}\tau_{y'z'}\tau_{z'x'}
\end{aligned}
\tag{1.8}
$$

where I_1, I_2, and I_3 are the first, second, and third invariants of stress, respectively. If the $Oxyz$ coordinate system is selected coincident with the principal directions, Eqs. (1.8) reduce to

$$
I_1 = \sigma_1 + \sigma_2 + \sigma_3 \qquad I_2 = \sigma_1\sigma_2 + \sigma_2\sigma_3 + \sigma_3\sigma_1 \qquad I_3 = \sigma_1\sigma_2\sigma_3 \tag{1.9}
$$

1.7 MAXIMUM SHEAR STRESS

In developing equations for maximum shear stresses, the special case will be considered in which $\tau_{xy} = \tau_{yz} = \tau_{zx} = 0$. No loss in generality is introduced by considering this special case since it involves only a reorientation of the reference axes to coincide with the principal directions. In the following development n_1, n_2, and n_3 will be used to denote the principal directions. In Sec. 1.3 the resultant stress on an oblique plane was given by

$$
T_n^2 = T_{nx}^2 + T_{ny}^2 + T_{nz}^2 \tag{a}
$$

Substitution of values for T_{nx}, T_{ny}, and T_{nz} from Eqs. (1.2) with principal normal stresses and zero shearing stresses yields

$$
T_n^2 = \sigma_1^2 \cos^2(n, n_1) + \sigma_2^2 \cos^2(n, n_2) + \sigma_3^2 \cos^2(n, n_3) \tag{b}
$$

Also from Eq. (1.5a)

$$
\sigma_n = \sigma_1 \cos^2(n, n_1) + \sigma_2 \cos^2(n, n_2) + \sigma_3 \cos^2(n, n_3) \tag{c}
$$

Since $\tau_n^2 = T_n^2 - \sigma_n^2$, an expression for the shear stress τ_n on the oblique plane is obtained from Eqs. (b) and (c) after substituting $l = \cos(n, n_1)$, $m = \cos(n, n_2)$, and

$n = \cos(n, n_3)$ as

$$\tau_n^2 = \sigma_1^2 l^2 + \sigma_2^2 m^2 + \sigma_3^2 n^2 - (\sigma_1 l^2 + \sigma_2 m^2 + \sigma_3 n^2)^2 \qquad (d)$$

The planes on which maximum and minimum shearing stresses occur can be obtained from Eq. (d) by differentiating with respect to the direction cosines l, m, and n. One of the direction cosines, n for example, in Eq. (d) can be eliminated by solving the expression

$$l^2 + m^2 + n^2 = 1 \qquad (e)$$

for l and substituting into Eq. (d). Thus

$$\tau_n^2 = (\sigma_1^2 - \sigma_3^2)l^2 + (\sigma_2^2 - \sigma_3^2)m^2 + \sigma_3^2 - [(\sigma_1 - \sigma_3)l^2 + (\sigma_2 - \sigma_3)m^2 + \sigma_3]^2 \quad (f)$$

By taking the partial derivatives of Eq. (f), first with respect to l and then with respect to m, and equating to zero, the following equations are obtained for determining the direction cosines associated with planes having maximum and minimum shearing stresses:

$$l[\tfrac{1}{2}(\sigma_1 - \sigma_3) - (\sigma_1 - \sigma_3)l^2 - (\sigma_2 - \sigma_3)m^2] = 0 \qquad (g)$$

$$m[\tfrac{1}{2}(\sigma_2 - \sigma_3) - (\sigma_1 - \sigma_3)l^2 - (\sigma_2 - \sigma_3)m^2] = 0 \qquad (h)$$

One solution of these equations is obviously $l = m = 0$. Then from Eq. (e), $n = \pm 1$ (a principal plane with zero shear). Solutions different from zero are also possible for this set of equations. Consider first that $m = 0$; then from Eq. (g), $l = \pm(\tfrac{1}{2})^{1/2}$ and from Eq. (e), $n = \pm(\tfrac{1}{2})^{1/2}$. Also if $l = 0$, then from Eq. (h), $m = \pm(\tfrac{1}{2})^{1/2}$ and from Eq. (e), $n = \pm(\tfrac{1}{2})^{1/2}$. Repeating the above procedure by eliminating l and m in turn from Eq. (f) yields other values for the direction cosines which make the shearing stresses maximum or minimum. Substituting the values $l = \pm(\tfrac{1}{2})^{1/2}$ and $n = \pm(\tfrac{1}{2})^{1/2}$ into Eq. (d) yields

$$\tau_n^2 = \tfrac{1}{2}\sigma_1^2 + 0 + \tfrac{1}{2}\sigma_3^2 - (\tfrac{1}{2}\sigma_1 + 0 + \tfrac{1}{2}\sigma_3)^2$$

from which

$$\tau_n = \tfrac{1}{2}(\sigma_1 - \sigma_3)$$

Similarly, using the other values for the direction cosines which make the shearing stresses maximum gives

$$\tau_n = \tfrac{1}{2}(\sigma_1 - \sigma_2) \qquad \text{and} \qquad \tau_n = \tfrac{1}{2}(\sigma_2 - \sigma_3)$$

Of these three possible results, the largest magnitude will be obtained from $\sigma_1 - \sigma_3$ if the principal stresses are ordered such that $\sigma_1 \geq \sigma_2 \geq \sigma_3$. Thus

$$\tau_{max} = \tfrac{1}{2}(\sigma_{max} - \sigma_{min}) = \tfrac{1}{2}(\sigma_1 - \sigma_3) \qquad (1.10)$$

A useful aid for visualizing the complete state of stress at a point is the three-dimensional Mohr's circle shown in Fig. 1.11. This representation, which is similar to the familiar two-dimensional Mohr's circle, shows the three principal stresses, the maximum shearing stresses, and the range of values within which the normal- and shear-stress components must lie for a given state of stress.

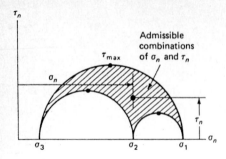

FIGURE 1.11
Mohr's circle for the three-dimensional state of stress.

1.8   THE TWO-DIMENSIONAL STATE OF STRESS

For two-dimensional stress fields where $\sigma_{zz} = \tau_{zx} = \tau_{yz} = 0$, z' is coincident with z, and θ is the angle between x and x', Eqs. (1.6a) to (1.6f) reduce to

$$\sigma_{x'x'} = \sigma_{xx} \cos^2 \theta + \sigma_{yy} \sin^2 \theta + 2\tau_{xy} \sin \theta \cos \theta$$

$$= \frac{\sigma_{xx} + \sigma_{yy}}{2} + \frac{\sigma_{xx} - \sigma_{yy}}{2} \cos 2\theta + \tau_{xy} \sin 2\theta \tag{1.11a}$$

$$\sigma_{y'y'} = \sigma_{yy} \cos^2 \theta + \sigma_{xx} \sin^2 \theta - 2\tau_{xy} \sin \theta \cos \theta$$

$$= \frac{\sigma_{yy} + \sigma_{xx}}{2} + \frac{\sigma_{yy} - \sigma_{xx}}{2} \cos 2\theta - \tau_{xy} \sin 2\theta \tag{1.11b}$$

$$\tau_{x'y'} = \sigma_{yy} \cos \theta \sin \theta - \sigma_{xx} \cos \theta \sin \theta + \tau_{xy}(\cos^2 \theta - \sin^2 \theta)$$

$$= \frac{\sigma_{yy} - \sigma_{xx}}{2} \sin 2\theta + \tau_{xy} \cos 2\theta \tag{1.11c}$$

$$\sigma_{z'z'} = \tau_{z'x'} = \tau_{y'z'} = 0 \tag{1.11d}$$

The relationships between stress components given in Eqs. (1.11) can be graphically represented by using Mohr's circle of stress, as indicated in Fig. 1.12. In this diagram, normal-stress components σ are plotted horizontally, while shear-stress components τ are plotted vertically. Tensile stresses are plotted to the right of the τ axis. Compressive stresses are plotted to the left. Shear-stress components which tend to produce a clockwise rotation of a small element surrounding the point are plotted above the σ axis. Those tending to produce a counterclockwise rotation are plotted below. When plotted in this manner, the stress components associated with each plane through the point are represented by a point on the circle. The diagram thus gives an excellent visual picture of the state of stress at a point. Mohr's circle and Eqs. (1.11) are often used in experimental stress-analysis work when stress components are transformed from one coordinate system to another. These relationships will be used frequently in later sections of this text, where strain gages and photoelasticity methods of analysis are discussed. Since two-dimensional stress systems are often considered in subsequent chapters, it will be useful to consider the principal stresses which

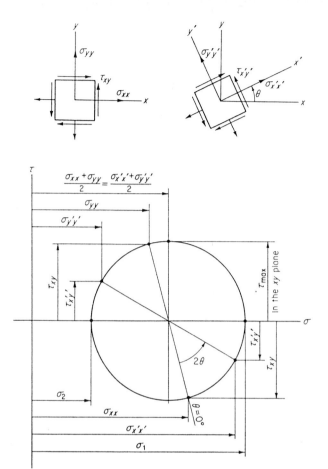

FIGURE 1.12
Mohr's circle of stress.

occur in a two-dimensional stress system. If a coordinate system is chosen so that $\sigma_{zz} = \tau_{zx} = \tau_{yz} = 0$, then a state of plane stress exists and Eq. (1.7) reduces to

$$\sigma_n[\sigma_n^2 - (\sigma_{xx} + \sigma_{yy})\sigma_n + (\sigma_{xx}\sigma_{yy} - \tau_{xy}^2)] = 0 \qquad (a)$$

Solving this equation for the three principal stresses yields

$$\sigma_1, \sigma_2 = \frac{\sigma_{xx} + \sigma_{yy}}{2} \pm \sqrt{\left(\frac{\sigma_{xx} - \sigma_{yy}}{2}\right)^2 + \tau_{xy}^2} \qquad \sigma_3 = 0 \qquad (1.12)$$

The two direction cosines which define the two principal planes can be determined from Eq. (1.11c), which gives $\tau_{x'y'}$ in terms of σ_{xx}, σ_{yy}, τ_{xy} and the angle θ between x and x'. If x' and y' are selected so that $x' = n_1$ and $y' = n_2$, then $\tau_{x'y'}$

must vanish since no shearing stresses can exist on principal planes. Thus the following equation can be written:

$$\frac{\sigma_{yy} - \sigma_{xx}}{2} \sin 2(n_1, x) + \tau_{xy} \cos 2(n_1, x) = 0 \tag{1.13}$$

Dividing through by $\cos 2(n_1, x)$ and simplifying gives

$$\tan 2(n_1, x) = \frac{2\tau_{xy}}{\sigma_{xx} - \sigma_{yy}} \tag{1.14a}$$

and hence

$$\cos 2(n_1, x) = \frac{\sigma_{xx} - \sigma_{yy}}{\sqrt{(\sigma_{xx} - \sigma_{yy})^2 + 4\tau_{xy}^2}} \tag{1.14b}$$

$$\sin 2(n_1, x) = \frac{2\tau_{xy}}{\sqrt{(\sigma_{xx} - \sigma_{yy})^2 + 4\tau_{xy}^2}} \tag{1.14c}$$

Equations (1.14) are used in solving for the direction of n_1 if the cartesian stress components τ_{xy}, σ_{xx}, σ_{yy} are known.

1.9   STRESSES RELATIVE TO A PRINCIPAL COORDINATE SYSTEM

If the coordinate system $Oxyz$ is selected to coincide with the three principal directions n_1, n_2, n_3, then $\sigma_1 = \sigma_{xx}$, $\sigma_2 = \sigma_{yy}$, $\sigma_3 = \sigma_{zz}$, and $\tau_{xy} = \tau_{yz} = \tau_{zx} = 0$. This reduces the six components of stress to three, which permits a considerable simplification in some of the previous results. Equations (1.2) become

$$T_{nx} = \sigma_1 \cos (n, x) \qquad T_{ny} = \sigma_2 \cos (n, y) \qquad T_{nz} = \sigma_3 \cos (n, z) \tag{1.15}$$

and Eqs. (1.6) reduce to

$$\begin{aligned}
\sigma_{x'x'} &= \sigma_1 \cos^2 (x', x) + \sigma_2 \cos^2 (x', y) + \sigma_3 \cos^2 (x', z) \\
\sigma_{y'y'} &= \sigma_1 \cos^2 (y', x) + \sigma_2 \cos^2 (y', y) + \sigma_3 \cos^2 (y', z) \\
\sigma_{z'z'} &= \sigma_1 \cos^2 (z', x) + \sigma_2 \cos^2 (z', y) + \sigma_3 \cos^2 (z', z) \\
\tau_{x'y'} &= \sigma_1 \cos (x', x) \cos (y', x) + \sigma_2 \cos (x', y) \cos (y', y) \\
&\quad + \sigma_3 \cos (x', z) \cos (y', z) \\
\tau_{y'z'} &= \sigma_1 \cos (y', x) \cos (z', x) + \sigma_2 \cos (y', y) \cos (z', y) \\
&\quad + \sigma_3 \cos (y', z) \cos (z', z) \\
\tau_{z'x'} &= \sigma_1 \cos (z', x) \cos (x', x) + \sigma_2 \cos (z', y) \cos (x', y) \\
&\quad + \sigma_3 \cos (z', z) \cos (x', z)
\end{aligned} \tag{1.16}$$

Often experimental methods yield principal stresses directly, and in these cases Eqs. (1.16) are frequently used to obtain the stresses acting on other planes.

1.10 SPECIAL STATES OF STRESS

Two states of stress occur so frequently in practice that they have been classified. They are the state of pure shearing stress and the hydrostatic state of stress. Both are defined below.

1. A state of pure shear stress exists if one particular set of axes $Oxyz$ can be found such that $\sigma_{xx} = \sigma_{yy} = \sigma_{zz} = 0$. It can be shown that this particular set of axes $Oxyz$ exists if and only if the first invariant of stress $I_1 = 0$. The proof of this condition is beyond the scope of this text. Two of the infinite number of arrays which represent a state of pure shearing stress are given below.

$$\begin{Vmatrix} 0 & \tau_{xy} & \tau_{xz} \\ \tau_{xy} & 0 & \tau_{yz} \\ \tau_{xz} & \tau_{yz} & 0 \end{Vmatrix} \quad \text{or} \quad \begin{Vmatrix} \sigma_{xx} & \tau_{xy} & \tau_{xz} \\ \tau_{xy} & \sigma_{yy} = -\sigma_{xx} & \tau_{yz} \\ \tau_{xz} & \tau_{yz} & 0 \end{Vmatrix}$$

Pure shear   Can be converted to the form shown on the left by a suitable rotation of the coordinate system

2. A state of stress is said to be hydrostatic if $\sigma_{xx} = \sigma_{yy} = \sigma_{zz} = -p$ and all the shearing stresses vanish. In photoelastic work a hydrostatic state of stress is often called an isotropic state of stress. The stress array for this case is

$$\begin{Vmatrix} -p & 0 & 0 \\ 0 & -p & 0 \\ 0 & 0 & -p \end{Vmatrix}$$

One particularly important property of these two states of stress is that they can be combined to form a general state of stress. Of more importance, however, is the fact that any state of stress can be separated into a state of pure shear plus a hydrostatic state of stress. This is easily seen from the three arrays shown below:

$$\begin{Vmatrix} \sigma_{xx} & \tau_{xy} & \tau_{xz} \\ \tau_{xy} & \sigma_{yy} & \tau_{yz} \\ \tau_{xz} & \tau_{yz} & \sigma_{zz} \end{Vmatrix} = \begin{Vmatrix} -p & 0 & 0 \\ 0 & -p & 0 \\ 0 & 0 & -p \end{Vmatrix} + \begin{Vmatrix} \sigma_{xx} + p & \tau_{xy} & \tau_{xz} \\ \tau_{xy} & \sigma_{yy} + p & \tau_{yz} \\ \tau_{xz} & \tau_{yz} & \sigma_{zz} + p \end{Vmatrix} \quad (1.17)$$

or

General state of stress = hydrostatic state of stress
+ state of pure shearing stress

It is immediately clear that the array on the left represents a general state of stress and that the center array represents a hydrostatic state of stress; however, the right-hand array represents a state of pure shear if and only if its first stress invariant is zero. This fact implies that

$$(\sigma_{xx} + p) + (\sigma_{yy} + p) + (\sigma_{zz} + p) = 0$$

Hence
$$p = -\tfrac{1}{3}(\sigma_{xx} + \sigma_{yy} + \sigma_{zz}) \tag{1.18}$$

If the p represented in the hydrostatic state of stress satisfies Eq. (1.18), then the separation of the state of stress given in Eq. (1.17) is valid. In the study of plasticity, the effect of the hydrostatic stresses is usually neglected; consequently, the principle illustrated above is quite important.

EXERCISES

1.1. Prepare a scale drawing representing the stresses associated with a resultant stress $\mathbf{T}_n$ where $|\mathbf{T}_n| = 10$ units and its direction is described by $\cos(n, x) = 0.6$ and $\cos(n, y) = 0.7$. The normal to the surface upon which $\mathbf{T}_n$ acts is given by $n = \tfrac{1}{3}\mathbf{i} + \tfrac{2}{3}\mathbf{j} + \tfrac{2}{3}\mathbf{k}$. In this drawing show σ_n and τ_n to scale.

1.2. Derive the relation

$$l^2 + m^2 + n^2 = 1$$

where $l = \cos(n, x)$, $m = \cos(n, y)$, and $n = \cos(n, z)$.

1.3. At a point in a stressed body, the cartesian components of stress are $\sigma_{xx} = 70$ MPa, $\sigma_{yy} = -35$ MPa, $\sigma_{zz} = 35$ MPa, $\tau_{xy} = 40$ MPa, $\tau_{yz} = 0$, and $\tau_{zx} = 0$. Determine the normal and shear stresses on a plane whose outer normal has the direction cosines

$$\cos(n, x) = \tfrac{15}{35} \qquad \cos(n, y) = \tfrac{18}{35} \qquad \cos(n, z) = \tfrac{26}{35}$$

1.4. At a point in a stressed body, the cartesian components of stress are $\sigma_{xx} = 75$ MPa, $\sigma_{yy} = 60$ MPa, $\sigma_{zz} = 50$ MPa, $\tau_{xy} = 25$ MPa, $\tau_{yz} = -25$ MPa, and $\tau_{zx} = 30$ MPa. Determine the normal and shear stresses on a plane whose outer normal has the direction cosines

$$\cos(n, x) = \tfrac{12}{25} \qquad \cos(n, y) = \tfrac{15}{25} \qquad \cos(n, z) = \tfrac{16}{25}$$

1.5. At a point in a stressed body, the cartesian components of stress are $\sigma_{xx} = 45$ MPa, $\sigma_{yy} = 90$ MPa, $\sigma_{zz} = 45$ MPa, $\tau_{xy} = 108$ MPa, $\tau_{yz} = 54$ MPa, and $T_{zx} = 36$ MPa. Determine (a) the normal and shear stresses on a plane whose outer normal has the direction cosines

$$\cos(n, x) = \tfrac{4}{9} \qquad \cos(n, y) = \tfrac{4}{9} \qquad \cos(n, z) = \tfrac{7}{9}$$

and (b) the angle between $\mathbf{T}_n$ and the outer normal n.

1.6. At a point in a stressed body, the cartesian components of stress are $\sigma_{xx} = 60$ MPa, $\sigma_{yy} = -40$ MPa, $\sigma_{zz} = 20$ MPa, $\tau_{xy} = -40$ MPa, $\tau_{yz} = 20$ MPa, and $\tau_{zx} = 30$ MPa. Determine (a) the normal and shear stresses on a plane whose outer normal has the direction cosines

$$\cos(n, x) = 0.429 \qquad \cos(n, y) = 0.514 \qquad \cos(n, z) = 0.743$$

and (b) the angle between $\mathbf{T}_n$ and the outer normal n.

1.7. Determine the normal and shear stresses on a plane whose outer normal makes equal angles with the x, y, and z axes if the cartesian components of stress at the point are $\sigma_{xx} = \sigma_{yy} = \sigma_{zz} = 0$, $\tau_{xy} = 75$ MPa, $\tau_{yz} = 0$, and $\tau_{zx} = 100$ MPa.

1.8. At a point in a stressed body, the cartesian components of stress are $\sigma_{xx} = \sigma_{yy} = 75$ MPa, $\sigma_{zz} = -30$ MPa, $\tau_{xy} = 0$, $\tau_{yz} = 45$ MPa, and $\tau_{zx} = 75$ MPa. Determine the normal and shear stresses on a plane whose outer normal has the direction cosines

$$\cos(n, x) = \tfrac{2}{3} \qquad \cos(n, y) = \tfrac{2}{3} \qquad \cos(n, z) = \tfrac{1}{3}$$

1.9. The following stress distribution has been determined for a machine component:

$$\sigma_{xx} = 3x^2 - 3y^2 - z \qquad \sigma_{yy} = 3y^2 \qquad \sigma_{zz} = 3x + y - z + \tfrac{5}{4}$$

$$\tau_{xy} = z - 6xy - \tfrac{3}{4} \qquad \tau_{yz} = 0 \qquad \tau_{zx} = x + y - \tfrac{3}{2}$$

Is equilibrium satisfied in the absence of body forces?

1.10. If the state of stress at any point in a body is given by the equations

$$\sigma_{xx} = ax + by + cz \qquad \sigma_{yy} = dx^2 + ey^2 + fz^2 \qquad \sigma_{zz} = gx^3 + hy^3 + iz^3$$

$$\tau_{xy} = k \qquad \tau_{yz} = ly + mz \qquad \tau_{zx} = nx^2 + pz^2$$

What equations must the body-force intensities F_x, F_y, and F_z satisfy?

1.11. For a two-dimensional state of stress with $\sigma_{zz} = \tau_{yz} = \tau_{zx} = 0$ and no body forces, show that

$$\int \frac{\partial \sigma_{xx}}{\partial x}\, dy = \int \frac{\partial \sigma_{yy}}{\partial y}\, dx$$

1.12. At a point in a stressed body, the cartesian components of stress are $\sigma_{xx} = 100$ MPa, $\sigma_{yy} = 50$ MPa, $\sigma_{zz} = 30$ MPa, $\tau_{xy} = 30$ MPa, $\tau_{yz} = -30$ MPa, and $\tau_{zx} = 60$ MPa. Transform this set of cartesian stress components into a new set of cartesian stress components relative to an $Ox'y'z'$ set of coordinates where the $Ox'y'z'$ axes are defined as:

θ	Case 1	Case 2	Case 3	Case 4
$x - x'$	$\pi/4$	$\pi/2$	0	$\pi/2$
$y - y'$	$\pi/4$	$\pi/2$	$\pi/2$	0
$z - z'$	0	0	$\pi/2$	$\pi/2$

1.13. At a point in a stressed body, the cartesian components of stress are $\sigma_{xx} = 100$ MPa, $\sigma_{yy} = 60$ MPa, $\sigma_{zz} = 50$ MPa, $\tau_{xy} = 40$ MPa, $\tau_{yz} = -30$ MPa, and $\tau_{zx} = 60$ MPa. Transform this set of cartesian stress components into a new set of cartesian stress components relative to an $Ox'y'z'$ set of coordinates where the direction angles associated with the $Ox'y'z'$ axes are:

	x	y	z
x'	$\pi/3$	$\pi/3$	$\pi/4$
y'	$3\pi/4$	$\pi/4$	$\pi/2$
z'	$\pi/3$	$\pi/3$	$3\pi/4$

1.14. At a point in a stressed body, the cartesian components of stress are $\sigma_{xx} = 90$ MPa, $\sigma_{yy} = 54$ MPa, $\sigma_{zz} = -36$ MPa, $\tau_{xy} = 27$ MPa, $\tau_{yz} = -36$ MPa, and $\tau_{zx} = 45$ MPa. Transform this set of cartesian stress components into a new set of cartesian stress components relative to an $Ox'y'z'$ set of coordinates where the $Ox'y'z'$ axes are defined by the following direction cosines:

<table>
<tr><td></td><td colspan="3">(a)</td></tr>
<tr><td></td><td>x</td><td>y</td><td>z</td></tr>
<tr><td>x'</td><td>$\frac{2}{3}$</td><td>$\frac{2}{3}$</td><td>$-\frac{1}{3}$</td></tr>
<tr><td>y'</td><td>$-\frac{2}{3}$</td><td>$\frac{1}{3}$</td><td>$-\frac{2}{3}$</td></tr>
<tr><td>z'</td><td>$-\frac{1}{3}$</td><td>$\frac{2}{3}$</td><td>$\frac{2}{3}$</td></tr>
</table>

<table>
<tr><td></td><td colspan="3">(b)</td></tr>
<tr><td></td><td>x</td><td>y</td><td>z</td></tr>
<tr><td>x'</td><td>$\frac{1}{9}$</td><td>$-\frac{8}{9}$</td><td>$\frac{4}{9}$</td></tr>
<tr><td>y'</td><td>$\frac{4}{9}$</td><td>$\frac{4}{9}$</td><td>$\frac{7}{9}$</td></tr>
<tr><td>z'</td><td>$-\frac{8}{9}$</td><td>$\frac{1}{9}$</td><td>$\frac{4}{9}$</td></tr>
</table>

1.15. At a point in a stressed body, the cartesian components of stress are $\sigma_{xx} = 90$ MPa, $\sigma_{yy} = 60$ MPa, $\sigma_{zz} = -30$ MPa, $\tau_{xy} = 45$ MPa, $\tau_{yz} = -15$ MPa, and $\tau_{zx} = 45$ MPa. Transform this set of cartesian stress components into a new set of cartesian stress components relative to an $Ox'y'z'$ set of coordinates where the $Ox'y'z'$ axes are defined by the following direction cosines:

<table>
<tr><td></td><td colspan="3">(a)</td></tr>
<tr><td></td><td>x</td><td>y</td><td>z</td></tr>
<tr><td>x'</td><td>$\frac{2}{15}$</td><td>$\frac{10}{15}$</td><td>$\frac{11}{15}$</td></tr>
<tr><td>y'</td><td>$\frac{14}{15}$</td><td>$-\frac{5}{15}$</td><td>$\frac{2}{15}$</td></tr>
<tr><td>z'</td><td>$-\frac{5}{15}$</td><td>$-\frac{10}{15}$</td><td>$\frac{10}{15}$</td></tr>
</table>

<table>
<tr><td></td><td colspan="3">(b)</td></tr>
<tr><td></td><td>x</td><td>y</td><td>z</td></tr>
<tr><td>x'</td><td>$\frac{2}{11}$</td><td>$\frac{6}{11}$</td><td>$\frac{9}{11}$</td></tr>
<tr><td>y'</td><td>$\frac{6}{11}$</td><td>$\frac{7}{11}$</td><td>$-\frac{6}{11}$</td></tr>
<tr><td>z'</td><td>$\frac{9}{11}$</td><td>$-\frac{6}{11}$</td><td>$\frac{2}{11}$</td></tr>
</table>

1.16. At a point in a stressed body, the cartesian components of stress are $\sigma_{xx} = 100$ MPa, $\sigma_{yy} = 50$ MPa, $\sigma_{zz} = -25$ MPa, $\tau_{xy} = 50$ MPa, $\tau_{yz} = -25$ MPa, and $\tau_{zx} = 75$ MPa. Transform this set of cartesian stress components into a new set of cartesian stress components relative to an $Ox'y'z'$ set of coordinates where the $Ox'y'z'$ axes are defined by the following direction cosines:

<table>
<tr><td></td><td colspan="3">(a)</td></tr>
<tr><td></td><td>x</td><td>y</td><td>z</td></tr>
<tr><td>x'</td><td>$\frac{1}{17}$</td><td>$\frac{12}{17}$</td><td>$\frac{12}{17}$</td></tr>
<tr><td>y'</td><td>$\frac{12}{17}$</td><td>$-\frac{9}{17}$</td><td>$\frac{8}{17}$</td></tr>
<tr><td>z'</td><td>$\frac{12}{17}$</td><td>$\frac{8}{17}$</td><td>$-\frac{9}{17}$</td></tr>
</table>

<table>
<tr><td></td><td colspan="3">(b)</td></tr>
<tr><td></td><td>x</td><td>y</td><td>z</td></tr>
<tr><td>x'</td><td>$\frac{9}{25}$</td><td>$\frac{12}{25}$</td><td>$\frac{20}{25}$</td></tr>
<tr><td>y'</td><td>$\frac{12}{25}$</td><td>$\frac{16}{25}$</td><td>$-\frac{15}{25}$</td></tr>
<tr><td>z'</td><td>$\frac{20}{25}$</td><td>$-\frac{15}{25}$</td><td>0</td></tr>
</table>

1.17. For the state of stress at the point of Exercise 1.3, determine the principal stresses and the maximum shear stress at the point.

1.18. For the state of stress at the point of Exercise 1.4, determine the principal stresses and the maximum shear stress at the point.

1.19. For the state of stress at the point of Exercise 1.5, determine the principal stresses and the maximum shear stress at the point.

1.20. For the state of stress at the point of Exercise 1.6, determine the principal stresses and the maximum shear stress at the point.

1.21. For the state of stress at the point of Exercise 1.12, determine the principal stresses and the maximum shear stress at the point.

1.22. For the state of stress at the point of Exercise 1.14, determine the principal stresses and the maximum shear stress at the point.

1.23. Determine the principal stresses and the maximum shear stress at the point $x = \frac{1}{2}$, $y = 1$, $z = \frac{3}{4}$ for the stress distribution given in Exercise 1.9.

1.24. At a point in a stressed body, the cartesian components of stress are $\sigma_{xx} = 50$ MPa, $\sigma_{yy} = 60$ MPa, $\sigma_{zz} = 70$ MPa, $\tau_{xy} = 100$ MPa, $\tau_{yz} = 75$ MPa, and $\tau_{zx} = 50$ MPa. Determine (a) the principal stresses and the maximum shear stress at the point and (b) the orientation of the plane on which the maximum tensile stress acts.

1.25. At a point in a stressed body, the cartesian components of stress are $\sigma_{xx} = \sigma_{yy} = \sigma_{zz} = 0$, $\tau_{xy} = 75$ MPa, $\tau_{yz} = 0$, and $\tau_{zx} = -75$ MPa. Determine (a) the principal stresses and the maximum shear stress at the point and (b) the orientation of the plane on which the maximum tensile stress acts.

1.26. At a point in a stressed body, the cartesian components of stress are $\sigma_{xx} = \sigma_{yy} = \sigma_{zz} = 50$ MPa, $\tau_{xy} = 200$ MPa, $\tau_{yz} = 0$, and $\tau_{zx} = 150$ MPa. Determine the principal stresses and the associated principal directions. Check on the invariance of I_1, I_2, and I_3.

1.27. At a point in a stressed body, the cartesian components of stress are $\sigma_{xx} = \sigma_{yy} = \sigma_{zz} = 0$, $\tau_{xy} = \tau_{yz} = \tau_{zx} = 60$ MPa. Determine the principal stresses and the associated principal directions. Check on the invariance of I_1, I_2, and I_3.

1.28. A machine component is subjected to loads which produce the following stress field in a region where an oil hole must be drilled: $\sigma_{xx} = 100$ MPa, $\sigma_{yy} = -60$ MPa, $\sigma_{zz} = -30$ MPa, $\tau_{xy} = 50$ MPa, $\tau_{yz} = 0$, and $\tau_{zx} = 0$. To minimize the effects of stress concentrations, the hole must be drilled along a line parallel to the direction of the maximum tensile stress in the region. Determine the direction cosines associated with the centerline of the hole with respect to the reference $Oxyz$ coordinate system.

1.29. A two-dimensional state of stress ($\sigma_{zz} = \tau_{zx} = \tau_{zy} = 0$) exists at a point on the free surface of a machine component. The remaining cartesian components of stress are $\sigma_{xx} = 90$ MPa, $\sigma_{yy} = -80$ MPa, and $\tau_{xy} = -30$ MPa. Determine (a) the principal stresses and their associated directions at the point and (b) the maximum shear stress at the point.

1.30. A two-dimensional state of stress ($\sigma_{zz} = \tau_{zx} = \tau_{zy} = 0$) exists at a point on the surface of a loaded member. Determine the principal stresses and the maximum shear stress at the point if the remaining cartesian components of stress are $\sigma_{xx} = 45$ MPa, $\sigma_{yy} = 30$ MPa, and $\tau_{xy} = 20$ MPa.

1.31. A two-dimensional state of stress ($\sigma_{zz} = \tau_{zx} = \tau_{zy} = 0$) exists at a point on the surface of a loaded member. The remaining cartesian components of stress are $\sigma_{xx} = 100$ MPa, $\sigma_{yy} = 50$ MPa, and $\tau_{xy} = -20$ MPa. Determine the principal stresses and the maximum shear stress at the point.

1.32. A two-dimensional state of stress ($\sigma_{zz} = \tau_{zx} = \tau_{zy} = 0$) exists at a point on the surface of a loaded member. The remaining cartesian components of stress are $\sigma_{xx} = 90$ MPa, $\sigma_{yy} = 40$ MPa, and $\tau_{xy} = 60$ MPa. Determine the principal stresses and the maximum shear stress at the point.

1.33. Solve Exercise 1.31 by means of Mohr's circle.

1.34. Solve Exercise 1.32 by means of Mohr's circle.

1.35. Write a computer program to determine (a) the principal stresses, (b) the principal directions, (c) the maximum shear stresses, and (d) the von Mises stress if the cartesian components of stresses are given.

1.36. At the point of Exercise 1.31, determine the normal and shear stresses on a plane whose outer normal has the direction cosines

$$\cos (n, x) = \tfrac{3}{5} \qquad \cos (n, y) = \tfrac{4}{5} \qquad \cos (n, z) = 0$$

1.37. At the point of Exercise 1.32, determine the normal and shear stresses on a plane whose outer normal has the direction cosines

$$\cos (n, x) = \tfrac{1}{3} \qquad \cos (n, y) = \tfrac{2}{3} \qquad \cos (n, z) = \tfrac{2}{3}$$

1.38. At a point in a machine part, the principal stresses are $\sigma_1 = 100$ MPa, $\sigma_2 = 60$ MPa, and $\sigma_3 = 30$ MPa. Determine the normal and shear stresses on a plane whose outer normal has the direction cosines

$$\cos (n, n_1) = \frac{\sqrt{3}}{2} \qquad \cos (n, n_2) = 0 \qquad \cos (n, n_3) = \tfrac{1}{2}$$

1.39. If the three principal stresses relative to the $Oxyz$ reference system are $\sigma_1 = \sigma_{xx} = 50$ MPa, $\sigma_2 = \sigma_{yy} = 40$ MPa, $\sigma_3 = \sigma_{zz} = -20$ MPa, determine the six cartesian components of stress relative to the $Ox'y'z'$ reference system where $Ox'y'z'$ is defined as:

θ	Case 1	Case 2	Case 3	Case 4
$x - x'$	$\pi/4$	$\pi/2$	0	$\pi/4$
$y - y'$	$\pi/4$	$\pi/2$	$\pi/4$	0
$z - z'$	0	0	$\pi/4$	$\pi/4$

1.40. If the three principal stresses relative to the $Oxyz$ reference system are $\sigma_1 = \sigma_{xx} = 75$ MPa, $\sigma_2 = \sigma_{yy} = 60$ MPa, $\sigma_3 = \sigma_{zz} = 50$ MPa, determine the six cartesian components of stress relative to an $Ox'y'z'$ reference system where the direction angles associated with the $Ox'y'z'$ axes are

	x	y	z
x'	$\pi/3$	$\pi/3$	$\pi/4$
y'	$3\pi/4$	$\pi/4$	$\pi/2$
z'	$\pi/3$	$\pi/3$	$3\pi/4$

1.41. If the three principal stresses relative to the $Oxyz$ reference system are $\sigma_1 = \sigma_{xx} = 135$ MPa, $\sigma_2 = \sigma_{yy} = 90$ MPa, $\sigma_3 = \sigma_{zz} = 45$ MPa, determine the six cartesian components of stress relative to an $Ox'y'z'$ reference system where the $Ox'y'z'$ axes are defined by the following direction cosines:

<table>
<tr><td align="center">(a)</td><td></td><td></td><td></td><td align="center">(b)</td><td></td><td></td><td></td></tr>
<tr><td></td><td align="center">x</td><td align="center">y</td><td align="center">z</td><td></td><td align="center">x</td><td align="center">y</td><td align="center">z</td></tr>
<tr><td align="center">x</td><td align="center">$\frac{2}{3}$</td><td align="center">$\frac{2}{3}$</td><td align="center">$-\frac{1}{3}$</td><td align="center">x'</td><td align="center">$\frac{1}{9}$</td><td align="center">$-\frac{8}{9}$</td><td align="center">$\frac{4}{9}$</td></tr>
<tr><td align="center">y'</td><td align="center">$-\frac{2}{3}$</td><td align="center">$\frac{1}{3}$</td><td align="center">$-\frac{2}{3}$</td><td align="center">y'</td><td align="center">$\frac{4}{9}$</td><td align="center">$\frac{4}{9}$</td><td align="center">$\frac{7}{9}$</td></tr>
<tr><td align="center">z'</td><td align="center">$-\frac{1}{3}$</td><td align="center">$\frac{2}{3}$</td><td align="center">$\frac{2}{3}$</td><td align="center">z'</td><td align="center">$-\frac{8}{9}$</td><td align="center">$\frac{1}{9}$</td><td align="center">$\frac{4}{9}$</td></tr>
</table>

1.42. Resolve the general state of stress given in Exercise 1.3 into a hydrostatic state of stress and a state of pure shearing stress.

1.43. Resolve the general state of stress given in Exercise 1.4 into a hydrostatic state of stress and a state of pure shearing stress.

1.44. Resolve the general state of stress given in Exercise 1.5 into a hydrostatic state of stress and a state of pure shearing stress.

1.45. Resolve the general state of stress given in Exercise 1.6 into a hydrostatic state of stress and a state of pure shearing stress.

1.46. Determine the octahedral normal and shearing stresses associated with the principal stresses σ_1, σ_2, and σ_3. Octahedral normal and shearing stresses occur on planes whose outer normal makes equal angles with the principal directions n_1, n_2, and n_3.

REFERENCES

1. Boresi, A. P., and P. P. Lynn: *Elasticity in Engineering Mechanics*, chap. 3, Prentice-Hall, Englewood Cliffs, N.J., 1974.
2. Chou, P. C., and N. J. Pagano: *Elasticity*, chap. 1, Van Nostrand, Princeton, N.J., 1967.
3. Durelli, A. J., E. A. Phillips, and C. H. Tsao: *Introduction to the Theoretical and Experimental Analysis of Stress and Strain*, chap. 1, McGraw-Hill, New York, 1958.
4. Love, A. E. H.: *A Treatise on the Mathematical Theory of Elasticity*, chap. 2, Dover, New York, 1944.
5. Sechler, E. E.: *Elasticity in Engineering*, chaps. 2 and 3, John Wiley & Sons, New York, 1952.
6. Sokolnikoff, I. S.: *Mathematical Theory of Elasticity*, 2d ed., chap. 2, McGraw-Hill, New York, 1956.
7. Southwell, R. V.: *An Introduction to the Theory of Elasticity*, chap. 8, Oxford University Press, Fair Lawn, N.J., 1953.
8. Timoshenko, S. P., and J. N. Goodier: *Theory of Elasticity*, 2d ed., chaps. 1, 8, and 9, McGraw-Hill, New York, 1951.

CHAPTER
2

STRAIN
AND THE
STRESS–STRAIN
RELATIONSHIPS

2.1 INTRODUCTION

In Chap. 1 the state of stress which develops at an arbitrary point within a body as a result of surface- or body-force loadings was discussed. The relationships obtained were based on the conditions of equilibrium, and since no assumptions were made regarding body deformations or physical properties of the material of which the body was composed, the results are valid for any material and for any amount of body deformation. In this chapter the subject of body deformation and associated strain will be discussed. Since strain is a pure geometric quantity, no restrictions on body material will be required. However, in order to obtain linear equations relating displacement to strain, restrictions must be placed on the allowable deformations. In a later section, when the stress-strain relations are developed, the elastic constants of the body material must be considered.

2.2 DEFINITIONS OF DISPLACEMENT AND STRAIN

If a given body is subjected to a system of forces, individual points of the body will, in general, move. This movement of an arbitrary point is a vector quantity known as a *displacement*. If the various points in the body undergo different movements, each can be represented by its own unique displacement vector. Each vector can be resolved into components parallel to a set of cartesian coordinate axes such that u, v, and w are the displacement components in the x, y, and z directions, respectively.

Motion of the body may be considered as the sum of two parts:

1. A translation and/or rotation of the body as a whole
2. The movement of the points of the body relative to each other

The translation or rotation of the body as a whole is known as *rigid-body motion*. This type of motion is applicable to either the idealized rigid body or the real deformable body. The movement of the points of the body relative to each other is known as a *deformation* and is obviously a property of real bodies only. Rigid-body motions can be large or small. Deformations, in general, are small except when rubberlike materials or specialized structures such as long, slender beams are involved.

Strain is a geometric quantity which depends on the relative movements of two or three points in the body and therefore is related only to the deformation displacements. Since rigid-body displacements do not produce strains, they will be neglected in all further developments in this chapter. In the preceding chapter two types of stress were discussed: normal stress and shear stress. This same classification will be used for strains. A normal strain is defined as the change in length of a line segment between two points divided by the original length of the line segment. A shearing strain is defined as the angular change between two line segments which were originally perpendicular. The relationships between strains and displacements can be determined by considering the deformation of an arbitrary cube in a body as a system of loads is applied. This deformation is illustrated in Fig. 2.1, in which a general point P is moved through a distance u in the x direction, v in the y direction, and w in the z direction. The other corners of the cube are also displaced and, in general, they will be displaced by amounts which differ from those at point P. For example the displacements u^*, v^*, and w^* associated with point Q can be expressed in terms of the displacements u, v, and w at point P by means of a Taylor-series expansion. Thus

$$u^* = u + \frac{\partial u}{\partial x} \Delta x + \frac{\partial u}{\partial y} \Delta y + \frac{\partial u}{\partial z} \Delta z + \cdots$$

$$v^* = v + \frac{\partial v}{\partial x} \Delta x + \frac{\partial v}{\partial y} \Delta y + \frac{\partial v}{\partial z} \Delta z + \cdots \tag{2.1}$$

$$w^* = w + \frac{\partial w}{\partial x} \Delta x + \frac{\partial w}{\partial y} \Delta y + \frac{\partial w}{\partial z} \Delta z + \cdots$$

The terms shown in the above expressions are the only significant terms if it is assumed that the cube is sufficiently small for higher-order terms such as $(\Delta x)^2$, $(\Delta y)^2$, $(\Delta z)^2, \ldots$ to be neglected. Under these conditions, planes will remain plane

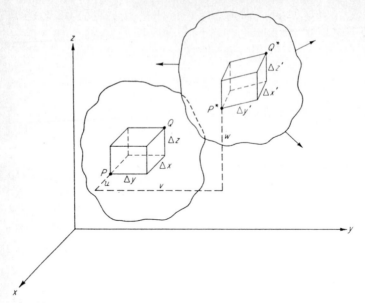

FIGURE 2.1
The distortion of an arbitrary cube in a body due to the application of a system of forces.

and straight lines will remain straight lines in the deformed cube, as shown in Fig. 2.1.

The average normal strain along an arbitrary line segment was previously defined as the change in length of the line segment divided by its original length. This normal strain can be expressed in terms of the displacements experienced by points at the ends of the segment. For example, consider the line PQ originally oriented parallel to the x axis, as shown in Fig. 2.2. Since y and z are constant along PQ, Eqs. (2.1) yield the following displacements for point Q if the displacements for point P are u, v, and w:

$$u^* = u + \frac{\partial u}{\partial x}\Delta x \qquad v^* = v + \frac{\partial v}{\partial x}\Delta x \qquad w^* = w + \frac{\partial w}{\partial x}\Delta x$$

From the definition of normal strain,

$$\epsilon_{xx} = \frac{\Delta x' - \Delta x}{\Delta x} \qquad (a)$$

which is equivalent to

$$\Delta x' = (1 + \epsilon_{xx})\Delta x \qquad (b)$$

As shown in Fig. 2.2, the deformed length $\Delta x'$ can be expressed in terms of the displacement gradients as

$$(\Delta x')^2 = \left[\left(1 + \frac{\partial u}{\partial x}\right)\Delta x\right]^2 + \left(\frac{\partial v}{\partial x}\Delta x\right)^2 + \left(\frac{\partial w}{\partial x}\Delta x\right)^2 \qquad (c)$$

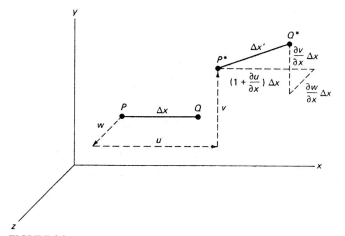

FIGURE 2.2
Displacement gradients associated with the normal strain ϵ_{xx}.

Squaring Eq. (*b*) and substituting Eq. (*c*) yields

$$(1 + \epsilon_{xx})^2(\Delta x)^2 = \left[1 + 2\frac{\partial u}{\partial x} + \left(\frac{\partial u}{\partial x}\right)^2 + \left(\frac{\partial v}{\partial x}\right)^2 + \left(\frac{\partial w}{\partial x}\right)^2\right](\Delta x)^2$$

or

$$\epsilon_{xx} = \sqrt{1 + 2\frac{\partial u}{\partial x} + \left(\frac{\partial u}{\partial x}\right)^2 + \left(\frac{\partial v}{\partial x}\right)^2 + \left(\frac{\partial w}{\partial x}\right)^2} - 1 \qquad (2.2a)$$

In a similar manner considering line segments originally oriented parallel to the y and z axes leads to

$$\epsilon_{yy} = \sqrt{1 + 2\frac{\partial v}{\partial y} + \left(\frac{\partial v}{\partial y}\right)^2 + \left(\frac{\partial w}{\partial y}\right)^2 + \left(\frac{\partial u}{\partial y}\right)^2} - 1 \qquad (2.2b)$$

$$\epsilon_{zz} = \sqrt{1 + 2\frac{\partial w}{\partial z} + \left(\frac{\partial w}{\partial z}\right)^2 + \left(\frac{\partial u}{\partial z}\right)^2 + \left(\frac{\partial v}{\partial z}\right)^2} - 1 \qquad (2.2c)$$

The shear-strain components can also be related to the displacements by considering the changes in right angle experienced by the edges of the cube during deformation. For example, consider lines PQ and PR, as shown in Fig. 2.3. The angle θ^* between P^*Q^* and P^*R^* in the deformed state can be expressed in terms of the displacement gradients since the cosine of the angle between any two intersecting lines in space is the sum of the pairwise products of the direction cosines of the lines with respect to the same set of reference axes. Thus

$$\cos\theta^* = \left[\left(1 + \frac{\partial u}{\partial x}\right)\frac{\Delta x}{\Delta x'}\right]\left(\frac{\partial u}{\partial y}\frac{\Delta y}{\Delta y'}\right) + \left(\frac{\partial v}{\partial x}\frac{\Delta x}{\Delta x'}\right)\left[\left(1 + \frac{\partial v}{\partial y}\right)\frac{\Delta y}{\Delta y'}\right]$$

$$+ \left(\frac{\partial w}{\partial x}\frac{\Delta x}{\Delta x'}\right)\left(\frac{\partial w}{\partial y}\frac{\Delta y}{\Delta y'}\right) \qquad (d)$$

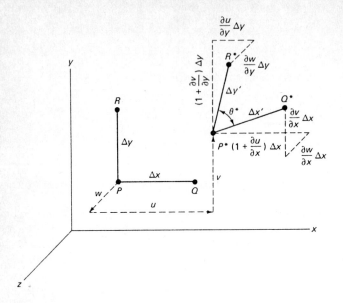

FIGURE 2.3
Displacement gradients associated with the shear strain γ_{xy}.

From the definition of shear strain

$$\gamma_{xy} = \left(\frac{\pi}{2} - \theta^*\right) \tag{e}$$

therefore

$$\sin \gamma_{xy} = \sin \left(\frac{\pi}{2} - \theta^*\right) = \cos \theta^* \tag{f}$$

Substituting Eq. (d) into Eq. (f) and simplifying yields

$$\sin \gamma_{xy} = \left[\left(1 + \frac{\partial u}{\partial x}\right)\frac{\partial u}{\partial y} + \left(1 + \frac{\partial v}{\partial x}\right)\frac{\partial v}{\partial y} + \frac{\partial w}{\partial x}\frac{\partial w}{\partial y}\right]\left(\frac{\Delta x \; \Delta y}{\Delta x' \; \Delta y'}\right)$$

From Eq. (b)

$$\Delta x' = (1 + \epsilon_{xx}) \, \Delta x \qquad \text{and} \qquad \Delta y' = (1 + \epsilon_{yy}) \, \Delta y$$

therefore

$$\gamma_{xy} = \arcsin \frac{\dfrac{\partial u}{\partial y} + \dfrac{\partial v}{\partial x} + \dfrac{\partial u}{\partial x}\dfrac{\partial u}{\partial y} + \dfrac{\partial v}{\partial x}\dfrac{\partial v}{\partial y} + \dfrac{\partial w}{\partial x}\dfrac{\partial w}{\partial y}}{(1 + \epsilon_{xx})(1 + \epsilon_{yy})} \tag{2.3a}$$

In a similar manner by considering two line segments originally oriented parallel to the y and z axes and the z and x axes

$$\gamma_{yz} = \arcsin \frac{\dfrac{\partial v}{\partial z} + \dfrac{\partial w}{\partial y} + \dfrac{\partial v}{\partial y}\dfrac{\partial v}{\partial z} + \dfrac{\partial w}{\partial y}\dfrac{\partial w}{\partial z} + \dfrac{\partial u}{\partial y}\dfrac{\partial u}{\partial z}}{(1 + \epsilon_{yy})(1 + \epsilon_{zz})} \tag{2.3b}$$

$$\gamma_{zx} = \arcsin \frac{\dfrac{\partial w}{\partial x} + \dfrac{\partial u}{\partial z} + \dfrac{\partial w}{\partial z}\dfrac{\partial w}{\partial x} + \dfrac{\partial u}{\partial z}\dfrac{\partial u}{\partial x} + \dfrac{\partial v}{\partial z}\dfrac{\partial v}{\partial x}}{(1 + \epsilon_{zz})(1 + \epsilon_{xx})} \tag{2.3c}$$

Equations (2.2) and (2.3) represent a common engineering description of strain in terms of positions of points in a body before and after deformation. In the development of these equations, no limitations were imposed on the magnitudes of the strains. One restriction was introduced, however, when the higher-order terms in the Taylor-series expansion for displacement were neglected. This restriction has the effect of limiting the length of the line segment (gage length) used for strain determinations unless displacement gradients ($\delta u/\delta x$, $\delta u/\delta y$, ...) in the region of interest are essentially constant. If displacement gradients change rapidly with position in the region of interest, very short gage lengths will be required for accurate strain measurements.

In a wide variety of engineering problems, the displacements and strains produced by the applied loads are very small. Under these conditions, it can be assumed that products and squares of displacement gradients will be small with respect to the displacement gradients and therefore can be neglected. With this assumption Eqs. (2.2) and (2.3) reduce to the strain-displacement equations frequently encountered in the theory of elasticity. The reduced form of the equations is

$$\epsilon_{xx} = \frac{\partial u}{\partial x} \tag{2.4a}$$

$$\epsilon_{yy} = \frac{\partial v}{\partial y} \tag{2.4b}$$

$$\epsilon_{zz} = \frac{\partial w}{\partial z} \tag{2.4c}$$

$$\gamma_{xy} = \frac{\partial v}{\partial x} + \frac{\partial u}{\partial y} \tag{2.4d}$$

$$\gamma_{yz} = \frac{\partial w}{\partial y} + \frac{\partial v}{\partial z} \tag{2.4e}$$

$$\gamma_{zx} = \frac{\partial u}{\partial z} + \frac{\partial w}{\partial x} \tag{2.4f}$$

Equations (2.4) indicate that it is a simple matter to convert a displacement field into a strain field. However, as will be emphasized later, an entire displacement field is rarely determined experimentally. Usually, strains are determined at a number of small areas on the surface of the body through the use of strain gages. In certain problems, however, the displacement field can be computed analytically, and in these instances Eqs. (2.4) become very important.

2.3 STRAIN EQUATIONS OF TRANSFORMATION

Now that the normal and shearing strains in the x, y, z directions have been determined, consider the normal strain in an arbitrary direction. Refer to Fig. 2.4 and consider the elongation of the diagonal PQ. By definition the strain along PQ is

$$\epsilon_{PQ} = \frac{P^*Q^* - PQ}{PQ} \tag{a}$$

From geometric considerations, as illustrated in Fig. 2.4,

$$(PQ)^2 = (\Delta x)^2 + (\Delta y)^2 + (\Delta z)^2 \tag{b}$$

$$(P^*Q^*)^2 = (\Delta x^*)^2 + (\Delta y^*)^2 + (\Delta z^*)^2 \tag{c}$$

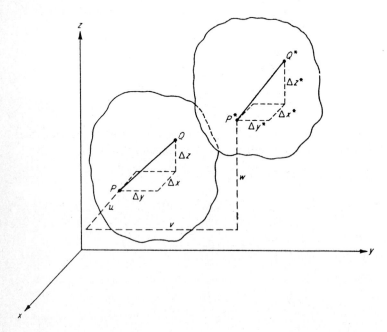

FIGURE 2.4
Displacements of points P and Q in a body which result from the application of a system of loads.

In general, the component Δx^* will have a different length than the component Δx because of the deformation of the body in the x direction. From Fig. 2.4 it can easily be seen that

$$\Delta x^* = \left(1 + \frac{\partial u}{\partial x}\right)\Delta x + \frac{\partial u}{\partial y}\Delta y + \frac{\partial u}{\partial z}\Delta z$$

$$\Delta y^* = \frac{\partial v}{\partial x}\Delta x + \left(1 + \frac{\partial v}{\partial y}\right)\Delta y + \frac{\partial v}{\partial z}\Delta z \qquad (d)$$

$$\Delta z^* = \frac{\partial w}{\partial x}\Delta x + \frac{\partial w}{\partial y}\Delta y + \left(1 + \frac{\partial w}{\partial z}\right)\Delta z$$

If Eqs. (d) are substituted into Eq. (c), the length of the deformed line segment P^*Q^* can be computed. In the substitution, since the deformations are extremely small, the products and squares of derivatives can be neglected. Thus

$$(P^*Q^*)^2 = \left(1 + 2\frac{\partial u}{\partial x}\right)(\Delta x)^2 + \left(1 + 2\frac{\partial v}{\partial y}\right)(\Delta y)^2$$

$$+ \left(1 + 2\frac{\partial w}{\partial z}\right)(\Delta z)^2 + 2\left(\frac{\partial u}{\partial y} + \frac{\partial v}{\partial x}\right)\Delta x\,\Delta y$$

$$+ 2\left(\frac{\partial v}{\partial z} + \frac{\partial w}{\partial y}\right)\Delta y\,\Delta z + 2\left(\frac{\partial w}{\partial x} + \frac{\partial u}{\partial z}\right)\Delta z\,\Delta x \qquad (e)$$

Equation (a) can be rearranged in the following form:

$$\epsilon_{PQ} = \frac{P^*Q^*}{PQ} - 1$$

or
$$(\epsilon_{PQ} + 1)^2 = \left(\frac{P^*Q^*}{PQ}\right)^2$$

If Eqs. (b) and (e) are substituted into this rearranged form of Eq. (a), the following equation can be obtained after some rearrangement of terms:

$$(\epsilon_{PQ} + 1)^2 = \cos^2(x, PQ) + \cos^2(y, PQ) + \cos^2(z, PQ)$$

$$+ 2\frac{\partial u}{\partial x}\cos^2(x, PQ) + 2\frac{\partial v}{\partial y}\cos^2(y, PQ) + 2\frac{\partial w}{\partial z}\cos^2(z, PQ)$$

$$+ 2\left(\frac{\partial u}{\partial y} + \frac{\partial v}{\partial x}\right)\cos(x, PQ)\cos(y, PQ)$$

$$+ 2\left(\frac{\partial v}{\partial z} + \frac{\partial w}{\partial y}\right)\cos(y, PQ)\cos(z, PQ)$$

$$+ 2\left(\frac{\partial w}{\partial x} + \frac{\partial u}{\partial z}\right)\cos(z, PQ)\cos(x, PQ) \qquad (f)$$

If the left-hand side of Eq. (f) is expanded, the ϵ_{PQ}^2 term can be neglected since it is of the same order of magnitude as the products and squares of displacement derivatives which were neglected in a previous step of this development. Recall also that

$$\cos^2(x, PQ) + \cos^2(y, PQ) + \cos^2(z, PQ) = 1$$

Thus the basic equation for the strain along an arbitrary line segment is

$$\epsilon_{PQ} = \frac{\partial u}{\partial x}\cos^2(x, PQ) + \frac{\partial v}{\partial y}\cos^2(y, PQ) + \frac{\partial w}{\partial z}\cos^2(z, PQ)$$

$$+ \left(\frac{\partial u}{\partial y} + \frac{\partial v}{\partial x}\right)\cos(x, PQ)\cos(y, PQ)$$

$$+ \left(\frac{\partial v}{\partial z} + \frac{\partial w}{\partial y}\right)\cos(y, PQ)\cos(z, PQ)$$

$$+ \left(\frac{\partial w}{\partial x} + \frac{\partial u}{\partial z}\right)\cos(z, PQ)\cos(x, PQ) \qquad (2.5a)$$

Equations (2.4) and (2.5a) can be used to determine $\epsilon_{x'x'}$ by choosing the direction of PQ parallel to the x' axis. Then

$$\begin{aligned}
\epsilon_{x'x'} = {} & \epsilon_{xx}\cos^2(x, x') + \epsilon_{yy}\cos^2(y, x') \\
& + \epsilon_{zz}\cos^2(z, x') + \gamma_{xy}\cos(x, x')\cos(y, x') \\
& + \gamma_{yz}\cos(y, x')\cos(z, x') \\
& + \gamma_{zx}\cos(z, x')\cos(x, x')
\end{aligned} \qquad (2.6a)$$

In a similar manner $\epsilon_{y'y'}$ and $\epsilon_{z'z'}$ can be determined by choosing the direction of PQ parallel to the y' and z' axes, respectively:

$$\begin{aligned}
\epsilon_{y'y'} = {} & \epsilon_{yy}\cos^2(y, y') + \epsilon_{zz}\cos^2(z, y') \\
& + \epsilon_{xx}\cos^2(x, y') + \gamma_{yz}\cos(y, y')\cos(z, y') \\
& + \gamma_{zx}\cos(z, y')\cos(x, y') \\
& + \gamma_{xy}\cos(x, y')\cos(y, y')
\end{aligned} \qquad (2.6b)$$

$$\begin{aligned}
\epsilon_{z'z'} = {} & \epsilon_{zz}\cos^2(z, z') + \epsilon_{xx}\cos^2(x, z') \\
& + \epsilon_{yy}\cos^2(y, z') + \gamma_{zx}\cos(z, z')\cos(x, z') \\
& + \gamma_{xy}\cos(x, z')\cos(y, z') \\
& + \gamma_{yz}\cos(y, z')\cos(z, z')
\end{aligned} \qquad (2.6c)$$

A similar but somewhat more involved derivation can be used to establish the shearing strains. Consider the angular change in an arbitrary right angle formed by two line segments PQ_1 and PQ_2. The shearing strain γ_{PQ_1, PQ_2} can be shown

[3, p. 44]† to be given by

$$\gamma_{PQ_1, PQ_2} = 2\epsilon_{xx} \cos(x, PQ_1) \cos(x, PQ_2)$$
$$+ 2\epsilon_{yy} \cos(y, PQ_1) \cos(y, PQ_2)$$
$$+ 2\epsilon_{zz} \cos(z, PQ_1) \cos(z, PQ_2)$$
$$+ \gamma_{xy}[\cos(x, PQ_1) \cos(y, PQ_2) + \cos(x, PQ_2) \cos(y, PQ_1)]$$
$$+ \gamma_{yz}[\cos(y, PQ_1) \cos(z, PQ_2) + \cos(y, PQ_2) \cos(z, PQ_1)]$$
$$+ \gamma_{zx}[\cos(z, PQ_1) \cos(x, PQ_2) + \cos(z, PQ_2) \cos(x, PQ_1)] \quad (2.5b)$$

By choosing PQ_1 parallel to x' and PQ_2 parallel to y', an expression for $\gamma_{x'y'}$ is obtained as follows:

$$\gamma_{x'y'} = 2\epsilon_{xx} \cos(x, x') \cos(x, y')$$
$$+ 2\epsilon_{yy} \cos(y, x') \cos(y, y')$$
$$+ 2\epsilon_{zz} \cos(z, x') \cos(z, y')$$
$$+ \gamma_{xy}[\cos(x, x') \cos(y, y') + \cos(x, y') \cos(y, x')]$$
$$+ \gamma_{yz}[\cos(y, x') \cos(z, y') + \cos(y, y') \cos(z, x')]$$
$$+ \gamma_{zx}[\cos(z, x') \cos(x, y') + \cos(z, y') \cos(x, x')] \quad (2.6d)$$

Similarly

$$\gamma_{y'z'} = 2\epsilon_{yy} \cos(y, y') \cos(y, z')$$
$$+ 2\epsilon_{zz} \cos(z, y') \cos(z, z')$$
$$+ 2\epsilon_{xx} \cos(x, y') \cos(x, z')$$
$$+ \gamma_{yz}[\cos(y, y') \cos(z, z') + \cos(y, z') \cos(z, y')]$$
$$+ \gamma_{zx}[\cos(z, y') \cos(x, z') + \cos(z, z') \cos(x, y')]$$
$$+ \gamma_{xy}[\cos(x, y') \cos(y, z') + \cos(x, z') \cos(y, y')] \quad (2.6e)$$

$$\gamma_{z'x'} = 2\epsilon_{zz} \cos(z, z') \cos(z, x')$$
$$+ 2\epsilon_{xx} \cos(x, z') \cos(x, x')$$
$$+ 2\epsilon_{yy} \cos(y, z') \cos(y, x')$$
$$+ \gamma_{zx}[\cos(z, z') \cos(x, x') + \cos(z, x') \cos(x, z')]$$
$$+ \gamma_{xy}[\cos(x, z') \cos(y, x') + \cos(x, x') \cos(y, z')]$$
$$+ \gamma_{yz}[\cos(y, z') \cos(z, x') + \cos(y, x') \cos(z, z')] \quad (2.6f)$$

Equations (2.6a) to (2.6f) are the strain equations of transformation and can be used to transform the six cartesian components of strain $\epsilon_{xx}, \epsilon_{yy}, \epsilon_{zz}, \gamma_{xy}, \gamma_{yz}, \gamma_{zx}$ relative to the $Oxyz$ reference system to six other cartesian components of strain relative to the $Ox'y'z'$ reference system.

---

† Numbers in brackets refer to numbered references at the end of the chapter.

A comparison of Eqs. (2.6) with the stress equations of transformation [Eq. (1.6)] shows remarkable similarities:

$$\sigma_{xx} \leftrightarrow \epsilon_{xx} \qquad 2\tau_{xy} \leftrightarrow \gamma_{xy}$$
$$\sigma_{yy} \leftrightarrow \epsilon_{yy} \qquad 2\tau_{yz} \leftrightarrow \gamma_{yz} \qquad (2.7)$$
$$\sigma_{zz} \leftrightarrow \epsilon_{zz} \qquad 2\tau_{zx} \leftrightarrow \gamma_{zx}$$

Here the symbol $\leftrightarrow$ indicates an interchange. This interchange is important since many of the derivations given in the preceding chapter for stresses can be converted directly into strains. Some of these conversions are indicated in Sec. 2.4.

2.4 PRINCIPAL STRAINS

From the similarity between the laws of stress and strain transformation it can be concluded that there exist at most three distinct principal strains with their three associated principal directions. By substituting the conversions indicated by Eqs. (2.7) into Eq. (1.7), the cubic equation whose roots give the principal strains is obtained:

$$\epsilon_n^3 - (\epsilon_{xx} + \epsilon_{yy} + \epsilon_{zz})\epsilon_n^2$$
$$+ \left(\epsilon_{xx}\epsilon_{yy} + \epsilon_{yy}\epsilon_{zz} + \epsilon_{zz}\epsilon_{xx} - \frac{\gamma_{xy}^2}{4} - \frac{\gamma_{yz}^2}{4} - \frac{\gamma_{zx}^2}{4} \right)\epsilon_n$$
$$- \left(\epsilon_{xx}\epsilon_{yy}\epsilon_{zz} - \epsilon_{xx}\frac{\gamma_{yz}^2}{4} - \epsilon_{yy}\frac{\gamma_{zx}^2}{4} - \epsilon_{zz}\frac{\gamma_{xy}^2}{4} + \frac{\gamma_{xy}\gamma_{yz}\gamma_{zx}}{4} \right) = 0 \qquad (2.8)$$

As with principal stresses, three situations exist:

$$\epsilon_1 \neq \epsilon_2 \neq \epsilon_3 \qquad \epsilon_1 = \epsilon_2 \neq \epsilon_3 \qquad \epsilon_1 = \epsilon_2 = \epsilon_3$$

The significance of these three cases is determined from the discussion in Sec. 1.6.

Similarly, there are three strain invariants which are analogous to the three stress invariants. By substituting Eqs. (2.7) into Eqs. (1.8), the following expressions are obtained for the strain invariants:

$$J_1 = \epsilon_{xx} + \epsilon_{yy} + \epsilon_{zz}$$
$$J_2 = \epsilon_{xx}\epsilon_{yy} + \epsilon_{yy}\epsilon_{zz} + \epsilon_{zz}\epsilon_{xx} - \frac{\gamma_{xy}^2}{4} - \frac{\gamma_{yz}^2}{4} - \frac{\gamma_{zx}^2}{4} \qquad (2.9)$$
$$J_3 = \epsilon_{xx}\epsilon_{yy}\epsilon_{zz} - \frac{\epsilon_{xx}\gamma_{yz}^2}{4} - \frac{\epsilon_{yy}\gamma_{zx}^2}{4} - \frac{\epsilon_{zz}\gamma_{xy}^2}{4} + \frac{\gamma_{xy}\gamma_{yz}\gamma_{zx}}{4}$$

It is clear that other equations derived in Chap. 1 for stresses could easily be converted into equations in terms of strains. A few more will be covered in the exercises at the end of the chapter, and others will be converted as the need arises.

2.5 COMPATIBILITY

From a given displacement field, i.e., three equations expressing u, v, and w as functions of x, y, and z, a unique strain field can be determined by using Eqs. (2.4). However, an arbitrary strain field may yield an impossible displacement field, i.e., one in which the body might contain voids after deformation. A valid displacement field can be ensured only if the body under consideration is simply connected and if the strain field satisfies a set of equations known as the *compatibility relations*. The six equations of compatibility which must be satisfied are

$$\frac{\partial^2 \gamma_{xy}}{\partial x\, \partial y} = \frac{\partial^2 \epsilon_{xx}}{\partial y^2} + \frac{\partial^2 \epsilon_{yy}}{\partial x^2} \tag{2.10a}$$

$$\frac{\partial^2 \gamma_{yz}}{\partial y\, \partial z} = \frac{\partial^2 \epsilon_{yy}}{\partial z^2} + \frac{\partial^2 \epsilon_{zz}}{\partial y^2} \tag{2.10b}$$

$$\frac{\partial^2 \gamma_{zx}}{\partial z\, \partial x} = \frac{\partial^2 \epsilon_{zz}}{\partial x^2} + \frac{\partial^2 \epsilon_{xx}}{\partial z^2} \tag{2.10c}$$

$$2\frac{\partial^2 \epsilon_{xx}}{\partial y\, \partial z} = \frac{\partial}{\partial x}\left(-\frac{\partial \gamma_{yz}}{\partial x} + \frac{\partial \gamma_{zx}}{\partial y} + \frac{\partial \gamma_{xy}}{\partial z} \right) \tag{2.10d}$$

$$2\frac{\partial^2 \epsilon_{yy}}{\partial z\, \partial x} = \frac{\partial}{\partial y}\left(\frac{\partial \gamma_{yz}}{\partial x} - \frac{\partial \gamma_{zx}}{\partial y} + \frac{\partial \gamma_{xy}}{\partial z} \right) \tag{2.10e}$$

$$2\frac{\partial^2 \epsilon_{zz}}{\partial x\, \partial y} = \frac{\partial}{\partial z}\left(\frac{\partial \gamma_{yz}}{\partial x} + \frac{\partial \gamma_{zx}}{\partial y} - \frac{\partial \gamma_{xy}}{\partial z} \right) \tag{2.10f}$$

In order to derive Eq. (2.10a), begin by recalling

$$\gamma_{xy} = \frac{\partial u}{\partial y} + \frac{\partial v}{\partial x} \tag{a}$$

Differentiating γ_{xy} once with respect to x and then again with respect to y gives

$$\frac{\partial^2 \gamma_{xy}}{\partial x\, \partial y} = \frac{\partial^3 u}{\partial x\, \partial y^2} + \frac{\partial^3 v}{\partial x^2\, \partial y} \tag{b}$$

Note that

$$\frac{\partial^2 \epsilon_{xx}}{\partial y^2} = \frac{\partial^3 u}{\partial x\, \partial y^2} \quad \text{and} \quad \frac{\partial^2 \epsilon_{yy}}{\partial x^2} = \frac{\partial^3 v}{\partial x^2\, \partial y} \tag{c}$$

Substituting Eqs. (c) into Eq. (b) gives

$$\frac{\partial^2 \gamma_{xy}}{\partial x\, \partial y} = \frac{\partial^2 \epsilon_{xx}}{\partial y^2} + \frac{\partial^2 \epsilon_{yy}}{\partial x^2} \tag{d}$$

which establishes Eq. (2.10a), and by the same methods Eqs. (2.10b) and (2.10c) could be verified. The proof of Eq. (2.10d) is obtained by considering four identities:

$$\frac{\partial^2 \epsilon_{xx}}{\partial y\, \partial z} = \frac{\partial^3 u}{\partial x\, \partial y\, \partial z} \tag{e}$$

$$\frac{\partial^2 \gamma_{xy}}{\partial x\, \partial z} = \frac{\partial^3 u}{\partial x\, \partial y\, \partial z} + \frac{\partial^3 v}{\partial x^2\, \partial z} \tag{f}$$

$$\frac{\partial^2 \gamma_{zx}}{\partial x\, \partial y} = \frac{\partial^3 w}{\partial x^2\, \partial y} + \frac{\partial^3 u}{\partial x\, \partial y\, \partial z} \tag{g}$$

$$\frac{\partial^2 \gamma_{yz}}{\partial x^2} = \frac{\partial^3 w}{\partial x^2\, \partial y} + \frac{\partial^3 v}{\partial x^2\, \partial z} \tag{h}$$

Now by forming

$$2(e) = (f) + (g) - (h) \tag{i}$$

thus obtaining

$$2\frac{\partial^2 \epsilon_{xx}}{\partial y\, \partial z} = \frac{\partial}{\partial x}\left(\frac{\partial \gamma_{xy}}{\partial z} + \frac{\partial \gamma_{zx}}{\partial y} - \frac{\partial \gamma_{yz}}{\partial x} \right) \tag{j}$$

Eq. (2.10d) is verified. The remaining two compatibility relations can be established in an identical manner.

In order to gain a better physical understanding of the compatibility relations, consider a two-dimensional body made up of a large number of small, square elements. When the body is loaded, the elements deform. By measuring angle changes and length changes, the strains which develop in each element can be determined. This procedure is accomplished theoretically by differentiating the displacement field. Consider now the inverse problem. Suppose a large number of small, deformed elements are given which must be fitted together to form a body free of voids and discontinuities. If and only if each element is properly strained can the body be reassembled without voids. The deformed elements correspond to the case of the prescribed strain field. The check to determine whether the elements are all properly strained and hence compatible with each other represents the compatibility relations. If these relations are satisfied, the elements will fit together properly, thus guaranteeing a satisfactory displacement field.

2.6 EXAMPLE OF A DISPLACEMENT FIELD COMPUTED FROM A STRAIN FIELD

If a circular shaft of radius a is loaded in torsion, the following strain field is produced:

$$\gamma_{zx} = -ay \qquad \gamma_{yz} = ax \qquad \epsilon_{xx} = \epsilon_{yy} = \epsilon_{zz} = \gamma_{xy} = 0 \tag{a}$$

where the z axis of the reference system is coincident with the centerline of the shaft. The first step in solving for the displacement field is to check the compatibility conditions:

1. The body must be simply connected, a condition which is obviously satisfied in this case.
2. The strain relations given in Eqs. (*a*) must satisfy all the compatibility relations given in Eqs. (2.10). It is clear that this linear system of strains does satisfy this requirement.

Next substitute Eqs. (*a*) into Eqs. (2.4) and integrate:

$$\epsilon_{xx} = \frac{\partial u}{\partial x} = 0 \qquad u = f(y, z)$$

$$\epsilon_{yy} = \frac{\partial v}{\partial y} = 0 \qquad v = g(x, z)$$

$$\epsilon_{zz} = \frac{\partial w}{\partial z} = 0 \qquad w = h(x, y)$$

(*b*)

$$\gamma_{xy} = \frac{\partial u}{\partial y} + \frac{\partial v}{\partial x} = 0 = \frac{\partial f(y, z)}{\partial y} + \frac{\partial g(x, z)}{\partial x}$$

The last of Eqs. (*b*) can be satisfied only if both right-hand terms are functions of *z* alone; hence

$$\frac{\partial f(y, z)}{\partial y} = -\frac{\partial g(x, z)}{\partial x} = F(z) \tag{c}$$

Integrating Eq. (*c*) gives

$$f(z) = yF(z) + C_1 = u \qquad g(z) = -xF(z) + C_2 = v \tag{d}$$

Recall the value of γ_{yz} from Eqs. (*a*), the definition of γ_{yz} from Eq. (2.4*e*), and the functional relation for *w* from the third of Eqs. (*b*) and form

$$\gamma_{yz} = \frac{\partial w}{\partial y} + \frac{\partial v}{\partial z} = ax = \frac{\partial h(x, y)}{\partial y} - x\frac{dF(z)}{dz} \tag{e}$$

Equation (*e*) can be satisfied if and only if both the right-hand terms are functions of *x* alone; hence

$$\frac{\partial h(x)}{\partial y} = H(x) \qquad \frac{dF(z)}{dz} = C_3 \tag{f}$$

Substituting Eqs. (*f*) into Eq. (*e*) gives

$$H(x) - C_3 x = ax$$
$$H(x) = (a + C_3)x \tag{g}$$

Integrating the second of Eqs. (f) yields

$$F(z) = C_3 z + C_4 \qquad (h)$$

Substituting Eq. (h) into Eqs. (d) gives

$$u = y(C_3 z + C_4) + C_1 \qquad v = -x(C_3 z + C_4) + C_2 \qquad (i)$$

By Eqs. (b), (f), and (g),

$$w = h(x, y) = \int H(x)\, \partial y = \int (a + C_3)x\, \partial y = (a + C_3)xy + C_5 \qquad (j)$$

Thus far five constants of integration have been introduced, and five of the six cartesian components of strain which were given have been used. By employing the last strain relation, the arbitrary constant C_3 can be evaluated. Recall from Eqs. (a):

$$\gamma_{zx} = \frac{\partial w}{\partial x} + \frac{\partial u}{\partial z} = (a + C_3)y + C_3 y = -ay \qquad C_3 = -a \qquad (k)$$

Substituting Eqs. (k) into Eqs. (i) and (j) gives

$$u = -ayz + C_4 y + C_1 \qquad v = axz - C_4 x + C_2 \qquad w = C_5 \qquad (l)$$

The constants C_1, C_2, and C_5 indicate rigid-body translation of the shaft. The constant C_4 indicates rigid-body rotation of the shaft. It is clear upon differentiation that C_1, C_2, C_3, and C_4 do not enter into the strains produced in the shaft and hence are not a part of the displacement due to deformation. The deformation displacements are given by

$$u = -ayz \qquad v = axz \qquad w = 0 \qquad (m)$$

As indicated by this simple example, the process by which the displacement field is calculated from a given strain field is quite lengthy. On the other hand, it is a very simple matter to go from a complex displacement field to a strain field.

2.7   VOLUME DILATATION

Consider a small, rectangular element in a deformed body which has its edges oriented along the principal axes. The length of each side of the block may have changed; however, the element will not be distorted since there are no shearing strains acting on the faces. The change in volume of such an element divided by the initial volume is, by definition, the volume dilatation D; that is,

$$D = \frac{V^* - V}{V}$$

where V is the initial volume, equal to the product of the three sides of the element, a_1, a_2, a_3, before deformation, and V^* is the final volume after straining, equal to the product of the three sides a_1^*, a_2^*, a_3^*, after deformation. Since

$$a_1^* = a_1(1 + \epsilon_1) \qquad a_2^* = a_2(1 + \epsilon_2) \qquad a_3^* = a_3(1 + \epsilon_3)$$

it follows that

$$D = \frac{a_1 a_2 a_3 (1 + \epsilon_1)(1 + \epsilon_2)(1 + \epsilon_3) - a_1 a_2 a_3}{a_1 a_2 a_3}$$

If the higher-order strain terms are neglected,

$$D = \epsilon_1 + \epsilon_2 + \epsilon_3 = J_1 \tag{2.11}$$

Equation (2.11) indicates that the volume dilatation D is equal to the first invariant of strain. Since the first invariant of strain is independent of the coordinate system being used, the volume dilatation of an element is independent of the reference frame forming its sides. Volume dilatation is thus a coordinate-independent concept.

2.8 STRESS-STRAIN RELATIONS

Thus far stress and strain have been discussed individually, and no assumptions have been required regarding the behavior of the material except that it was a continuous medium.[†] In this section, stress will be related to strain; therefore, certain restrictive assumptions regarding the body material must be introduced. The first of these assumptions regards linearity of the stress versus strain in the body. With a linear stress-strain relationship it is possible to write the general stress-strain expressions as follows:

$$\sigma_{xx} = K_{11}\epsilon_{xx} + K_{12}\epsilon_{yy} + K_{13}\epsilon_{zz} + K_{14}\gamma_{xy} + K_{15}\gamma_{yz} + K_{16}\gamma_{zx}$$

$$\sigma_{yy} = K_{21}\epsilon_{xx} + K_{22}\epsilon_{yy} + K_{23}\epsilon_{zz} + K_{24}\gamma_{xy} + K_{25}\gamma_{yz} + K_{26}\gamma_{zx}$$

$$\sigma_{zz} = K_{31}\epsilon_{xx} + K_{32}\epsilon_{yy} + K_{33}\epsilon_{zz} + K_{34}\gamma_{xy} + K_{35}\gamma_{yz} + K_{36}\gamma_{zx}$$

$$\tau_{xy} = K_{41}\epsilon_{xx} + K_{42}\epsilon_{yy} + K_{43}\epsilon_{zz} + K_{44}\gamma_{xy} + K_{45}\gamma_{yz} + K_{46}\gamma_{zx} \tag{2.12}$$

$$\tau_{yz} = K_{51}\epsilon_{xx} + K_{52}\epsilon_{yy} + K_{53}\epsilon_{zz} + K_{54}\gamma_{xy} + K_{55}\gamma_{yz} + K_{56}\gamma_{zx}$$

$$\tau_{zx} = K_{61}\epsilon_{xx} + K_{62}\epsilon_{yy} + K_{63}\epsilon_{zz} + K_{64}\gamma_{xy} + K_{65}\gamma_{yz} + K_{66}\gamma_{zx}$$

where K_{11} to K_{66} are the coefficients of elasticity of the material and are independent of the magnitudes of both the stress and the strain, provided the elastic limit of the material is not exceeded. If the elastic limit is exceeded, the linear relationship between stress and strain no longer holds, and Eqs. (2.12) are not valid.

[†] Actually most metals are not strictly continuous since they are composed of a large number of rather small grains. However, the grains are in almost all cases small enough in comparison with the size of the body for the body to behave as if it were a continuous medium.

There are 36 coefficients of elasticity in Eqs. (2.12); however, they are not all independent. By strain energy considerations, which are beyond the scope of this book, the number of independent coefficients of elasticity can be reduced to 21. This reduction is quite significant; however, even with 21 constants, Eqs. (2.12) may be considered rather long and involved. By assuming that the material is isotropic, i.e., that the elastic constants are the same in all directions and hence independent of the choice of a coordinate system, the 21 coefficients of elasticity reduce to two constants. The stress-strain relationships then reduce to

$$\sigma_{xx} = \lambda J_1 + 2\mu\epsilon_{xx} \qquad \sigma_{yy} = \lambda J_1 + 2\mu\epsilon_{yy} \qquad \sigma_{zz} = \lambda J_1 + 2\mu\epsilon_{zz}$$

$$\tau_{xy} = \mu\gamma_{xy} \qquad \tau_{yz} = \mu\gamma_{yz} \qquad \tau_{zx} = \mu\gamma_{zx}$$

(2.13)

where J_1 = first invariant of strain $(\epsilon_{xx} + \epsilon_{yy} + \epsilon_{zz})$
λ = Lamé's constant
μ = shear modulus

Equations (2.13) can be solved to give the strains as a function of stress:

$$\epsilon_{xx} = \frac{\lambda + \mu}{\mu(3\lambda + 2\mu)}\sigma_{xx} - \frac{\lambda}{2\mu(3\lambda + 2\mu)}(\sigma_{yy} + \sigma_{zz})$$

$$\epsilon_{yy} = \frac{\lambda + \mu}{\mu(3\lambda + 2\mu)}\sigma_{yy} - \frac{\lambda}{2\mu(3\lambda + 2\mu)}(\sigma_{xx} + \sigma_{zz})$$

(2.14)

$$\epsilon_{zz} = \frac{\lambda + \mu}{\mu(3\lambda + 2\mu)}\sigma_{zz} - \frac{\lambda}{2\mu(3\lambda + 2\mu)}(\sigma_{yy} + \sigma_{xx})$$

$$\gamma_{xy} = \frac{1}{\mu}\tau_{xy} \qquad \gamma_{yz} = \frac{1}{\mu}\tau_{yz} \qquad \gamma_{zx} = \frac{1}{\mu}\tau_{zx}$$

The elastic coefficients μ and λ shown in Eqs. (2.13) and (2.14) arise from a mathematical treatment of the general linear stress-strain relations. In experimental work, Lamé's constant λ is rarely used since it has no physical significance; however, as will be shown later, the shear modulus has physical significance and can easily be measured.

Consider a two-dimensional case of pure shear where

$$\sigma_{xx} = \sigma_{yy} = \sigma_{zz} = \tau_{zx} = \tau_{yz} = 0 \qquad \tau_{xy} = \text{applied shearing stress}$$

From Eqs. (2.14),

$$\mu = \frac{\tau_{xy}}{\gamma_{xy}}$$

(2.15a)

Hence, the shear modulus μ is the ratio of the shearing stress to the shearing strain in a two-dimensional state of pure shear.

In a conventional tension test which is often used to determine the mechanical properties of materials, a long, slender bar is subjected to a state of uniaxial stress in, say, the x direction. In this instance

$$\sigma_{yy} = \sigma_{zz} = \tau_{xy} = \tau_{yz} = \tau_{zx} = 0 \qquad \sigma_{xx} = \text{applied normal stress}$$

From Eqs. (2.14),

$$\epsilon_{xx} = \frac{\lambda + \mu}{\mu(3\lambda + 2\mu)} \sigma_{xx} \tag{a}$$

$$\epsilon_{yy} = \epsilon_{zz} = -\frac{\lambda}{2\mu(3\lambda - 2\mu)} \sigma_{xx} \tag{b}$$

In elementary strength-of-materials texts, the stress-strain relations for the case of uniaxial stress are often written

$$\epsilon_{xx} = \frac{1}{E} \sigma_{xx} \tag{c}$$

$$\epsilon_{yy} = \epsilon_{zz} = -\frac{v}{E} \sigma_{xx} \tag{d}$$

By equating the coefficients in Eqs. (a) and (b) to those in Eqs. (c) and (d),

$$E = \frac{\mu(3\lambda + 2\mu)}{\lambda + \mu} \tag{2.15b}$$

$$v = \frac{\lambda}{2(\lambda + \mu)} \tag{2.15c}$$

where E is the modulus of elasticity and v is Poisson's ratio, defined as

$$v = -\frac{\epsilon_{yy}}{\epsilon_{xx}} \tag{2.15d}$$

Equations (2.15b) and (2.15c) indicate the conversion from Lamé's constant λ and the shear modulus μ to the more commonly used modulus of elasticity E and Poisson's ratio v.

To establish the definition and physical significance of a fifth elastic constant, consider a state of hydrostatic stress where

$$\sigma_{xx} = \sigma_{yy} = \sigma_{zz} = -p \qquad \tau_{xy} = \tau_{yz} = \tau_{zx} = 0$$

where p is the uniform pressure acting on the body.

Adding together the first three of Eqs. (2.13) gives

$$-3p = (3\lambda + 2\mu)J_1$$

or

$$p = -\frac{3\lambda + 2\mu}{3} J_1 = -KJ_1 = -KD$$

Thus

$$K = \frac{3\lambda + 2\mu}{3} = -\frac{p}{D} \tag{2.15e}$$

The constant K is known as the *bulk modulus* and is the ratio of the applied hydrostatic pressure to the volume dilatation.

Five elastic constants λ, μ, E, v, and K have been discussed. The constant λ has no physical significance and is employed because it simplifies, mathematically speaking, the stress-strain relations. The constant μ has both mathematical and physical significance. It is used extensively in torsional problems. The constants E and v are the most widely recognized of the five constants considered and are used in almost all areas of stress analysis. The rather specialized bulk modulus K is used primarily for computing volume changes in a given body subjected to hydrostatic pressure. As indicated previously, there are two and only two independent elastic constants. The five constants discussed are related to each other as shown in Table 2.1.

TABLE 2.1
Relationships between the elastic constants

	λ equals	μ equals	E equals	v equals	K equals
λ, μ			$\dfrac{\mu(3\lambda + 2\mu)}{\lambda + \mu}$	$\dfrac{\lambda}{2(\lambda + \mu)}$	$\dfrac{3\lambda + 2\mu}{3}$
λ, E		$\dfrac{A^\dagger + (E - 3\lambda)}{4}$		$\dfrac{A^\dagger - (E + \lambda)}{4\lambda}$	$\dfrac{A^\dagger + (3\lambda + E)}{6}$
λ, v		$\dfrac{\lambda(1 - 2v)}{2v}$	$\dfrac{\lambda(1 + v)(1 - 2v)}{v}$		$\dfrac{\lambda(1 + v)}{3v}$
λ, K		$\dfrac{3(K - \lambda)}{2}$	$\dfrac{9K(K - \lambda)}{3K - \lambda}$	$\dfrac{\lambda}{3K - \lambda}$	
μ, E	$\dfrac{\mu(2\mu - E)}{E - 3\mu}$			$\dfrac{E - 2\mu}{2\mu}$	$\dfrac{\mu E}{3(3\mu - E)}$
μ, v	$\dfrac{2\mu v}{1 - 2v}$		$2\mu(1 + v)$		$\dfrac{2\mu(1 + v)}{3(1 - 2v)}$
μ, K	$\dfrac{3K - 2\mu}{3}$		$\dfrac{9K\mu}{3K + \mu}$	$\dfrac{3K - 2\mu}{2(3K + \mu)}$	
E, v	$\dfrac{vE}{(1 + v)(1 - 2v)}$	$\dfrac{E}{2(1 + v)}$			$\dfrac{E}{3(1 - 2v)}$
K, E	$\dfrac{3K(3K - E)}{9K - E}$	$\dfrac{3EK}{9K - E}$		$\dfrac{3K - E}{6K}$	
v, K	$\dfrac{3Kv}{1 + v}$	$\dfrac{3K(1 - 2v)}{2(1 + v)}$	$3K(1 - 2v)$		

$^\dagger A = \sqrt{E^2 + 2\lambda E + 9\lambda^2}$.

Since the constants E and v will be used almost exclusively throughout the remainder of this text, Eqs. (2.15b) and (2.15c) have been substituted into Eqs. (2.13) and (2.14) to obtain expressions for strain in terms of stress and the constants

$$\epsilon_{xx} = \frac{1}{E}[\sigma_{xx} - v(\sigma_{yy} + \sigma_{zz})]$$

$$\epsilon_{yy} = \frac{1}{E}[\sigma_{yy} - v(\sigma_{xx} + \sigma_{zz})]$$

$$\epsilon_{zz} = \frac{1}{E}[\sigma_{zz} - v(\sigma_{yy} + \sigma_{xx})]$$

$$\gamma_{xy} = \frac{2(1 + v)}{E}\tau_{xy} \qquad \gamma_{yz} = \frac{2(1 + v)}{E}\tau_{yz} \qquad \gamma_{zx} = \frac{2(1 + v)}{E}\tau_{zx}$$

(2.16)

and for stress in terms of strain and the constants

$$\sigma_{xx} = \frac{E}{(1 + v)(1 - 2v)}[(1 - v)\epsilon_{xx} + v(\epsilon_{yy} + \epsilon_{zz})]$$

$$\sigma_{yy} = \frac{E}{(1 + v)(1 - 2v)}[(1 - v)\epsilon_{yy} + v(\epsilon_{xx} + \epsilon_{zz})]$$

$$\sigma_{zz} = \frac{E}{(1 + v)(1 - 2v)}[(1 - v)\epsilon_{zz} + v(\epsilon_{xx} + \epsilon_{yy})]$$

$$\tau_{xy} = \frac{E}{2(1 + v)}\gamma_{xy} \qquad \tau_{yz} = \frac{E}{2(1 + v)}\gamma_{yz} \qquad \tau_{zx} = \frac{E}{2(1 + v)}\gamma_{zx}$$

(2.17)

2.9 STRAIN-TRANSFORMATION EQUATIONS AND STRESS-STRAIN RELATIONS FOR A TWO-DIMENSIONAL STATE OF STRESS

Simplified forms of the strain-transformation equations and the stress-strain relations, which will be extremely useful in later chapters when brittle coating and electrical-resistance strain-gage analyses are discussed, are the equations applicable to the strain field associated with a two-dimensional state of stress ($\sigma_{zz} = \tau_{zx} = \tau_{zy} = 0$).

The strain-transformation equations can be obtained from Eqs. (2.6) by selecting z' coincident with z and noting from Eqs. (2.16) that $\gamma_{zx} = \gamma_{yz} = 0$. The notation can also be simplified by denoting the angle between x' and x as θ. The equations obtained are

$$\epsilon_{x'x'} = \epsilon_{xx} \cos^2 \theta + \epsilon_{yy} \sin^2 \theta + \gamma_{xy} \sin \theta \cos \theta$$

$$\epsilon_{y'y'} = \epsilon_{yy} \cos^2 \theta + \epsilon_{xx} \sin^2 \theta - \gamma_{xy} \sin \theta \cos \theta$$

$$\gamma_{x'y'} = 2(\epsilon_{yy} - \epsilon_{xx}) \sin \theta \cos \theta + \gamma_{xy}(\cos^2 \theta - \sin^2 \theta) \qquad (2.18)$$

$$\epsilon_{z'z'} = \epsilon_{zz} \qquad \gamma_{y'z'} = \gamma_{z'x'} = 0$$

The stress-strain relations for a two-dimensional state of stress are obtained by substituting $\sigma_{zz} = \tau_{zx} = \tau_{yz} = 0$ into Eqs. (2.16). Thus

$$\epsilon_{xx} = \frac{1}{E}(\sigma_{xx} - \nu\sigma_{yy}) \qquad \epsilon_{yy} = \frac{1}{E}(\sigma_{yy} - \nu\sigma_{xx}) \qquad \epsilon_{zz} = -\frac{\nu}{E}(\sigma_{xx} + \sigma_{yy})$$

$$\gamma_{xy} = \frac{2(1 + \nu)}{E}\tau_{xy} \qquad \gamma_{yz} = \gamma_{zx} = 0 \qquad (2.19)$$

In a similar manner the equations for stress in terms of strain for the two-dimensional state of stress are obtained from Eqs. (2.17). Thus

$$\sigma_{xx} = \frac{E}{1 - \nu^2}(\epsilon_{xx} + \nu\epsilon_{yy}) \qquad \sigma_{yy} = \frac{E}{1 - \nu^2}(\epsilon_{yy} + \nu\epsilon_{xx})$$

$$\sigma_{zz} = \tau_{zx} = \tau_{yz} = 0 \qquad \tau_{xy} = \frac{E}{2(1 + \nu)}\gamma_{xy} \qquad (2.20)$$

One additional relationship which relates the strain ϵ_{zz} to the measured strains ϵ_{xx} and ϵ_{yy} in experimental analyses is obtained from Eq. (2.17) by substituting $\sigma_{zz} = 0$. Thus

$$\epsilon_{zz} = -\frac{\nu}{1 - \nu}(\epsilon_{xx} + \epsilon_{yy}) \qquad (2.21)$$

This equation can be used to establish the magnitude of the third principal strain associated with a two-dimensional state of stress. This information is useful for maximum shear-strain determinations.

EXERCISES

2.1. Given the displacement field

$$u = (3x^4 + 2x^2y^2 + x + y + z^3 + 3)(10^{-3})$$

$$v = (3xy + y^3 + y^2z + z^2 + 1)(10^{-3})$$

$$w = (x^2 + xy + yz + zx + y^2 + z^2 + 2)(10^{-3})$$

compute the associated strains at point (1, 1, 1). Compare the results obtained by using Eqs. (2.2) and (2.3) with those obtained by using Eqs. (2.4).

2.2. Repeat Exercise 2.1 but change the multiplier from 10^{-3} to 10^{-1}.

2.3. Given the displacement field

$$u = (x^2 + y^4 + 2y^2z + yz)(10^{-3})$$

$$v = (xy + xz + 3x^2z)(10^{-3})$$

$$w = (y^4 + 4y^3 + 2z^2)(10^{-3})$$

compute the associated strains at point (2, 2, 2). Compare the results obtained by using Eqs. (2.2) and (2.3) with those obtained by using Eqs. (2.4).

2.4. Repeat Exercise 2.3 but change the multiplier from 10^{-3} to 10^{-2}.

2.5. Transform the set of cartesian strain components

$$\epsilon_{xx} = 300 \ \mu\epsilon \qquad \epsilon_{yy} = 200 \ \mu\epsilon \qquad \epsilon_{zz} = 100 \ \mu\epsilon$$

$$\gamma_{xy} = 200 \ \mu\epsilon \qquad \gamma_{yz} = 100 \ \mu\epsilon \qquad \gamma_{zx} = 150 \ \mu\epsilon$$

into a new set of cartesian strain components relative to an $Ox'y'z'$ set of coordinates, where the direction angles associated with the $Ox'y'z'$ axes are

θ	Case 1	Case 2	Case 3	Case 4
$x - x'$	$\pi/4$	$\pi/2$	0	$\pi/2$
$y - y'$	$\pi/4$	$\pi/2$	$\pi/2$	0
$z - z'$	0	0	$\pi/2$	$\pi/2$

2.6. Transform the set of cartesian strain components

$$\epsilon_{xx} = 400 \ \mu\epsilon \qquad \epsilon_{yy} = 250 \ \mu\epsilon \qquad \epsilon_{zz} = 125 \ \mu\epsilon$$

$$\gamma_{xy} = 275 \ \mu\epsilon \qquad \gamma_{yz} = 175 \ \mu\epsilon \qquad \gamma_{zx} = 225 \ \mu\epsilon$$

into a new set of cartesian strain components relative to an $Ox'y'z'$ set of coordinates, where the direction angles associated with the $Ox'y'z'$ axes are

	x	y	z
x'	$\pi/3$	$\pi/3$	$\pi/4$
y'	$3\pi/4$	$\pi/4$	$\pi/2$
z'	$\pi/3$	$\pi/3$	$3\pi/4$

2.7. At a point in a stressed body, the cartesian components of strain are

$$\epsilon_{xx} = 300 \ \mu\epsilon \qquad \epsilon_{yy} = 450 \ \mu\epsilon \qquad \epsilon_{zz} = 300 \ \mu\epsilon$$

$$\gamma_{xy} = 600 \ \mu\epsilon \qquad \gamma_{yz} = 375 \ \mu\epsilon \qquad \gamma_{zx} = 450 \ \mu\epsilon$$

Transform this set of cartesian components into a new set of cartesian strain components relative to an $Ox'y'z'$ set of coordinates where the $Ox'y'z'$ axes are defined by the following direction cosines:

(a)

	x	y	z
x	$\frac{2}{3}$	$\frac{2}{3}$	$-\frac{1}{3}$
y'	$-\frac{2}{3}$	$\frac{1}{3}$	$-\frac{2}{3}$
z'	$-\frac{1}{3}$	$\frac{2}{3}$	$\frac{2}{3}$

(b)

	x	y	z
x'	$\frac{1}{9}$	$-\frac{8}{9}$	$\frac{4}{9}$
y'	$\frac{4}{9}$	$\frac{4}{9}$	$\frac{7}{9}$
z'	$-\frac{8}{9}$	$\frac{1}{9}$	$\frac{4}{9}$

2.8. At a point in a stressed body, the cartesian components of strain are

$$\epsilon_{xx} = 450 \ \mu\epsilon \qquad \epsilon_{yy} = 300 \ \mu\epsilon \qquad \epsilon_{zz} = 150 \ \mu\epsilon$$

$$\gamma_{xy} = 150 \ \mu\epsilon \qquad \gamma_{yz} = 150 \ \mu\epsilon \qquad \gamma_{zx} = 300 \ \mu\epsilon$$

Transform this set of cartesian strain components into a new set of cartesian strain components relative to an $Ox'y'z'$ set of coordinates where the $Ox'y'z'$ axes are defined by the following direction cosines:

(a)

	x	y	z
x'	$\frac{1}{17}$	$\frac{12}{17}$	$\frac{12}{17}$
y'	$\frac{12}{17}$	$-\frac{9}{17}$	$\frac{8}{17}$
z'	$\frac{12}{17}$	$\frac{8}{17}$	$-\frac{9}{17}$

(b)

	x	y	z
x'	$\frac{9}{25}$	$\frac{12}{25}$	$\frac{20}{25}$
y'	$\frac{12}{25}$	$\frac{16}{25}$	$-\frac{15}{25}$
z'	$\frac{20}{25}$	$-\frac{15}{25}$	0

2.9. At a point in a stressed body, the cartesian components of strain are

$$\epsilon_{xx} = 990 \ \mu\epsilon \qquad \epsilon_{yy} = 825 \ \mu\epsilon \qquad \epsilon_{zz} = 550 \ \mu\epsilon$$

$$\gamma_{xy} = 330 \ \mu\epsilon \qquad \gamma_{yz} = 660 \ \mu\epsilon \qquad \gamma_{zx} = 500 \ \mu\epsilon$$

Transform this set of cartesian strain components into a new set of cartesian strain components relative to an $Ox'y'z'$ set of coordinates where the $Ox'y'z'$ axes are defined by the following direction cosines:

(a)

	x	y	z
x'	$\frac{2}{15}$	$\frac{10}{15}$	$\frac{11}{15}$
y'	$\frac{14}{15}$	$-\frac{5}{15}$	$\frac{2}{15}$
z'	$-\frac{5}{15}$	$-\frac{10}{15}$	$\frac{10}{15}$

(b)

	x	y	z
x'	$\frac{2}{11}$	$\frac{6}{11}$	$\frac{9}{11}$
y'	$\frac{6}{11}$	$\frac{7}{11}$	$-\frac{6}{11}$
z'	$\frac{9}{11}$	$-\frac{6}{11}$	$\frac{2}{11}$

2.10. Write a computer program to determine the cartesian strain components relative to the $Ox'y'z'$ axes in terms of the cartesian strain components referred to $Oxyz$ and the appropriate direction cosines.

2.11. Determine the three principal strains and the maximum shearing strain at the point having the cartesian strain components given in Exercise 2.5. Check the three strain invariants.

2.12. Determine the three principal strains and the maximum shearing strain at the point having the cartesian strain components given in Exercise 2.6. Check the three strain invariants.

2.13. Determine the three principal strains and the maximum shearing strain at the point having the cartesian strain components given in Exercise 2.7.

2.14. Determine the three principal strains and the maximum shearing strain at the point having the cartesian strain components given in Exercise 2.8.

2.15. Determine whether the following strain fields are compatible:

(a) $\epsilon_{xx} = 2x^2 + 3y^2 + z + 1$

$\epsilon_{yy} = 2y^2 + x^2 + 3z + 2$

$\epsilon_{zz} = 3x + 2y + z^2 + 1$

$\gamma_{xy} = 10xy$

$\gamma_{yz} = 0$

$\gamma_{zx} = 0$

(b) $\epsilon_{xx} = 3y^2 + xy$

$\epsilon_{yy} = 2y + 4z + 3$

$\epsilon_{zz} = 3zx + 2xy + 3yz + 2$

$\gamma_{xy} = 5xy$

$\gamma_{yz} = 2x + z$

$\gamma_{zx} = 2y + x$

2.16. Determine whether the following strain fields are compatible:

(a) $\epsilon_{xx} = 3x^2 + 4xy - 4y^2$

$\epsilon_{yy} = x^2 + xy + 3y^2$

$\epsilon_{zz} = 0$

$\gamma_{xy} = -x^2 - 6xy - 4y^2$

$\gamma_{yz} = 2x + y$

$\gamma_{zx} = z + 3$

(b) $\epsilon_{xx} = 12x^2 - 6y^2 - 4z$

$\epsilon_{yy} = 12y^2 - 6x^2 + 4z$

$\epsilon_{zz} = 12x + 4y - z + 5$

$\gamma_{xy} = 4z - 24xy - 3$

$\gamma_{yz} = y + z - 4$

$\gamma_{zx} = 4x + 4y - 6$

2.17. Given the strain field

$$\epsilon_{xx} = ay \qquad \epsilon_{yy} = by \qquad \epsilon_{zz} = by$$

$$\gamma_{xy} = 0 \qquad \gamma_{yz} = 0 \qquad \gamma_{zx} = 0$$

compute the displacement fields. What physical problem does this strain field represent?

2.18. Determine the volume dilatation for the strain field in Exercise 2.5.

2.19. Determine the volume dilatation for the strain field in Exercise 2.6.

2.20. Determine the volume dilatation for the strain field in Exercise 2.7.

2.21. Determine the volume dilatation for the strain field in Exercise 2.8.

2.22. Determine the volume dilatation at point $(2, 1, 2)$ of the displacement field given in Exercise 2.3.

2.23. Determine λ, μ, and K for steel, brass, aluminum, plastic, and magnesium by using the following values for E and v:

Material	E, GPa	v
Steel	207	0.30
Brass	106	0.33
Aluminum	71	0.33
Plastic	3	0.40
Magnesium	45	0.35

2.24. A cube of steel ($E = 207$ GPa and $v = 0.30$) is loaded with a uniformly distributed pressure of 300 MPa on the four faces having outward normals in the x and y directions. Rigid constraints limit the total deformation of the cube in the z direction to 0.05 mm. Determine the normal stress, if any, which develops in the z direction. The length of a side of the cube is 190 mm.

2.25. Determine the change in volume of a 10-mm cube of aluminum ($E = 71$ GPa and $v = 0.33$) when dropped a distance of 8 km to the ocean floor.

2.26. Determine the stresses at a point in a steel ($E = 207$ GPa and $v = 0.30$) machine component if the cartesian components of strain at the point are as listed in Exercise 2.5.

2.27. Determine the stresses at a point in an aluminum ($E = 71$ GPa, $v = 0.33$) machine component if the cartesian components of strain at the point are as listed in Exercise 2.6.

2.28. The cartesian components of stress at a point in a steel ($E = 207$ GPa and $v = 0.30$) machine part are:

$$\sigma_{xx} = 220 \text{ MPa} \qquad \sigma_{yy} = 77 \text{ MPa} \qquad \sigma_{zz} = 154 \text{ MPa}$$

$$\tau_{xy} = 110 \text{ MPa} \qquad \tau_{yz} = 55 \text{ MPa} \qquad \tau_{zx} = 66 \text{ MPa}$$

Determine the principal strains at the point.

2.29. At a point on the free surface of an alloy steel ($E = 207$ GPa and $v = 0.30$) machine part normal strains of 1000 $\mu\epsilon$, 2000 $\mu\epsilon$, and 1200 $\mu\epsilon$ were measured at angles of $0°$, $60°$, and $120°$, respectively, relative to the x axis. Design considerations limit the maximum normal stress to 510 MPa, the maximum shearing stress to 275 MPa, the maximum normal strain to 2200 $\mu\epsilon$, and the maximum shearing strain to 2500 $\mu\epsilon$. What is your evaluation of the design?

2.30. A thick-walled cylindrical pressure vessel will be used to store gas under a pressure of 100 MPa. During initial pressurization of the vessel, axial and hoop components of strain were measured on the inside and outside surfaces. On the inside surface, the axial strain was 500 $\mu\epsilon$ and the hoop strain was 750 $\mu\epsilon$. On the outside surface, the axial strain was 500 $\mu\epsilon$, and the hoop strain was 100 $\mu\epsilon$. Determine the axial and hoop components of stress associated with these strains if $E = 207$ GPa and $v = 0.30$.

2.31. The cartesian components of stress at a point in a steel ($E = 207$ GPa and $v = 0.30$) machine part are as follows

$$\sigma_{xx} = 280 \text{ MPa} \qquad \sigma_{yy} = -120 \text{ MPa} \qquad \sigma_{zz} = 140 \text{ MPa}$$

$$\tau_{xy} = 280 \text{ MPa} \qquad \tau_{yz} = 0 \qquad \tau_{zx} = 0$$

Determine the three principal strains, the principal-strain directions, and the maximum shearing strain.

2.32. Mohr's circle for stress and Mohr's circle for strain are convenient graphical methods for visualizing three-dimensional states of stress and strain at points in a stressed body. At a particular point in a body fabricated from steel ($E = 207$ GPa and $v = 0.30$), the three principal stresses are:

$$\sigma_1 = 120 \text{ MPa} \qquad \sigma_2 = 60 \text{ MPa} \qquad \sigma_3 = -40 \text{ MPa}$$

(a) Sketch the three-dimensional Mohr's circle for stress at the point.
(b) Sketch the three-dimensional Mohr's circle for strain at the point.
(c) On a plane through the point, the shearing stress $\tau = 70$ MPa. What normal stress must exist on this plane?
(d) On another plane through the point, the shearing stress $\tau = 50$ MPa. Within what range of values must the normal stress associated with this plane fall?
(e) Along a line through the point (say the x axis), the normal strain is zero. What range of values may the shearing strain γ assume for different orientation of the x axis?

2.33. A thin rubber membrane is stretched in such a manner that the following uniform strain field is produced:

$$\epsilon_{xx} = 5000 \ \mu\epsilon \qquad \epsilon_{yy} = -6000 \ \mu\epsilon \qquad \gamma_{xy} = 2000 \ \mu\epsilon$$

A rectangle is drawn on the membrane before stretching. How should the rectangle be oriented if the angles are to remain 90° during stretching?

2.34. A thin rectangular aluminum ($E = 71$ GPa and $v = 0.33$) plate 75×100 mm is acted upon by a two-dimensional stress distribution which produces the following uniform distribution of strains in the plate:

$$\epsilon_{xx} = 2000 \ \mu\epsilon \qquad \epsilon_{yy} = -500 \ \mu\varepsilon \qquad \gamma_{xy} = 2000 \ \mu\varepsilon$$

(a) Determine the changes in length of the diagonals of the plate.
(b) Determine the maximum shearing strain in the plate. Indicate on a sketch two of the initially perpendicular lines in the plate associated with this maximum shearing strain.

2.35. For an aluminum ($E = 71$ GPa and $v = 0.33$) body under plane-stress conditions with $\sigma_{zz} = \tau_{yz} = \tau_{zx} = 0$, strains on the surface of the body at a given point are:

$$\epsilon_{xx} = 1200 \ \mu\epsilon \qquad \text{and} \qquad \epsilon_{yy} = 900 \ \mu\epsilon$$

Determine the strain ε_{zz}.

2.36. Determine the stresses σ_{xx}, σ_{yy}, and σ_{zz} in a material with $v = \frac{1}{2}$ if

$$\epsilon_{xx} = \epsilon_{yy} = \epsilon_{zz} = -1000 \ \mu\epsilon$$

Explain your results.

2.37. Write a computer program using the relations given in Table 2.1 so that all five elastic constants λ, μ, E, v, and K can be determined if any two of them are given.

REFERENCES

1. Boresi, A. P., and P. P. Lynn: *Elasticity in Engineering Mechanics*, chap. 2, Prentice-Hall, Englewood Cliffs, N.J., 1974.
2. Chou, P. C., and N. J. Pagano: *Elasticity*, chaps. 2 and 3, Van Nostrand, Princeton, N.J., 1967.
3. Durelli, A. J., E. A. Phillips, and C. H. Tsao: *Introduction to the Theoretical and Experimental Analysis of Stress and Strain*, chaps. 2 and 4, McGraw-Hill, New York, 1958.
4. Love, A. E. H.: *A Treatise on the Mathematical Theory of Elasticity*, chaps. 1 and 3, Dover, New York, 1944.
5. Sechler, E. E.: *Elasticity in Engineering*, chaps. 4 and 5, John Wiley & Sons, New York, 1952.
6. Sokolnikoff, I. S.: *Mathematical Theory of Elasticity*, 2d ed., chaps. 1 and 3, McGraw-Hill, New York, 1956.
7. Southwell, R. V.: *An Introduction to the Theory of Elasticity*, chaps. 9 and 10, Oxford University Press, Fair Lawn, N.J., 1953.
8. Timoshenko, S. P., and J. N. Goodier: *Theory of Elasticity*, 2d ed., chaps. 1, 8, and 9, McGraw-Hill, New York, 1951.

CHAPTER
3

BASIC
EQUATIONS AND
PLANE-ELASTICITY
THEORY

3.1 FORMULATION OF THE PROBLEM

In the general three-dimensional elasticity problem there are 15 unknown quantities which must be determined at every point in the body, namely, the 6 cartesian components of stress, the 6 cartesian components of strain, and the 3 components of displacement. Attempts can be made to obtain a solution to a given problem after the following quantities have been adequately defined:

1. The geometry of the body
2. The boundary conditions
3. The body-force field as a function of position
4. The elastic constants

In order to solve for the above-mentioned 15 unknown quantities, 15 independent equations are required. Three are provided by the stress equations of equilibrium [Eqs. (1.3)], six are provided by the strain-displacement relations [Eqs. (2.4)], and the remaining six can be obtained from the stress-strain expressions [Eqs. (2.16)].

A solution to an elasticity problem, in addition to satisfying these 15 equations, must also satisfy the boundary conditions. In other words, the stresses acting over the surface of the body must produce tractions which are equivalent

55

to the loads being applied to the body. Boundary conditions are often classified to define the four different types of boundary-value problem listed below:

Type 1. If the displacements are prescribed over the entire boundary, the problem is classified as a type 1 boundary-value problem. As an example, consider a long, slender rod which is given an axial displacement, say, u and transverse displacements v and w. In this instance displacements are prescribed over the entire boundary of the rod.

Type 2. The most frequently encountered boundary-value problem is the type where normal and shearing forces are given over the entire surface. For instance, a sphere subjected to a uniform hydrostatic pressure has zero shearing stress and a normal stress equal to $-p$ on the surface and hence is a type 2 boundary-value problem.

Type 3. This is a mixed boundary-value problem where the normal and shearing forces are given over a portion of the boundary and the displacements are given over the remainder of the body. To illustrate this type of problem, consider the shrinking of a sleeve over a shaft. In the shrinking process a radial displacement is given to the sleeve at the interface between the shaft and the sleeve. On all other surfaces of the sleeve, both the normal and the shearing components of stress are zero.

Type 4. This type of boundary-value problem is the most general of the four considered. Over a portion of the surface, displacements are prescribed. Over a second portion of the surface, normal and shearing stresses are prescribed. Over a third portion of the surface, the normal component of displacement and the shearing component of stress are prescribed. Over a fourth portion of the surface, the shearing component of displacement and the normal component of stress are prescribed. Obviously, the first three types of problem can be regarded as special cases of this general fourth type.

One of the most difficult problems encountered in any experimental study is the design and construction of the loading fixture for applying the required displacements or tractions to the model being studied. The classifications given previously should be kept in mind when one designs the fixture. In general, it has been found that tractions cannot be adequately simulated by applying a displacement field to the model and vice versa. Type 1 and type 2 boundary-value problems are usually the easiest to approach experimentally. In general, type 3 and type 4 problems offer more difficulties in properly loading the model.

3.2   FIELD EQUATIONS

Thus far in the development four sets of field equations have been discussed, namely, the stress equations of equilibrium, the strain-displacement relations, the stress-strain expressions, and the equations of compatibility. Quite often two or more of these sets of equations can be combined to give a new set which may be more applicable to a specific problem. As an example, consider the six stress-displacement equations which can be obtained from the six stress-strain relations and the six strain-displacement equations by substituting Eqs. (2.4) into Eqs. (2.16):

$$\frac{\partial u}{\partial x} = \frac{1}{E}[\sigma_{xx} - v(\sigma_{yy} + \sigma_{zz})]$$

$$\frac{\partial v}{\partial y} = \frac{1}{E}[\sigma_{yy} - v(\sigma_{zz} + \sigma_{xx})]$$

$$\frac{\partial w}{\partial z} = \frac{1}{E}[\sigma_{zz} - v(\sigma_{xx} + \sigma_{yy})]$$

$$\frac{\partial u}{\partial y} + \frac{\partial v}{\partial x} = \frac{1}{\mu}\tau_{xy} \qquad \frac{\partial v}{\partial z} + \frac{\partial w}{\partial y} = \frac{1}{\mu}\tau_{yz} \qquad \frac{\partial w}{\partial x} + \frac{\partial u}{\partial z} = \frac{1}{\mu}\tau_{zx}$$

(3.1)

It is interesting to note that the set of equations consisting of the stress equations of equilibrium [Eqs. (1.3)] and the stress-displacement relations [Eqs. (3.1)] are expressed as nine equations in terms of nine unknowns. The reduction in the number of unknowns from 15 to 9 was made possible by eliminating the strains.

The problem can be reduced further (from nine to three unknowns) if the stress equations of equilibrium [Eqs. (1.3)] are combined with the stress-displacement equations [Eqs. (3.1)]. The displacement equations of equilibrium obtained can be written as follows:

$$\nabla^2 u + \frac{1}{1 - 2v}\frac{\partial}{\partial x}\left(\frac{\partial u}{\partial x} + \frac{\partial v}{\partial y} + \frac{\partial w}{\partial z}\right) + \frac{1}{\mu}F_x = 0$$

$$\nabla^2 v + \frac{1}{1 - 2v}\frac{\partial}{\partial y}\left(\frac{\partial u}{\partial x} + \frac{\partial v}{\partial y} + \frac{\partial w}{\partial z}\right) + \frac{1}{\mu}F_y = 0$$

(3.2)

$$\nabla^2 w + \frac{1}{1 - 2v}\frac{\partial}{\partial z}\left(\frac{\partial u}{\partial x} + \frac{\partial v}{\partial y} + \frac{\partial w}{\partial z}\right) + \frac{1}{\mu}F_z = 0$$

where ∇^2 is the operator $\partial^2/\partial x^2 + \partial^2/\partial y^2 + \partial^2/\partial z^2$.

It is clear that a solution of the displacement equations of equilibrium will yield the three displacements u, v, w. Once the displacements are known, the six strains and the six stresses can easily be obtained by using Eqs. (2.4) to obtain the strains and Eqs. (2.16) to obtain the stresses.

Analytical solutions for three-dimensional elasticity problems are quite difficult to obtain, and the number of problems which have been solved in an exact fashion to date is surprisingly small. The most successful approach to date has been through the use of the Boussinesq-Popkovich stress functions, which are defined so as to satisfy Eq. (3.2). The development of this approach is somewhat involved and is therefore beyond the scope and objectives of this elementary treatment of the theory of elasticity. The interested student should consult the selected references at the end of the chapter for a detailed development of the Boussinesq-Popkovich stress-function approach.

Before this section is completed, the stress equations of compatibility will be developed since they are the basis for an important theorem regarding the dependence of stresses on the elastic constants. If the stress-strain relations [Eqs. (2.16)], the stress equations of equilibrium [Eqs. (1.3)], and the strain compatibility equations [Eqs. (2.10)] are combined, the six stress equations of compatibility are obtained as follows:

$$\nabla^2 \sigma_{xx} + \frac{1}{1+v} \frac{\partial^2}{\partial x^2} I_1 = -\frac{v}{1-v} \left(\frac{\partial F_x}{\partial x} + \frac{\partial F_y}{\partial y} + \frac{\partial F_z}{\partial z} \right) - 2 \frac{\partial F_x}{\partial x}$$

$$\nabla^2 \sigma_{yy} + \frac{1}{1+v} \frac{\partial^2}{\partial y^2} I_1 = -\frac{v}{1-v} \left(\frac{\partial F_x}{\partial x} + \frac{\partial F_y}{\partial y} + \frac{\partial F_z}{\partial z} \right) - 2 \frac{\partial F_y}{\partial y}$$

$$\nabla^2 \sigma_{zz} + \frac{1}{1+v} \frac{\partial^2}{\partial z^2} I_1 = -\frac{v}{1-v} \left(\frac{\partial F_x}{\partial x} + \frac{\partial F_y}{\partial y} + \frac{\partial F_z}{\partial z} \right) - 2 \frac{\partial F_z}{\partial z}$$

$$\nabla^2 \tau_{xy} + \frac{1}{1+v} \frac{\partial^2}{\partial x \, \partial y} I_1 = -\left(\frac{\partial F_x}{\partial y} + \frac{\partial F_y}{\partial x} \right)$$

(3.3)

$$\nabla^2 \tau_{yz} + \frac{1}{1+v} \frac{\partial^2}{\partial y \, \partial z} I_1 = -\left(\frac{\partial F_y}{\partial z} + \frac{\partial F_z}{\partial y} \right)$$

$$\nabla^2 \tau_{zx} + \frac{1}{1+v} \frac{\partial^2}{\partial z \, \partial x} I_1 = -\left(\frac{\partial F_x}{\partial z} + \frac{\partial F_z}{\partial x} \right)$$

where I_1 is the first invariant of stress $\sigma_{xx} + \sigma_{yy} + \sigma_{zz}$ and F_x, F_y, F_z are the body-force intensities in the x, y, z directions, respectively.

If this system of six equations is solved for the six cartesian stress components, and if the boundary conditions are satisfied, the problem can be considered solved. Of great importance to the experimentalist is the appearance of elastic constants in Eqs. (3.3). Recall that equations of stress equilibrium did not contain elastic constants. Since only Poisson's ratio v appears in Eqs. (3.3), it follows that the stresses are independent of the modulus of elasticity E of the model material and can at most depend upon Poisson's ratio alone. Of course, this is true only for a simply connected body since the strain compatibility equations are valid only for this condition.

This independence of the stresses on the elastic modulus is very important in three-dimensional photoelasticity, where a low-modulus plastic model is used to simulate a metal prototype. Only the difference in Poisson's ratio between the model and the prototype is a source of error. The very large difference between the moduli of elasticity of the model and the prototype does not produce any significant errors in the determination of stresses using a three-dimensional photoelastic approach, provided the strains induced in the photoelastic model remain sufficiently small.

3.3 THE PLANE ELASTIC PROBLEM

In the theory of elasticity there exists a special class of problems, known as *plane problems*, which can be solved more readily than the general three-dimensional problem since certain simplifying assumptions can be made in their treatment. The geometry of the body and the nature of the loading on the boundaries which permit a problem to be classified as a plane problem are as follows:

By definition a plane body consists of a region of uniform thickness bounded by two parallel planes and by any closed lateral surface B_L, as indicated by Fig. 3.1. Although the thickness of the body must be uniform, it need not be limited. It may be very thick or very thin; in fact, these two extremes represent the most desirable cases for this approach, as will be pointed out later.

In addition to the restrictions on the geometry of the body, the following restrictions are imposed on the loads applied to the plane body.

1. Body forces, if they exist, cannot vary through the thickness of the region; that is, $F_x = F_x(x, y)$ and $F_y = F_y(x, y)$. Furthermore, the body force in the z direction must equal zero.
2. The surface tractions or loads on the lateral boundary B_L must be in the plane of the model and must be uniformly distributed across the thickness, i.e., constant in the z direction. Hence, $T_x = T_x(x, y)$, $T_y = T_y(x, y)$, and $T_z = 0$.
3. No loads can be applied on the parallel planes bounding the top and bottom surfaces; that is, $\mathbf{T}_n = 0$ on $z = \pm t$.

Once the geometry and loading have been defined, stresses can be determined by using either the plane-strain or the plane-stress approach. Usually the plane-strain approach is used when the body is very thick relative to its lateral dimensions. The plane-stress approach is employed when the body is relatively thin in relation to its lateral dimensions.

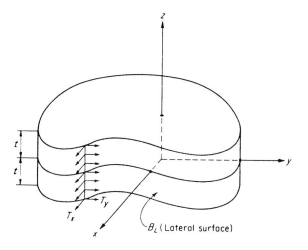

FIGURE 3.1
A body which may be considered for the plane-elasticity approach is bounded on the top and bottom by two parallel planes and is bounded laterally by any surface which is normal to the top and bottom planes.

3.4 THE PLANE-STRAIN APPROACH

If it is assumed that the strains in the body are plane, i.e., the strains in the x and y directions are functions of x and y alone, and also that the strains in the z directions are equal to zero, the strain-displacement relation [Eqs. (2.4)] can be simplified as follows:

$$\epsilon_{xx} = \frac{\partial u}{\partial x} \qquad \epsilon_{yy} = \frac{\partial v}{\partial y} \qquad \epsilon_{zz} = \frac{\partial w}{\partial z} = 0$$

$$\gamma_{xy} = \frac{\partial u}{\partial y} + \frac{\partial v}{\partial x} \qquad \gamma_{yz} = \frac{\partial v}{\partial z} + \frac{\partial w}{\partial y} = 0 \qquad \gamma_{zx} = \frac{\partial w}{\partial x} + \frac{\partial u}{\partial z} = 0$$

$$(3.4)$$

Similarly, if Eqs. (3.4) are substituted into Eqs. (2.13), a reduced form of the stress-strain relations for the case of plane strain is obtained:

$$\sigma_{xx} = \lambda J_1 + 2\mu\epsilon_{xx} \qquad \sigma_{yy} = \lambda J_1 + 2\mu\epsilon_{yy} \qquad \sigma_{zz} = \lambda J_1$$

$$\tau_{xy} = \mu\gamma_{xy} \qquad \tau_{yz} = \tau_{zx} = 0$$

$$(3.5)$$

where $J_1 = \epsilon_{xx} + \epsilon_{yy}$. In addition, the stress equations of equilibrium [Eqs. (1.3)] reduce to

$$\frac{\partial\sigma_{xx}}{\partial x} + \frac{\partial\tau_{xy}}{\partial y} + F_x = 0 \qquad \frac{\partial\tau_{xy}}{\partial x} + \frac{\partial\sigma_{yy}}{\partial y} + F_y = 0 \qquad (3.6)$$

Any solution for a plane-strain problem must satisfy Eqs. (3.4) to (3.6) in addition to the boundary conditions on the lateral boundary B_L and the bounding planes. The boundary conditions on B_L can be expressed in terms of the stresses by referring to Eqs. (1.2), which give the x, y, and z components of the resultant-stress vector in terms of the cartesian components of stress. Thus, on B_L the following relations must be satisfied:

$$T_{nx} = \sigma_{xx} \cos(n, x) + \tau_{xy} \cos(n, y)$$

$$T_{ny} = \tau_{xy} \cos(n, x) + \sigma_{yy} \cos(n, y) \qquad (3.7)$$

$$T_{nz} = 0$$

where T_{nx}, T_{ny}, T_{nz} are the x, y, z components of the stresses applied to the body on surface B_L. Finally, on the two parallel bounding planes,

$$\mathbf{T}_n = 0 \qquad (3.8)$$

i.e., no tractions are applied to these surfaces; hence, τ_{yz}, τ_{zx}, σ_{zz} must be zero on these surfaces.

It is clear from Eqs. (3.5) that σ_{zz} will be equal to zero, as demanded by Eq. (3.8), only when the dilatation J_1 is equal to zero. In most problems J_1 will not be equal to zero; therefore, the solution will not be exact since the boundary conditions on the parallel planes are violated. In many problems this violation of the boundary conditions can be cleared by superimposing an equal and opposite distribution of σ_{zz} (residual solution) onto the original solution.

It is possible to obtain an exact solution to the residual problem only when σ_{zz} is a linear function of x and y. When σ_{zz} is nonlinear, an approximate solution based on Saint-Venant's principle[†] is often utilized. When the nonlinear distribution of σ_{zz} on the parallel boundaries is replaced by a linear distribution which is statically equivalent, the solution will be valid only in regions well removed from the parallel bounding planes. Thus, it is clear that the plane-strain approach is necessarily limited to the central regions of bodies such as shafts or dams which are very long, i.e., thick, relative to their lateral dimensions. In the central region of such a long body, the stresses σ_{xx}, σ_{yy}, and τ_{xy} can be found from the solution of the original problem since the superposition of the residual solution onto the original problem does not influence these stresses but only serves to make σ_{zz} vanish.

In this section the plane-strain approach has been discussed without indicating a method for solving for σ_{xx}, σ_{yy}, and τ_{xy}. This problem will be treated later, in Sec. 3.6, when the Airy's-stress-function approach is discussed. In this plane-strain section it is important for the student to understand the plane-strain assumption, why it usually leads to a violation of the boundary conditions on the two parallel planes, and finally how these undesired stresses can be removed from the planes by superimposing a statically equivalent linear stress system. Also quite important is Saint-Venant's principle, since an experimentalist in simulating loads often relies on this principle to permit simplification in the design of the loading fixtures.

3.5 PLANE STRESS

In Sec. 3.4 it was noted that the plane-strain method is limited to very long or thick bodies. In those cases where the body thickness is small relative to its lateral dimensions, it is advantageous to assume that

$$\sigma_{zz} = \tau_{yz} = \tau_{zx} = 0 \qquad (3.9)$$

throughout the thickness of the plate. With this assumption the stress equations of equilibrium again reduce to

$$\frac{\partial \sigma_{xx}}{\partial x} + \frac{\partial \tau_{xy}}{\partial y} + F_x = 0 \qquad \frac{\partial \tau_{xy}}{\partial x} + \frac{\partial \sigma_{yy}}{\partial y} + F_y = 0 \qquad (3.10)$$

and the stress-strain relations [Eqs. (2.13)] become

$$\sigma_{xx} = \lambda J_1 + 2\mu \epsilon_{xx} \qquad \sigma_{yy} = \lambda J_1 + 2\mu \epsilon_{yy} \qquad \sigma_{zz} = \lambda J_1 + 2\mu \epsilon_{zz} = 0$$
$$\tau_{xy} = \mu \gamma_{xy} \qquad \tau_{yz} = \mu \gamma_{yz} = 0 \qquad \tau_{zx} = \mu \gamma_{zx} = 0 \qquad (3.11)$$

[†] Saint-Venant's principle states that a system of forces acting over a small region of the boundary can be replaced by a statically equivalent system of forces without introducing appreciable changes in the distribution of stresses in regions well removed from the area of load application.

From the third of Eqs. (3.11) the following relationship can be obtained:

$$\epsilon_{zz} = -\frac{\lambda}{\lambda + 2\mu}(\epsilon_{xx} + \epsilon_{yy}) \tag{a}$$

With this value of ϵ_{zz} the first strain invariant J_1 becomes

$$J_1 = \frac{2\mu}{\lambda + 2\mu}(\epsilon_{xx} + \epsilon_{yy}) \tag{b}$$

Substituting the value for J_1 given in Eq. (b) into Eqs. (3.11) yields

$$\sigma_{xx} = \frac{2\lambda\mu}{\lambda + 2\mu}(\epsilon_{xx} + \epsilon_{yy}) + 2\mu\epsilon_{xx}$$

$$\sigma_{yy} = \frac{2\lambda\mu}{\lambda + 2\mu}(\epsilon_{xx} + \epsilon_{yy}) + 2\mu\varepsilon_{yy} \tag{3.12}$$

$$\tau_{xy} = \mu\gamma_{xy} \qquad \sigma_{zz} = \tau_{yz} = \tau_{zx} = 0$$

Unfortunately, in the general case σ_{xx}, σ_{yy}, and τ_{xy} are not independent of z, and thus the boundary conditions imposed on the boundary B_L cannot be rigorously satisfied. To overcome this difficulty, average stresses and displacements over the thickness are commonly used. If the body is relatively thin, these averages closely approximate the true boundary conditions on B_L. Average values for the stresses and displacements over the thickness of the body are obtained as follows:

$$\tilde{\sigma}_{xx} = \frac{1}{2t}\int_{-t}^{t}\sigma_{xx}\,dz \qquad \tilde{\sigma}_{yy} = \frac{1}{2t}\int_{-t}^{t}\sigma_{yy}\,dz \qquad \tilde{\tau}_{xy} = \frac{1}{2t}\int_{-t}^{t}\tau_{xy}\,dz$$
$$\tilde{u} = \frac{1}{2t}\int_{-t}^{t}u\,dz \qquad \tilde{v} = \frac{1}{2t}\int_{-t}^{t}v\,dz \tag{3.13}$$

The symbol $\sim$ over the stresses and displacements indicates average values. Substituting the average values of the stresses into Eqs. (1.2) gives the boundary conditions which must be satisfied on B_L:

$$T_{nx} = \tilde{\sigma}_{xx}\cos(n, x) + \tilde{\tau}_{xy}\cos(n, y)$$
$$T_{ny} = \tilde{\tau}_{xy}\cos(n, x) + \tilde{\sigma}_{yy}\cos(n, y) \tag{3.14}$$

If the equations which the plane-strain and the plane-stress solutions must satisfy are compared, it can be observed that they are identical except for the comparison between Eqs. (3.5) and (3.11). An examination of a typical equation from each of these sets,

$$\sigma_{xx} = \begin{cases} \lambda(\epsilon_{xx} + \epsilon_{yy}) + 2\mu\epsilon_{xx} & \text{plane strain} \\ \dfrac{2\lambda\mu}{\lambda + 2\mu}(\epsilon_{xx} + \epsilon_{yy}) + 2\mu\epsilon_{xx} & \text{plane stress} \end{cases}$$

indicates that they are identical except for the coefficients of the $\epsilon_{xx} + \epsilon_{yy}$ term. Since all other equations for the plane-stress and plane-strain solutions are identical, results from plane strain can be transformed into plane stress by letting

$$\lambda \to \frac{2\lambda\mu}{\lambda + 2\mu}$$

which is equivalent to letting

$$\frac{v}{1 - v} \to v \tag{3.15}$$

In a similar manner a plane-stress solution can be transformed into a plane-strain solution by letting

$$\frac{2\lambda\mu}{\lambda + 2\mu} \to \lambda$$

or

$$v \to \frac{v}{1 - v} \tag{3.16}$$

In the plane-stress approach it is generally assumed that

$$\sigma_{zz} = \tau_{yz} = \tau_{zx} = 0$$

and the unknown stresses σ_{xx}, σ_{yy}, and τ_{xy} will have a z dependence. As a result of this z dependence, the boundary conditions on B_L are violated. This difficulty can be eliminated and an approximate solution to the problem can be obtained by using average values for the stresses and displacements. Finally, it was shown that plane-stress and plane-strain solutions can be transformed from one case into the other by a simple replacement involving Poisson's ratio, as indicated in Eqs. (3.15) and (3.16).

3.6 AIRY'S STRESS FUNCTION

In the plane problem three unknowns σ_{xx}, σ_{yy}, and τ_{xy} must be determined which will satisfy the required field equations and boundary conditions. The most convenient sets of field equations to use in this determination are the two equations of equilibrium and one stress equation of compatibility.

The equilibrium equations in two dimensions are

$$\frac{\partial \sigma_{xx}}{\partial x} + \frac{\partial \tau_{xy}}{\partial y} + F_x = 0 \tag{3.17a}$$

$$\frac{\partial \tau_{xy}}{\partial x} + \frac{\partial \sigma_{yy}}{\partial y} + F_y = 0 \tag{3.17b}$$

The stress compatibility equation for the case of plane strain is

$$\nabla^2(\sigma_{xx} + \sigma_{yy}) = -\frac{2(\lambda + \mu)}{\lambda + 2\mu}\left(\frac{\partial F_x}{\partial x} + \frac{\partial F_y}{\partial y}\right) \tag{3.17c}$$

Suppose the body-force field is defined by $\Omega(x, y)$ so that the body-force intensities are given by

$$F_x = -\frac{\partial \Omega}{\partial x} \qquad F_y = -\frac{\partial \Omega}{\partial y} \qquad (3.18)$$

Then by substituting Eqs. (3.18) into Eqs. (3.17) and noting that $2(\lambda + \mu)/(\lambda + 2\mu) = 1/(1 - v)$, it is apparent that

$$\frac{\partial \sigma_{xx}}{\partial x} + \frac{\partial \tau_{xy}}{\partial y} = \frac{\partial \Omega}{\partial x} \qquad \frac{\partial \tau_{xy}}{\partial x} + \frac{\partial \sigma_{yy}}{\partial y} = \frac{\partial \Omega}{\partial y}$$

$$\nabla^2 \left(\sigma_{xx} + \sigma_{yy} - \frac{\Omega}{1 - v} \right) = 0 \qquad (3.19)$$

Equations (3.19) represent the three field equations which σ_{xx}, σ_{yy}, and τ_{xy} must satisfy.

Assume that the stresses can be represented by a stress function ϕ such that

$$\sigma_{xx} = \frac{\partial^2 \phi}{\partial y^2} + \Omega \qquad \sigma_{yy} = \frac{\partial^2 \phi}{\partial x^2} + \Omega \qquad \tau_{xy} = -\frac{\partial^2 \phi}{\partial x\, \partial y} \qquad (3.20)$$

If Eqs. (3.20) are substituted into Eqs. (3.19), it can be seen that the two equations of equilibrium are exactly satisfied, and the last of Eqs. (3.19) gives

$$\nabla^4 \phi = -\frac{1 - 2v}{1 - v} \nabla^2 \Omega \qquad (3.21)$$

Thus, equilibrium and compatibility are immediately satisfied if ϕ satisfies Eq. (3.21). The expression ϕ is known as *Airy's stress function*. If Eq. (3.21) is solved for ϕ, an expression containing x, y, and a number of constants will be obtained. The constants are evaluated from the boundary conditions given in Eqs. (3.17), and the stresses are computed from ϕ according to Eqs. (3.20). Of course, evaluation of ϕ from Eq. (3.21) produces stresses for the plane-strain case. Stresses for the plane-stress case can be obtained by letting $v/(1 - v) \to v$, as indicated in Eq. (3.15). This substitution leads to

$$\nabla^4 \phi = -(1 - v) \nabla^2 \Omega \qquad (3.22)$$

which is valid for plane-stress problems.

It is important to note that if the body-force intensities are zero or constant, such as those encountered in a gravitational field, then

$$\nabla^2 \Omega = 0$$

and Eqs. (3.21) and (3.22) both become

$$\nabla^4 \phi = 0 \qquad (3.23a)$$

This is a biharmonic equation, which can also be written in the form

$$\frac{\partial^4 \phi}{\partial x^4} + 2 \frac{\partial^4 \phi}{\partial x^2\, \partial y^2} + \frac{\partial^4 \phi}{\partial y^4} = 0 \qquad (3.23b)$$

Examination of this equation shows that ϕ and thus σ_{xx}, σ_{yy}, and τ_{xy} are independent of the elastic constants. This consideration is very important in two-dimensional photoelasticity since it indicates that the stresses obtained from a plastic model are identical to those in a metal prototype if the model is simply connected and subjected to a zero or a uniform body-force field. Differences in the values of the modulus of elasticity and Poisson's ratio between model and prototype do not influence the results for the stresses. There are exceptions to the simply connected restriction, however, which will be covered in Chap. 13, on applied photoelasticity.

3.7 AIRY'S STRESS FUNCTION IN CARTESIAN COORDINATES

Any Airy's stress function used in the solution of a plane problem must satisfy Eqs. (3.23a) and (3.23b) and provide stresses via Eqs. (3.20) which satisfy the defined boundary conditions. Some Airy's stress functions commonly used are polynomials in x and y. In this section, polynomials from the first to the fifth degree will be considered.

3.7.1 Airy's Stress Function in Terms of a First-Degree Polynomial $\phi_1 = a_1 x + b_1 y$

It is clear from Eqs. (3.20) that

$$\sigma_{xx} = \sigma_{yy} = \tau_{xy} = 0 \tag{3.24}$$

and that Eqs. (3.23a) and (3.23b) are satisfied. This function is suitable only for indicating a stress-free field and therefore is of little use in the solution of any problem.

3.7.2 Airy's Stress Function in Terms of a Second-Degree Polynomial $\phi_2 = a_2 x^2 + b_2 xy + c_2 y^2$

From Eqs. (3.20) the stresses are

$$\sigma_{xx} = 2c_2 \qquad \sigma_{yy} = 2a_2 \qquad \tau_{xy} = -b_2 \tag{3.25}$$

Note that Eqs. (3.23a) and (3.23b) are satisfied and that the stress function ϕ_2 gives a uniform stress field over the entire body which is independent of x and y.

3.7.3 Airy's Stress Function in Terms of a Third-Degree Polynomial

$$\phi_3 = a_3 x^3 + b_3 x^2 y + c_3 xy^2 + d_3 y^3$$

Again, by use of Eqs. (3.20) the stresses are given by

$$\sigma_{xx} = 2c_3 x + 6d_3 y \qquad \sigma_{yy} = 6a_3 x + 2b_3 y \qquad \tau_{xy} = -2b_3 x - 2c_3 y \tag{3.26}$$

Equations (3.23a) and (3.23b) are satisfied unconditionally, and the stress function ϕ_3 provides a linearly varying stress field over the body.

3.7.4 Airy's Stress Function in Terms of a Fourth-Degree Polynomial

$$\phi_4 = a_4 x^4 + b_4 x^3 y + c_4 x^2 y^2 + d_4 xy^3 + e_4 y^4$$

From Eqs. (3.20) it is apparent that

$$\sigma_{xx} = 2c_4 x^2 + 6d_4 xy + 12e_4 y^2$$
$$\sigma_{yy} = 12a_4 x^2 + 6b_4 xy + 2c_4 y^2 \tag{3.27}$$
$$\tau_{xy} = -3b_4 x^2 - 4c_4 xy - 3d_4 y^2$$

When ϕ_4 is substituted into Eq. (3.23b), it should be noted that it is not unconditionally satisfied. In order for $\nabla^4 \phi = 0$ it is necessary that

$$e_4 = -\left(a_4 + \frac{c_4}{3}\right)$$

Substituting this equation into the relations for the stresses gives

$$\sigma_{xx} = 2c_4 x^2 + 6d_4 xy - 12a_4 y^2 - 4c_4 y^2$$

and σ_{yy} and τ_{xy} are unchanged. Thus, ϕ_4 yields a stress field which is a second-degree polynomial in x and y.

3.7.5 Airy's Stress Function in Terms of a Fifth-Degree Polynomial

$$\phi_5 = a_5 x^5 + b_5 x^4 y + c_5 x^3 y^2 + d_5 x^2 y^3 + e_5 xy^4 + f_5 y^5$$

Employing Eqs. (3.20) to solve for the stresses gives

$$\sigma_{xx} = 2c_5 x^3 + 6d_5 x^2 y + 12e_5 xy^2 + 20f_5 y^3$$
$$\sigma_{yy} = 20a_5 x^3 + 12b_5 x^2 y + 6c_5 xy^2 + 2d_5 y^3$$
$$\tau_{xy} = -4b_5 x^3 - 6c_5 x^2 y - 6d_5 xy^2 - 4e_5 y^3$$

Again, note that ϕ_5 must be subjected to certain conditions involving the constants e_5 and f_5. For Eqs. (3.23a) and (3.23b) to be satisfied, these conditions are

$$e_5 = -(5a_5 + c_5) \qquad f_5 = -\tfrac{1}{5}(b_5 + d_5)$$

Subject to the restrictive conditions listed above, the cartesian stress components become

$$\sigma_{xx} = 2c_5 x^3 + 6d_5 x^2 y - 12(5a_5 + c_5)xy^2 - 4(b_5 + d_5)y^3$$
$$\sigma_{yy} = 20a_5 x^3 + 12b_5 x^2 y + 6c_5 xy^2 + 2d_5 y^3 \tag{3.28}$$
$$\tau_{xy} = -4b_5 x^3 - 6c_5 x^2 y - 6d_5 xy^2 + 4(5a_5 + c_5)y^3$$

Thus, it is clear that ϕ_5 yields a stress field which is a third-degree polynomial in x and y. It is possible to continue this procedure to ϕ_6, ϕ_7, etc., as long as Eqs. (3.23a) and (3.23b) are satisfied. It is also possible to add together two or more stress functions to form another, for example, $\phi^* = \phi_2 + \phi_3$. Thus, by simply adding terms or by eliminating terms from the stress function it is theoretically possible to build up any stress field that can be expressed as a function of x and y.

3.8   EXAMPLE PROBLEM

Airy's stress function expressed in cartesian coordinates can be employed to solve a particular class of two-dimensional problems where the boundaries of the body can be adequately represented by the cartesian reference frame. As an example, consider the simply supported beam with uniform loads shown in Fig. 3.2. An examination of the loading conditions indicates that

$$\sigma_{yy} = \begin{cases} \tau_{xy} = 0 & \text{at } y = \dfrac{-h}{2} \\[3mm] -q & \text{at } y = \dfrac{+h}{2} \end{cases} \tag{a}$$

$$\tau_{xy} = 0 \qquad \text{at } y = \dfrac{+h}{2} \tag{b}$$

Also at $x = \pm L/2$

$$\int_{-h/2}^{h/2} \tau_{xy} \, dy = R = \frac{qL}{2} \tag{c}$$

$$\int_{-h/2}^{h/2} \sigma_{xx} \, dy = 0 \tag{d}$$

$$\int_{-h/2}^{h/2} \sigma_{xx} y \, dy = 0 \tag{e}$$

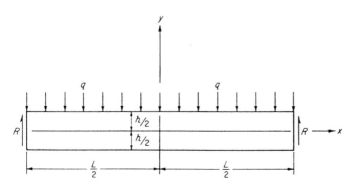

FIGURE 3.2
Simply supported beam of length L, height h, and unit depth subjected to a uniformly distributed load.

Note that the bending moment (and consequently σ_{xx}) is a maximum at position $x = 0$ and decreases with a change in x in either the positive or the negative direction. This is possible only if the stress function contains even functions of x. Note also that σ_{yy} varies from zero at $y = -h/2$ to a maximum value of $-q$ at $y = +h/2$; thus the stress function must contain odd functions of y. From the stress functions listed in Sec. 3.7, the following even and odd functions can be selected to form a new stress function ϕ which satisfies the previously listed conditions.

$$\phi = a_2 x^2 + b_3 x^2 y + d_3 y^3 + a_4 x^4 + b_5 x^4 y + d_5 x^2 y^3 + f_5 y^5 \qquad (f)$$

This stress function ϕ must satisfy the equation $\nabla^4 \phi = 0$; hence

$$a_4 = 0 \qquad f_5 = -\tfrac{1}{5}(b_5 + d_5) \qquad (g)$$

From Eqs. (3.20) the cartesian stress components are

$$\sigma_{xx} = 6d_3 y + 6d_5 x^2 y - 4(b_5 + d_5)y^3$$
$$\sigma_{yy} = 2a_2 + 2b_3 y + 12b_5 x^2 y + 2d_5 y^3 \qquad (h)$$
$$\tau_{xy} = -2b_3 x - 4b_5 x^3 - 6d_5 xy^2$$

Examination of the boundary conditions shown in Eq. (a) indicates that σ_{yy} must be independent of x; hence the coefficient $b_5 = 0$. Consequently, Eqs. (h) reduce to

$$\sigma_{xx} = 6d_3 y + 6d_5 x^2 y - 4d_5 y^3$$
$$\sigma_{yy} = 2a_2 + 2b_3 y + 2d_5 y^3 \qquad (i)$$
$$\tau_{xy} = -2b_3 x - 6d_5 xy^2$$

The problem can be solved if the coefficients a_2, b_3, d_3, and d_5 can be selected so that the boundary conditions given in Eqs. (a) to (e) are satisfied. From Eqs. (a)

$$\sigma_{yy} = 0 = 2a_2 + 2b_3\left(-\frac{h}{2}\right) + 2d_5\left(-\frac{h}{2}\right)^3 \qquad a_2 - \frac{b_3 h}{2} - \frac{d_5 h^3}{8} = 0 \qquad (j)$$

and from Eqs. (b)

$$\sigma_{yy} = -q = 2a_2 + 2b_3\frac{h}{2} + 2d_5\left(\frac{h}{2}\right)^3 \qquad a_2 + \frac{b_3 h}{2} + \frac{d_5 h^3}{8} = -\frac{q}{2} \qquad (k)$$

Adding Eqs. (j) and (k) gives

$$a_2 = -\frac{q}{4} \qquad (l)$$

From Eqs. (a) and (b)

$$\tau_{xy} = 0 = -2x\left[b_3 + 3d_5\left(\pm\frac{h}{2}\right)^2\right] \qquad b_3 = -\frac{3}{4}h^2 d_5 \qquad (m)$$

Substituting Eqs. (*m*) into Eqs. (*j*),

$$\frac{d_5 h^3}{8} - \frac{3h^3 d_5}{8} = -\frac{q}{4} \qquad d_5 = \frac{q}{h^3} \tag{n}$$

and

$$b_3 = -\frac{3}{4}\frac{q}{h} \tag{o}$$

With the values of a_2, b_3, and d_5 given by Eqs. (*l*), (*o*), and (*n*), respectively, Eqs. (*c*) and (*d*) are identically satisfied. Equation (*e*) can be used to solve for the remaining unknown d_3.

$$\int_{-h/2}^{h/2} \left(6d_3 y^2 + \frac{3}{2}\frac{q}{h^3} L^2 y^2 - 4\frac{qy^4}{L^3} \right) dy = 0$$

$$\left[2d_3 y^3 + \frac{qL^2 y^3}{2h^3} - \frac{4qy^5}{5h^3} \right]_{-h/2}^{+h/2} = 0 \tag{p}$$

Solving Eq. (*p*) for d_3 gives

$$d_3 = \frac{q}{240I} (2h^2 + 5L^2) \tag{q}$$

where $I = h^3/12$ is the moment of inertia of the unit-width beam. Substituting Eqs. (*q*), (*o*), (*n*), and (*l*) into Eqs. (*i*) gives the final equations for the cartesian components of stress:

$$\sigma_{xx} = \frac{q}{8I}(4x^2 - L^2)y + \frac{q}{60I}(3h^2 y - 20y^3)$$

$$\sigma_{yy} = \frac{q}{24I}(4y^3 - 3h^2 y - h^3) \tag{r}$$

$$\tau_{xy} = \frac{qx}{8I}(h^2 - 4y^2)$$

The conventional strength-of-materials solution for this problem, namely, that $\sigma_{xx} = My/I$, gives

$$\sigma_{xx} = \frac{q}{8I}(4x^2 - L^2)y \tag{s}$$

which is identical with the first term of the relation given for σ_{xx} in Eqs. (*r*). The second term, $(q/60I)(3h^2 y - 20y^3)$, is a correction term for the strength-of-materials solution. In the strength-of-materials approach, recall that it is assumed that plane sections remain plane after bending. This is not exactly true, and as a consequence the solution obtained lacks the correction term shown above. It is clear that the correction term is small when $L \gg h$, and the strength-of-materials solution will be sufficiently accurate.

This simple example illustrates how elementary elasticity theory can be employed to extend the student's understanding of the distribution of stresses in simple two-dimensional problems. Other examples are included in the exercises at the end of this chapter.

3.9 TWO-DIMENSIONAL PROBLEMS IN POLAR COORDINATES

In Sec. 3.6 the Airy's-stress-function approach to the solution of two-dimensional elasticity problems in cartesian coordinates was developed. This method was then applied to solve an elementary problem which was well suited to the cartesian reference frame. In many problems, however, the geometry of the body does not lend itself to the use of a cartesian coordinate system, and it is more expeditious to work with a different system. A large class of problems (such as circular rings, curved beams, and half-planes) can be solved by employing a commonly used system, the polar coordinate system. In any elasticity problem the proper choice of the coordinate system is extremely important since this choice establishes the complexity of the mathematical expressions employed to satisfy the field equations and the boundary conditions.

In order to solve two-dimensional elasticity problems by employing a polar-coordinate reference frame, the equations of equilibrium, the definition of Airy's stress function, and one of the stress equations of compatibility must be reestablished in terms of polar coordinates. On the following pages the equations of equilibrium will be derived by considering a polar element instead of a cartesian element. The equations for the polar components of stress in terms of Airy's stress function as well as the stress equation of compatibility will be transformed from cartesian to polar coordinates. Finally, a set of stress functions is developed which satisfies the stress equation of compatibility.

The stress equations of equilibrium in polar coordinates can be derived from the free-body diagram of the polar element shown in Fig. 3.3. The element is

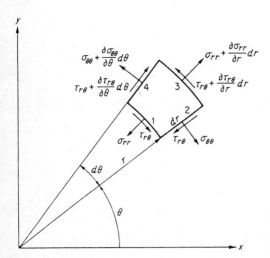

FIGURE 3.3
Polar element of unit depth showing the stresses acting on the four faces.

assumed to be very small. The average values of the normal and shearing stresses which act on surface 1 are denoted by σ_{rr} and $\tau_{r\theta}$, respectively. Since the stresses may vary as a function of r, values of the normal and shearing stresses on surface 3 are given by $\sigma_{rr} + (\partial\sigma_{rr}/\partial r)\, dr$ and $\tau_{r\theta} + (\partial\tau_{r\theta}/\partial r)\, dr$. Similarly, the average values of the normal and shearing stresses which act on surface 2 are given by $\sigma_{\theta\theta}$ and $\tau_{r\theta}$. Since the stresses may also vary as a function of θ, values of the normal and shearing stresses on surface 4 are $\sigma_{\theta\theta} + (\partial\sigma_{\theta\theta}/\partial\theta)\, d\theta$ and $\tau_{r\theta} + (\partial\tau_{r\theta}/\partial\theta)\, d\theta$.

For a polar element of unit thickness to be in a state of equilibrium the sum of all forces in the radial r and tangential θ directions must equal zero. Summing forces first in the radial direction and considering the body-force intensity F_r gives the equation of equilibrium

$$\left(\sigma_{rr} + \frac{\partial\sigma_{rr}}{\partial r}\, dr\right)(r + dr)\, d\theta - \sigma_{rr}r\, d\theta - \left[\sigma_{\theta\theta}\, dr + \left(\sigma_{\theta\theta} + \frac{\partial\sigma_{\theta\theta}}{\partial\theta}\, d\theta\right) dr\right]\frac{d\theta}{2}$$

$$+ \left(\tau_{r\theta} + \frac{\partial\tau_{r\theta}}{\partial\theta}\, d\theta - \tau_{r\theta}\right) dr + F_r r\, d\theta\, dr = 0 \qquad (a)$$

Dividing Eq. (a) by $dr\, d\theta$ and simplifying gives

$$\frac{\partial\sigma_{rr}}{\partial r}\, dr - \frac{\partial\sigma_{\theta\theta}}{\partial\theta}\frac{d\theta}{2} + \sigma_{rr} + \frac{\partial\sigma_{rr}}{\partial r}\, r - \sigma_{\theta\theta} + \frac{\partial\tau_{r\theta}}{\partial\theta} + F_r r = 0 \qquad (b)$$

If the element is made infinitely small by permitting dr and $d\theta$ each to approach zero, the first two terms in Eq. (b) also approach zero and the expression can be rewritten as

$$\frac{\partial\sigma_{rr}}{\partial r} + \frac{1}{r}\frac{\partial\tau_{r\theta}}{\partial\theta} + \frac{1}{r}(\sigma_{rr} - \sigma_{\theta\theta}) + F_r = 0 \qquad (3.29a)$$

The equation of equilibrium in the tangential direction can be derived in the same manner if the forces acting in the θ direction on the polar element are summed and set equal to zero. Hence

$$\frac{1}{r}\frac{\partial\sigma_{\theta\theta}}{\partial\theta} + \frac{\partial\tau_{r\theta}}{\partial r} + \frac{2\tau_{r\theta}}{r} + F_\theta = 0 \qquad (3.29b)$$

Equations (3.29a) and (3.29b) represent the equations of equilibrium in polar coordinates. They are analogous to the equations of equilibrium in cartesian coordinates presented in Eqs. (3.17a) and (3.17b). Any solution to an elasticity problem must satisfy these field equations.

3.10 TRANSFORMATION OF THE EQUATION $\nabla^4\phi = 0$ INTO POLAR COORDINATES

In the coverage of Airy's stress function given in Sec. 3.6 it was shown that the stress function ϕ had to satisfy the biharmonic equation $\nabla^4\phi = 0$, provided the body forces are zero or constants. In polar coordinates the stress function must satisfy this same equation; however, the definition of the ∇^4 operator must be modified to suit the polar-coordinate system. This modification may be accomplished by transforming the ∇^4 operator from the cartesian system to the polar system.

In transforming from cartesian coordinates to polar coordinates, recall that

$$r^2 = x^2 + y^2 \qquad \theta = \arctan \frac{y}{x} \qquad (3.30)$$

where r and θ are defined in Fig. 3.3.

Differentiating Eqs. (3.30) gives

$$\frac{\partial r}{\partial x} = \frac{x}{r} = \cos \theta \qquad \frac{\partial r}{\partial y} = \frac{y}{r} = \sin \theta$$

$$\frac{\partial \theta}{\partial x} = -\frac{y}{r^2} = -\frac{\sin \theta}{r} \qquad \frac{\partial \theta}{\partial y} = \frac{x}{r^2} = \frac{\cos \theta}{r} \qquad (3.31)$$

The form of the ∇^4 operator in cartesian coordinates is

$$\nabla^4 \phi = \left(\frac{\partial^2}{\partial x^2} + \frac{\partial^2}{\partial y^2} \right) \left(\frac{\partial^2 \phi}{\partial x^2} + \frac{\partial^2 \phi}{\partial y^2} \right)$$

Individual elements of this expression can be transformed by employing Eqs. (3.30) and (3.31) as follows. If it is assumed that ϕ is a function of r and θ,

$$\frac{\partial \phi}{\partial x} = \frac{\partial \phi}{\partial r} \frac{\partial r}{\partial x} + \frac{\partial \phi}{\partial \theta} \frac{\partial \theta}{\partial x} \qquad (a)$$

$$\frac{\partial^2 \phi}{\partial x^2} = \frac{\partial \phi}{\partial r} \frac{\partial^2 r}{\partial x^2} + \left(\frac{\partial r}{\partial x} \right)^2 \frac{\partial^2 \phi}{\partial r^2} + 2 \frac{\partial^2 \phi}{\partial r \partial \theta} \frac{\partial r}{\partial x} \frac{\partial \theta}{\partial x} + \frac{\partial \phi}{\partial \theta} \frac{\partial^2 \theta}{\partial x^2} + \left(\frac{\partial \theta}{\partial x} \right)^2 \frac{\partial^2 \phi}{\partial \theta^2} \qquad (b)$$

Substituting the equalities given in Eqs. (3.31) into Eq. (b) yields

$$\frac{\partial^2 \phi}{\partial x^2} = \frac{\sin^2 \theta}{r} \frac{\partial \phi}{\partial r} + \cos^2 \theta \frac{\partial^2 \phi}{\partial r^2} - \frac{\sin 2\theta}{r} \frac{\partial^2 \phi}{\partial r \partial \theta} + \frac{\sin 2\theta}{r^2} \frac{\partial \phi}{\partial \theta} + \frac{\sin^2 \theta}{r^2} \frac{\partial^2 \phi}{\partial \theta^2} \qquad (3.32a)$$

Following the same procedure makes it clear that

$$\frac{\partial^2 \phi}{\partial y^2} = \frac{\cos^2 \theta}{r} \frac{\partial \phi}{\partial r} + \sin^2 \theta \frac{\partial^2 \phi}{\partial r^2} + \frac{\sin 2\theta}{r} \frac{\partial^2 \phi}{\partial r \partial \theta} - \frac{\sin 2\theta}{r^2} \frac{\partial \phi}{\partial \theta} + \frac{\cos^2 \theta}{r^2} \frac{\partial^2 \phi}{\partial \theta^2} \qquad (3.32b)$$

$$\frac{\partial^2 \phi}{\partial x \partial y} = -\frac{\sin \theta \cos \theta}{r} \frac{\partial \phi}{\partial r} + \sin \theta \cos \theta \frac{\partial^2 \phi}{\partial r^2} + \frac{\cos 2\theta}{r} \frac{\partial^2 \phi}{\partial r \partial \theta}$$

$$- \frac{\cos 2\theta}{r^2} \frac{\partial \phi}{\partial \theta} - \frac{\sin \theta \cos \theta}{r^2} \frac{\partial^2 \phi}{\partial \theta^2} \qquad (3.32c)$$

Adding Eqs. (3.32a) and (3.32b) gives

$$\frac{\partial^2 \phi}{\partial x^2} + \frac{\partial^2 \phi}{\partial y^2} = \frac{\partial^2 \phi}{\partial r^2} + \frac{1}{r}\frac{\partial \phi}{\partial r} + \frac{1}{r^2}\frac{\partial^2 \phi}{\partial \theta^2} \tag{3.33}$$

Furthermore, it is easily seen that

$$\nabla^4 \phi = \left(\frac{\partial^2}{\partial x^2} + \frac{\partial^2}{\partial y^2}\right)\left(\frac{\partial^2 \phi}{\partial x^2} + \frac{\partial^2 \phi}{\partial y^2}\right)$$

$$= \left(\frac{\partial^2}{\partial r^2} + \frac{1}{r}\frac{\partial}{\partial r} + \frac{1}{r^2}\frac{\partial^2}{\partial \theta^2}\right)\left(\frac{\partial^2 \phi}{\partial r^2} + \frac{1}{r}\frac{\partial \phi}{\partial r} + \frac{1}{r^2}\frac{\partial^2 \phi}{\partial \theta^2}\right) = 0 \tag{3.34}$$

Equation (3.34) is the stress equation of compatibility in terms of Airy's stress function referred to a polar coordinate system.

3.11 POLAR COMPONENTS OF STRESS IN TERMS OF AIRY'S STRESS FUNCTION

By referring to the two-dimensional equations of stress transformation [Eqs. (1.11)], expressions can be obtained which relate the polar stress components σ_{rr}, $\sigma_{\theta\theta}$, and $\tau_{r\theta}$ to the cartesian stress components σ_{xx}, σ_{yy}, and τ_{xy} as follows:

$$\sigma_{rr} = \sigma_{xx} \cos^2 \theta + \sigma_{yy} \sin^2 \theta + \tau_{xy} \sin 2\theta$$

$$\sigma_{\theta\theta} = \sigma_{yy} \cos^2 \theta + \sigma_{xx} \sin^2 \theta - \tau_{xy} \sin 2\theta \tag{3.35}$$

$$\tau_{r\theta} = (\sigma_{yy} - \sigma_{xx}) \sin \theta \cos \theta + \tau_{xy} \cos 2\theta$$

If Eqs. (3.20) are substituted into Eqs. (3.35) and Ω set equal to zero (which is equivalent to setting both F_x and F_y equal to zero), then

$$\sigma_{rr} = \frac{\partial^2 \phi}{\partial y^2} \cos^2 \theta + \frac{\partial^2 \phi}{\partial x^2} \sin^2 \theta - \frac{\partial^2 \phi}{\partial x\,\partial y} \sin 2\theta$$

$$\sigma_{\theta\theta} = \frac{\partial^2 \phi}{\partial x^2} \cos^2 \theta + \frac{\partial^2 \phi}{\partial y^2} \sin^2 \theta + \frac{\partial^2 \phi}{\partial x\,\partial y} \sin 2\theta \tag{3.36}$$

$$\tau_{r\theta} = \left(\frac{\partial^2 \phi}{\partial x^2} - \frac{\partial^2 \phi}{\partial y^2}\right) \sin \theta \cos \theta - \frac{\partial^2 \phi}{\partial x\,\partial y} \cos 2\theta$$

If the results from Eqs. (3.32a) to (3.32c) are substituted into Eqs. (3.36), the polar components of stress in terms of Airy's stress function are obtained:

$$\sigma_{rr} = \frac{1}{r}\frac{\partial \phi}{\partial r} + \frac{1}{r^2}\frac{\partial^2 \phi}{\partial \theta^2} \qquad \sigma_{\theta\theta} = \frac{\partial^2 \phi}{\partial r^2}$$

$$\tau_{r\theta} = \frac{1}{r^2}\frac{\partial \phi}{\partial \theta} - \frac{1}{r}\frac{\partial^2 \phi}{\partial r\,\partial \theta} \tag{3.37}$$

When Airy's stress function ϕ in polar coordinates has been established, these relations can be employed to determine the stress field as a function of r and θ.

3.12 FORMS OF AIRY'S STRESS FUNCTION IN POLAR COORDINATES

The equation $\nabla^4 \phi = 0$ is a fourth-order biharmonic partial differential equation which can be reduced to an ordinary fourth-order differential equation by using a separation-of-variables technique, where

$$\phi^{(n)} = R_n(r) \begin{Bmatrix} \cos n\theta \\ \sin n\theta \end{Bmatrix}$$

The resulting differential equation is an Euler type which yields four different stress functions upon solution. These stress functions are tabulated, together with the stress and displacement distributions which they provide, on the following pages.

One of the stress functions obtained can be expressed in the following form:

$$\phi^{(0)}(r) = a_0 + b_0 \ln r + c_0 r^2 + d_0 r^2 \ln r \qquad (3.38a)$$

By using Eqs. (3.37), the stresses associated with this particular stress function can be expressed as

$$\sigma_{rr} = \frac{b_0}{r^2} + 2c_0 + d_0(1 + 2 \ln r)$$

$$\sigma_{\theta\theta} = -\frac{b_0}{r^2} + 2c_0 + d_0(3 + 2 \ln r) \qquad \tau_{r\theta} = 0$$

$$(3.38b)$$

The displacements associated with this function can be determined by integrating the stress displacement relations, giving

$$u_r = \frac{1}{E} \left[-(1 + v)\frac{b_0}{r} + 2(1 - v)c_0 r + 2(1 - v)d_0 r \ln r - (1 + v)d_0 r \right]$$

$$+ \alpha_2 \cos \theta + \alpha_3 \sin \theta \qquad (3.38c)$$

$$u_\theta = \frac{1}{E} (4d_0 r\theta) - \alpha_1 r - \alpha_2 \sin \theta + \alpha_3 \cos \theta$$

where u_r and u_θ are the displacements in the radial and circumferential directions, respectively. The terms containing α_1, α_2, and α_3 are associated with rigid-body displacements.

It should be noted that the stresses in this solution are independent of θ; hence, the stress function $\phi^{(0)}$ should be employed to solve problems which have rotational symmetry.

One of the other stress functions and the stresses and displacements associated with it can be expressed as follows:

$$\phi^{(1)} = \left(a_1 r + \frac{b_1}{r} + c_1 r^3 + d_1 r \ln r\right)\begin{Bmatrix} \sin\theta \\ \cos\theta \end{Bmatrix}$$

$$\sigma_{rr} = \left(-\frac{2b_1}{r^3} + 2c_1 r + \frac{d_1}{r}\right)\begin{Bmatrix} \sin\theta \\ \cos\theta \end{Bmatrix}$$

$$\sigma_{\theta\theta} = \left(\frac{2b_1}{r^3} + 6c_1 r + \frac{d_1}{r}\right)\begin{Bmatrix} \sin\theta \\ \cos\theta \end{Bmatrix}$$

$$\tau_{r\theta} = \left(-\frac{2b_1}{r^3} + 2c_1 r + \frac{d_1}{r}\right)\begin{Bmatrix} -\cos\theta \\ \sin\theta \end{Bmatrix}$$

$$\begin{aligned}
u_r = \frac{1}{E}\Bigg\{ & \left[(1+v)\frac{b_1}{r^2} + (1-3v)c_1 r^2 - (1+v)d_1 \right. \\
& \left. + (1-v)d_1 \ln r\right]\begin{Bmatrix} \sin\theta \\ \cos\theta \end{Bmatrix} - (2d_1\theta)\begin{Bmatrix} \cos\theta \\ -\sin\theta \end{Bmatrix}\Bigg\} + \alpha_2 \cos\theta + \alpha_3 \sin\theta
\end{aligned}$$

$$\begin{aligned}
u_\theta = \frac{1}{E}\Bigg\{ & \left[-(1+v)\frac{b_1}{r^2} - (5+v)c_1 r^2 + (1-v)d_1 \ln r\right]\begin{Bmatrix} \cos\theta \\ -\sin\theta \end{Bmatrix} \\
& + (2d_1\theta)\begin{Bmatrix} \sin\theta \\ \cos\theta \end{Bmatrix}\Bigg\} - \alpha_1 r - \alpha_2 \sin\theta + \alpha_3 \cos\theta
\end{aligned}$$

(3.39)

The third stress function of interest and its associated stresses and displacements are as follows:

$$\phi^{(n)} = (a_n r^n + b_n r^{-n} + c_n r^{2+n} + d_n r^{2-n})\begin{Bmatrix} \sin n\theta \\ \cos n\theta \end{Bmatrix}$$

$$\begin{aligned}
\sigma_{rr} = [& a_n(n-n^2)r^{n-2} - b_n(n+n^2)r^{-n-2} + c_n(2+n-n^2)r^n \\
& + d_n(2-n-n^2)r^{-n}]\begin{Bmatrix} \sin n\theta \\ \cos n\theta \end{Bmatrix}
\end{aligned}$$

$$\begin{aligned}
\sigma_{\theta\theta} = [& a_n(n^2-n)r^{n-2} + b_n(n^2+n)r^{-n-2} + c_n(2+3n+n^2)r^n \\
& + d_n(2-3n+n^2)r^{-n}]\begin{Bmatrix} \sin n\theta \\ \cos n\theta \end{Bmatrix}
\end{aligned}$$

$$\begin{aligned}
\tau_{r\theta} = [& a_n(n^2-n)r^{n-2} - b_n(n+n^2)r^{-n-2} + c_n(n+n^2)r^n \\
& + d_n(n-n^2)r^{-n}]\begin{Bmatrix} -\cos n\theta \\ \sin n\theta \end{Bmatrix}
\end{aligned}$$

(3.40)

$$\begin{aligned}
u_r = \frac{1}{E}\Big\{ & -a_n(1+v)nr^{n-1} + b_n(1+v)nr^{-n-1} \\
& + c_n[4-(1+v)(2+n)]r^{n+1} \\
& + d_n[4-(1+v)(2-n)]r^{-n+1}\Big\}\begin{Bmatrix} \sin n\theta \\ \cos n\theta \end{Bmatrix} + \alpha_2 \cos\theta + \alpha_3 \sin\theta
\end{aligned}$$

$$u_\theta = \frac{1}{E} \{ -a_n(1+v)nr^{n-1} - b_n(1+v)nr^{-n-1}$$

$$- c_n[4+(1+v)n]r^{n+1} + d_n[4-(1+v)n]r^{-n+1} \} \begin{Bmatrix} \cos n\theta \\ -\sin n\theta \end{Bmatrix}$$

$$- \alpha_1 r - \alpha_2 \sin\theta + \alpha_3 \cos\theta$$

For the stress function $\phi^{(n)}$ the value of n can be greater than or equal to 2 (that is, $n \geq 2$).

The fourth stress function of interest and the associated stresses and displacements are expressed as follows:

$$\phi^{(*)} = a_* \theta + b_* r^2 \theta + c_* r\theta \sin\theta + d_* r\theta \cos\theta$$

$$\sigma_{rr} = 2b_* \theta + 2c_* \frac{\cos\theta}{r} - 2d_* \frac{\sin\theta}{r}$$

$$\sigma_{\theta\theta} = 2b_* \theta$$

$$\tau_{r\theta} = \frac{a_*}{r^2} - b_*$$

$$u_r = \frac{1}{E} [2(1-v)b_* r\theta + (1-v)c_* \theta \sin\theta + 2c_* \ln r \cos\theta$$

$$+ (1-v)d_* \theta \cos\theta - 2d_* \ln r \sin\theta] + \alpha_2 \cos\theta + \alpha_3 \sin\theta \qquad (3.41)$$

$$u_\theta = \frac{1}{E} \left[-(1+v)\frac{a_*}{r} + (3-v)b_* r - 4b_* r \ln r \right.$$

$$- (1+v)c_* \sin\theta - 2c_* \ln r \sin\theta + (1-v)c_* \theta \cos\theta$$

$$\left. - (1+v)d_* \cos\theta - 2d_* \ln r \cos\theta - (1-v)d_* \theta \sin\theta \right]$$

$$- \alpha_1 r - \alpha_2 \sin\theta + \alpha_3 \cos\theta$$

In the example problems which follow, the stress functions previously listed will be employed to determine the stresses and displacements for problems which lend themselves to polar coordinates. As the selection of the stress function is often the most difficult phase of the problem, particular emphasis will be placed on the reasoning behind the selection.

3.13 STRESSES AND DISPLACEMENTS IN A CIRCULAR CYLINDER SUBJECTED TO INTERNAL AND EXTERNAL PRESSURE

Consider the long hollow cylinder shown in Fig. 3.4, which is subjected to an internal pressure p_i and an external pressure p_o. The inside and outside radii of the cylinder are denoted as a and b, respectively.

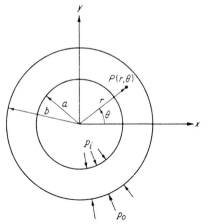

FIGURE 3.4
Circular cylinder subjected to internal and external pressures.

As stated previously, the first step in the solution of an elasticity problem after the geometry of the body has been defined is to establish the boundary conditions. For the problem under consideration these conditions can be listed as follows:

$$\sigma_{rr} = -p_i \qquad \tau_{r\theta} = 0 \qquad \text{at } r = a$$
$$\sigma_{rr} = -p_o \qquad \tau_{r\theta} = 0 \qquad \text{at } r = b \qquad (a)$$

An examination of the boundary conditions indicates that they are independent of θ; hence the four stress functions $\phi^{(0)}$, $\phi^{(1)}$, $\phi^{(n)}$, and $\phi^{(*)}$ should be inspected to determine which will provide a stress field independent of θ. The stress function $\phi^{(0)}$ given in Eq. (3.38a) yields stresses which satisfy this requirement, as shown below:

$$\sigma_{rr} = \frac{b_0}{r^2} + 2c_0 + d_0(1 + 2 \ln r)$$

$$\sigma_{\theta\theta} = -\frac{b_0}{r^2} + 2c_0 + d_0(3 + 2 \ln r) \qquad (b)$$

$$\tau_{r\theta} = 0$$

Equations (b) will provide the desired solution to the problem if the constants b_0, c_0, and d_0 can be determined so that the boundary conditions given in Eqs. (a) are satisfied.

An examination of Eqs. (b) indicates that the condition $\tau_{r\theta} = 0$ throughout the body satisfies part of the boundary conditions. From symmetry considerations it is also obvious that both u_r and u_θ must be independent of θ. This condition can be satisfied only if $d_0 = 0$ in Eqs. (3.38c). The two remaining constants b_0 and

c_0 can be evaluated by using the remaining boundary conditions in Eqs. (a):

$$\sigma_{rr} = -p_i = \frac{b_0}{a^2} + 2c_0 \qquad \sigma_{rr} = -p_o = \frac{b_0}{b^2} + 2c_0 \qquad (c)$$

Solving Eqs. (c) for b_0 and c_0 yields

$$b_0 = \frac{a^2 b^2 (p_o - p_i)}{b^2 - a^2} \qquad c_0 = \frac{a^2 p_i - b^2 p_o}{2(b^2 - a^2)}$$

These values when substituted into Eqs. (3.38) provide the required solution.

$$\sigma_{rr} = \frac{a^2 b^2 (p_o - p_i)}{(b^2 - a^2) r^2} + \frac{a^2 p_i - b^2 p_o}{b^2 - a^2}$$

$$\sigma_{\theta\theta} = -\frac{a^2 b^2 (p_o - p_i)}{(b^2 - a^2) r^2} + \frac{a^2 p_i - b^2 p_o}{b^2 - a^2}$$

$$\tau_{r\theta} = 0 \qquad (3.42)$$

$$u_r = \frac{1}{E}\left[-(1 + v)\frac{a^2 b^2 (p_o - p_i)}{(b^2 - a^2) r} + (1 - v)\frac{a^2 p_i - b^2 p_o}{b^2 - a^2} r \right]$$

$$u_\theta = 0$$

Equations (3.42) give the stresses and displacements at a point $P(r, \theta)$ in the cylinder if the two pressures, the radii, and the elastic constants are known. Three special cases of this problem are of interest.

CASE 1: EXTERNAL PRESSURE EQUALS ZERO. Setting $p_o = 0$ in Eqs. (3.42) leads to

$$\sigma_{rr} = \frac{a^2 p_i}{b^2 - a^2}\left(1 - \frac{b^2}{r^2}\right) \qquad \sigma_{\theta\theta} = \frac{a^2 p_i}{b^2 - a^2}\left(1 + \frac{b^2}{r^2}\right) \qquad \tau_{r\theta} = 0$$

$$\qquad (3.43)$$

$$u_r = \frac{a^2 p_i}{Er(b^2 - a^2)}[(1 + v)b^2 + (1 - v)r^2] \qquad u_\theta = 0$$

This special case is often encountered when dealing with stresses in piping systems or pressure vessels.

CASE 2: INTERNAL PRESSURE EQUALS ZERO. Setting $p_i = 0$ in Eqs. (3.42) leads to

$$\sigma_{rr} = \frac{b^2 p_o}{b^2 - a^2}\left(\frac{a^2}{r^2} - 1\right) \qquad \sigma_{\theta\theta} = -\frac{b^2 p_o}{b^2 - a^2}\left(\frac{a^2}{r^2} + 1\right) \qquad \tau_{r\theta} = 0$$

$$u_r = -\frac{b^2 p_o}{Er(b^2 - a^2)}[(1 + v)a^2 + (1 - v)r^2] \qquad u_\theta = 0 \qquad (3.44)$$

When external pressure is applied to a cylindrical shell, the problem of buckling should also be considered.

CASE 3: EXTERNAL PRESSURE ON A SOLID CIRCULAR CYLINDER. When one sets $a = 0$ in Eqs. (3.44), the hole in the cylinder vanishes and the stresses become

$$\sigma_{rr} = \sigma_{\theta\theta} = -p_o \qquad \tau_{r\theta} = 0$$

$$u_r = -\frac{1-v}{E}\, p_o r \qquad u_\theta = 0$$

(3.45)

3.14 STRESS DISTRIBUTION IN A THIN, INFINITE PLATE WITH A CIRCULAR HOLE SUBJECTED TO UNIAXIAL TENSILE LOADS

A thin plate of infinite length and width with a circular hole is shown in Fig. 3.5. The plate is subjected to a uniform tensile-type load which produces a uniform stress σ_0 in the y direction at $r = \infty$. The distribution of the stresses about the hole, along the x axis, and along the y axis can be determined by using the Airy's-stress-function approach.

The boundary conditions which must be satisfied are

$$\sigma_{rr} = \tau_{r\theta} = 0 \qquad \text{at } r = a \qquad (a)$$

$$\sigma_{yy} = \sigma_0 \qquad \text{at } r \to \infty$$

$$\sigma_{xx} = \tau_{xy} = 0 \qquad \text{at } r \to \infty \qquad (b)$$

Selection of a stress function for this particular problem is difficult since none of the four functions previously tabulated is satisfactory. In order to overcome this difficulty, a method of superposition is commonly used which

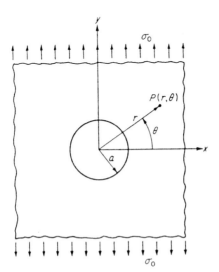

FIGURE 3.5
Thin, infinite plate with a circular hole subjected to a uniaxial tensile stress σ_0.

employs two different stress functions. The first function is selected such that the stresses associated with it satisfy the boundary conditions at $r \to \infty$ but in general violate the conditions on the boundary of the hole. The second stress function must have associated stresses which cancel the stresses on the boundary of the hole without influencing the stresses at $r \to \infty$. An illustration of this superposition process is presented in Fig. 3.6.

The boundary conditions at $r \to \infty$ can be satisfied by the uniform stress field associated with the stress function ϕ_2 in Eqs. (3.25). For the case of uniaxial tension in the y direction, ϕ_2 reduces to

$$\phi_2 = a_2 x^2 = \frac{\sigma_0 x^2}{2} \tag{c}$$

The stresses throughout the plate for a plate without a hole are

$$\sigma_{yy} = \sigma_0 \qquad \sigma_{xx} = \tau_{xy} = 0 \tag{d}$$

If an imaginary hole of radius a is cut into the plate, the stresses σ_{rr}, $\sigma_{\theta\theta}$, and $\tau_{r\theta}$ on the boundary of the imaginary hole can be computed from Eqs. (3.31) as follows:

$$\sigma_{rr}^I = \sigma_0 \sin^2 \theta = \frac{\sigma_0}{2} (1 - \cos 2\theta)$$

$$\sigma_{\theta\theta}^I = \sigma_0 \cos^2 \theta = \frac{\sigma_0}{2} (1 + \cos 2\theta) \tag{e}$$

$$\tau_{r\theta}^I = \sigma_0 \sin \theta \cos \theta = \frac{\sigma_0}{2} \sin 2\theta$$

In the original problem the boundary conditions at $r = a$ were

$$\sigma_{rr} = \tau_{r\theta} = 0$$

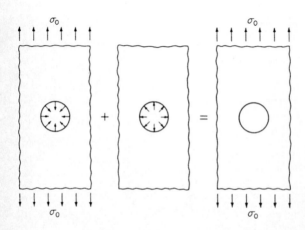

FIGURE 3.6
The method of superposition.

The boundary conditions to be satisfied by the stresses associated with the second stress function are therefore

$$\sigma_{rr} = -\sigma_0 \sin^2 \theta = -\frac{\sigma_0}{2}(1 - \cos 2\theta) \qquad \text{at } r = a$$

$$\sigma_{rr} = \tau_{r\theta} = \sigma_{\theta\theta} = 0 \qquad \text{at } r \to \infty \qquad (f)$$

$$\tau_{r\theta} = -\sigma_0 \sin \theta \cos \theta = -\frac{\sigma_0}{2} \sin 2\theta \qquad \text{at } r = a$$

From Eqs. (f) it is apparent that the stresses σ'_{rr} and $\tau_{r\theta}$ are functions of sin 2θ and cos 2θ, which suggests $\phi^{(2)}$ given by Eqs. (3.40) as a possible stress function. Inspection of Eqs. (3.40) indicates, however, that this function can satisfy the boundary conditions only for $\tau_{r\theta}$. From Eqs. (3.38), however, it can be seen that the stresses associated with $\phi^{(0)}$ can satisfy the boundary conditions for σ_{rr} without influencing $\tau_{r\theta}$. Thus, the stress function $\phi^{(0)} + \phi^{(2)}$ may be applicable. From Eqs. (3.38) and (3.40),

$$\phi^{(0)} + \phi^{(2)} = a_0 + b_0 \ln r + c_0 r^2 + d_0 r^2 \ln r$$
$$+ (a_2 r^2 + b_2 r^{-2} + c_2 r^4 + d_2) \cos 2\theta \qquad (g)$$

$$\sigma_{rr} = \frac{b_0}{r^2} + 2c_0 + d_0(1 + 2 \ln r)$$

$$- \left(2a_2 + \frac{6b_2}{r^4} + \frac{4d_2}{r^2} \right) \cos 2\theta \qquad (h)$$

$$\sigma_{\theta\theta} = -\frac{b_0}{r^2} + 2c_0 + d_0(3 + 2 \ln r)$$

$$+ \left(2a_2 + \frac{6b_2}{r^4} + 10c_2 r^2 \right) \cos 2\theta \qquad (i)$$

$$\tau_{r\theta} = \left(2a_2 - \frac{6b_2}{r^4} + 6c_2 r^2 - \frac{2d_2}{r^2} \right) \sin 2\theta \qquad (j)$$

Equations (h) to (j) contain seven unknowns: b_0, c_0, d_0, a_2, b_2, c_2, and d_2. Since $\sigma_{\theta\theta} = \sigma_{rr} = \tau_{r\theta} = 0$ as $r \to \infty$,

$$c_0 = d_0 = a_2 = c_2 = 0 \qquad (k)$$

and Eqs. (h) to (j) reduce to

$$\sigma_{rr} = \frac{1}{r^2} \left[b_0 - \left(\frac{6b_2}{r^2} + 4d_2 \right) \cos 2\theta \right] \qquad (l)$$

$$\sigma_{\theta\theta} = \frac{1}{r^2} \left(-b_0 + \frac{6b_2}{r^2} \cos 2\theta \right) \qquad (m)$$

$$\tau_{r\theta} = -\frac{1}{r^2} \left[\left(\frac{6b_2}{r^2} + 2d_2 \right) \sin 2\theta \right] \qquad (n)$$

From the boundary conditions at $r = a$,

$$\tau_{r\theta} = -\frac{1}{a^2}\left[\left(\frac{6b_2}{a^2} + 2d_2\right)\sin 2\theta\right] = -\frac{\sigma_0}{2}\sin 2\theta \qquad (o)$$

$$\sigma_{rr} = \frac{1}{a^2}\left[b_0 - \left(\frac{6b_2}{a^2} + 4d_2\right)\cos 2\theta\right] = -\frac{\sigma_0}{2}(1 - \cos 2\theta) \qquad (p)$$

Solving Eqs. (o) and (p) for the coefficients gives

$$b_0 = -\frac{\sigma_0 a^2}{2} \qquad b_2 = \frac{\sigma_0 a^4}{4} \qquad d_2 = -\frac{\sigma_0 a^2}{2} \qquad (q)$$

Substituting Eqs. (q) into Eqs. (l) to (n) gives

$$\sigma_{rr}^{II} = -\frac{\sigma_0 a^2}{2r^2}\left[1 + \left(\frac{3a^2}{r^2} - 4\right)\cos 2\theta\right]$$

$$\sigma_{\theta\theta}^{II} = \frac{\sigma_0 a^2}{2r^2}\left(1 + \frac{3a^2}{r^2}\cos 2\theta\right)$$

$$\tau_{r\theta}^{II} = -\frac{\sigma_0 a^2}{2r^2}\left[\left(\frac{3a^2}{r^2} - 2\right)\sin 2\theta\right]$$

The required solution for the original problem is obtained by superposition as follows:

$$\sigma_{rr} = \sigma_{rr}^{I} + \sigma_{rr}^{II} = \frac{\sigma_0}{2}\left\{\left(1 - \frac{a^2}{r^2}\right)\left[1 + \left(\frac{3a^2}{r^2} - 1\right)\cos 2\theta\right]\right\}$$

$$\sigma_{\theta\theta} = \sigma_{\theta\theta}^{I} + \sigma_{\theta\theta}^{II} = \frac{\sigma_0}{2}\left[\left(1 + \frac{a^2}{r^2}\right) + \left(1 + \frac{3a^4}{r^4}\right)\cos 2\theta\right] \qquad (3.46)$$

$$\tau_{r\theta} = \tau_{r\theta}^{I} + \tau_{r\theta}^{II} = \frac{\sigma_0}{2}\left[\left(1 + \frac{3a}{r^2}\right)\left(1 - \frac{a^2}{r^2}\right)\sin 2\theta\right]$$

Equations (3.46) give the polar components of stress at any point in the body defined by r, θ. Through the use of Eqs. (3.46) the stresses along the x axis, along the y axis, and about the boundary of the hole can easily be computed.

The stresses along the x axis can be obtained by setting $\theta = 0$ and $r = x$ in Eqs. (3.46):

$$\sigma_{rr} = \sigma_{xx} = \frac{\sigma_0}{2}\left(1 - \frac{a^2}{x^2}\right)\frac{3a^2}{x^2}$$

$$\sigma_{\theta\theta} = \sigma_{yy} = \frac{\sigma_0}{2}\left(2 + \frac{a^2}{x^2} + \frac{3a^4}{x^4}\right) \qquad (3.47)$$

$$\tau_{r\theta} = \tau_{xy} = 0$$

The distribution of σ_{xx}/σ_0 and σ_{yy}/σ_0 is plotted as a function of position along the x axis in Fig. 3.7. An examination of this figure clearly indicates that the presence of the hole in the infinite plate under uniaxial tension increases the stress σ_{yy} by a factor of 3. This factor is often called a *stress concentration factor*. In Chap. 13, it will be shown how photoelasticity can be effectively employed to determine stress concentration factors.

In a similar manner the stresses along the y axis can be obtained by setting $\theta = \pi/2$ and $r = y$ in Eqs. (3.46):

$$\sigma_{rr} = \sigma_{yy} = \frac{\sigma_0}{2}\left(2 - \frac{5a^2}{y^2} + \frac{3a^4}{y^4}\right)$$

$$\sigma_{\theta\theta} = \sigma_{xx} = \frac{\sigma_0}{2}\left(\frac{a^2}{y^2} - \frac{3a^4}{y^4}\right) \tag{3.48}$$

$$\tau_{r\theta} = \tau_{xy} = 0$$

A distribution of σ_{xx}/σ_0 and σ_{yy}/σ_0 is plotted as a function of position along the y axis in Fig. 3.8. In this figure it can be noted that $\sigma_{xx}/\sigma_0 = -1$ at the boundary of the hole; thus the influence of the hole not only produces a concentration of the stresses but in this case also produces a change in the sign of the stresses.

The distribution of $\sigma_{\theta\theta}$ about the boundary of the hole is obtained by setting $r = a$ into Eqs. (3.46):

$$\sigma_{rr} = \tau_{r\theta} = 0 \qquad \sigma_{\theta\theta} = \sigma_0(1 + 2\cos 2\theta) \tag{3.49}$$

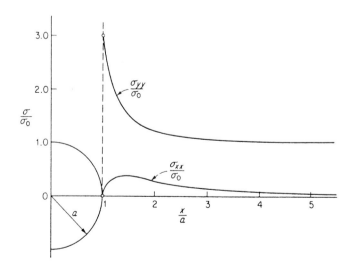

FIGURE 3.7
Distribution of σ_{xx}/σ_0 and σ_{yy}/σ_0 along the x axis.

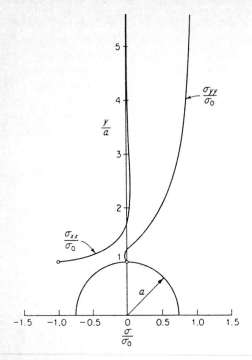

FIGURE 3.8
Distribution of σ_{xx}/σ_0 and σ_{yy}/σ_0 along the y axis.

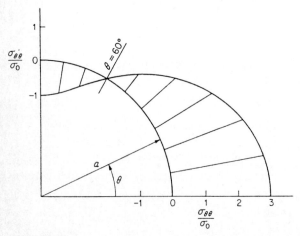

FIGURE 3.9
Distribution of $\sigma_{\theta\theta}/\sigma_0$ about the boundary of the hole.

The distribution of $\sigma_{\theta\theta}/\sigma_0$ about the boundary of the hole is shown in Fig. 3.9. The maximum $\sigma_{\theta\theta}/\sigma_0$ occurs at the x axis ($\sigma_{\theta\theta}/\sigma_0 = 3$), and the minimum occurs at the y axis ($\sigma_{\theta\theta}/\sigma_0 = -1$). At the point defined by $\theta = 60°$ on the boundary of the hole, all stresses are zero. This type of point is commonly referred to as a *singular point*.

EXERCISES

3.1. Give an example of a type 1 boundary-value problem other than the one cited in Sec. 3.1.

3.2. Give an example of a type 2 boundary-value problem other than the one cited in Sec. 3.1.

3.3. Give an example of a type 3 boundary-value problem other than the one cited in Sec. 3.1.

3.4. Give an example of a type 4 boundary-value problem other than the one cited in Sec. 3.1.

3.5. Prepare a list of the names of all of the field equations. How many field quantities exist in the elastic problem?

3.6. What is the advantage of using the stress equations of compatibility, Eqs. (3.3), in the solution of elasticity problems?

3.7. Discuss the influence of the elastic constants on determining stresses from scale models fabricated from plastics by using Eqs. (3.3) as the basis for your discussion.

3.8. Describe the key features which differentiate three-dimensional, plane-stress, and plane-strain formulations in elasticity theory.

3.9. Determine the stress-strain and the strain-stress relations from Eqs. (2.16) and (2.17) with the assumptions of the plane-stress problem.

3.10. Determine the stress-strain and the strain-stress relations from Eqs. (2.16) and (2.17) with the assumptions of the plane-strain problem.

3.11. Verify Eq. (3.17c).

3.12. In Eqs. (3.18), the body-force intensities were defined as

$$F_x = -\frac{\partial\Omega}{\partial x} \quad \text{and} \quad F_y = -\frac{\partial\Omega}{\partial y}$$

Determine Ω for a gravitational field of intensity q in the x direction.

3.13. Verify Eqs. (3.26) and Eqs. (3.28).

3.14. Verify Eqs. (r) on page 69.

3.15. Establish Eqs. (3.29).

3.16. Determine the stresses in the uniformly loaded cantilever beam shown in Fig. E3.16, by using the Airy's-stress-function approach. Assume a unit thickness.

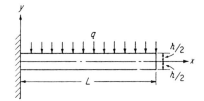

FIGURE E3.16

3.17. Determine the stresses in the cantilever beam shown in Fig. E3.17 by using the Airy's-stress-function approach.

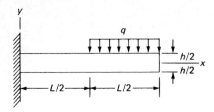

FIGURE E3.17

3.18. For the built-in triangular plate shown in Fig. E3.18, verify that the applicable stress function is

$$\phi = \frac{q \cot \alpha}{2(1 - \alpha \cot \alpha)} \left[-x^2 \tan \alpha + xy + (x^2 + y^2)\left(\alpha - \tan^{-1} \frac{y}{x} \right) \right]$$

For the particular case of $\alpha = 45°$ and a plate of unit thickness, determine the normal stress distribution along the line $x = L/2$ and compare the solution with the results determined by using elementary beam theory.

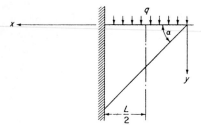

FIGURE E3.18

3.19. Show that

$$\frac{\partial^2 \phi}{\partial y^2} = \frac{\partial \phi}{\partial r} \left(\frac{\partial^2 r}{\partial y^2} \right) + \frac{\partial^2 \phi}{\partial r^2} \left(\frac{\partial r}{\partial y} \right)^2 + 2 \frac{\partial^2 \phi}{\partial r \partial \theta} \left(\frac{\partial r}{\partial y} \right)\left(\frac{\partial \theta}{\partial y} \right) + \frac{\partial \phi}{\partial \theta} \left(\frac{\partial^2 \theta}{\partial y^2} \right) + \frac{\partial^2 \phi}{\partial \theta^2} \left(\frac{\partial \theta}{\partial y} \right)^2$$

and verify Eq. (3.32b).

3.20. Show that

$$\frac{\partial^2 \phi}{\partial x \partial y} = \frac{\partial \phi}{\partial r} \left(\frac{\partial^2 r}{\partial x \partial y} \right) + \frac{\partial^2 \phi}{\partial r^2} \left(\frac{\partial r}{\partial x} \right)\left(\frac{\partial r}{\partial y} \right) + \frac{\partial^2 \phi}{\partial r \partial \theta} \left(\frac{\partial r}{\partial x} \right)\left(\frac{\partial \theta}{\partial y} \right)$$

$$+ \frac{\partial \phi}{\partial \theta} \left(\frac{\partial^2 \theta}{\partial x \partial y} \right) + \frac{\partial^2 \phi}{\partial \theta \partial r} \left(\frac{\partial \theta}{\partial x} \right)\left(\frac{\partial r}{\partial y} \right) + \frac{\partial^2 \phi}{\partial \theta^2} \left(\frac{\partial \theta}{\partial x} \right)\left(\frac{\partial \theta}{\partial y} \right)$$

and verify Eq. (3.32c).

3.21. Verify Eqs. (3.37).

3.22. Determine the polar components of the stresses σ_{rr}, $\sigma_{\theta\theta}$ in a thick-walled cylindrical vessel whose inside diameter is 1000 mm and outside diameter is 1400 mm if the vessel is subjected to an internal pressure of 35 MPa. Determine the radial displacement of the outside diameter if the vessel is made from steel ($E = 207$ GPa and $v = 0.30$).

3.23. Show how the radial displacement of the inside and outside surfaces of a cylindrical pressure vessel subjected to internal pressure can be used to determine E and v.

3.24. Determine the stresses and displacements in the curved beam shown in Fig. E3.24 when subjected to the moment load M.

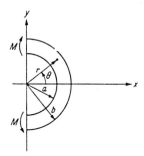

FIGURE E3.24

3.25. Determine the stresses and displacements in the curved beam shown in Fig. E3.25 when subjected to the shear load V.

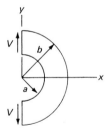

FIGURE E3.25

3.26. In Exercise 3.24, let a increase while holding $b - a$ constant and compare the values of the maximum stress $\sigma_{\theta\theta}$ from curved-beam theory and from straight-beam theory. Draw conclusions regarding the influence of the radius of curvature on $\sigma_{\theta\theta}$.

3.27. Determine the stresses in a semi-infinite plate due to a normal load acting on its edge as shown in Fig. E3.27. *Hint:* Try the stress function ϕ^*.

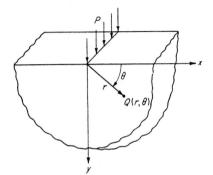

FIGURE E3.27

3.28. Determine the stresses in a semi-infinite plate due to a shear load acting on its edge as shown in Fig. E3.28. *Hint*: Use the stress function ϕ^*.

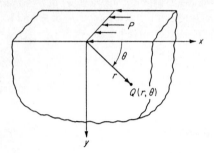

FIGURE E3.28

3.29. Determine the stresses in a semi-infinite plate due to an inclined load acting on its edge as shown in Fig. E3.29.

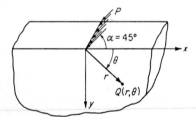

FIGURE E3.29

3.30. Determine the stresses in a semi-infinite body due to a distributed normal load (Fig. E3.30).

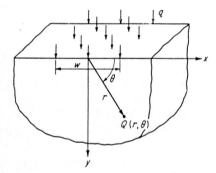

FIGURE E3.30

3.31. A steel ($E = 207$ GPa and $v = 0.30$) ring is shrunk onto another steel ring, as shown in Fig. E3.31. Determine the maximum interference possible without yielding one of

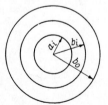

FIGURE E3.31

the rings if the yield strength of both rings is 900 MPa and the dimensions of the rings are as follows:

θ	Case 1	Case 2	Case 3	Case 4
a_i, mm	100	100	100	0
b_i, mm	125	115	150	100
b_o, mm	150	200	175	150

3.32. Discuss the possibility of fabricating gun tubes by shrink-fitting two or more long cylinders together to form the tube.

3.33. Using the solution of Exercise 3.30, determine σ_1, σ_2, and τ_{max}. Prepare a drawing illustrating the contour lines for σ_1, σ_2, and τ_{max}. Also prepare a drawing of the isostatics (stress trajectories) for σ_1 and σ_2.

REFERENCES

1. Boresi, A. P., and P. P. Lynn: *Elasticity in Engineering Mechanics*, chaps. 4–6, Prentice-Hall, Englewood Cliffs, N.J., 1974.
2. Chou, P. C., and N. J. Pagano: *Elasticity*, chaps. 4 and 5, Van Nostrand, Princeton, N.J., 1967.
3. Durelli, A. J., E. A. Phillips, and C. H. Tsao: *Introduction to the Theoretical and Experimental Analysis of Stress and Strain*, chaps. 6 and 9, McGraw-Hill, New York, 1958.
4. Sechler, E. E.: *Elasticity in Engineering*, chaps. 6 and 8, John Wiley & Sons, New York, 1952.
5. Timoshenko, S. P., and J. N. Goodier: *Theory of Elasticity*, 2d ed., chap. 4, McGraw-Hill, New York, 1951.
6. Kirsch, G.: Die Theorie der Elasticität und die Bedürfnisse der Festigkeitlehre, *Z. Ver. Dtsch. Ing.*, vol. 32, pp. 797–807, 1898.
7. Sternberg, E., and M. Sadowsky: Three-Dimensional Solution for the Stress Concentration around a Circular Hole in a Plate of Arbitrary Thickness, *J. Appl. Mech.*, vol. 16, pp. 27–38, 1949.
8. Howland, R. C. J.: On the Stresses in the Neighborhood of a Circular Hole in a Strip under Tension, *Trans. R. Soc.*, vol. A229, pp. 49–86, 1929.
9. Inglis, C. E.: Stresses in a Plate Due to the Presence of Cracks and Sharp Corners, *Proc. Inst. Nav. Arch.*, vol. 55, part 1, pp. 219–230, 1913.
10. Greenspan, M.: Effect of a Small Hole on the Stresses in a Uniformly Loaded Plate, *Q. Appl. Math.*, vol. 2, pp. 60–71, 1944.
11. Jeffrey, G. B.: Plane Stress and Plane Strain in Bi-Polar Co-Ordinates, *Phil. Trans. R. Soc.*, vol. A-221, pp. 265–293, 1920.
12. Mindlin, R. D.: Stress Distribution around a Hole near the Edge of a Plate under Tension, *Proc. SESA*, vol. V, no. 2, pp. 56–68, 1948.
13. Ling, C. B.: Stresses in a Notched Strip under Tension, *J. Appl. Mech.*, vol. 14, pp. 275–280, 1947.

CHAPTER
4

ELEMENTARY
FRACTURE
MECHANICS

4.1 INTRODUCTION

In the first three chapters, the theory of elasticity was introduced to show procedures for determining stresses and strains in bodies free of flaws. However, when flaws such as cracks exist in the body, elasticity theory is not sufficient to completely predict the onset of failure. The difficulty is due to the geometry of the crack tip. The crack tip is sharp with a radius of curvature approaching zero and this sharpness produces local stresses σ_{xx}, σ_{yy}, and τ_{xy} which tend to infinity as one approaches the crack tip. Since the stresses go to infinity for any loading of the body, the theories of failure, such as Tresca or Von Mises, cannot be applied and the load required to produce either localized yielding or the onset of crack propagation cannot be predicted.

To treat bodies containing cracks, it is necessary to introduce a method which deals with the singular state of stress at the crack tip. Fracture mechanics, developed by Irwin [1] from the earlier work of Inglis [2], Griffith [3], and Westergaard [4], treats singular stress fields by introducing a quantity known as a stress intensity factor K_I defined as

$$K_I = \lim_{r \to 0} (\sqrt{2\pi r}\, \sigma_{yy}) \tag{4.1}$$

where the coordinate system is as shown in Fig. 4.1 and σ_{yy} is evaluated in the limit along the $\theta = 0$ line. The limit process gives a stress intensity factor K_I that is a linear function of the loads applied to the body. The stress intensity factor

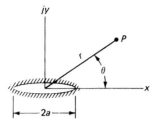

FIGURE 4.1
Coordinate system defining a double-ended crack ($z = x + jy = re^{j\theta}$).

remains finite and provides a basis for determining the critical load. The critical condition in fracture mechanics is the onset of crack initiation, where the crack extends suddenly at high velocity in cleavage failure or where the crack extends at low velocity by tearing in a shear-rupture-type failure. In either type of crack extension, the structure has failed. Clearly the introduction of the stress intensity factor provides a means, which is dependent upon the applied state of stress (i.e., σ_{yy}) and the crack length, of judging the criticality of a crack.

The second and equally important aspect of fracture mechanics is the characterization of the material property that defines the onset of crack initiation. This property is termed *toughness* and it is determined using well-defined testing procedures. For example, the plane-strain crack initiation toughness K_{Ic} defines the resistance of the material to crack initiation. When the applied stress intensity factor

$$K_I \geq K_{Ic} \tag{4.2}$$

the crack initiates and extends into the structure. Since K_I is a linear function of the applied load, say P, the failure criterion given in Eq. (4.2) provides an approach for predicting the critical load P_{cr} associated with crack initiation. This approach is illustrated in Fig. 4.2.

Before beginning a study of fracture mechanics, it is important to understand singular stress fields, where all of the cartesian components of stress tend to infinity at a point while the loads applied to the body remain finite. To show this singular stress state, the Inglis solution for a plate under uniaxial load with an elliptical hole will be introduced in Sec. 4.2.

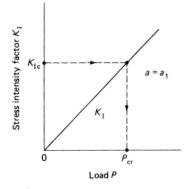

FIGURE 4.2
Approach using fracture mechanics for predicting the critical load P_{cr} associated with crack initiation.

4.2 STRESSES DUE TO AN ELLIPTICAL HOLE IN A UNIFORMLY LOADED PLATE

Elasticity problems involving elliptical or hyperbolic boundaries are treated using an elliptical coordinate system, defined in Fig. 4.3a, where

$$x = c \cosh \xi \cos \eta \qquad \text{and} \qquad y = c \sinh \xi \sin \eta \qquad (4.3a)$$

with c constant. Eliminating η from Eq. (4.3a) gives

$$\frac{x^2}{c^2 \cosh^2 \xi} + \frac{y^2}{c^2 \sinh^2 \xi} = 1 \qquad (4.3b)$$

When $\xi = \xi_0$, this is the equation of an ellipse with major and minor axes a and b given by

$$a = c \cosh \xi_0 \qquad \text{and} \qquad b = c \sinh \xi_0 \qquad (4.4)$$

The foci of the ellipse are at $x = \pm c$. It is clear that the aspect ratio of the ellipse varies as a function of ξ_0. If ξ_0 is very large (approaching infinity), the ellipse approaches a circle with $a = b$. However, if $\xi_0 \to 0$, the ellipse becomes a line of length $2c$ and represents a crack.

Inglis considered the infinite plate with an elliptical hole subjected to uniaxial loading, as shown in Fig. 4.3a, and found that the stresses σ_η about the hole are given by the equation

$$\sigma_\eta = \sigma_0 e^{2\xi_0} \left[\frac{\sinh 2\xi_0 (1 - e^{-2\xi_0})}{\cosh 2\xi_0 - \cos 2\eta} - 1 \right] \qquad (4.5)$$

The boundary stress σ_η is a maximum at the ends of the major axis where $\cos 2\eta = 1$. Substituting this value of η into Eq. (4.5) leads to

$$(\sigma_\eta)_{max} = \sigma_0 \left(1 + \frac{2a}{b} \right) \qquad (4.6)$$

It is instructive to examine the results of Eq. (4.6) for the two limit cases. First, when $a = b$ (ξ_0 large), the elliptical hole becomes circular and $(\sigma_\eta)_{max} = 3\sigma_0$. This result confirms the stress concentration for a circular hole in an infinite plate with a uniaxial load as determined by Kirsch [5] and described by Eq. (3.49). Second, when $b = 0$ ($\xi_0 = 0$), the elliptical hole becomes a flat opening representing a crack. In this case Eq. (4.6) shows that $[\sigma_\eta]_{max} \to \infty$ as $b \to 0$. Note that the maximum stress at the tip of the crack located at the ends of the major axis of the ellipse goes to infinity regardless of the magnitude of the applied stress σ_0. This fact indicates that localized yielding will occur at the crack tips for any nonzero load. The commonly employed failure theories such as Tresca and Von Mises predict yielding for any level of load and, therefore, do not provide useful information regarding the stability of the crack as the applied stress σ_0 is increased from zero to the value where crack initiation occurs.

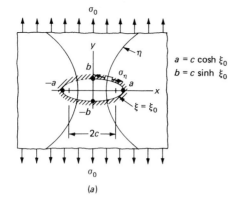

$a = c \cosh \xi_0$
$b = c \sinh \xi_0$

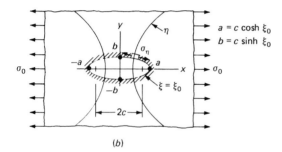

$a = c \cosh \xi_0$
$b = c \sinh \xi_0$

FIGURE 4.3
Elliptical hole in an infinite plate: (*a*) with uniaxial loading σ_0 perpendicular to *x*; (*b*) with uniaxial loading σ_0 parallel to *x*.

When the applied stress σ_0 is parallel to the major axis of the elliptical hole, as shown in Fig. 4.3*b*, the maximum value of σ_η on the boundary of the hole is at the ends of the minor axes, and is

$$(\sigma_\eta)_{\max} = \sigma_0 \left(1 + \frac{2b}{a}\right) \tag{4.7}$$

In the limit when $b \to 0$ and the ellipse represents a crack the stress $(\sigma_\eta)_{\max} = \sigma_0$. At the ends of the major axis of the elliptical hole Eq. (4.7) does not apply but $\sigma_\eta = -\sigma_0$ for any value of b/a.

The solution of Inglis for the elliptical hole in the plate provides, in the limit as $b \to 0$, the stress distribution for a crack in a plate. It is evident that the stresses at the tip of the crack are singular when the crack is perpendicular to the applied stress σ_0. The fact that the stresses at the crack tip are singular indicates local yielding will occur for any nonzero stress σ_0 and that methods for predicting structural stability based on, say, Tresca or Von Mises theories for yielding are inadequate. One must introduce concepts of crack stability based on fracture mechanics to predict if the crack will initiate at a specified applied stress σ_0.

4.3 THE WESTERGAARD STRESS FUNCTION

Westergaard introduced a complex stress function $Z(z)$ that is related to the Airy's stress function ϕ by the equation

$$\phi = \text{Re } \bar{\bar{Z}} + y \text{ Im } \bar{Z} \tag{4.8}$$

Since Z is a complex function, it is clear that

$$Z(z) = \text{Re } Z + j \text{ Im } Z \tag{4.9a}$$

where z is defined as

$$z = x + jy = re^{j\theta} \tag{4.9b}$$

Note that Z is analytic over the region of interest, and the Cauchy-Riemann conditions lead to

$$\nabla^2 \text{ Re } Z = \nabla^2 \text{ Im } Z = 0 \tag{4.10}$$

This result shows that the Westergaard stress function automatically satisfies Eq. (3.23b). The bars over the stress function Z in Eq. (4.8) indicate integration. Thus

$$\frac{d\bar{\bar{Z}}}{dz} = \bar{Z} \quad \text{or} \quad \bar{\bar{Z}} = \int \bar{Z} \, dz$$

$$\frac{d\bar{Z}}{dz} = Z \quad \text{or} \quad \bar{Z} = \int Z \, dz \tag{4.11}$$

$$\frac{dZ}{dz} = Z' \quad \text{or} \quad Z = \int Z' \, dz$$

where the bar and the prime represents integration and differentiation, respectively. Substituting Eq. (4.8) into Eqs. (3.20) gives the cartesian components of stress in terms of the real and imaginary parts of the Westergaard stress function as

$$\sigma_{xx} = \text{Re } Z - y \text{ Im } Z'$$

$$\sigma_{yy} = \text{Re } Z + y \text{ Im } Z' \tag{4.12}$$

$$\tau_{xy} = -y \text{ Re } Z'$$

Equations (4.12) will yield stresses for functions $Z(z)$ that are analytic; however, the stress function must be selected to satisfy boundary conditions corresponding to the problem being investigated. The formulation given by Eqs. (4.12), as originally proposed by Westergaard, correctly accounts for the stress singularity at the crack tip; however, additional terms must be added to adequately represent the stress field in regions adjacent to the crack tip. These additional terms will be introduced in later sections which deal with experimental methods for measuring K_I.

The classic problem in fracture mechanics, shown in Fig. 4.4a, is an infinite plate with a central crack of length $2a$. The plate is subjected to biaxial tension.

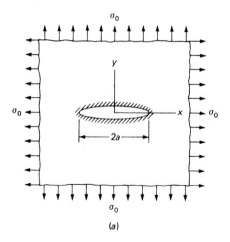

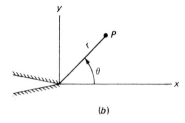

FIGURE 4.4
(a) A crack of length $2a$ in an infinite plate subjected to biaxial stress σ_0; (b) coordinates defined with the origin at the crack tip.

The stress function Z which is applied to solve this problem is

$$Z = \frac{\sigma_0 z}{\sqrt{z^2 - a^2}} \tag{4.13}$$

Substitution of Eq. (4.13) into Eqs. (4.12) with $z \to \infty$ yields $\sigma_{xx} = \sigma_{yy} = \sigma_0$ and $\tau_{xy} = 0$, as required to satisfy the far-field boundary conditions. On the crack surface where $y = 0$ and $z = x$, for $-a \le x \le a$, Re $Z = 0$ and $\sigma_{yy} = \tau_{xy} = 0$. Clearly, the stress function Z given in Eq. (4.13) satisfies the boundary conditions over the free surfaces of the crack.

It is more convenient to relocate the origin of the coordinate system and place it at the tip of the crack, as shown in Fig. 4.4b. The translation of the origin requires that z in Eq. (4.13) be replaced by $z + a$, and the new stress function becomes

$$Z = \frac{\sigma_0(z + a)}{\sqrt{z(z + 2a)}} \tag{4.14}$$

Next, consider a small region near the crack tip where $z \ll a$. Then, Eq. (4.14) reduces to

$$Z = \sqrt{\frac{a}{2}}\, \sigma_0 z^{-1/2} \tag{4.15}$$

Substituting Eq. (4.9b) into Eq. (4.15) yields

$$Z = \sqrt{\frac{a}{2r}}\,\sigma_0 e^{-j\theta/2} \tag{4.16}$$

Recall the identity

$$e^{\pm j\theta} = \cos\theta \pm j\sin\theta \tag{4.17}$$

Substituting Eq. (4.17) into Eq. (4.16) shows that the real part of Z is

$$\mathrm{Re}\,Z = \sqrt{\frac{a}{2r}}\,\sigma_0 \cos\frac{\theta}{2} \tag{4.18}$$

Along the crack line where θ and y are both equal to zero, Eqs. (4.18) and (4.12) give

$$\sigma_{yy} = \mathrm{Re}\,Z = \sqrt{\frac{a}{2r}}\,\sigma_0 \tag{4.19}$$

This result shows that the stress $\sigma_{yy} \to \infty$ and is singular with order $1/\sqrt{r}$ as one approaches the crack tip along the x axis. Finally, Eq. (4.19) can be substituted into Eq. (4.1) to obtain the stress intensity factor K_I as

$$K_I = \sqrt{\pi a}\,\sigma_0 \tag{4.20}$$

This result shows that the stress intensity factor K_I varies as a linear function of the applied stress σ_0 and increases with the crack length as a function of the $\sqrt{a}$, as illustrated in Fig. 4.5.

The units of K_I are $\mathrm{psi}\cdot\sqrt{\mathrm{in}}$ or $\mathrm{MPa}\cdot\sqrt{\mathrm{m}}$.

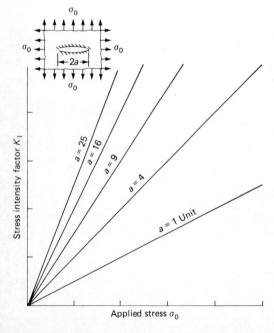

FIGURE 4.5
Stress intensity factor K_I as a function of applied stress σ_0 with crack length a as a parameter.

4.4   STRESS INTENSITY FACTORS FOR SELECT GEOMETRIES

Cracks in plane bodies of finite size are important since the crack poses a threat to the stability and safety of the entire structure even though only one plate may be cracked. It is important to determine the stress intensity factor for the specific geometry and loading involved to assess the safety factor for the cracked plate. Many solutions for boundary-value problems with cracks exist and a few of the more commonly encountered geometries and loadings will be reviewed in this section. A much more complete listing of solutions is given by Tada, Paris, and Irwin [6].

First, consider the tension strip with a centrally located crack, as shown in Fig. 4.6. The loading at the ends of the strip, well removed from the crack, is a uniaxial stress σ_0. An approximate solution for K_I, due to Irwin [7], is

$$K_I = \sqrt{\frac{W}{\pi a} \tan \frac{\pi a}{W}} \sqrt{\pi a}\, \sigma_0 \qquad (4.21)$$

Note that Eq. (4.21) may be written as

$$K_I = \alpha K_{I\infty} \qquad (4.22)$$

where $\alpha = \sqrt{W/(\pi a) \tan (\pi a/W)}$ is a multiplier and the term $K_{I\infty} = \sqrt{\pi a}\, \sigma_0$ corresponds to the solution for the infinite plate under biaxial loading. With K_I expressed in the form shown in Eq. (4.22), the effect of the finite boundaries of the body is given in terms of a multiplying factor α that acts on $K_{I\infty}$. The magnitude of α depends on a/W and ranges from 1 for $(a/W) \to 0$ to 4 for $(a/W) = 0.487$.

Next, consider single- and double-edge cracks in a tension strip, as shown in Fig. 4.7a and b. Series solutions for these two problems yield the following expressions for the multiplier α in terms of (a/W).

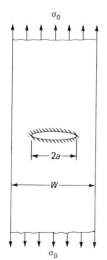

FIGURE 4.6
Tension strip with a central crack.

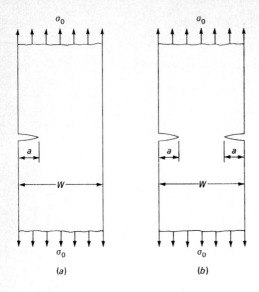

FIGURE 4.7
Edge-cracked tension strips: (a) single-edge crack; (b) double-edge crack.

For the single-edge crack (SEC):

$$\alpha = 1.12 - 0.231\frac{a}{W} + 10.55\left(\frac{a}{W}\right)^2$$

$$- 21.71\left(\frac{a}{W}\right)^3 + 30.38\left(\frac{a}{W}\right)^4 \tag{4.23a}$$

For the double-edge crack (DEC):

$$\alpha = 1.12 - 0.429\frac{a}{W} - 4.78\left(\frac{a}{W}\right)^2 + 15.44\left(\frac{a}{W}\right)^3 \tag{4.23b}$$

A listing of multipliers for different a/W ratios is presented in Table 4.1.

As the third example, consider the beam with a crack subjected to three- and four-point bending, as shown in Fig. 4.8a and b. With bending, the form of the solution changes and the multiplying factor α is not used. Instead, the stress intensity factor K_I is expressed in terms of the beam parameters S, W, and B, as defined in Fig. 4.8, and a series in terms of the (a/W) ratio.

For three-point bending:

$$K_I = \frac{PS}{BW^{3/2}}\left[2.9\left(\frac{a}{W}\right)^{1/2} - 4.6\left(\frac{a}{W}\right)^{3/2} + 21.8\left(\frac{a}{W}\right)^{5/2}\right.$$

$$\left. - 37.6\left(\frac{a}{W}\right)^{7/2} + 38.7\left(\frac{a}{W}\right)^{9/2}\right] \tag{4.24a}$$

TABLE 4.1
Multiplier α as a function of a/W for single- and double-edge cracked tension strips

a/W	SEC	DEC
0	1.12	1.12
0.1	1.18	1.13
0.2	1.37	1.14
0.3	1.66	1.24
0.4	2.10	1.52
0.45	2.42	1.75

For the beam subjected to pure bending:

$$K_{\mathrm{I}} = \frac{6M}{BW^2} \sqrt{\pi a} \left[1.122 - 1.4 \frac{a}{W} + 7.33 \left(\frac{a}{W} \right)^2 - 13.08 \left(\frac{a}{W} \right)^3 + 14.0 \left(\frac{a}{W} \right)^4 \right]$$

(4.24b)

As the final example, consider the compact tension specimen, defined in Fig. 4.9, that is loaded through pins positioned over the crack. The stress intensity factor K_{I} is given as a series in terms of (a/W). Thus,

$$K_{\mathrm{I}} = \frac{P}{BW^{1/2}} \left[29.6 \left(\frac{a}{W} \right)^{1/2} - 185.5 \left(\frac{a}{W} \right)^{3/2} + 655.7 \left(\frac{a}{W} \right)^{5/2} \right.$$
$$\left. - 1017 \left(\frac{a}{W} \right)^{7/2} + 638.9 \left(\frac{a}{W} \right)^{9/2} \right]$$

(4.25)

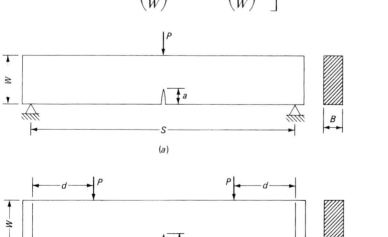

FIGURE 4.8
Cracked beams in bending: (a) symmetric three-point bending; (b) four-point bending.

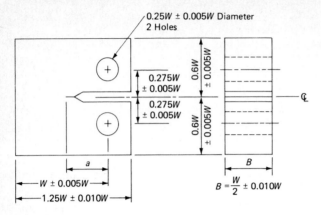

FIGURE 4.9
Compact tension specimen.

4.5 THE THREE MODES OF LOADING

A crack in a body can be subjected to three different types of loading. The opening mode of loading, illustrated in Fig. 4.10a, involves loads that produce displacements of the crack surfaces perpendicular to the plane of the crack. In this case the crack line or the x axis, as shown in Fig. 4.4b, is a principal axis. The subscript I (pronounced "one") on K_I indicates that the stress intensity factor is due to opening-mode loading.

The shearing mode of loading, illustrated in Fig. 4.10b, is due to in-plane shear loads which cause the two crack surfaces to slide on one another. The displacement of the crack surfaces is in the plane of the crack and perpendicular to the leading edge of the crack. The subscript II (pronounced "two") on K_{II} implies that the stress intensity factor is due to shear-mode loading. Of course, opening- and shear-mode loading conditions can occur together. In this case the loading is defined as mixed-mode, where both K_I and K_{II} exist in the region near the crack tip.

The tearing mode of loading, shown in Fig. 4.10c, is due to out-of-plane shear loadings. The displacements of the crack surfaces are in the plane of the crack and parallel to the leading edge of the crack. The subscript III on K_{III} is used to depict the tearing mode. While the superposition of the three modes of

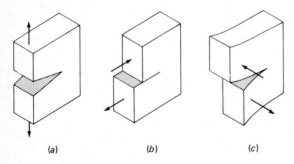

(a)   (b)   (c)

FIGURE 4.10
Three modes of crack loading: (a) opening; (b) shearing; (c) tearing.

loading gives the most general loading condition, one usually is interested in mode I or mixed-mode loading since they occur more frequently in engineering applications.

4.6 CRACK-TIP PLASTICITY

The elasticity solutions of Inglis show that a stress singularity exists at the tip of the crack. This fact implies that the material, at the tip of the crack and in a small region near the tip, has exceeded the yield strength of the specimen material. In this region, plastic deformation occurs and, because the stresses are limited by yielding, a stress singularity cannot occur. This region of plastic deformation is called the *crack-tip plastic zone.*

The size of the plastic zone is important because most of the developments in fracture mechanics are valid only if the size of the plastic zone is small relative to the length of the ligament from the crack tip to the nearest boundary. To study the size of the plastic zone, consider first the magnitude of σ_{yy} [given by Eq. (4.19)] along the x axis where $\theta = 0$. Setting $\sigma_{yy} = \sigma_{ys}$, where σ_{ys} is the yield strength of the material, gives

$$\sigma_{yy} = \frac{\sqrt{a}}{2r_p^*}\sigma_0 = \sigma_{ys} \tag{4.26}$$

where r_p^* is the apparent distance along the x axis in the plastic zone. Solving Eq. (4.26) for r_p^* gives

$$r_p^* = \frac{a}{2}\left(\frac{\sigma_0}{\sigma_{ys}}\right)^2 \tag{4.27}$$

The apparent distance r_p^* is shown in Fig. 4.11a, where the distribution of σ_{yy} along the x axis is presented. The shaded area shown in Fig. 4.11a indicates load shedding P_s due to yielding. Thus,

$$P_s = B\int_0^{r_p^*}(\sigma_{yy} - \sigma_{ys})\,dx \tag{4.28}$$

The result of load shedding is to increase the size of the plastic zone since P_s must still be carried by the body.

Irwin [7] has shown that the effect of the plastic zone is to artificially extend the crack by a distance δ, as shown in Fig. 4.11b. The coordinate system is translated by δ to the extended crack tip and the elastic stresses σ_{yy} are again limited by the yield stress σ_{ys}. Two areas, A_1 and A_2, are shown in Fig. 4.11b. Area A_1 is related to load shedding as given by Eq. (4.28); however, area A_2 is related to load loss by the crack extension. To maintain equilibrium, these two areas must be equal. Modifying Eq. (4.26) to account for changes due to the artificial extension of the crack tip gives

$$\beta = \frac{a + \delta}{2}\left(\frac{\sigma_0}{\sigma_{ys}}\right)^2 \tag{4.29}$$

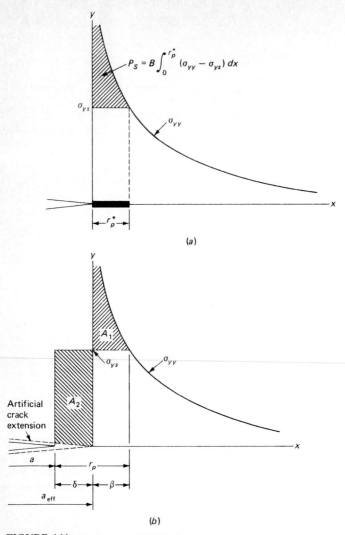

FIGURE 4.11
(a) Yielding at the tip in front of the crack; (b) an improved yield model incorporating artificial crack extension.

Since $\delta \ll a$, Eq. (4.29) shows that $\beta \approx r_p^*$. The equilibrium requirement that $A_1 = A_2$ together with Eqs. (4.28) and (4.26) can be used to give

$$\sigma_0 \int_0^{r_p^*} \sqrt{\frac{a + \delta}{2x}} \, dx = \sigma_{ys}(r_p^* + \delta) \tag{a}$$

Integrating and simplifying Eq. (a) yields

$$(r_p^* + \delta)^2 = 2(a + \delta)(r_p^*)\left(\frac{\sigma_0}{\sigma_{ys}}\right)^2 \tag{b}$$

Since δ is small when compared to the original crack length a, Eqs. (b) and (4.27) can be used to obtain

$$r_p^* + \delta = 2r_p^* \qquad (c)$$

which gives $r_p^* = \delta$ and the true length r_p of the plastic zone in front of the crack as

$$r_p = \delta + \beta = 2r_p^* \qquad (4.30)$$

To account for the effect of the plastic zone, an effective crack length a_{eff} is introduced such that

$$a_{\text{eff}} = a + r_p^* \qquad (4.31)$$

The value of a_{eff} is used to determine K_I and relations like Eq. (4.20) are modified to give

$$K_I = \sqrt{\pi(a + r_p^*)}\,\sigma_0 \qquad (4.20a)$$

Note that $r_p^* \approx \beta$ and Eq. (4.29) gives

$$r_p^* = \frac{a}{2}\left(\frac{\sigma_0}{\sigma_{\text{ys}}}\right)^2 \qquad (4.32)$$

Substituting Eq. (4.32) into Eq. (4.20a) yields K_I corrected for the plastic deformation at the crack tip as

$$K_I = \sqrt{\pi a \left[1 + \frac{1}{2}\left(\frac{\sigma_0}{\sigma_{\text{ys}}}\right)^2\right]}\,\sigma_0 \qquad (4.20b)$$

Similar corrections are made to Eqs. (4.21) through (4.25) to determine K_I for ductile materials where yielding can occur.

The value of r_p is the measure of the size of the plastic zone along the x axis. A more complete description of the plastic zone size and shape is presented in Sec. 4.8. Apart from correcting the determination of K_I, r_p is important in establishing if the body is in a state of small-scale yielding or large-scale yielding. When r_p is small compared to the length of the uncracked ligament, the body is subjected to small-scale yielding. In this state it responds in nearly an elastic manner with strains and displacements, outside the small plastic enclave, varying linearly with respect to load. At some critical load, the crack becomes unstable and propagates at a very high velocity (500 to 1000 m/s in steel). The failure is a cleavage type, often with little or no evidence of ductility.

In large-scale yielding, r_p is large compared to the length of the unbroken ligament. The body does not respond elastically due to the significant redistribution of stress caused by large-scale yielding. With large plastic deformation, the crack tip becomes blunted. In some materials, the crack will become unstable and extend at relatively low velocities by tearing. In other materials, the size of the plastic zone increases until it reaches a boundary. The specimen then fails by plastic instability. For failure by plastic instability, limit load analysis is the preferred approach for predicting the critical load. Failure by crack tearing is a current research topic that it is not possible to cover within the scope of this text.

4.7 THE FIELD EQUATIONS FOR THE REGION ADJACENT TO THE CRACK TIP

While it is possible in many cases to determine the stress intensity factors analytically (Sec. 4.6) or with numerical methods, in other instances it is necessary to measure K_I or K_{II} in carefully controlled experiments. Any experimental method for determining the stress intensity factors depends upon a complete knowledge of the field equations which are valid near the tip of the crack. In this treatment, the field adjacent to the crack tip is divided into three regions, as shown in Fig. 4.12. The field quantities (stresses, strains, and displacements) are represented in series form. For example, the stresses are

$$\sigma_{ij} = \sum_{n=0}^{N} A_n r^{(n-1/2)} f_n(\theta) \sum_{m=0}^{M} B_m r^m g_m(\theta) \tag{4.33}$$

where A_n and B_m are unknown coefficients and $f_n(\theta)$ and $g_m(\theta)$ are trigonometric functions.

Examination of Eq. (4.33) shows that for $r \to 0$, the only term in the series which contributes significantly to σ_{ij} is the $n = 0$ term. All of the other terms vanish. Thus, the very-near-field region, region (1) in Fig. 4.12, is defined as that area adjoining the crack tip where a single term in the series representation is sufficient to determine the field quantity of interest.

As one moves away from the crack tip, the nonsingular terms become significant and a single-term representation of the field quantities is not adequate. Additional terms in the series are required to improve the accuracy of the determination of the field quantity. The near-field region, region (2) in Fig. 4.12, is defined as that area beyond region (1) where the field quantities can be accurately represented with a small number of terms, up to say six, of the series.

For still larger values of r, a very large number of terms are required in Eq. (4.33) to accurately describe the stress field. This area is depicted as the far-field region, region (3) in Fig. 4.12. The far-field region is avoided in measuring field quantities in an attempt to determine the stress intensity factor, because the large

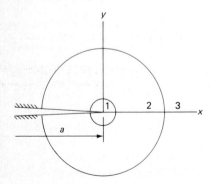

FIGURE 4.12
The field adjacent to the crack tip is divided into (1) the very near field, (2) the near field, (3) the far field.

number of unknown coefficients in the series requires that large amounts of data be taken with very high accuracy.

4.7.1 The Very-Near-Field Equations—Opening Mode

Consider first region (1) where r is small and a single term in the series representation of the field quantities is adequate. In this case the stress function Z may be written as

$$Z = A_0 z^{-1/2} = A_0 r^{-1/2} e^{-j(\theta/2)} \tag{4.34}$$

where

$$A_0 = \frac{K_1}{\sqrt{2\pi}} \tag{4.35}$$

By taking the real and imaginary parts of Z and Z' and substituting into Eqs. (4.12), the relations for the stresses are obtained as

$$\sigma_{xx} = A_0 r^{-1/2} \cos\frac{\theta}{2}\left(1 - \sin\frac{\theta}{2}\sin\frac{3\theta}{2}\right)$$

$$\sigma_{yy} = A_0 r^{-1/2} \cos\frac{\theta}{2}\left(1 + \sin\frac{\theta}{2}\sin\frac{3\theta}{2}\right) \tag{4.36}$$

$$\tau_{xy} = A_0 r^{-1/2} \cos\frac{\theta}{2}\sin\frac{\theta}{2}\cos\frac{3\theta}{2}$$

Next, consider a state of plane stress where $\sigma_{zz} = 0$ and substitute Eqs. (4.36) into Eqs. (2.19) to determine the strains. Thus,

$$E\epsilon_{xx} = A_0 r^{-1/2} \cos\frac{\theta}{2}\left[(1 - \nu) - (1 + \nu)\sin\frac{\theta}{2}\sin\frac{3\theta}{2}\right]$$

$$E\epsilon_{yy} = A_0 r^{-1/2} \cos\frac{\theta}{2}\left[(1 - \nu) + (1 + \nu)\sin\frac{\theta}{2}\sin\frac{3\theta}{2}\right] \tag{4.37}$$

$$2\mu\gamma_{xy} = A_0 r^{-1/2} \sin\theta\cos\frac{3\theta}{2}$$

Finally, the displacements are given by substituting Eqs. (4.37) into Eqs. (2.4) and integrating to obtain

$$\mu u = A_0 r^{1/2} \cos\frac{\theta}{2}\left(\frac{1 - \nu}{1 + \nu} + \sin^2\frac{\theta}{2}\right)$$

$$\mu v = A_0 r^{1/2} \sin\frac{\theta}{2}\left(\frac{2}{1 + \nu} - \cos^2\frac{\theta}{2}\right) \tag{4.38}$$

$$Ew = -2\nu A_0 r^{-1/2} \cos\frac{\theta}{2} B$$

where B is the thickness of the two-dimensional specimen.

Note from Eqs. (4.38) that only the displacement w is singular as $r \to 0$. The displacements u and v go to zero at the crack tip.

The very-near-field equations describe the field quantities in region (1) that is located in the local region of the crack tip. Chona [8] has shown that this region is very small and extends only a distance of $0.02a$ from the tip of the crack. Since the stress distribution in this region is three-dimensional, neither plane-stress nor plain-strain assumptions are valid. For this reason the very-near-field equations should only be used in a qualitative sense to give first approximations for A_0 or K_I. A more complete set of relations will be provided in Sec. 4.7.3 which permit improved accuracy in determining A_0 or K_I from the field quantities.

4.7.2 The Very-Near-Field Equations—Shear Mode

Again consider region (1) with a single-term representation of the plane-stress state very near the crack tip. The loading is due to in-plane shear stress τ applied remote to the crack tip. The Airy's stress function for the shear mode is

$$\phi = -y \, \text{Re} \, \bar{Z} \tag{4.39}$$

Let Z be represented by

$$Z = C_0 z^{-1/2} = C_0 r^{-1/2} e^{-j(\theta/2)} \tag{4.40}$$

where

$$C_0 = \frac{K_{II}}{\sqrt{2\pi}} \tag{4.41}$$

Substituting Eq. (4.39) into Eqs. (3.20) yields

$$\sigma_{xx} = y \, \text{Re} \, Z' + 2 \, \text{Im} \, Z$$

$$\sigma_{yy} = -y \, \text{Re} \, Z' \tag{4.42}$$

$$\tau_{xy} = \text{Re} \, Z - y \, \text{Im} \, Z'$$

Substituting Eq. (4.40) into Eqs. (4.42) gives

$$\sigma_{xx} = -C_0 r^{-1/2} \sin \frac{\theta}{2} \left(2 + \cos \frac{\theta}{2} \cos \frac{3\theta}{2} \right)$$

$$\sigma_{yy} = C_0 r^{-1/2} \sin \frac{\theta}{2} \cos \frac{\theta}{2} \cos \frac{3\theta}{2} \tag{4.43}$$

$$\tau_{xy} = C_0 r^{-1/2} \cos \frac{\theta}{2} \left(1 - \sin \frac{\theta}{2} \sin \frac{3\theta}{2} \right)$$

Again taking the plane-stress state and substituting Eqs. (4.43) into Eqs. (2.19) gives the strains as

$$E\epsilon_{xx} = -C_0 r^{-1/2} \sin\frac{\theta}{2}\left[2 + (1 + v)\cos\frac{\theta}{2}\cos\frac{3\theta}{2}\right]$$

$$E\epsilon_{yy} = C_0 r^{-1/2} \sin\frac{\theta}{2}\left[2v + (1 + v)\cos\frac{\theta}{2}\cos\frac{3\theta}{2}\right] \qquad (4.44)$$

$$\mu\gamma_{xy} = C_0 r^{-1/2} \cos\frac{\theta}{2}\left(1 - \sin\frac{\theta}{2}\sin\frac{3\theta}{2}\right)$$

The displacements are determined by substituting Eqs. (4.44) into Eqs. (2.4) and integrating. Thus,

$$\mu u = C_0 r^{1/2} \sin\frac{\theta}{2}\left(\frac{2}{1 + v} + \cos^2\frac{\theta}{2}\right)$$

$$\mu v = C_0 r^{1/2} \cos\frac{\theta}{2}\left(-\frac{1 - v}{1 + v} + \sin^2\frac{\theta}{2}\right) \qquad (4.45)$$

$$Ew = 2vC_0 r^{-1/2} \sin\frac{\theta}{2} B$$

These relations which describe the field quantities in terms of C_0 or K_{II} are approximate because the plane-stress state in the very near field (region 1) is not valid.

4.7.3 The Near-Field Equations—Opening Mode

Consider region (2) of Fig. 4.12 where r is sufficiently large to be outside the zone where the stress state is three-dimensional. Of course r is limited so that the field quantities can be expressed with reasonable accuracy (2 to 5 percent) with a small number of terms (say three to six) in the series expansion. In region (2) the Westergaard equations must be modified, with additional terms added, to account for all of the nonsingular terms [9]. In the modified form, one adds to Eqs. (4.12) and expresses the stresses in terms of two stress functions Z and Y as:

$$\sigma_{xx} = \text{Re } Z - y \text{ Im } Z' - y \text{ Im } Y' + 2 \text{ Re } Y$$

$$\sigma_{yy} = \text{Re } Z + y \text{ Im } Z' + y \text{ Im } Y' \qquad (4.46)$$

$$\tau_{xy} = -y \text{ Re } Z' - y \text{ Re } Y' - \text{Im } Y$$

where the stress functions are given as series relations in terms of z as

$$Z(z) = \sum_{n=0}^{N} A_n z^{(n-1/2)}$$

$$\qquad (4.47)$$

$$Y(z) = \sum_{m=0}^{M} B_m z^m$$

The number of terms used in the series $(N + M + 2)$ will depend on r, the proximity of the boundaries relative to the crack tip, and the influence of the loads applied at finite distances from the crack tip. Substituting Eqs. (4.47) into Eqs. (4.46) gives

$$
\begin{aligned}
\sigma_{xx} = &\; A_0 r^{-1/2} \cos\frac{\theta}{2}\left(1 - \sin\frac{\theta}{2}\sin\frac{3\theta}{2}\right) + 2B_0 \\
&+ A_1 r^{1/2} \cos\frac{\theta}{2}\left(1 + \sin^2\frac{\theta}{2}\right) + 2B_1 r \cos\theta \\
&+ A_2 r^{3/2}\left(\cos\frac{3\theta}{2} - \frac{3}{2}\sin\theta\sin\frac{\theta}{2}\right) + 2B_2 r^2(1 - 3\sin^2\theta)
\end{aligned}
$$

$$
\begin{aligned}
\sigma_{yy} = &\; A_0 r^{-1/2} \cos\frac{\theta}{2}\left(1 + \sin\frac{\theta}{2}\sin\frac{3\theta}{2}\right) \\
&+ A_1 r^{1/2} \cos\frac{\theta}{2}\left(1 - \sin^2\frac{\theta}{2}\right) \\
&+ A_2 r^{3/2}\left(\cos\frac{3\theta}{2} + \frac{3}{2}\sin\theta\sin\frac{\theta}{2}\right) + 2B_2 r^2 \sin^2\theta
\end{aligned}
\tag{4.48}
$$

$$
\begin{aligned}
\tau_{xy} = &\; A_0 r^{1/2} \cos\frac{\theta}{2}\sin\frac{\theta}{2}\cos\frac{3\theta}{2} \\
&- A_1 r^{1/2} \sin\frac{\theta}{2}\cos^2\frac{\theta}{2} - 2B_1 r \sin\theta \\
&- 3A_2 r^{3/2} \sin\frac{\theta}{2}\cos^2\frac{\theta}{2} - 2B_2 r^2 \sin 2\theta
\end{aligned}
$$

Substituting Eqs. (4.48) into the stress-strain relations gives

$$
\begin{aligned}
E\epsilon_{xx} = &\; A_0 r^{-1/2} \cos\frac{\theta}{2}\left[(1 - v) - (1 + v)\sin\frac{\theta}{2}\sin\frac{3\theta}{2}\right] + 2B_0 \\
&+ A_1 r^{1/2} \cos\frac{\theta}{2}\left[(1 - v) + (1 + v)\sin^2\frac{\theta}{2}\right] + 2B_1 r \cos\theta \\
&+ \frac{A_2}{2} r^{3/2}\left[2(1 - v)\cos\frac{3\theta}{2} - 3(1 + v)\sin\theta\sin\frac{\theta}{2}\right] \\
&+ 2B_2 r^2[1 - (3 + v)\sin^2\theta]
\end{aligned}
$$

$$E\epsilon_{yy} = A_0 r^{-1/2} \cos\frac{\theta}{2}\left[(1-v)+(1+v)\sin\frac{\theta}{2}\sin\frac{3\theta}{2}\right]-2vB_0$$

$$+ A_1 r^{1/2} \cos\frac{\theta}{2}\left[(1-v)-(1+v)\sin^2\frac{\theta}{2}\right]-2vB_1 r\cos\theta$$

$$+\frac{A_2}{2} r^{3/2}\left[2(1-v)\cos\frac{3\theta}{2}+3(1+v)\sin\theta\sin\frac{\theta}{2}\right]$$

$$+ 2B_2 r^2[-v+(3v+1)\sin^2\theta] \tag{4.49}$$

$$\mu\gamma_{xy} = \frac{A_0}{2} r^{-1/2}\left(\sin\theta\cos\frac{3\theta}{2}\right)$$

$$-\frac{A_1}{2} r^{1/2}\left(\sin\theta\cos\frac{\theta}{2}\right)-2B_1 r\sin\theta$$

$$-\frac{3A_2}{2} r^{3/2}\sin\theta\cos\frac{\theta}{2}-2B_2 r^2\sin2\theta$$

4.7.4 The Near-Field Equations—Shearing Mode

The stress field near a single-ended crack tip loaded in the shearing mode in terms of stress functions Z and Y is

$$\sigma_{xx} = \text{Im } Y + y\,\text{Re } Y' + y\,\text{Re } Z' + 2\,\text{Im } Z$$
$$\sigma_{yy} = \text{Im } Y - y\,\text{Re } Y' - y\,\text{Re } Z' \tag{4.50}$$
$$\tau_{xy} = -y\,\text{Im } Y' - y\,\text{Im } Z' + \text{Re } Z$$

For the shearing mode, the stress functions are expressed as

$$Z(z) = \sum_{n=0}^{N} C_n z^{(n-1/2)}$$
$$Y(z) = \sum_{m=0}^{M} D_m z^m \tag{4.51}$$

Note that the stress intensity factor K_{II} is related to C_0 as indicated in Eq. (4.41). From Eqs. (4.50) and (4.51) it is evident that

$$\sigma_{xx} = \sum_{n=0}^{N} C_n r^{(n-1/2)}[(n-\tfrac{1}{2})\sin\theta\cos(n-\tfrac{3}{2})\theta + 2\sin(n-\tfrac{1}{2})\theta]$$

$$+ \sum_{m=0}^{M} D_m r^m[\sin m\theta + m\sin\theta\cos(m-1)\theta]$$

$$\sigma_{yy} = \sum_{n=0}^{N} C_n r^{(n-1/2)}[-(n-\tfrac{1}{2})\sin\theta\cos(n-\tfrac{3}{2})\theta]$$

$$+ \sum_{m=0}^{M} D_m r^m[\sin m\theta - m\sin\theta\cos(m-1)\theta] \qquad (4.52)$$

$$\tau_{xy} = \sum_{n=0}^{N} r^{(n-1/2)}[\cos(n-\tfrac{1}{2})\theta - (n-\tfrac{1}{2})\sin\theta\sin(n-\tfrac{3}{2})\theta]$$

$$+ \sum_{m=0}^{M} D_m r^m[-m\sin\theta\sin(m-1)\theta]$$

Next, the strains are obtained by using the stress-strain relations. Thus,

$$E\epsilon_{xx} = -C_0 r^{-1/2}\sin\frac{\theta}{2}\left[(1+v)\cos\frac{\theta}{2}\cos\frac{3\theta}{2}+2\right]$$

$$+ C_1 r^{1/2}\sin\frac{\theta}{2}\left[2+(1+v)\cos^2\frac{\theta}{2}\right]+2D_1 r\sin\theta$$

$$= + C_2 r^{3/2}\left[2\sin\frac{3\theta}{2}+\frac{3}{2}(1+v)\sin\theta\cos\frac{\theta}{2}\right]+2D_2 r^2\sin 2\theta$$

$$E\epsilon_{yy} = C_0 r^{-1/2}\sin\frac{\theta}{2}\left[2v+(1+v)\cos\frac{\theta}{2}\cos\frac{3\theta}{2}\right]$$

$$\qquad (4.53)$$

$$- C_1 r^{1/2}\sin\frac{\theta}{2}\left[2v+(1+v)\cos^2\frac{\theta}{2}\right]-2D_1 rv\sin\theta$$

$$- C_2 r^{3/2}\left[2v\sin\frac{3\theta}{2}+\frac{3}{2}(1+v)\sin\theta\cos\frac{\theta}{2}\right]-2D_2 r^2 v\sin 2\theta$$

$$\mu\gamma_{xy} = C_0 r^{-1/2}\cos\frac{\theta}{2}\left(1-\sin\frac{\theta}{2}\sin\frac{3\theta}{2}\right)+C_1 r^{1/2}\cos\frac{\theta}{2}\left(\sin^2\frac{\theta}{2}+1\right)$$

$$+ C_2 r^{3/2}\left(\cos\frac{3\theta}{2}-\frac{3}{2}\sin\theta\sin\frac{\theta}{2}\right)-2D_2 r^2\sin^2\theta$$

where the series representing Eqs. (4.52) were evaluated at $N = M = 2$. Inspection of Eqs. (4.52) and (4.53) shows that D_0 does not occur in either the stress or strain field for mode II loading.

When the specimen is loaded with tractions or boundary displacements which produce both mode I and II fields, a mixed-mode loading exists. In these cases, the equations representing the stresses or the strains are superimposed. The coefficients in the series A_n, B_m, C_n, and D_m are determined to satisfy field quantities measured in the near-field region. The stress intensity factors K_I and/or K_{II} are then determined from the coefficients A_0 and C_0 from Eqs. (4.35) and (4.41).

4.8 SHAPE OF THE PLASTIC ZONE

In Sec. 4.6 the development of a plastic zone at the crack tip was introduced and the size of the zone was determined as r_p, a distance measured along the x axis. Consider next the size and shape of the plastic zone in the very-near-field region where the boundary parameters r_p and θ can be defined in terms of the stress intensity factor. The shape of the plastic zone will depend on the yield criterion applicable for the specimen material.

The Von Mises yield relation for the plane-stress state is given by

$$\sigma_1^2 - \sigma_1\sigma_2 + \sigma_2^2 = \sigma_{ys}^2 \tag{4.54}$$

Using the single-term representation of the stresses valid in the very near field, Eqs. (4.36), together with Eqs. (1.12), yields

$$\sigma_1 = A_0 r^{-1/2} \cos\frac{\theta}{2}\left(1 + \sin\frac{\theta}{2}\right)$$

$$\sigma_2 = A_0 r^{-1/2} \cos\frac{\theta}{2}\left(1 - \sin\frac{\theta}{2}\right) \tag{4.55}$$

$$\sigma_3 = 0$$

Now substitute Eqs. (4.55) and Eq. (4.35) into Eq. (4.54) to obtain the expression for the boundary of the plastic zone as a function of θ. Thus,

$$r_p = \cos^2\frac{\theta}{2}\left(1 + 3\sin^2\frac{\theta}{2}\right)\left(\frac{1}{2\pi}\right)\left(\frac{K_I}{\sigma_{ys}}\right)^2 \tag{4.56}$$

The shape of the plastic zone given by $r_p/(K_I/\sigma_{ys})^2$ for mode I loading is presented in Fig. 4.13 as a function of θ. If one takes $\theta = 0$ in Eq. (4.56) and

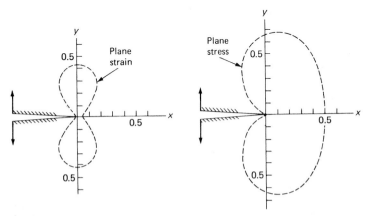

FIGURE 4.13
Plastic zone shape for mode I loading for a material yielding according to the Von Mises criterion.

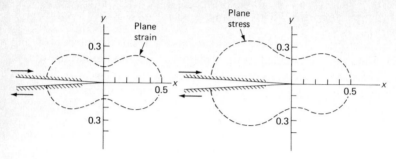

FIGURE 4.14
Plastic zone shape for mode II loading for a material yielding according to the Von Mises criterion.

substitutes Eq. (4.20) into the result for K_I with $r_p \ll a$, then it is clear that

$$r_p = \frac{a}{2}\left(\frac{\sigma_0}{\sigma_{ys}}\right)^2 = r_p^*$$

It is evident then that along the x axis the Von Mises yield criterion gives the same size r_p of the plastic zone as the elementary theory used to develop Eq. (4.27).

Following similar procedures it is easy to show that the shape of the plastic zone for mode II loading is

$$r_p = \frac{1}{2\pi}\left(\frac{K_{II}}{k}\right)^2 = \frac{a}{2}\left(\frac{\tau_0}{k}\right)^2 \tag{4.57}$$

where τ_0 is the value of τ_{xy} remote from the crack and k is the stress required to yield the material in shear. The shape of the yield zone for mode II loading is given in Fig. 4.14 for both the plane-stress and the plane-strain states of stress.

4.9 FRACTURE-ENERGY APPROACHES

4.9.1 The Energy Release Rate or Crack Extension Force $\mathscr{G}$

The earliest contribution to the fracture-mechanics approach made by Griffith [3] was based on the concept of an energy balance where the energy consumed to extend a crack equals the energy provided by the system. The system includes the plate with a crack and the loading frame. Consider first the case where the loading frame is perfectly rigid and the ends of the stretched plate are fixed. Since the ends are fixed, no work is added by the frame and the energy W consumed by the crack during an extension da is provided by a decrease in the strain energy U stored in the plate. The energy equation for the fixed-end condition for a plate of unit thickness is

$$\frac{d}{da}(U + W) = 0 \quad \text{or} \quad \frac{dW}{da} = -\frac{dU}{da} \tag{4.58}$$

The right-hand side of Eq. (4.58) is defined as

$$\mathcal{G} = -\frac{dU}{da} \tag{4.59}$$

and is called either the *energy release rate* or the *crack extension force*. The left-hand side of Eq. (4.58) is defined as

$$\mathcal{R} = \frac{dW}{da} \tag{4.60}$$

and is called the *crack growth resistance*. For a stationary crack,

$$\mathcal{G} \leq \mathcal{R} \tag{4.61a}$$

for a given material. For an unstable crack that initiates and propagates,

$$\mathcal{G} > \mathcal{R} \tag{4.61b}$$

It is possible to show a relationship between the crack extension force $\mathcal{G}$ and the stress intensity factor K_1 by considering the work released near the crack tip as the crack extends some small distance δ. Again take the cracked plate with fixed ends and a crack of length a. Now apply forces to the crack faces over a distance δ which will close the crack, as shown in Fig. 4.15. The work due to the closure forces, in the limit as $\delta \to 0$, is the same as the crack extension force $\mathcal{G}$. Writing the expression for this work gives

$$\mathcal{G} = \lim_{\delta \to 0} \frac{1}{\delta} \int_0^\delta \sigma_{yy} v \, dx \tag{4.62}$$

With the origin of the coordinate system at the center of the crack Westergaard [4] showed that the crack opening v is given by the expression

$$v = \frac{2\sigma_0}{E} \sqrt{a^2 - x^2} \tag{4.63}$$

Substituting $K_1 = \sqrt{\pi a}\sigma_0$ into Eq. (4.63) gives

$$v = \frac{2K_1}{\sqrt{\pi}E}\left(a - \frac{x^2}{a}\right)^{1/2} \tag{a}$$

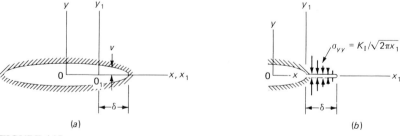

(a)   (b)

FIGURE 4.15
Application of forces distributed in proportion to σ_{yy} to close the crack over a length δ: (a) coordinate systems Oxy and $O_1x_1y_1$ and the crack opening v; (b) closing the crack over the increment length δ.

Consider now a second coordinate system $O_1 x_1 y_1$ where O_1 is located at the position $x = a - \delta$. It is clear from Fig. 4.15 that

$$x = a - \delta + x_1 \quad \text{or} \quad x = r + a - \delta \tag{b}$$

Now substitute Eq. (b) into Eq. (a) and note that $r \ll a$ and $\delta \ll a$. If second-order terms are neglected,

$$v = \frac{2K_I}{\sqrt{\pi E}} \sqrt{2(\delta - r)} \tag{4.64}$$

Next, recall Eqs. (4.35) and (4.36) which give σ_{yy} along the x_1 axis near the crack tip where $r \approx x_1$ as

$$\sigma_{yy} = \frac{K_I}{\sqrt{2\pi r}} \tag{c}$$

If Eq. (4.64) and Eq. (c) are substituted into Eq. (4.62),

$$\mathscr{G} = \lim_{\delta \to 0} \frac{2K_I^2}{\pi E \delta} \int_0^\delta \left[\frac{1 - r/\delta}{r/\delta} \right]^{1/2} dr \tag{d}$$

Integration of Eq. (d) gives

$$\mathscr{G}_I = \frac{K_I^2}{E} \tag{4.65}$$

where the subscript I is added to $\mathscr{G}$ to indicate opening-mode loading. Recall that plane stress conditions prevail. In a similar fashion $\mathscr{G}_{II}$ is expressed in terms of K_{II} as

$$\mathscr{G}_{II} = \frac{K_{II}^2}{E} \tag{4.66}$$

For elastic bodies, including small-scale yielding, $\mathscr{G}$ and K are interchangeable by using Eq. (4.65) for mode I and Eq. (4.66) for mode II loadings. For mixed-mode loading conditions, the energies from the different modes are added to give

$$\mathscr{G} = \mathscr{G}_I + \mathscr{G}_{II} = \frac{K_I^2 + K_{II}^2}{E} \tag{4.67}$$

4.9.2 Strain Energy Density

The strain energy density W is defined as the strain energy per unit volume, which can be expressed as

$$W = \frac{1}{E} \left[\frac{1}{2} (\sigma_{xx}^2 + \sigma_{yy}^2 + \sigma_{zz}^2) - v(\sigma_{xx}\sigma_{yy} + \sigma_{yy}\sigma_{zz} + \sigma_{zz}\sigma_{xx}) \right.$$

$$\left. + (1 + v)(\tau_{xy}^2 + \tau_{yz}^2 + \tau_{zx}^2) \right] \tag{4.68}$$

Under plane-stress conditions where $\sigma_{zz} = \tau_{zx} = \tau_{zy} = 0$, Eq. (4.68) reduces to

$$W = \frac{1}{2E} \left[\sigma_{xx}^2 + \sigma_{yy}^2 - 2\nu\sigma_{xx}\sigma_{yy} + 2(1 + \nu)\tau_{xy}^2 \right] \qquad (4.69)$$

The strain energy density in the near-field region about the tip of the crack is of the form

$$W = \frac{S}{r} + h(r, \theta) \qquad (4.70)$$

where $h(r, \theta)$ are nonsingular terms and S, the coefficient of $1/r$, is called the *strain energy density factor*. Clearly $W \to \infty$ as $r \to 0$ and the strain energy density is singular at the tip of the crack.

Sih [10] has proposed a critical value of the strain energy density W_c evaluated at a critical distance r_c as a criterion for crack initiation and propagation. If the nonsingular terms in Eq. (4.70) are neglected, it is clear that

$$S_c = r_c W_c \qquad (4.71)$$

If Eq. (4.70) is substituted into Eq. (4.71), the stresses at the critical distance r_c are

$$\sigma_{xx}^2 + \sigma_{yy}^2 - 2\nu\sigma_{xx}\sigma_{yy} + 2(1 + \nu)\tau_{xy}^2 = \frac{2ES_c}{r_c} \qquad (4.72)$$

4.9.3 The *J* Integral

The quantities K and $\mathscr{G}$ describe the stress state near the crack tip when the field is elastic with a relatively small plastic zone. The plastic enclave is surrounded by elastic material and the plate behaves in an essentially linear elastic manner. However, for tough ductile materials, the plastic zone becomes very large and is not small relative to the crack length a. In these cases K and $\mathscr{G}$ do not provide adequate descriptions of the elastic-plastic behavior of the tough specimens.

To determine an effective energy release rate for specimens where the plasticity effects must be considered, Rice [11] introduced a contour integral taken about the crack tip, as shown in Fig. 4.16a. The contour integral J was defined originally by Eshelby [12] as

$$J = \int_\Gamma \left(W \, dy - \bar{T} \cdot \frac{\partial \bar{u}}{\partial x} \, ds \right) \qquad (4.73)$$

where W = the strain energy density defined in Eq. (4.66)
$\bar{u}$ = the displacement vector
ds = an element along the contour Γ
$\bar{T}$ = the tension vector on Γ in the direction of the outer normal n to Γ and is given by $T_i = \sigma_{ij}n$

The J integral vanishes ($J = 0$) along any closed contour. Also, the integral is path-independent and J_1 determined along Γ_1 of Fig. 4.16b is the same as J_2

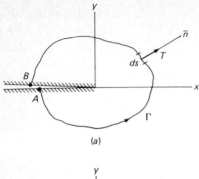

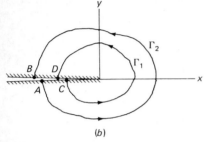

FIGURE 4.16
Contour integration about the path Γ to obtain J: (a) contour about a crack tip; (b) two contours about the crack tip.

determined along Γ_2. It is easy to prove that $J_1 = J_2$ since $dy = 0$ and $T = 0$ along the crack line. Thus, the crack line may be included in the contour without contributing to the value of J. For this reason points A and B in Fig. 4.16 do not need to coincide.

When the contour Γ encircles the crack tip, the J integral is the change in elastic energy U for a virtual crack extension da and is written as

$$J = -\frac{\partial U}{\partial a} \tag{4.74}$$

For an elastic stress state, Eq. (4.74) is the same as the definition of $\mathscr{G}$ given in Eq. (4.59); hence,

$$\mathscr{G}_1 = J_1 = \frac{K_I^2}{2E} \tag{4.75}$$

These relations between $\mathscr{G}_1$, J_1, and K_I hold as long as the yielding at the crack tip is "small scale." When r_p becomes large relative to the crack length a, Eq. (4.75) does not hold, the concept of either $\mathscr{G}_1$ or K_I is not valid, and elasto-plastic characterization of the field quantities at the crack tip, such as the J integral, is required.

4.10 CRITERIA FOR CRACK INSTABILITY

Griffith [3] first proposed a criterion for crack instability based on the energy consumed in forming the crack extension. For a perfectly elastic and brittle material, which does not form a plastic zone at the crack tip, the incremental energy dW necessary to produce the two new crack surfaces involved in an extension da of the crack is

$$\frac{dW}{da} = 2\gamma = \mathscr{R} \qquad (4.76)$$

where γ is the surface energy of the material. Using Eqs. (4.65), (4.61), and (4.20) with Eq. (4.76) leads to the Griffith equation for the critical stress σ_c:

$$\sigma_c = \sqrt{\frac{2\gamma E}{\pi a}} \qquad (4.77)$$

Although this relation for σ_c is limited to a narrow class of very brittle materials, it indicates two important concepts in fracture mechanics. First, for a given crack length a, the crack will remain stable if $\sigma_0 < \sigma_c$ in spite of the fact that the stresses at the crack tip are singular. Second, it indicates that growth of the crack by any mechanism (i.e., fatigue extension, stress-corrosion cracking, etc.) decreases the critical stress required to initiate the crack.

More than 20 years later Irwin [13] extended the Griffith approach by recasting Eq. (4.76) as

$$\frac{dW}{da} = 2\gamma + \frac{dW_p}{da} \qquad (4.78)$$

where the additional term dW_p/da accounts for the work necessary to form the plastic zone in front of the tip of the crack during an extension. This concept is valid if the plastic zone size remains constant as the crack advances. Experimental observations indicate that the plastic zone size is nearly constant in many different materials when fabricated as thick plates subjected to plane-strain conditions. With the constants imposed by plane strain,

$$\frac{dW}{da} = \mathscr{R} = \mathscr{G}_{1c} \qquad (4.79)$$

where $\mathscr{G}_{1c}$ is the critical strain-energy release rate which is a material property. Test procedures for establishing $\mathscr{G}_{1c}$ are described in Ref. 14.

Replacing 2γ with $\mathscr{G}_{1c}$ in Eq. (4.78) leads to the more modern representation of the critical stress as

$$\sigma_c = \sqrt{\frac{E\mathscr{G}_{1c}}{\pi a}} \qquad (4.80)$$

Before interpreting Eq. (4.80) it is useful to make a final modification of the relation for the critical stress. Note that the strain-energy release rate in terms of the stress intensity factor K is

For plane-stress conditions:

$$\mathscr{G}_I = \frac{K_I^2}{E} \qquad (4.65)$$

For plane-strain conditions:

$$\mathscr{G}_I = \frac{(1 - v^2)K_I^2}{E} \tag{4.81}$$

Finally, substituting Eq. (4.81) into Eq. (4.80) gives

$$\sigma_c = K_{Ic}\sqrt{\frac{1 - v^2}{\pi a}} \tag{4.82}$$

where K_{Ic} is the crack initiation toughness which is a material property determined under plane-strain testing conditions [14].

One may use either $\mathscr{G}_{Ic}$ or K_{Ic} in a crack stability analysis where the plastic zone size is considered small with $r_p \ll a$. For a given material with a toughness property specified in terms of, say, $\mathscr{G}_{Ic}$, the influence of the applied stress level σ_0 is illustrated in Fig. 4.17. This graph shows $\mathscr{G}_I$ for a wide (thick) plate with a central crack of length $2a$ where a is the independent variable. Four values of σ_0 are represented. The intersection of the $G_I - a$ lines with the material constant line corresponding to $\mathscr{G}_{Ic}$ gives the critical crack length a_{c1}, to a_{c4} for the different applied stress levels σ_0.

Another graphical display of the stability criteria for small-scale yielding is shown in Fig. 4.18. For the same problem, K_I is shown as a function of the far-field stress σ_0 with the crack length a as a parameter. The intersection of the $K_I - \sigma_0$ lines with the material constant line corresponding to K_{Ic} gives the critical stress $\sigma_{c1} - \sigma_{c3}$ necessary to initiate the cracks. It is of interest to note that the line for $a = 1$ unit does not intersect the K_{Ic} line before $\sigma_0 > \sigma_{ys}$. This fact implies the plate will yield and fail by excessive plastic deformation while the crack remains stationary.

For large-scale plasticity, the plastic zone size is large and the crack stability analyses using Eqs. (4.80) and (4.82) are not valid. In these cases a critical value, the J integral J_{Ic}, is determined. When the applied stress σ_0 produces J, which is

$$J > J_{Ic} \tag{4.83}$$

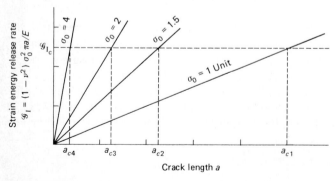

FIGURE 4.17
Graphical representation of Eq. (4.78) showing critical crack length for different applied stress levels.

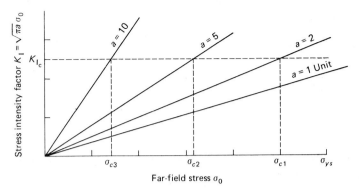

FIGURE 4.18
The influence of crack length a on the far-field stress σ_0 required to initiate a crack in a material with a crack initiation toughness K_{Ic}.

the crack will begin to grow in a stable manner at very low velocity. Indeed, the stress must be increased to maintain the tearing velocity in this stable-growth regime of fracture.

4.11 FRACTURE CONTROL

Large, complex structures usually have inherent flaws of one type or another. These flaws are often introduced in fabrication during the welding process where incomplete welds, embedded slag, holes, inadequate fusion bonding, and shrinkage cracks are common. Flaws are sometimes introduced during service. Mechanisms which produce flaws include fatigue, stress-corrosion cracking, and impact damage. Regardless of the source of the flaws, they are common and the history of technology contains many examples of dramatic and catastrophic failures which caused significant losses of life and property. Indeed, as recently as 1972 a 584-ft-long tank barge [15] broke almost completely in half while in port with calm seas. The vessel was only 1 year old, indicating that modern methods of design, material procurement, and welding procedures do not always ensure safe structures.

Since flaws in structures and components are common, it is essential in the design, fabrication, and maintenance of a structure to establish a fracture-control procedure. If properly implemented, the fracture-control plan can ensure the safety of the structure during its service life. Fracture control involves first the recognition that flaws exist, and second that the flaws must be maintained in a stable state. To describe the technical aspects of fracture control, consider a simple structure, namely a large plate, with the stress intensity factor given by

$$K_I = \sqrt{\pi a}\,\sigma_0 \tag{4.20}$$

The crack in this structure will remain stable if

$$K_{Ic} > K_I \tag{4.2}$$

Combining these two equations gives the stability relation

$$\sqrt{\pi a}\,\sigma_0 < K_{Ic} \qquad\qquad (4.84)$$

for this simple structure. This stability relation illustrates the three elements in all fracture-control procedures, namely,

1. Flaw size, a
2. Load or stress imposed, σ_0
3. Material toughness, K_{Ic}

Clearly, to ensure the integrity of a structure requires that the variables in the stability relation be maintained so that equality in Eq. (4.84) never occurs during the life of a structure.

A typical fracture-control plan usually considers material resistance (i.e., the crack initiation toughness K_{Ic}) as the first step in maintaining cracks in a stable state. The material selection sets the upper limit for K_{Ic} and it is controlled by the designer in the specification of the material for the structure. Typical values of K_{Ic} for common engineering materials are presented in Table 4.2.

To ensure that the material meets the specification, the supplier provides a certification indicating that tests have been conducted using samples drawn from the "heat" and that the test results exceed the minimum required strength and toughness.

With welded structures, selection and certification of the base plate is only the first step in ensuring the specified toughness. Weld materials and weld processes must be selected and then certified. The weld material is often undermatched (lower yield strength) relative to the base plate with a higher toughness. The difficulty is in certifying the weld process since this certification requires extensive testing of specimens containing the weld and the heat-affected zone. This testing is expensive

TABLE 4.2
Initiation toughness for some engineering materials

Material	K_{Ic}		σ_{ys}	
	MPa $\cdot \sqrt{m}$	ksi $\cdot \sqrt{in}$	MPa	ksi
Aluminum				
2024-T851	26	24	455	66
7075-T651	24	22	495	72
Titanium				
Ti-6AL-4V	115	105	910	132
Ti-6AL-4V*	55	50	1035	150
Steel				
4340	99	90	860	125
52100	14	13	2070	300

but necessary, because experience has shown that welding often degrades the toughness of the base plate in the heat-affected zone.

In some applications K_{Ic} decreases with service life. For instance, in reactor pressure vessels, radiation damage due to high neutron flux occurs and the toughness degrades with extended service. Also, some alloys which age-harden will exhibit losses in K_{Ic} with service life which is measured in decades.

The second step in a fracture-control plan involves inspection to determine the flaw size a. It is evident from Eq. (4.84) that the critical crack length a_{cr} is

$$a_{cr} = \frac{1}{\pi} \left(\frac{K_{Ic}}{\sigma_0} \right)^2 \tag{4.85}$$

Inspection with x-rays, eddy current sensors, and ultrasonics is employed to locate cracks. Large cracks which approach or exceed a_{cr} are repaired. However, small cracks with a length a_d below the minimum detectable length or with a length a_r below the repairable limit remain in the structure. These cracks may grow during service due to fatigue loading or stress-corrosion cracking, and they degrade the strength of the structure as shown in Fig. 4.19. The growth period represents the time interval during which one or more cracks extends from a_r to a_{cr}. This period is usually long and provides the opportunity for periodic inspections during the service life. It is important in the fracture-control plan to specify inspection intervals which are sufficiently short relative to the growth period to

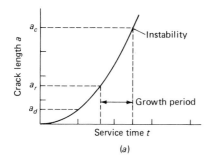

(a)

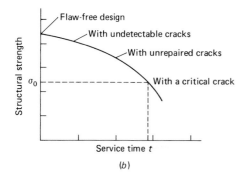

(b)

FIGURE 4.19
(a) Crack growth with time in service; (b) loss of structural strength with time and crack growth.

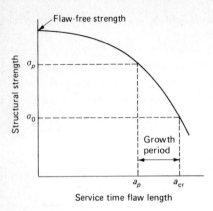

FIGURE 4.20

Proof testing at a stress σ_p to ensure that $a < a_p$.

provide two or more opportunities to detect the cracks before they achieve critical length. When cracks of length $a \to a_{cr}$ are detected, they are repaired (in the field) and the growth period is extended.

The third step in the fracture-control plan involves the applied loads which determine the stress σ_0. From Fig. 4.20, it is evident that the structural strength will exceed σ_0 provided $a < a_{cr}$. In many types of structures, periodic inspections and field repair ensure that crack growth is controlled and that the cracks do not approach critical length. In these structures, structural stability is achieved simply by maintaining the loads so that $\sigma \le \sigma_0$. However, in some structures, accessibility is limited and complete inspection is difficult if not impossible. In these instances, a proof-testing procedure can be applied to the structure to ensure safety. In proof testing, loads are applied to the structure which are larger than the normal loading which produces σ_0. These proof loads produce an elevated stress σ_p and, if the structure does not fail, the cracks in the structure are $a \le a_p$, as shown in Fig. 4.20. The single cycle of proof load provides confidence that the cracks in the structure are short and stable. Indeed, depending upon the ratio σ_p/σ_0, a growth period can be predicted where the safety of the structure can be ensured. It is important to divide the growth period into two or more parts to determine the proof-test interval. One repeats proof testing at these intervals to continue to extend the growth period. In this sense repeated proof testing is similar to repeated inspections. However, with an inspection procedure, the cracks which grow with time in service can be repaired; but with a proof-test procedure, the cracks continue to grow and when $a \ge a_p$ the structure will fail during the proof test.

EXERCISES

4.1. Your manager directs you to perform a stress analysis of a machine component containing a well-defined crack, and to employ the Von Mises theory of yielding to predict when the structure will fail. How do you respond to this directive?

4.2. If the ratio of the major axis to the minor axis of an ellipse is $a/b = 3$, determine the coordinate ξ_0 to form an ellipse of these proportions in an elliptical coordinate system. If the dimension $b = 10$ mm, determine the scaling constant c. Determine the coordinate η which locates a point at the intersection of the x axis with ξ_0 and another η which locates the intersection of the y axis with ξ_0.

4.3. Repeat Exercise 4.2 for an ellipse with $a/b = 10$.

4.4. Write an equation for the stress σ_η about the boundary of an elliptical hole with $a/b = 5$, if $\sigma_0 = 100$. Evaluate this equation for η ranging from 0 to $\pi/2$. Prepare a graph showing the distribution of the stresses about the perimeter of the first quadrant of the elliptical hole.

4.5. Repeat Exercise 4.4 for an elliptical hole with $a/b = 20$, and $a/b = 1$. Compare the stress concentration factor $K = (\sigma_\eta/\sigma_0)_{\eta=0}$ for these two cases.

4.6. Determine the stress concentration at the tip of a crack of length $2a$ when a uniaxial stress σ_0 is applied in a direction parallel to the length of the crack.

4.7. Verify the conversion [from Eq. (4.13) to Eq. (4.14)] of the expression for the stress function Z when the origin is shifted from the center of the crack to the crack tip.

4.8. Beginning with Eq. (4.14) verify Eqs. (4.15) to (4.19).

4.9. A crack 60 mm long develops in a very large plate which is to support a stress $\sigma_0 = 70$ MPa. Determine the stress intensity factor K_I. What assumption did you make in this determination?

4.10. If the crack in Exercise 4.9 grows at a rate of 10 mm/month, determine the service life remaining before crack initiation if $K_{Ic} = 40$ MPa $\cdot \sqrt{m}$.

4.11. Prepare a graph showing α as a function of a/W for a tension strip with a centrally located crack.

4.12. A steel strap 1 mm thick and 20 mm wide with a central crack 4 mm long is loaded to failure. Determine the critical load if K_{Ic} for the strap material is 80 MPa $\cdot \sqrt{m}$.

4.13. Determine the failure stress σ_0 applied to the strap of Exercise 4.12. Compare this stress to the yield and tensile strengths of 4340 alloy steel with a hardness $R_c = 50$.

4.14. Write a program to compute the multiplier α for SEC and DEC cracked tension strips. Evaluate α for $0 < a/W < 0.49$ in increments of 0.01. Display the results in a listing similar to Table 4.1.

4.15. A steel tension bar 8 mm thick and 50 mm wide with a single-edge crack 10 mm long is subjected to a uniaxial stress $\sigma_0 = 140$ MPa. Determine the stress intensity factor K_I. If K_{Ic} for this steel is 60 MPa $\cdot \sqrt{m}$, is the crack stable?

4.16. Determine the critical crack length for the steel bar in Exercise 4.15.

4.17. Determine the critical load for the steel bar in Exercise 4.15.

4.18. A steel tension bar 8 mm thick and 50 mm wide with double-edge cracks each 5 mm long is subjected to a uniaxial stress $\sigma_0 = 140$ MPa. Determine the stress intensity factor K_I. If K_{Ic} for this steel is 60 MPa $\cdot \sqrt{m}$, is the crack stable?

4.19. Determine the critical crack length for the steel bar in Exercise 4.18.

4.20. Determine the critical load for the steel bar in Exercise 4.18.

4.21. Determine the normalized stress intensity factor $K_I BW^{3/2}/PS$ as a function of a/W for a beam subjected to three-point bending.

4.22. Determine the normalized stress intensity factor $K_I BW^{3/2}/M$ as a function of a/W for a beam subjected to pure bending.

4.23. Find the critical load that can be applied to a beam subjected to three-point bending if $S = 1$ m, $W = 100$ mm, $B = 40$ mm, $a = 20$ mm, and $K_{Ic} = 100$ MPa $\cdot \sqrt{m}$.

4.24. Find the critical moment that can be applied to a beam in pure bending if $S = 1$ m, $W = 100$ mm, $B = 40$ mm, $a = 20$ mm, and $K_{Ic} = 100$ MPa $\cdot \sqrt{m}$.

4.25. Determine the normalized stress intensity factor $K_1 BW^{1/2}/P$ as a function of a/W for the compact tension specimen.

4.26. Determine the critical load for a compact tension specimen if $B = 50$ mm, $W = 250$ mm, $a = 100$ mm, and $K_{1c} = 120$ MPa $\cdot \sqrt{m}$.

4.27. Give an example of
   (a) Opening-mode loading
   (b) In-plane shear-mode loading
   (c) Out-of-plane shear-mode loading
   (d) Mixed-mode (I and II) loading.

4.28. Derive Eq. (4.31), which gives the effective crack length to account for plastic deformation at the crack tip.

4.29. Verify Eq. (4.20b).

4.30. Determine r_p at the critical load for the beam in Exercise 4.23 if $\sigma_{ys} = 700$ MPa. Compare r_p to the crack length and the ligament length. Comment on this comparison.

4.31. For the beam in Exercise 4.23, adjust the crack length to accommodate the effect of yielding and determine the new critical load. Compare this result to that found in the original solution for Exercise 4.23.

4.32. Justify the rationale in dividing the field adjacent to the crack tip into three regions. Explain how you would describe the stresses in each of these three regions.

4.33. Verify Eqs. (4.36) beginning with Eq. (4.34).

4.34. Verify Eqs. (4.37) beginning with Eq. (4.36).

4.35. Use the strains defined in Eqs. (4.37) and integrate to determine the displacement field given by Eqs. (4.38).

4.36. Verify Eqs. (4.43) beginning with Eq. (4.39).

4.37. Verify Eqs. (4.44) beginning with Eqs. (4.43).

4.38. Use the strains defined in Eqs. (4.44) and integrate to determine the displacement field given in Eqs. (4.45).

4.39. Derive Eqs. (4.48) from Eqs. (4.46) with definitions of Z and Y given by Eqs. (4.47).

4.40. Verify
   (a) The first of Eqs. (4.49)
   (b) The second of Eqs. (4.49)
   (c) The third of Eqs. (4.49)

4.41. Derive Eqs. (4.52) from Eqs. (4.50) with definitions of Z and Y given by Eqs. (4.51).

4.42. Verify
   (a) The first of Eqs. (4.53)
   (b) The second of Eqs. (4.53)
   (c) The third of Eqs. (4.53)

4.43. Beginning with Eq. (4.54) verify Eq. (4.56).

4.44. Beginning with Eq. (4.62) derive Eq. (4.65).

4.45. Construct a graph showing the crack opening displacement v as a function of position from the origin to $x = a$ for a central crack if $(K_1/E) = 0.5 \times 10^{-3} \sqrt{m}$.

4.46. Derive Eq. (4.69).

4.47. At the point defined by $r = r_c$ and $\theta = \pi/2$, find the relation between K_{1c} and the critical strain-energy-density factor S_c. Let $a = 20$ mm.

4.48. Repeat Exercise 4.47 for the point defined by $r = r_c$ and $\theta = 0$. Let $v = \frac{1}{3}$.

4.49. Determine the J integral for the contour a, b, c, d shown in Fig. E4.49. In this determination, find the contribution along each of the four line segments.

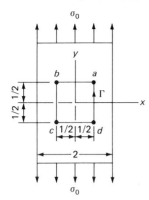

FIGURE E4.49

4.50. Derive Eq. (4.77).

4.51. The critical value of K_{Ic} for a steel is 70 MPa $\cdot \sqrt{m}$. Determine the critical value of the strain energy release rate under
(*a*) Plane stress
(*b*) Plane strain.

4.52. The critical value of K_{Ic} for a steel is 70 MPa $\cdot \sqrt{m}$. If the steel is used in a plane-strain application, determine the critical stress σ_c for cracks ranging in length from 1 to 100 mm.

4.53. For an alloy steel with $K_{Ic} = 90$ MPa $\cdot \sqrt{m}$ and $\sigma_{ys} = 900$ MPa, determine the crack length $2a$ where $\sigma_c = \sigma_{ys}$. Assume the crack is centrally located in a thick plate loaded uniaxially with σ_0.

4.54. Describe the three steps involved in fracture-control plans.

4.55. Construct a graph showing a_{cr} as a function of K_{Ic}/σ_0 over the range from 0.01 to 0.15 $\sqrt{m}$.

4.56. Write a fracture-control plan for a large natural gas storage tank to be located adjacent to a densely populated metropolitan area.

4.57. Why is proof testing used to ensure structural integrity?

REFERENCES

1. Irwin, G. R.: "Fracture I," in S. Flügge (ed.), *Handbuch der Physik VI*, pp. 558–590, Springer-Verlag, New York, 1958.
2. Inglis, C. E.: Stresses in a Plate due to the Presence of Cracks and Sharp Corners, *Trans. Inst. Nav. Arch.*, vol. 55, pp. 219–241, 1913.
3. Griffith, A. A.: The Phenomena of Rupture and Flow in Solids, *Phil. Trans. R. Soc.*, vol. 221, pp. 163–197, 1921.
4. Westergaard, H. M.: Bearing Pressures and Cracks, *J. Appl. Mech.*, vol. 61, pp. A49–A53, 1939.

5. Kirsch, G.: Die Theorie der Elasticität und die Bedürfnisse der Festigkeitlehre, *Z. Ver. Dtsch. Ing.*, vol. 32, pp. 797–807, 1898.
6. Tada, H., P. Paris, and G. R. Irwin: *The Stress Analysis of Cracks Handbook*, Del Research, Hellertown, Pa., 1973.
7. Irwin, G. R.: Plastic Zone Near a Crack and Fracture Toughness, *Proc. 7th Sagamore Conf.*, p. IV-63, 1960.
8. Chona, R.: "Non-Singular Stress Effects in Fracture Test Specimens—A Photoelastic Study," M.S. thesis, University of Maryland, College Park, August 1985.
9. Sanford, R. J.: A Critical Re-Examination of the Westergaard Method for Solving Opening-Mode Crack Problems, *Mech. Res. Commun.*, vol. 6, no. 5, pp. 289–294, 1979.
10. Sih, G. C.: Some Basic Problems in Fracture Mechanics and New Concepts, *Eng. Fract. Mech.*, vol. 5, pp. 365–377, 1973.
11. Rice, J. R.: A Path Independent Integral and the Approximate Analysis of Strain Concentrations by Notches and Cracks, *J. Appl. Mech.*, vol. 35, pp. 379–386, 1968.
12. Eshelby, J. D.: Calculation of Energy Release Rate, in *Prospects of Fracture Mechanics*, (eds.), G. C. Sih, H. C. Von Elst, and D. Broek. Noordhoff, Groningen, 1974, pp. 69–84.
13. Irwin, G. R.: *Fracture Dynamics, Fract. of Met.*, ASM, 1948; pp. 147–166.
14. Brown, W. F., Jr., and J. E. Srawley: "Fracture Toughness Testing," in *Fracture Toughness Testing and Its Applications*, ASTM publication, ASTM-STP-381, 1965, pp. 133–185.
15. U.S. Coast Guard: *Structural Failure of the Tank Barge I.O.S. 3301 Involving the Motor Vessel Martha R. Ingram on 10 January 1972 without Loss of Life*, Marine Casualty Report No. SDCG/NTSB, March 1974.

PART
II

STRAIN-MEASUREMENT METHODS AND RELATED INSTRUMENTATION

CHAPTER

5

INTRODUCTION TO STRAIN MEASUREMENTS

5.1 DEFINITION OF STRAIN AND ITS RELATION TO EXPERIMENTAL DETERMINATIONS

A state of strain may be characterized by its six cartesian strain components or, equally well, by its three principal strain components with the three associated principal directions. The six cartesian components of strain are defined in terms of the displacement field by the following set of equations when the strains are small (normally the case for elastic analyses):

$$
\epsilon_{xx} = \frac{\partial u}{\partial x} \qquad \epsilon_{yy} = \frac{\partial v}{\partial y} \qquad \epsilon_{zz} = \frac{\partial w}{\partial z}
$$

$$
\gamma_{xy} = \frac{\partial v}{\partial x} + \frac{\partial u}{\partial y} \qquad \gamma_{yz} = \frac{\partial w}{\partial y} + \frac{\partial v}{\partial z} \qquad \gamma_{zx} = \frac{\partial u}{\partial z} + \frac{\partial w}{\partial x}
$$

(2.4)

The ϵ components are normal strains, and ϵ_{xx}, for instance, is defined as the change in length of a line segment parallel to the x axis divided by its original length. The γ components are shearing strains, and γ_{xy}, for instance, is defined as the change in the right angle formed by the line segments parallel to the x and y axes.

  For the most part, strain-gage applications are confined to the free surfaces of a body. The two-dimensional state of stress existing on this surface can be expressed in terms of three cartesian strains ϵ_{xx}, ϵ_{yy}, γ_{xy}. Thus, if the two displacements u and v can be established over the surface of the body, the strains can be determined directly from Eqs. (2.4). In certain isolated cases, the most appropriate approach for establishing the stress and strain field is to determine the displacement field. As an example, consider the very simple problem of a transversely loaded beam. The deflections of the beam $w(x)$ along its longitudinal axis can be accurately determined with relatively simple experimental techniques. The strains and stresses are related to the deflection $w(x)$ of the beam by

$$\epsilon_{xx} = \frac{z}{\rho} = z\frac{d^2w}{dx^2} \qquad \sigma_{xx} = E\epsilon_{xx} = Ez\frac{d^2w}{dx^2} \tag{5.1}$$

where ρ is the radius of curvature of the beam and z is the distance from the neutral axis of the beam to the point of interest.

Measurement of the transverse displacements of plates can also be accomplished with relative ease, and stresses and strains computed by employing equations similar to (5.1). In the case of a more general body, however, the displacement field cannot readily be measured. Also, the conversion from displacements to strains requires a determination (by differentiation) of the gradients of experimentally determined displacements at many points on the surface of the specimen. Since the displacements are often difficult to obtain and the differentiation process is subject to large errors, it is advisable to employ a strain gage of one form or another to measure the surface strains directly.

Examination of Eqs. (2.4) shows that the strains ϵ_{xx}, ϵ_{yy}, and γ_{xy} are really the slopes of the displacement surfaces u and v. Moreover, these strains are not, in general, uniform; instead, they vary from point to point. The slopes of the displacement surfaces cannot be established unless the in-plane displacements u and v can be accurately established. Since the in-plane displacements are often exceedingly small in comparison with the transverse (out-of-plane) displacements mentioned previously, their direct measurement over the entire surface of a body is very difficult. To circumvent this difficulty, one displacement component is usually measured over a small region of the body along a short line segment, as illustrated in Fig. 5.1. This displacement measurement is converted to strain by the relationship

$$\epsilon_{xx} = \frac{l_x - l_0}{l_0} = \frac{\Delta u}{\Delta x} \tag{5.2}$$

where $\Delta u = l_x - l_0$ is the displacement in the x direction over the length of the line segment $l_0 = \Delta x$. Strain measured in this manner is not exact since the determination is made over some finite length l_0 and not at a point, as the definition for strain ϵ requires. The error involved in this approach depends upon the strain gradient and the length of the line segment l_0. If the strain determination is considered to represent the strain which occurs at the center of the line segment l_0, that is, point x_1 in Fig. 5.1, the error involved for various strain gradients is

Case 1, strain constant: $\epsilon_{xx} = k_1$ (no error is induced)

Case 2, strain linear: $\epsilon_{xx} = k_1x + k_2$ (no error is induced)

Case 3, strain quadratic: $\epsilon_{xx} = k_1x^2 + k_2x + k_3$

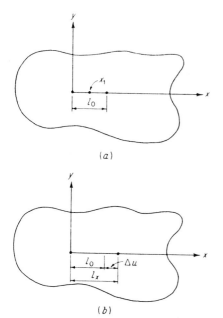

(a)

(b)

FIGURE 5.1
Strain measurement over a short line segment of length l_0: (a) before deformation; (b) after deformation.

In case 3, an error is involved since the strain at the midpoint x_1 is not equal to the average strain over the gage length l_0. The average strain ϵ_{avg} over the gage length l_0 can be computed as

$$\epsilon_{avg} = \frac{\int_0^{l_0} (k_1 x^2 + k_2 x + k_3)\, dx}{l_0} = \frac{k_1 l_0^2}{3} + \frac{k_2 l_0}{2} + k_3 \qquad (a)$$

and the strain at the midpoint $x_1 = l_0/2$ is given by

$$\epsilon_{xx}|_{x_1} = \frac{k_1 l_0^2}{4} + \frac{k_2 l_0}{2} + k_3 \qquad (b)$$

The difference between the average and midpoint strains represents the error $\mathscr{E}$ that is given by

$$\mathscr{E} = \frac{k_1 l_0^2}{12} \qquad (5.3)$$

In this example, the error involved depends upon the values of k_1 and l_0. If the gradient is sharp, the value of k_1 will be significant and the error induced will be large unless l_0 is reduced to an absolute minimum. Other examples corresponding to cubic and quartic strain distributions can also be analyzed; however, the fact that an error is induced is established by considering any strain distribution other than a linear one.

In view of the error introduced by the length of the line segment in certain strain fields, great effort has been expended in reducing the gage length l_0. Two

FIGURE 5.2
Berry-type strain gage.

factors complicate these efforts. (1) Mechanical difficulties are encountered when l_0 is reduced. Regardless of the type of gage employed to measure the strain, it must have a certain finite size and a certain number of parts. As the size is reduced, the parts become smaller and the dimensional tolerances required on each become prohibitive. (2) The strain to be measured is a very small quantity. Suppose, for example, that strain determinations are to be made with an accuracy of ± 1 $\mu\epsilon$ over a gage length of 0.1 in (2.5 mm). The strain gage must measure the corresponding displacement to an accuracy of $\pm 1 \times 10^{-6} \times 0.1 = \pm 1 \times 10^{-7}$ in (2.5 nm). These size and accuracy requirements place heavy demands on the talents of investigators in the area of strain-gage development.

The smallest gage developed and sold commercially to date is an electrical-resistance type. This gage is prepared from an ultrathin alloy foil which is photoetched to produce a grid with a gage length of only 0.008 in (0.2 mm). On the other hand, mechanical strain gages are still employed in civil engineering structural applications where the gage length $l_0 = 8$ in or 200 mm (Berry strain gage). These Berry gages are rugged, simple to use, and sufficiently accurate in structural applications where the strain distribution is essentially linear over the 8-in (200-mm) gage length. A Berry-type strain gage is shown in Fig. 5.2.

5.2 PROPERTIES OF STRAIN-GAGE SYSTEMS

Historically, the development of strain gages has followed many different paths, and gages have been developed which are based on mechanical, optical, electrical, acoustical, and pneumatic principles. No single gage system, regardless of the principle upon which it is based, has all the properties required of an optimum gage. Thus there is a need for a wide variety of gaging systems to meet the requirements of a wide range of different engineering problems involving strain measurement.

Some of the characteristics commonly used to judge the adequacy of a strain-gage system for a particular application are the following:

1. The calibration constant for the gage should be stable; it should not vary with either time, temperature, or other environmental factors.
2. The gage should be able to measure strains with an accuracy of $\pm 1 \mu\epsilon$ over a strain range of ± 10 percent.
3. The gage size, i.e., the gage length l_0 and width w_0, should be small so that strain at a point is approximated with small error.
4. The response of the gage, largely controlled by its inertia, should be sufficient to permit recording of dynamic strains with frequency components exceeding 100 kHz.
5. The gage system should permit on-location or remote readout.
6. The output from the gage during the readout period should be independent of temperature and other environmental parameters.
7. Both the gage and the associated auxiliary equipment should be low in cost to permit wide usage.
8. The gage system should be easy to install and operate.
9. The gage should exhibit a linear response to strain over a wide range.
10. The gage should be suitable for use as the sensing element in other transducer systems where an unknown quantity such as pressure is measured in terms of strain.

No single strain-gage system satisfies all of these characteristics. However, a strain-gage system for a particular application can be selected after proper consideration is given to each of these characteristics in terms of the requirements of the measurement to be made. Over the last 50 years a large number of systems, with wide variations in design, have been conceived, developed, and marketed. Each of these systems has four basic characteristics which deserve additional consideration, namely, the gage length l_0, the gage sensitivity, the range of strain, and the accuracy of the readout.

Strains cannot be measured at a point with any type of gage, and, as a consequence, nonlinear strain fields cannot be measured without some degree of error being introduced. In these cases, the error will definitely depend on the gage length l_0 in the manner described in Sec. 5.1 and may also depend on the gage width w_0. The gage size for a mechanical strain gage is characterized by the distance between the two knife edges in contact with the specimen (the gage length l_0) and by the width of the movable knife edge (the gage width w_0). The gage length and width of the metal-film resistance strain gage are determined by the size of the active area of the grid. In selecting a gage for a given application, gage length is one of the most important conditions.

A second basic characteristic of a strain gage is its sensitivity. Sensitivity is the smallest value of strain which can be read on the scale associated with the strain gage. The term "sensitivity" should not be mistaken for accuracy or precision, since very large values of magnification can be designed into a gage to increase its sensitivity; but friction, wear, noise, drift, etc., introduce large errors which limit the accuracy. In certain applications gages can be employed with sensitivities of less than 1 $\mu\epsilon$ if proper procedures are established. In other applications, where sensitivity is not important, 50 to 100 $\mu\epsilon$ is often sufficient. The choice of a gage is dependent upon the degree of sensitivity required, and quite often the selection of a gage with a very high sensitivity, when it is not really necessary, needlessly increases the complexity of the measuring method.

A third basic characteristic of a strain gage is its range. Range represents the maximum strain which can be recorded without rezeroing or replacing the strain gage. The range and sensitivity are interrelated since very sensitive gages respond to small strains with appreciable response and the range is usually limited to the full-scale deflection or count of the indicator. Often it is necessary to compromise between the range and sensitivity characteristics of a gage to obtain reasonable performance for both these categories.

The final basic consideration is the accuracy or precision. As was previously pointed out, sensitivity does not ensure accuracy. Usually the very sensitive instruments are quite prone to errors unless they are employed with the utmost care. In a mechanical strain gage, inaccuracies may result from lost motion due to wear or slippage, or deflection of the components. On all strain gages there is a readout error whether the output of the gage is manually recorded or displayed on a digital multimeter.

5.3 TYPES OF STRAIN GAGES

The problem encountered in measuring strain is to determine the displacement between two points some distance l_0 apart. The physical principles which have been employed to accomplish this task are very numerous, and a complete survey will not be attempted; however, a few of the more applicable methods will be covered briefly. The principles employed in strain-gage construction can be used as the basis for classifying the gages into the following four groups:

1. Mechanical
2. Optical
3. Electrical
4. Acoustical

5.3.1 Mechanical Strain Gages

Mechanical strain gages such as the Huggenberger tensometer or the Johansson mikrokator are rarely used today because the electrical-resistance strain gages are more accurate, lower in cost, and easier to use. Mechanical gages, often called *extensometers,* are still widely used today in material test systems. However, these extensometers utilize electrical devices such as displacement transformers or resistance strain gages for sensors. A typical extensometer, shown in Fig. 5.3*a*, is employed in the conventional tensile test where the stress-strain diagram is recorded. The extensometer is equipped with knife edges and a wire spring that forces the knife edges into the tension specimen. Elongation or compression of

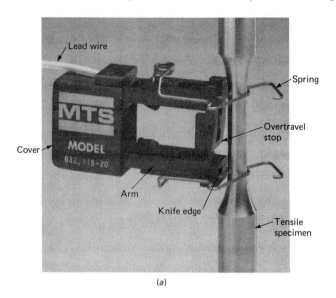

(a)

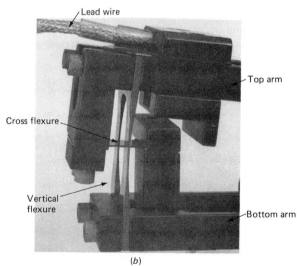

(b)

FIGURE 5.3
Detail of an extensometer: (*a*) extensometer attached to a tensile specimen (gage length—25 mm); (*b*) cover removed to show the vertical and cross-flexural plates which respond to motion of the arms. (*MTS Systems Corporation.*)

the specimen causes movement of the arms. As these arms move, they bend a small cross-flexural element (see Fig. 5.3b) ensuring center-point bending over the entire range of the extensometer. The cross-flexural member, which is the sensing element, also provides good lateral stability and requires low actuating forces (about 50 g). Electrical-resistance strain gages, bonded to the cross-flexural element, sense the bending strains and give a signal output that is proportional to the contraction or extension of the tensile specimen. The extensometer provides an accurate response to specimen strain with maximum nonlinearity of 0.3 percent of range and maximum hysteresis of only 0.1 percent of range.

5.3.2 Optical Strain Gages [1–4]

During the past 20 years, considerable research effort has been devoted to the area of optical methods of experimental stress analysis. The availability of gas and ruby lasers as monochromatic, collimated, and coherent light sources has led to several new developments in strain gages. Two of these developments—the diffraction strain gage and the interferometric strain gage—are described to indicate the capabilities of optical gages which use coherent light.

The diffraction strain gage is quite simple in construction; it consists of two blades that are bonded or welded to the component, as illustrated in Fig. 5.4. The two blades are separated by a distance b to form a narrow aperture and are fixed to the specimen along the edges (see Fig. 5.4) to give a gage length l. A beam of collimated monochromatic light from a helium-neon laser is directed onto the aperture to produce a diffraction pattern that can be observed as a line of dots on a screen a distance R from the aperture. An example of a diffraction pattern is illustrated in Fig. 5.5.

When the distance R to the screen is very large compared with the aperture width b, the distribution of the intensity I of light in the diffraction pattern (see Sec. 10.5) is

$$I = A_0^2 \frac{\sin^2 \beta}{\beta^2} \qquad (5.4)$$

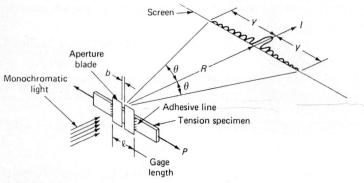

FIGURE 5.4
Arrangement of the diffraction-type strain gage. (*After T. R. Pryor and W. P. T. North.*)

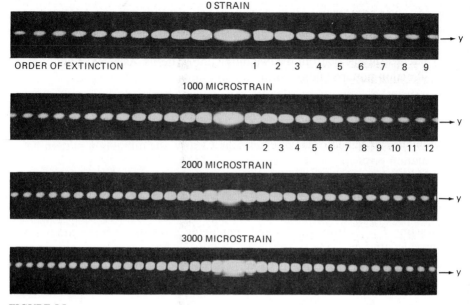

FIGURE 5.5
Diffractograms showing changes in the diffraction pattern with increasing strain. (*Courtesy of T. R. Pryor and W. P. T. North.*)

where A_0 is the amplitude of the light on the centerline of the pattern ($\theta = 0$) and

$$\beta = \frac{\pi b}{\lambda} \sin \theta \qquad (5.5)$$

where θ is defined in Fig. 5.4 and λ is the wavelength of the light. If the analysis of the diffraction pattern is limited to short distances y from the centerline of the system, $\sin \theta$ is small enough to be represented by y/R and Eq. (5.5) becomes

$$\beta = \frac{\pi b}{\lambda} \frac{y}{R} \qquad (5.6)$$

The intensity I vanishes according to Eq. (5.4) when $\sin \beta = 0$ or when $\beta = n\pi$, where $n = 1, 2, 3, \ldots$. By considering those points in the diffraction pattern where $I = 0$, it is possible to obtain a relationship between their location in the pattern and the aperture width b. Thus

$$b = \frac{\lambda R n}{y} \qquad (5.7)$$

where n is the order of extinction in the diffraction pattern at the position located by y.

As the specimen is strained, the deformation results in a change in the aperture width $\Delta b = \epsilon l$ and a corresponding change in the diffraction pattern, as indicated in Fig. 5.5. The magnitude of this strain ϵ can be determined from Eq. (5.7) and measurements from the two diffraction patterns. As an example, consider the diffraction pattern after deformation and note

$$b + \Delta b = \frac{\lambda R n^*}{y_1} \qquad (a)$$

where n^* is a specified order of extinction. Clearly, the diffraction pattern before deformation gives

$$b = \frac{\lambda R n^*}{y_0} \qquad (b)$$

Subtracting Eq. (b) from Eq. (a) and simplifying gives the average strain ϵ over the gage length l as

$$\epsilon = \frac{\Delta b}{l} = \frac{\lambda R n^*}{l} \frac{y_0 - y_1}{y_0 y_1} \qquad (5.8)$$

In practice, the order of extinction n^* is selected as high as possible consistent with the optical quality of the diffraction pattern. When the higher orders of extinction are used, the distance y can be measured with sufficient accuracy with calipers and an engineering scale and elaborate measuring devices can be avoided.

In an automated readout system, a linear array of photodiode cells replaces the screen in Fig. 5.4. The voltage from each cell is proportional to the light intensity and the value of y for each extinction order is determined by locating each cell which exhibits a minimum intensity. The output from the linear photodiode array is monitored with an online computer in real time.

The diffraction strain gage is extremely simple to install and use provided the component can be observed during the test. The method has many advantages for strain measurement at high temperatures since it is automatically temperature-compensated if the blades forming the aperture are constructed of the same material as the specimen.

A second optical method of strain measurement utilizes the interference patterns produced when coherent, monochromatic light from a source such as a helium-neon laser is reflected from two shallow V grooves ruled on a highly polished portion of the specimen surface. The V grooves are usually cut with a diamond to a depth of approximately 0.000040 in (0.001 mm) and are spaced approximately 0.005 in (0.125 mm) apart. An interference pattern from a pair of grooves having this depth and spacing and a groove angle of 110° is shown in Fig. 5.6.

When the grooves, which serve as reflective surfaces, are small enough to cause the light to diffract, and close enough together to permit the diffracted light rays to superimpose, an interference pattern is produced. The intensity of light

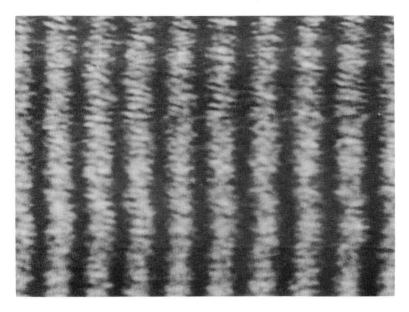

FIGURE 5.6
Interference fringe pattern produced by reflected light from two V-shaped grooves. (*Courtesy of W. N. Sharpe, Jr.*)

in the pattern is given by the expression

$$I = 4A_0^2 \frac{\sin^2 \beta}{\beta^2} \cos^2 \phi \qquad (5.9)$$

where $\beta = \dfrac{\pi b}{\lambda} \sin \theta$

$\phi = \dfrac{\pi d}{\lambda} \sin \theta$

b = width of groove
d = width between grooves
θ = angle from central maximum, as previously defined in Fig. 5.4.

As the light is reflected from the sides of the V grooves, two different interference patterns are formed, as indicated in Fig. 5.7. In an actual experimental situation, the fringe patterns are observed on screens located approximately 8 in (200 mm) from the grooves.

The intensity in the interference pattern goes to zero and a dark fringe is produced whenever $\beta = n\pi$, with $n = 1, 2, 3, \ldots$, or when $\phi = (m + \frac{1}{2})\pi$, with $m = 0, 1, 2, \ldots$. When the specimen is strained, both the distance d between grooves and the width b of the grooves change. These effects produce shifts in the fringes of the two interference patterns which can be related to the average strain

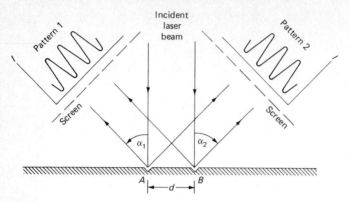

FIGURE 5.7
Schematic diagram showing the light rays which form the two interference patterns.

between the two grooves. The proof is beyond the scope of this presentation: however, it is shown in Ref. 4 that

$$\epsilon = \frac{(\Delta N_1 - \Delta N_2)\lambda}{2d \sin \alpha} \tag{5.10}$$

where ΔN_1 and ΔN_2 are the changes in fringe order in the two patterns produced by the strain and α is the angle between the incident light beam and the diffracted rays which produce the interference pattern, as shown in Fig. 5.7.

The interferometric gage offers a method for measuring strain without the actual use of a device or component, thus eliminating any reinforcing effects or bonding difficulties. Of course, it is necessary to polish the surface of the specimen and to diamond-scribe the grooves. Since no contact is made, the method can be employed on rotating parts or in hostile environments. Temperature compensation is automatic, and the method can be employed at very high temperatures. Also, photodiode arrays can be employed for automated readout.

5.3.3 Electrical Strain Gages

During the past 50 years, electrical strain gages have become so widely accepted that they now dominate the entire strain-gage field except for a few special applications. The most important electrical strain gage is the resistance type, which will be covered in much greater detail in Chaps. 6 and 7. Two less commonly employed electrical strain gages, the capacitance type and the inductance type, will be introduced in this section. Although these gages have limited use in conventional stress analysis, they are often employed in transducer applications and on occasion find special application in measuring strain.

THE CAPACITANCE STRAIN GAGE [5–6]. The capacitance of the parallel-plate capacitor illustrated in Fig. 5.8 can be computed from the relation

$$C = \begin{cases} 0.225 \dfrac{kA}{h} & \dfrac{A}{h} \text{ in inches} \\[3mm] 8.86(10^{-3}) \dfrac{kA}{h} & \dfrac{A}{h} \text{ in millimeters} \end{cases} \tag{5.11}$$

where C = capacitance, pF
$\quad k$ = dielectric constant of the medium between the two plates
$\quad A$ = cross-sectional area of the plates
$\quad h$ = distance between the two parallel plates

The flat-plate capacitor can be employed as a strain or displacement gage in one of three possible ways: (1) by changing the gap h between the plates; (2) by moving one plate in a transverse direction with respect to the other, thereby changing the area A between the two plates; and (3) by moving a body with a dielectric constant higher than air between the two plates.

A schematic drawing of a capacitor-type strain gage with a variable air gap is shown in Fig. 5.8. The change in capacitance ΔC that occurs with a change Δh in the air gap can be obtained from Eqs. (5.11) by noting that

$$C + \Delta C = \frac{kA}{h + \Delta h}$$

which can be simplified to

$$\frac{\Delta C}{C} = -\frac{\Delta h/h}{1 + \Delta h/h} \tag{5.12}$$

The nonlinear relationship between $\Delta h/h$ and $\Delta C/C$ given in Eq. (5.12) is important because in practice $\Delta h/h$ is not small enough when compared to 1 to be neglected.

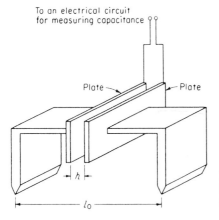

To an electrical circuit
for measuring capacitance

Plate

Plate

h

l_0

FIGURE 5.8
Schematic illustration of a parallel-plate capacitor strain gage with a variable air gap.

To avoid this nonlinear response of the capacitor sensor, consider the capacitive reactance $\mathscr{R}_c$ that is given by

$$\mathscr{R}_c = \frac{1}{\omega C} \tag{5.13}$$

where ω is the circular frequency of the ac voltage applied across the capacitor. If Eqs. (5.11) are substituted into Eq. (5.13),

$$\mathscr{R}_c = \frac{h}{k_1} \tag{5.14}$$

where $k_1 = 0.225kA\omega$ is a constant. From Eq. (5.14) it is evident that

$$\frac{\Delta \mathscr{R}_c}{\mathscr{R}_c} = \frac{\Delta h}{h} \tag{5.15}$$

Examination of Eq. (5.15) shows that the capacitive reactance varies linearly with the change in air gap Δh.

When the gage illustrated in Fig. 5.8 is mounted on a specimen, which in turn is loaded, the gage length changes by Δl and the air gap changes by $\Delta h = \Delta l$. Hence, the strain ϵ produces a change in the capacitive reactance

$$\frac{\Delta \mathscr{R}_c}{\mathscr{R}_c} = \frac{\epsilon l_0}{h} \tag{5.16}$$

For a capacitor gage with $l_0 = 1$ in (25 mm), $h = 0.010$ in (0.25 mm), and $\epsilon = 1 \ \mu\epsilon$, the value obtained for $\Delta C/C$ is 10^{-4}. Electrical circuits (see Ref. 7) can be employed to accurately measure very small changes in capacitance for both static applications and low-frequency dynamic applications. Thus, the sensitivity and accuracy of the capacitor gage are quite sufficient for application to the general problem of determining strain. The primary disadvantage of the capacitor gage is its relatively large size and its mechanical attachment through knife edges. The application of the capacitor principle to other transducer systems, however, does show promise since it can be employed at elevated temperatures. The dielectric constant k of air is nearly constant with temperature up to 1500°F (815°C), and this fact indicates that the calibration constant for a capacitive sensor will be stable over a wide range in temperatures.

THE INDUCTANCE STRAIN GAGE [8–9]. Of the many types of inductance measuring systems which could be employed to measure strain, the differential-transformer system will be considered here. The linear differential transformer is an excellent device for converting mechanical displacement into an electrical signal. It can be employed in a large variety of transducers, including strain, displacement, pressure, acceleration, force, and temperature. A schematic illustration of a linear differential transformer employed as a strain-gage transducer is shown in Fig. 5.9. Mechanical knife edges are displaced over the gage length l_0 by the strain induced

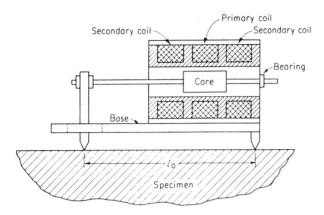

FIGURE 5.9
Schematic illustration of a linear differential transformer employed as a strain transducer.

in the specimen. This displacement is transmitted to the core, which moves relative to the coils, and an electrical output is produced across the coils.

A linear differential transformer has three coils, a primary coil and two secondary coils, on either side of the primary. A core of magnetic material supported on a shaft of nonmagnetic material is positioned in the center of the coils, as shown in the circuit drawing in Fig. 5.10. As the core moves within the coils, it varies the mutual inductance between the primary and each secondary winding, with one secondary becoming more tightly coupled to the primary and the other secondary becoming more loosely coupled. The two secondary coils are wired in series opposition, and consequently, the output voltage E_o is the difference between the voltages developed in each secondary, that is, $E_o = E_1 - E_2$. In a symmetrically constructed transformer, a null output occurs when the core is at the centerpoint between the two secondary coils. Movement of the core in one direction off the null position will induce an unbalance in E_1 and E_2, and some net E_o will be indicated. Displacement in the opposite direction will also produce an unbalance, with a resulting E_o which is 180° out of phase with the first output. If the direction of the displacement is required, it is determined from the phase angle.

The sensitivity of commercial differential transformers varies from 0.08 to 6.3 V/in (0.003 to 0.25 V/mm) of displacement per volt of excitation. Normal excitation supplied to the primary coil is 3 Vac with the frequency ranging from

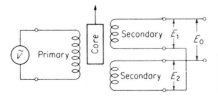

FIGURE 5.10
Schematic diagram of the linear-differential-transformer circuit.

50 Hz to 10 kHz. At rated voltage, 3 V, the most sensitive sensors provide an output of 18.9 mV/mm.

The range of the differential transformer varies with the design of the unit. With commercially available transformers, it is possible to obtain ranges which vary from ± 0.005 to ± 25 in (0.125 mm to 0.64 m). The greater the range of a particular transformer, the lower the sensitivity.

The errors of the linear differential transformer amount, in general, to about 0.5 percent of the maximum linear output. The deviation from linearity of the commercial units is also of the order 0.5 percent of the specified range. The dynamic response is limited by the mass of the core and supporting mechanical assembly. It is also limited electrically by the frequency of the applied ac voltage. This is a carrier frequency, which should be at least 10 times the maximum frequency being measured.

The linear differential transformer requires a very small driving force (a fraction of a gram) to move the core. The operation of the differential transformer can be severely affected, however, by the presence of metal masses or by stray magnetic fields. A magnetic shield is employed around the coil holder to minimize these effects.

In recent years the linear differential transformer has been supplied with a dc to ac power source and with a demodulator to condition the output voltage. When packaged as a single unit, the device is known as a *direct current differential transformer* (DCDT). It is small, rugged, and extremely useful in the laboratory. It can be used for many applications since it can be powered with an ordinary 6-V dry cell and the output can be displayed on a digital multimeter. Both these items are readily available in most laboratories. The use of the linear differential transformer or the DCDT for strain-gage applications has been limited because of the mechanical-attachment problem; however, it is one of the best displacement transducers available for general laboratory use.

5.3.4 Acoustical Strain Gages [10–12]

Acoustical strain gages have been employed in a variety of forms in several countries since the late 1920s. To date, they have been largely supplemented by the electrical-resistance strain gage. However, they are unique among all forms of strain gages in view of their long-term stability and freedom from drift over extended time periods. The acoustical gage described here, due to R. S. Jerrett, was developed in 1944 and is typical of the devices currently being employed. The strain-measuring system is based on the use of two identical gages identified as a test gage and a reference gage. The significant parts of a gage are shown schematically in Fig. 5.11.

In the figure it can be seen that the gage has the common knife-edge mounting provision. One knife edge is mounted to the main body, which is fixed, while the other knife edge is mounted in a bearing suspension and is free to elongate with the specimen. The gage length l_o is 3 in (76 mm). One end of a steel

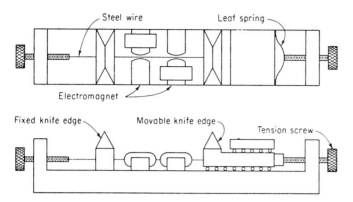

FIGURE 5.11
Schematic drawing showing the operation of the Jerrett acoustical strain gage.

wire is attached to the movable knife edge while the other end of the wire passes through a small hole in the fixed knife edge and is attached to a tension screw. The movable knife edge is connected to a second tension screw by a leaf spring. This design permits the initial tension in the wire to be applied without the transmission of load to the knife edges.

The wire passes between the pole pieces of two small electromagnets. One of these magnets is used to keep the wire vibrating at its natural frequency; the other is employed to pick up the frequency of the system. Electrically both magnets operate together in that the signal from the pickup magnet is amplified and fed back into the driving magnet to keep the string excited in its natural frequency.

The reference gage is identical to the test gage except that the knife edges are removed and a micrometer is used to tension the wire. A helical spring is employed in series with the wire to permit larger rotations of the micrometer head for small changes of stress in the wire.

To operate the system, the test gage is mounted and adjusted and the reference gage is placed near it to attempt to compensate for temperature effects. Both gages are energized, and each wire emits a musical note. If the frequency of vibrations from the two gages is not the same, beats will occur. The micrometer setting is varied on the reference gage until the beat frequency decreases to zero. The reading of the micrometer is then taken and the strain is applied to the test gage. The change in tension in the wire of the test gage produces a change in frequencies, and it is necessary to adjust the reference gage with the micrometer until the beats are eliminated. This new micrometer reading is proportional to the strain.

If the test gage is located in a remote position and the beat signals from the test and reference gages cannot be developed, it is possible to balance the two gages by using an oscilloscope. The voltage output from the pickup coils of each gage are displayed while operating the oscilloscope in the xy mode. The resulting

Lissajous figure provides the readout which permits adjustment of the micrometer on the reference gage to match the frequency of the test gage.

The natural frequency f of a wire held between two fixed points is given by the expression

$$f = \frac{1}{2L}\sqrt{\frac{\sigma}{\rho}} \tag{5.17}$$

where L = the length of wire between supports
σ = the stress in the wire
ρ = the density of the wire

In terms of strain in the wire, the frequency is governed by the following equation, which comes directly from Eq. (5.17):

$$f = \frac{1}{2L}\sqrt{\frac{E\epsilon}{\rho}} \tag{5.18}$$

where E is the modulus of elasticity.

The sensitivity of this instrument is very high, with possible determinations of displacements of the order of 0.1 μin (2.5 μm). The range is limited, in general, to about one-thousandth of the wire length before over- or understressing of the sensing wire becomes critical. The gage is temperature-sensitive unless the thermal coefficients of expansion of the base and wire are closely matched over the temperature range encountered during a test. Finally, the force required to drive the transducer is relatively large, and it should not be employed in high-compliance systems where the large driving force will be detrimental.

5.4 SEMICONDUCTOR STRAIN GAGES [13–17]

The development of semiconductor strain gages was an outgrowth of research at the Bell Telephone Laboratories which led to the introduction of the transistor in the early 1950s. The piezoresistive properties of semiconducting silicon and germanium were determined by Smith in 1954. Further development of semiconductor transducers by Mason and Thurston in 1957 eventually led to the commercial marketing of piezoresistive strain gages in 1960.

Basically, the semiconductor strain gage consists of a small, ultrathin rectangular filament of single-crystal silicon. The semiconducting materials exhibit a very high strain sensitivity S_A, with values ranging from 50 to 175 depending upon the type and amount of impurity diffused into the pure silicon crystal. The resistivity ρ of a single-crystal semiconductor with impurity concentrations of the order of 10^{16} to 10^{20} atoms/cm^3 is given by

$$\rho = \frac{1}{eN\mu} \tag{5.19}$$

where e = electron charge, which depends on the type of impurity

N = number of charge carriers, which depends on the concentration of the impurity

μ = mobility of the charge carriers, which depends on strain and its direction relative to the crystal axes.

Piezoresistive strain gages occupy a niche in the strain-gage market. Their advantage of a high sensitivity is balanced by several disadvantages which include high cost, limited range, and large temperature effects.

The importance of Eq. (5.19) can be better understood if it is considered in terms of the sensitivity S_A of a conductor to strain, which can be written as

$$S_A = 1 + 2v + \frac{d\rho/\rho}{\epsilon} \tag{5.20}$$

For metallic conductors, $1 + 2v \approx 1.6$ and $(d\rho/\rho)/\epsilon$ ranges from 0.4 to 2.0 for the common strain-gage alloys. For semiconductor materials $(d\rho/\rho)/\epsilon$ can be varied between -125 and $+175$ by selecting the type and concentration of the impurity. Thus, very high conductor sensitivities are possible where the resistance change is about 100 times larger than that obtained for the same strain with metallic-alloy gages. Also, negative gage factors are possible which permit large electrical outputs from Wheatstone-bridge circuits where multiple strain gages are employed.

In producing semiconductor strain gages, ultrapure single-crystal silicon is employed. Boron is used as the trace impurity in producing the P-type (positive gage factor) piezoresistive material. Arsenic is used to produce the N-type (negative gage factor) material.

The semiconductor materials have another advantage over metallic alloys for strain-gage applications. The resistivity of P-type silicon is of the order of 500 $\mu\Omega \cdot$ m which is 1000 times greater than the resistivity of Constantan, which is 0.49 $\mu\Omega \cdot$ m. Because of this very high resistivity, semiconductor strain gages do not utilize grids. They are usually very short single elements with leads, as shown in Fig. 5.12.

The very high sensitivity of semiconductor gages to strain and their high resistivity have led to their application in measuring extremely small strains, in miniaturized transducers, and in very-high-signal-output transducers.

5.4.1 Piezoresistive Properties of Semiconductors [17–18]

A crystal of a semiconducting material is electrically anisotropic; consequently, the relation between the potential gradient E and the current density i is formulated relative to the directions of the crystal axes to give components E_1, E_2, and E_3 of the vector E. Thus

$$\begin{aligned}
E_1 &= \rho_{11}i_1 + \rho_{12}i_2 + \rho_{13}i_3 \\
E_2 &= \rho_{21}i_1 + \rho_{22}i_2 + \rho_{23}i_3 \\
E_3 &= \rho_{31}i_1 + \rho_{32}i_2 + \rho_{33}i_3
\end{aligned} \tag{5.21}$$

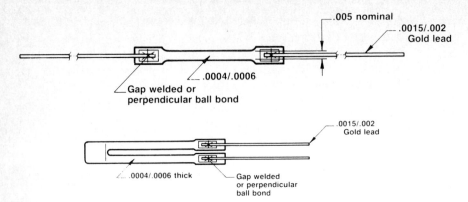

FIGURE 5.12
Semiconductor elements with lead wires. (*Micron Instruments.*)

where the first subscript of each resistivity coefficient indicates the component of the voltage field to which it contributes and the second identifies the component of current. The single crystal will permit isotropic conduction only if

$$\rho_{11} = \rho_{22} = \rho_{33} = \rho$$

and

$$\rho_{12} = \rho_{13} = \rho_{21} = \rho_{23} = \rho_{31} = \rho_{32} = 0$$

Since this situation exists for unstressed cubic crystals, Eqs. (5.21) reduce to

$$E_1 = \rho i_1 \qquad E_2 = \rho i_2 \qquad E_3 = \rho i_3 \qquad (5.22)$$

When a state of stress is imposed on the crystal, it responds by exhibiting a piezoresistive effect which can be described by the expression

$$\rho_{ij} = \delta_{ij}\rho + \pi_{ijkl}\tau_{kl} \qquad (5.23)$$

where subscripts $i, j, k,$ and l range from 1 to 3 and π_{ijkl} is a fourth-rank piezoresistivity tensor, which is a function of the crystal and the level and type of the impurities. A complete discussion of the results obtained from Eqs. (5.22) is beyond the scope of this text. Fortunately, several reductions in the complexity of Eqs. (5.22) are possible for the type of crystals used in the fabrication of semiconductor strain gages. Recall from Chap. 1 [Eqs. (1.4)] that the stress tensor τ_{kl} is symmetric. If the analysis is limited to cubic crystals, the resistivity tensor ρ_{ij} is also symmetric. Finally, due to the properties of a silicon crystal, the piezoresistive coefficients can be shown to reduce to

$$\pi_{1111} = \pi_{2222} = \pi_{3333} = \rho\pi_{11}$$

$$\pi_{1122} = \pi_{2211} = \pi_{1133} = \pi_{3311} = \pi_{2233} = \pi_{3322} = \rho\pi_{12}$$

$$\pi_{1212} = \pi_{2121} = \pi_{2323} = \pi_{3232} = \pi_{1313} = \pi_{3131} = \frac{\rho\pi_{44}}{2}$$

The simplified matrix representing the piezoresistive coefficients for silicon, since all other coefficients are zero, is then

$$
\begin{bmatrix}
\pi_{11} & \pi_{12} & \pi_{12} & 0 & 0 & 0 \\
\pi_{12} & \pi_{11} & \pi_{12} & 0 & 0 & 0 \\
\pi_{12} & \pi_{12} & \pi_{11} & 0 & 0 & 0 \\
0 & 0 & 0 & \pi_{44} & 0 & 0 \\
0 & 0 & 0 & 0 & \pi_{44} & 0 \\
0 & 0 & 0 & 0 & 0 & \pi_{44}
\end{bmatrix}
$$

Equation (5.23) can then be expressed in terms of the simplified notation as

$$
\rho_{11} = \rho[1 + \pi_{11}\sigma_{11} + \pi_{12}(\sigma_{22} + \sigma_{33})]
$$

$$
\rho_{22} = \rho[1 + \pi_{11}\sigma_{22} + \pi_{12}(\sigma_{33} + \sigma_{11})]
$$

$$
\rho_{33} = \rho[1 + \pi_{11}\sigma_{33} + \pi_{12}(\sigma_{11} + \sigma_{22})]
$$

$$
\rho_{12} = \rho\pi_{44}\tau_{12} \qquad \rho_{23} = \rho\pi_{44}\tau_{23} \qquad \rho_{31} = \rho\pi_{44}\tau_{31}
$$

(5.24)

Substitution of Eqs. (5.24) into Eqs. (5.21) yields

$$
\frac{E_1}{\rho} = i_1[1 + \pi_{11}\sigma_{11} + \pi_{12}(\sigma_{22} + \sigma_{33})] + \pi_{44}(i_2\tau_{12} + i_3\tau_{31})
$$

$$
\frac{E_2}{\rho} = i_2[1 + \pi_{11}\sigma_{22} + \pi_{12}(\sigma_{33} + \sigma_{11})] + \pi_{44}(i_3\tau_{23} + i_1\tau_{12})
$$

(5.25)

$$
\frac{E_3}{\rho} = i_3[1 + \pi_{11}\sigma_{33} + \pi_{12}(\sigma_{11} + \sigma_{22})] + \pi_{44}(i_1\tau_{31} + i_2\tau_{23})
$$

These results indicate that the voltage drop across a gage fabricated from a piezoelectric element will depend upon the current density, the imposed stresses, and the three piezoresistivity coefficients π_{11}, π_{12}, and π_{44}.

Consider now a strain gage consisting of a single conductor, which is long relative to its lateral dimensions and cut in an arbitrary direction from a cubic crystal, as illustrated in Fig. 5.13. The direction of the gage conductor relative to the crystal axes is given by the unit vector **g**. Consider that lead wires are attached to the ends of this element in the manner illustrated in Fig. 5.12 and that a current density i_g is applied along the gage axis. The strain transmitted to the gage after it is bonded to a component will produce a uniaxial stress σ_g along the axis of the element. Since **i** is a vector quantity, the gage current exhibits components along the crystal axes. Thus

$$
i_1 = i_g \cos(\mathbf{g}, 1) = l i_g
$$

$$
i_2 = i_g \cos(\mathbf{g}, 2) = m i_g
$$

$$
i_3 = i_g \cos(\mathbf{g}, 3) = n i_g
$$

where l, m, and n are the direction cosines associated with the unit vector **g**.

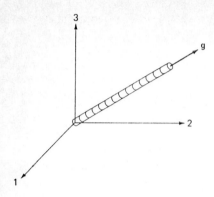

FIGURE 5.13
Orientation of strain-gage element relative to the crystal axes 1, 2, and 3.

The stresses along the crystal axes in terms of the stress along the axis of the gage are given by Eqs. (1.6) as

$$\sigma_{11} = l^2\sigma_g \qquad \sigma_{22} = m^2\sigma_g \qquad \sigma_{33} = n^2\sigma_g$$
$$\tau_{12} = lm\sigma_g \qquad \tau_{23} = mn\sigma_g \qquad \tau_{31} = nl\sigma_g \tag{5.26}$$

From the vector properties of the voltage gradient it is also obvious that

$$E_g = lE_1 + mE_2 + nE_3 \tag{5.27}$$

Substituting Eqs. (5.25) and (5.26) into Eq. (5.27) and simplifying gives

$$\frac{E_g}{\rho} = i_g\{1 + \sigma_g[\pi_{11} + 2(\pi_{12} + \pi_{44} - \pi_{11})(l^2m^2 + m^2n^2 + n^2l^2)]\} \tag{5.28}$$

which can be rewritten as

$$\frac{E_g}{\rho} = i_g(1 + \pi_g\sigma_g) \tag{5.29}$$

where π_g represents the sensitivity of the gage to the uniaxial stress σ_g and is related to the piezoresistive coefficients by the expression

$$\pi_g = A + B(l^2m^2 + m^2n^2 + n^2l^2) \tag{5.30}$$

where $A = \pi_{11}$ and $B = 2(\pi_{12} + \pi_{44} - \pi_{11})$.

Examination of Eq. (5.30) indicates that the sensitivity π_g of the gage to stress can be varied by changing the orientation **g** or by changing the dopant (N or P) and the level of doping of the silicon. The maximum sensitivity $\pi_g = A + B/3$ for P-type silicon with $l = m = n = \pm\sqrt{3}/3$. For N-type silicon the maximum sensitivity $\pi_g = A$ when $l = m = n = 0$.

From Eq. (5.29), the difference in voltage gradient ΔE_g across the semiconductor element before and after stressing is

$$\frac{\Delta E_g}{\rho} = i_g(1 + \pi_g\sigma_g) - i_g = i_g\pi_g\sigma_g$$

Normalizing this relationship with respect to the voltage gradient in the unstressed state gives

$$\frac{\Delta E_g}{E_g} = \frac{\Delta R_g}{R_g} = \pi_g \sigma_g \tag{5.31}$$

Since a uniaxial state of stress exists in the gage element,

$$\sigma_g = E\epsilon \tag{a}$$

where E is the modulus of elasticity of the semiconducting silicon and ϵ is the strain transmitted to the element from the specimen. Substituting Eq. (a) into Eq. (5.31) gives

$$\frac{\Delta R_g}{R_g} = \pi_g E\epsilon = S_{sc}\epsilon \tag{5.32}$$

where $S_{sc} = \pi_g E$ is the strain sensitivity of the piezoresistive material due to strain-induced changes in resistivity. It should be noted that S_{sc} is sufficiently large (≈ 100) that the effect of changing length and width on the sensitivity of the conducting element is very small (≈ 1.5).

5.4.2 Temperature Effects [19–20]

The results of Sec. 5.4.1 indicate that the response of a semiconductor strain gage is linear with respect to strain. Unfortunately, this is a simplification which is not true in the general case. For lightly doped semiconducting materials (10^{19} atoms/cm^3 or less), the sensitivity S_A is markedly dependent upon both strain and temperature with

$$S_A = \frac{T_0}{T}(S_A)_0 + C_1\left(\frac{T_0}{T}\right)^2\epsilon + C_2\left(\frac{T_0}{T}\right)^3\epsilon^2 + \cdots \tag{5.33}$$

where $(S_A)_0$ = room-temperature zero-strain sensitivity, as defined in Eq. (5.20)
$\quad\quad T$ = temperature, with $T_0 = 294$ K
$\quad C_1, C_2$ = constants, which depend on the type of impurity, the level of doping, and the orientation of the element with respect to the crystal axes

The variation of alloy sensitivity S_A as a function of doping level for P-type silicon is shown in Fig. 5.14. As the impurity concentration is increased from 10^{16} to 10^{20} atoms/cm^3, the sensitivity S_A decreases from 155 to 50. Two significant advantages related to the temperature effect compensate for this loss of sensitivity: (1) the effect of temperature on the sensitivity S_A is greatly diminished (since the temperature coefficient of sensitivity approaches zero) when the impurity concentration approaches 10^{20} atoms/cm^3, as shown in Fig. 5.14; (2) the temperature coefficient of resistance decreases from 0.009/°C for impurity concentrations of 10^{16} atoms/cm^3 to 0.00036/°C for impurity concentrations of 10^{19} atoms/cm^3.

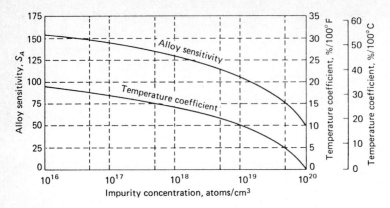

FIGURE 5.14
Alloy sensitivity and temperature coefficient of sensitivity as a function of impurity concentration of
P-type silicon.

Experiments with P-type strain gages indicate that the effects of temperature
are minimized with impurity concentrations of 10^{19} atom/cm^3. This concentration
gives $S_A = 105$.

Temperature compensation of single-element P-type semiconductor strain
gages is not possible since the gages respond to temperature and the gage factor
changes. As a result, a temperature-induced apparent strain occurs.

5.4.3 Linearity and Strain Limits [19–21]

It is clear from Eq. (5.33) that the response of semiconductor strain gages may be
nonlinear with respect to strain. For piezoresistive materials with a low impurity
concentration, the nonlinearities are significant. Increasing the impurity level to
10^{19} to 10^{20} atoms/cm^3 markedly improves the linearity. Unfortunately, most
commercial gages are P type with a gage factor of about 140, which corresponds
to an impurity level of 10^{17} atoms/cm^3. This level is below the optimum for
linearity. Typical linearity specifications are ± 0.25 percent to 600 $\mu\varepsilon$ and ± 1.5
percent to 1500 $\mu\varepsilon$.

The small elements used in fabricating semiconductor strain gages are
removed from a single crystal of silicon by a slicing procedure similar to the one
illustrated in Fig. 5.15. The single crystal is sectioned and then sliced with a
diamond saw to produce thin plates of P-type silicon with the orientation relative
to the [111] crystal axis defined in Fig. 5.15. The thin plates are lapped and etched
to eliminate flaws induced by sawing and to improve the surface finish. The small
sensing elements are then etched from the plate by using photoresist images to
protect the sensing elements.

The silicon material is glasslike and must be treated with care during

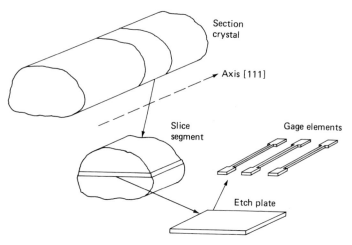

FIGURE 5.15
Sectioning the crystal to remove the strain-gage sensing elements.

mounting. On flat surfaces, installation presents no problems, and semiconductor gages can be subjected to approximately 3000 $\mu\epsilon$ before the gages begin to fail by brittle rupture. When semiconductor gages are installed in a fillet and bending stresses are imposed during installation, extreme care must be exercised to avoid rupture of the element. Fillet radii from 0.1 to 0.25 in (2.5 to 6 mm) can be gaged; however, the strains induced in the element during installation reduce its strain range for subsequent tests.

The fatigue life of semiconductor gages is rated at 10^7 cycles for cyclic strains of $\pm 500 \mu\varepsilon$. This is considerably less than the capability of metallic-foil-type gages; however, semiconductor strain gages are usually employed in low-magnitude strain fields, so that the relatively low strain limits placed on single-cycle and multiple-cycle measurements do not cause problems.

5.4.4 Summary

Semiconductor strain gages have the advantage of a high gage factor and a high resistivity, which leads to miniaturization of the sensing element. They have the disadvantage of exhibiting significant temperature effects.

The primary application of semiconductor gages is as sensors for very rigid miniature transducers which provide high output signals and very high frequency response. The semiconductor gages are less suited as sensors in the more common general-purpose high-accuracy transducers.

The use of semiconductor gages in experimental stress or fracture analysis is very rare. They are applied only to measure very small strains with high accuracy under stable thermal conditions.

5.5 GRID METHOD OF STRAIN ANALYSIS [22–27]

The grid method of strain analysis is one of the oldest techniques known to experimental stress analysis. The method requires placement of a grid (a series of well-defined parallel lines) on the surface of the specimen. Next, the grid is carefully photographed before and after loading the specimen to obtain negatives which will show the distortion of the grid. Measurements of the distance between the grid lines before and after deformation give lengths l_i and l_f, respectively. These lengths may be interpreted in terms of strain in several different ways:

Lagrangian strain:
$$\epsilon = \frac{l_f - l_i}{l_i} \tag{5.34}$$

Eulerian strain:
$$\epsilon = \frac{l_f - l_i}{l_f} \tag{5.35}$$

Natural strain:
$$\epsilon = \ln \frac{l_f}{l_i} \tag{5.36}$$

The exact form used to determine the strain will depend upon the purpose of the analysis and the amount of deformation experienced by the model.

If the grid consists of two series of orthogonal parallel lines over the entire surface of the specimen (see Fig. 5.16), the grid method will give strain components over the entire field. The strain component ϵ_{xx} in the horizontal direction is obtained by comparing the distances between the vertical grid lines before and after deformation. The vertical or ϵ_{yy} component of strain can be obtained from the horizontal array of grid lines, and the $\epsilon_{45°}$ component of strain can be obtained from measurements across the diagonals if the orthogonal grid array is square.

In general, two main difficulties are associated with the utilization of the grid method for measuring strain. First, the strains being measured are usually quite small, and in most instances the displacement readings cannot be made with sufficient precision to keep the accuracy of the strain determinations within reasonable limits. Also, the definition of the grid lines on the negatives is usually poor when magnified, so that appreciable errors are introduced in the displacement readings.

In order to effectively employ the grid method and avoid these two difficulties, the deformations applied to the model must be large enough to impose strain levels of the order of 5 percent. Usually, model deformations this large are to be avoided in an elastic stress analysis; hence other experimental methods such as the moiré method or electrical-resistance strain gages are preferred. However, when large deformations are associated with the stress-analysis problem, e.g., plastic deformations or deformations in rubberlike materials, the grid method is one of the most effective methods known. In fact, if the strains exceed 10 to 30 percent, the range of the commonly employed strain gages is exceeded, and only

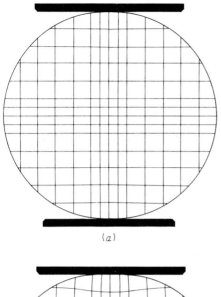

(a)

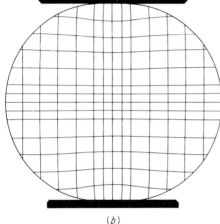

(b)

FIGURE 5.16
An embedded rubber-thread grid in a transparent urethane model of a circular disk under diametral compression: (a) before deformation; (b) after deformation.

the moiré method or the grid method is adequate for measuring these large strains.

5.5.1 The Replica Technique [27]

A modification of the grid method for determining static strains which eliminates the requirement for accurate placement of a grid array on the specimen has been developed by Hickson. With this method, a family of parallel scratches of uneven spacing and thickness is applied to a polished region of a specimen by drawing fine abrasive paper along the edge of a template. Lines equivalent to those produced by the best engraving techniques are present in such a scratch pattern. A second pattern applied perpendicular to the first produces the rectangular array of lines required for grid analyses of deformation and strain. A coarse pattern of broad reference lines is scribed over the scratch pattern to serve as datum lines and to help identify specific scratch lines being used for measurements.

Since it is extremely difficult to make precise measurements of displacement with the specimen under load, a replica technique is used to record the scratch patterns before and after loading. The replicas are made by using a thin film of fusible alloy on a stiffening platen. The alloy is capable of reproducing the finest detail in the scratch pattern. The platen serves to provide dimensional stability and temperature compensation if it is made of the same material as the specimen. Such replicas have been shown to be accurate to at least 40 μin (1 μm).

Displacement measurements are made by comparing a pair of replicas, taken before and after loading, with a microscope fitted with a micrometer eyepiece. The two replicas are mounted in a holder which permits the two scratch patterns to be viewed alternately in the same part of the microscope field. The identification of lines being used for strain determinations is simplified by use of the reference grid and the random nature of the lines in the scratch pattern.

5.5.2 Circular Spot Arrays [28–30]

With significant advancements in video cameras, low-cost memory, and image processing, a new approach to the grid method is being developed. Consider the grid produced with a periodic array of black spots on a white background as illustrated in Fig. 5.17. Note that the pattern includes a single T marking which can be used to indicate rigid-body rotations.

The array of circular dots is placed on the surface of a plane specimen and digital images of the pattern are made and stored during the loading process. Comparison of the digital images from two different loads, known as *pattern matching*, is simply the modern adaptation of the well-established grid method described in Sec. 5.5. What is new is the digital-image acquisition hardware and the image-analysis procedures used in tracking the displacement fields.

The digital-image acquisition hardware includes a video camera and a digitizing (A/D) board which converts the camera signals into digital format and stores the intensity level for each picture element (pixel). The size and shape of the pixels depend upon the video camera. For this discussion consider that the pixels are rectangular in shape and arranged in an imaging matrix, as illustrated

FIGURE 5.17
Array of black circular dots on a white background.

in Fig. 5.18. The photosensitive portion of the pixel is shown as the cross-hatched area A_s. The intensity I_{kl} registered at each pixel location is given by

$$I_{kl} = \frac{1}{A_s} \int i(x, y) \qquad (5.37)$$

where k and l locate the pixel in the image matrix and $i(x, y)$ is the intensity of light on the pixel.

This intensity I_{kl} is an analog signal which is, upon command, converted to a digital value called a *gray level*. The gray level G_{kl} is an integer number ranging from 0 to 255 for 8-bit A/D converters. The gray level G_{kl} is related to a registered intensity at each pixel location by the equation

$$G_{kl} = \mathrm{INT}(I_{kl}) \qquad (5.38)$$

where INT is an operator which rounds I_{kl} downward to the nearest integer between 0 and 255.

Now consider as the intensity of light $i(x, y)$ the grid spot projected over a region of pixels, as indicated in Fig. 5.18. For this configuration it is clear that

$$i(x, y) = \begin{cases} G_D & \text{for } r \leq R \\ 0 & \text{for } r > R \end{cases} \qquad (5.39)$$

where R = the radius of the spot
$r = \sqrt{x^2 + y^2}$
G_D = the gray level difference between the black spot and the white (0) background

If the spot completely covers the pixel, $I_{kl} = G_D$. If the pixel is outside the spot, $I_{kl} = 0$. However, if the spot partially covers the pixel, I_{kl} is obtained by integrating Eq. (5.37). The gray levels G_{kl} are stored for each pixel location. With typical 512×512 arrays, an 8-bit word corresponding to 256 gray levels is stored for each of the 262,144 locations.

The next step is to analyze this image to determine the position of the

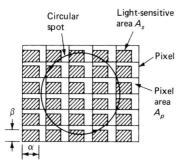

FIGURE 5.18
Imaging matrix showing pixel area A_p, the light-sensitive area A_s, the fill factor parameters λ and β, and a circular spot.

centroid for each circular spot. The centroid for a typical spot is determined from

$$\bar{y} = \frac{\sum\limits_{k=k_0}^{k_f} \sum\limits_{l=l_0}^{l_f} k(T - G_{kl})}{\sum\limits_{k=k_0}^{k_f} \sum\limits_{l=l_0}^{l_f} (T - G_{kl})}$$

$$\bar{x} = \frac{\sum\limits_{k=k_0}^{k_f} \sum\limits_{l=l_0}^{l_f} l(T - G_{kl})}{\sum\limits_{k=k_0}^{k_f} \sum\limits_{l=l_0}^{l_f} (T - G_{kl})}$$

(5.40)

where k_0, k_f, l_0, l_f define a window of pixels including the grid point and T is a threshold level. Only those pixels with $G_{kl} < T$ are included in computing the centroid locations given by Eqs. (5.40).

For a given circular spot, analyses of the two images yield

$$u = \bar{x}_2 - \bar{x}_1$$
$$v = \bar{y}_2 - \bar{y}_1$$

(5.41)

where subscripts 1 and 2 refer to two load levels. Since a large number of points can be employed over the surface of the body, the method can be classified as full field.

The use of Eqs. (5.41) assumes that rigid-body rotations do not occur between the two load levels. If rigid-body rotations do occur, it is necessary to analyze the T mark. In this image analysis, the direction of the principal inertia axis is determined for each load level to give θ_{y1} and θ_{y2}. The rigid-body rotation ω, defined in Fig. 5.19, is given by

$$\omega = \theta_{y2} - \theta_{y1}$$

(5.42)

If ω is not zero, it is necessary to adjust both $\bar{x}_2$ and $\bar{y}_2$ for each circular spot to account for the displacements due to rigid-body rotation.

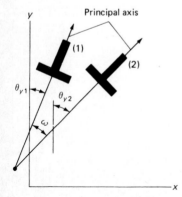

FIGURE 5.19
Image of T pattern to determine rigid-body rotation.

Sirkis has recently examined the parameters which affect the accuracy of this method, which include the fill factor, the size of the circular spot T, the gray level difference G_D, and signal-to-noise ratio.

The fill factor is defined as the ratio of the sensitive area A_s to the entire area A_p of the pixel. Currently $A_s/A_p \approx 0.6$ on typical video cameras and is not a variable. Clearly, as the sensing arrays in video cameras improve, the ratio A_s/A_p will increase and the error due to the fill factor will decrease.

The spot radius markedly affects the error, as indicated in Fig. 5.20. Large spot radii, 10 to 20 times the pixel size, are necessary to limit error. As the fill factor increases, the spot size required for a given error decreases.

The gray level difference G_D should be maximized. This is controlled by the operator in forming the grid (contrast), adjusting the lighting, adjusting the gain and offset of the A/D converter, and deciding on the number of bits for gray level distinction.

Noise, like fill factor, is dependent upon equipment; but the noise can be reduced with image averaging, electromagnetic isolation, and cooling the video camera.

With careful techniques giving a signal-to-noise ratio of 200, it is possible to limit error to ± 0.05 pixel. The corresponding surface displacement depends upon the magnification of the optical system used to project the image of the circular spots onto the sensing array of the camera. The time required to process an array of 1000 circular spots is about 100 s when an IBM-AT is used to process the image.

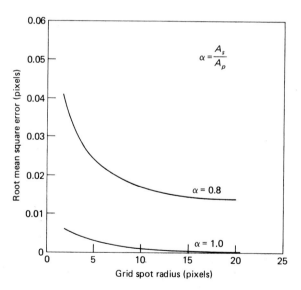

FIGURE 5.20
Error as a function of spot size. (*From Data by Sirkis.*)

EXERCISES

5.1. The stress distribution in a thin, wide, steel plate ($E = 207$ GPa and $v = 0.30$) with a central circular hole is given by Eqs. (3.46) when the plate is subjected to a uniaxial tensile or compressive load. Determine the error made in the determination of the maximum stress σ_{max} on the boundary of the hole if a strain gage having a gage length $l_0 = 3$ mm is used. The diameter of the hole in the plate is 10 mm.

5.2. Determine the error associated with measurements of ε_{yy} along the longitudinal axis of the plate of Exercise 5.1 if gages with $l_0/a = 0.1$ are located at $y/a = 1.5$, 2.0, and 3.0.

5.3. Determine the error associated with measurements of ϵ_{yy} along the transverse axis of the plate of Exercise 5.1 if gages with $w_0/a = 0.1$ are located at $x/a = 1.5, 2.0$, and 3.0.

5.4. For the plate of Exercise 5.1, determine the error associated with measurements of ϵ_{yy} along the transverse axis of the plate due to both gage width and gage length. Take $w_0/a = 0.1$ and $l_0/a = 0.1$ and consider gages located at $x/a = 1.5, 2.0$, and 3.0.

5.5. Consider a strain field given by the expression

$$\epsilon_{xx} = b \sin \frac{\pi x}{a}$$

Determine the error made in the measurement of ϵ_{xx} at $x = 0$, $a/4$, and $a/2$ as a function of l_0/a and plot this error for $0.01 \leq l_0/a \leq 1$.

5.6. Consider a strain field given by the expression

$$\epsilon_{xx} = b + b \sin \frac{\pi x}{a}$$

Determine the error made in the measurement of ϵ_{xx} at $x = a/2$ when $l_0/a = \frac{1}{4}$. Compare with the results from Exercise 5.5.

5.7. Consider a strain field given by the expression

$$\epsilon_{xx} = b + bx + b \sin \frac{\pi x}{a}$$

Determine the error made in the measurement of ϵ_{xx} at $x = a/2$ when $l_0/a = \frac{1}{4}$. Compare with the results from Exercises 5.5 and 5.6. What is the effect of adding constant and linear contributions to the strain field on the overall error?

5.8. In these days of digital data processing where varying calibration constants are easily programmed, why is it still important that a strain gage should respond linearly?

5.9. Justify with a one-paragraph engineering brief each of the ten characteristics listed on page 133 that are used to judge the adequacy of a strain gage system.

5.10. Describe the design features of the extensometer shown in Fig. 5.3 and indicate the importance of each of these features to an individual using the extensometer to determine stress-strain curves in a tension test of round bar specimens.

5.11. A helium-neon laser ($\lambda = 632.8$ nm) is used to illuminate a diffraction-type strain gage with a gage length 20 mm. The diffraction pattern is displayed on a screen which is located 3 m from the aperture. The initial aperture width b is 0.1 mm.
   (a) Determine the density n/y of the diffraction pattern.
   (b) If the gage is subjected to a strain of 1000 $\mu\varepsilon$, what is the new aperture width and the new density n/y_1?

(c) Suppose the diffraction pattern is sufficiently well defined for the $+8$ and -8 orders of extinction to be clearly observed before and after subjecting the gage to the strain. If distances y_0 and y_1 can be determined from scale measurements on the screen to ± 1 mm, estimate the percent error in the measurement of strain.

5.12. Repeat Exercise 5.12 if the initial aperture width $b = 0.05$ mm.

5.13. Derive Eq. (5.9).

5.14. Derive Eq. (5.10).

5.15. A capacitance strain gage (see Fig. 5.8) has a gage length $l_0 = 20$ mm and $h = 0.10$ mm. Construct a graph of capacitance C versus strain ϵ as the strain is varied from 0 to 100 percent.
(a) Is the output from the gage linear over the entire range?
(b) If not, what is the linear range?
(c) Why was the maximum value in part (b) selected as the limit of the linear range?

5.16. Repeat Exercise 5.15 but construct a graph of change in reactance $\Delta \mathscr{R}_c / \mathscr{R}_c$ with strain over the same range. Comment on the effect of measuring reactance $\mathscr{R}_c$ instead of capacitance C.

5.17. Design a strain extensometer, incorporating a commercially available DCDT, which can be utilized with an xy recorder to plot stress-strain curves automatically during tensile tests of standard ASTM specimens of 1020 carbon steel in a universal testing machine. Select the range of the DCDT, specify the input voltage, decide on the gage length of the extensometer, design the linkage (the specimen may fracture with the extensometer in place), determine the output voltage as a function of stress on the specimen, and specify the characteristics of the xy recorder to be used to plot the stress-strain curves automatically.

5.18. Design an acoustical strain gage which can be installed in a concrete dam and monitored over the life (estimated to exceed two centuries) of the dam. The gage will be monitored periodically to record the change in strain with rising and falling head, to record any change in effective modulus of the concrete due to cracking or other deterioration of the concrete, and to estimate damage to the structure after any natural occurrence such as an earthquake. Items to be considered in the design include selection of materials for the components of the gage, gage length, wire size, wire type, and frequency range during operation. Estimate the accuracy of the strain measurement and comment on the ability of the gage to detect structural damage or deterioration of the concrete.

5.19. Show the matrix which represents the 81 terms in the fourth-rank piezoresistivity tensor π_{ijkl} given in Eq. (5.23).

5.20. Show that Eq. (5.23) reduces Eqs. (5.24) when the tensors ρ_{ij} and τ_{ij} are symmetrical and that the piezoresistive coefficients are reduced to the three terms defined in the simplified π matrix.

5.21. Verify Eq. (5.28) by substituting Eqs. (5.25) and (5.26) into Eq. (5.27) and simplifying the resulting relationship.

5.22. One of the advantages of semiconductor strain gages is related to the fact that both positive and negative gage factors can be achieved with P- and N-type silicon. Show the gain in sensitivity, based on gage factor, achieved in the design of a beam-type deflection transducer when four gages (two of the P type and two of the N type) are employed.

5.23. Outline the two methods which can be used to compensate for temperature-induced apparent strains in semiconductor strain gages.

5.24. Describe the precautions which must be taken in installing and cycling semiconductor strain gages due to their inherently brittle characteristics.

5.25. Determine the power which can be dissipated by a semiconductor strain gage 1.5 mm long. If the gage resistance is 500 Ω, determine the current passing through the gage and the voltage drop across the gage when the maximum power is being dissipated.

REFERENCES

1. Weaver, P. R.: An Optical Strain Gage for Use at Elevated Temperatures, *Proc. SESA*, vol. IX, no. 1, pp. 159–162, 1951.
2. Pryor, T. R., and W. P. T. North: The Diffractographic Strain Gage, *Exp. Mech.*, vol. 11, no. 12, pp. 565–578, 1971.
3. Pryor, T. R., O. L. Hageniers, and W. P. T. North: Displacement Measurement along a Line by the Diffractographic Method, *Exp. Mech.*, vol. 12, no. 8, pp. 384–386, 1972.
4. Sharpe, W. N., Jr.: The Interferometric Strain Gage, *Exp. Mech.*, vol. 8, no. 4, pp. 164–170, 1968.
5. Carter, B. C., J. F. Shannon, and J. R. Forshaw: Measurements of Displacement and Strain by Capacity Methods, *Proc. Inst. Mech. Eng. (Lond.)*, vol. 152, pp. 215–221, 1945.
6. Harting, D. R.: "Evaluation of a Capacitive Strain Measuring System for Use to 1500°F," ISA ASI publication 75251, 1975, pp. 289–297.
7. Foster, R. L., and S. P. Wnuk, Jr.: "High Temperature Capacitive Displacement Sensing," Instrument Society of America, paper 85-0123, 0096-7238, 1985, pp. 245–252.
8. Schaevitz, H.: The Linear Variable Differential Transformer, *Proc. SESA*, vol. IV, no. 2, pp. 79–88, 1947.
9. Herceg, E. E.: *Handbook of Measurement and Control*, Schaevitz Engineering, Pennsauken, N.J., 1972.
10. Shepherd, R.: Strain Measurement Using Vibrating-Wire Gages, *Exp. Mech.*, vol. 4, no. 8, pp. 244–248, 1964.
11. Jerrett, R. S.: The Acoustic Strain Gage, *J. Sci. Instrum.*, vol. 22, no. 2, pp. 29–34, 1945.
12. Potocki, F. P.: Vibrating-Wire Strain Gauge for Long-Term Internal Measurements in Concrete, *Engineer*, vol. 206, pp. 964–967, 1958.
13. Mason, W. P., and R. N. Thurston: Piezoresistive Materials in Measuring Displacement, Force, and Torque, *J. Acoust. Soc. Am.*, vol. 29, no. 10, pp. 1096–1101, 1957.
14. Geyling, F. T., and J. J. Forst: Semiconductor Strain Transducers, *Bell Syst. Tech. J.*, vol. 39, pp. 705–707, 1960.
15. Mason, W. P.: Semiconductors in Strain Gages, *Bell Lab. Rec.*, vol. 37, no. 1, pp. 7–9, 1959.
16. Smith, C. S.: Piezoresistive Effect in Germanium and Silicon, *Phys. Rev.*, vol. 94, pp. 42–49, 1954.
17. Padgett, E. D., and W. V. Wright: "Silicon Piezoresistive Devices," in M. Dean and R. D. Douglas (eds.), *Semiconductor and Conventional Strain Gages*, Academic Press, New York, 1962, pp. 1–20.
18. O'Regan, R.: "Development of the Semiconductor Strain Gage and Some of Its Applications," in ibid., pp. 245–257.
19. Mason, W. P., J. J. Forst, and L. M. Tornillo: "Recent Developments in Semiconductor Strain Transducers," in ibid., pp. 109–120.
20. Kurtz, A. D.: "Adjusting Crystal Characteristics to Minimize Temperature Dependence," in ibid., pp. 259–272.
21. Sanchez, J. C., and W. V. Wright: "Recent Developments in Flexible Silicon Strain Gages," in ibid., pp. 307–346.
22. Durelli, A. J., and I. M. Daniel: A Nondestructive Three-dimensional Strain-Analysis Method, *J. Appl. Mech.*, vol. 28, ser. E, no. 1, pp. 83–86, 1961.
23. Durelli, A. J., J. W. Dally, and W. F. Riley: Developments in the Application of the Grid Method to Dynamic Problems, *J. Appl. Mech.*, vol. 26, no. 4, pp. 629–634, 1959.
24. Durelli, A. J., and W. F. Riley: Developments in the Grid Method of Experimental Stress Analysis, *Proc. SESA*, vol. XIV, no. 2, pp. 91–100, 1957.

25. Parks, V. J., and A. J. Durelli: On the Definitions of Strain and Their Use in Large Strain Analysis, *Exp. Mech*, vol. 7, no. 6, pp. 279–280, 1967.
26. Parks, V. J., and A. J. Durelli: Various Forms of Strain Displacement Relations Applied to Experimental Strain Analysis, *Exp. Mech.*, vol. 4, no. 2, pp. 37–47, 1964.
27. Hickson, V. M.: A Replica Technique for Measuring Static Strains, *J. Mech. Eng. Sci.*, vol. 1, no. 2, pp. 171–183, 1959.
28. Fail, W. F., and C. E. Taylor: An Application of Pattern Mapping to Plane Motion, *Exp. Mech.*, in press.
29. Pratt, W. K.: *Digital Image Processing*, John Wiley and Sons, New York, 1978.
30. Sirkis, J. S.: "Improved Grid Methods through Displacement Pattern Matching," *Proc. 1989 SEM Spring Conf. on Exp. Mech.*, pp. 439–444, 1989.

CHAPTER
6

ELECTRICAL-
RESISTANCE
STRAIN
GAGES

6.1 INTRODUCTION [1-3]

In Chap. 5, several different strain-measuring systems were introduced, and their performance characteristics such as range, sensitivity, gage length, and precision of measurement were discussed. None of these different systems, regardless of the principle upon which the gage is based, exhibits all the properties required for an optimum device; however, the electrical-resistance strain gage approaches the requirements for an optimum system. As such, the electrical-resistance strain gage is the most frequently used device in stress-analysis work throughout the world today. Electrical-resistance strain gages are also widely used as sensors in transducers designed to measure such quantities as load, torque, pressure, and acceleration.

The discovery of the principle upon which the electrical-resistance strain gage is based was made in 1856 by Lord Kelvin, who loaded copper and iron wires in tension and noted that their resistance increased with the strain applied to the wire. Furthermore, he observed that the iron wire showed a greater increase in resistance than the copper wire when they were both subjected to the same strain. Finally, Lord Kelvin employed a Wheatstone bridge to measure the resistance change. In this classic experiment he established three vital facts which have greatly aided the development of the electrical-resistance strain gage: (1) the resistance of the wire changes as a function of strain; (2) different materials have different sensitivities; and (3) the Wheatstone bridge can be used to measure these resistance changes accurately. It is indeed remarkable that 80 years passed before strain gages based on Lord Kelvin's experiments became commercially available.

Today, after more than 50 years of commercial development and extensive utilization by industrial and academic laboratories throughout the world, the bonded-foil gage monitored with a Wheatstone bridge has become a highly perfected measuring system. Precise results for surface strains can be obtained quickly using relatively simple methods and inexpensive gages and instrumentation systems. In spite of the relative ease in employing strain gages, there are many features of the gages which must be thoroughly understood to obtain optimum performance from the measuring system in applied stress analysis. In this chapter, the electrical-resistance strain gage will be examined in detail to illustrate each feature affecting its performance. Strain-gage circuits and recording instruments used in measuring the strain-related resistance changes will be treated in Chaps. 7 and 8.

6.2   STRAIN SENSITIVITY IN METALLIC ALLOYS [4–5]

Lord Kelvin noted that the resistance of a wire increases with increasing strain and decreases with decreasing strain. The question then arises whether this change in resistance is due to the dimensional change in the wire under strain or to the change in resistivity of the wire with strain. It is possible to answer this question by performing a very simple analysis and comparing the results with experimental data which have been compiled on the characteristics of certain metallic alloys.

The resistance R of a uniform conductor with a length L, cross-sectional area A, and specific resistance ρ is given by

$$R = \rho \frac{L}{A} \tag{6.1}$$

Differentiating Eq. (6.1) and dividing by the total resistance R leads to

$$\frac{dR}{R} = \frac{d\rho}{\rho} + \frac{dL}{L} - \frac{dA}{A} \tag{a}$$

The term dA represents the change in cross-sectional area of the conductor due to the transverse strain, which is equal to $-v\,dL/L$. If the diameter of the conductor before the application of the axial strain is noted as d_0, then the diameter after the strain is applied is given by

$$d_f = d_0\left(1 - v\frac{dL}{L}\right) \tag{b}$$

and from Eq. (*b*) it is clear that

$$\frac{dA}{A} = -2v\frac{dL}{L} + v^2\left(\frac{dL}{L}\right)^2 \approx -2v\frac{dL}{L} \tag{c}$$

Substituting Eq. (*c*) into Eq. (*a*) gives

$$\frac{dR}{R} = \frac{d\rho}{\rho} + \frac{dL}{L}(1 + 2v) \tag{6.2}$$

which can be rewritten as

$$S_A = \frac{dR/R}{\epsilon} = 1 + 2v + \frac{d\rho/\rho}{\epsilon} \tag{5.20}$$

where S_A is the sensitivity of the metallic alloy used in the conductor and is defined as the resistance change per unit of initial resistance divided by the applied strain.

Examination of Eq. (5.20) shows that the strain sensitivity of any alloy is due to two factors, namely, the change in the dimensions of the conductor, as expressed by the $1 + 2v$ term, and the change in specific resistance, as represented by the $(d\rho/\rho)/\epsilon$ term. Experimental results show that S_A varies from about 2 to 4 for most metallic alloys. For pure metals, the range is from -12.1 (nickel) to $+6.1$ (platinum). This fact implies that the change in specific resistance can be quite large for certain metals since $1 + 2v$ usually ranges between 1.4 and 1.7. The change in specific resistance is due to the number of free electrons and the variation of their mobility with applied strain.

A list of some metallic alloys commonly employed in commercial strain gages, together with their sensitivities, is presented in Table 6.1. It should be noted that the sensitivity depends upon the particular alloy being considered. Moreover, the values assigned to S_A in Table 6.1 are not necessarily constants. The value of the strain sensitivity S_A will depend upon the degree of cold working imparted to the conductor in its formation, the impurities in the alloy, and the range of strain over which the measurement of S_A is made.

Most electrical-resistance strain gages produced today are fabricated from the copper-nickel alloy known as Advance or Constantan. A typical curve showing

TABLE 6.1
Strain sensitivity S_A for common strain-gage alloys

Material	Composition, %	S_A
Advance or Constantan	45 Ni, 55 Cu	2.1
Nichrome V	80 Ni, 20 Cr	2.2
Isoelastic	36 Ni, 8 Cr, 0.5 Mo, 55.5 Fe	3.6
Karma	74 Ni, 20 Cr, 3 Al, 3 Fe	2.0
Armour D	70 Fe, 20 Cr, 10 Al	2.0
Alloy 479	92 Pt, 8 W	4.1

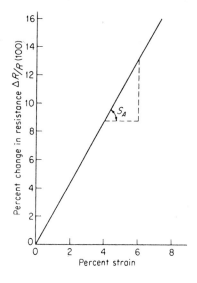

FIGURE 6.1
Percent change in resistance as a function of percent strain for Advance alloy.

the percent change in resistance $\Delta R/R$ as a function of percent strain for this alloy is given in Fig. 6.1. This alloy is useful in strain-gage applications for the following reasons:

1. The value of the strain sensitivity S_A is linear over a wide range of strain, and the hysteresis of bonded filaments is extremely small.
2. The value of S_A does not change significantly as the material goes plastic.
3. The alloy has a high specific resistance ($\rho = 0.49 \ \mu\Omega \cdot m$).
4. The alloy has excellent thermal stability and is not influenced appreciably by temperature changes when mounted on common structural materials.
5. The small temperature-induced changes in resistance of the alloy can be controlled with trace impurities or by heat treatment.

The first advantage of the Advance-type alloy over other alloys implies that the gage calibration constant will not vary with strain level; therefore, a single calibration constant is adequate for all levels of strain. The wide range of linearity with strain (even into the alloy's plastic region) indicates that it can be employed for measurements of both elastic and plastic strains in most structural materials. The high specific resistance of the alloy is useful when constructing a small gage with a relatively high resistance. Finally, the temperature characteristics of selected melts of the alloy permit the fabrication of temperature-compensating strain gages for each structural material. With temperature-compensating strain gages, the temperature-induced $\Delta R/R$ on a given material can be maintained at less than $10^{-6}/°C$.

The Isoelastic alloy is also employed in commercial gages because of its high sensitivity (3.6 for Isoelastic compared with 2.1 for Advance) and its high fatigue strength. The increased sensitivity is advantageous in dynamic applications where the strain-gage output must be amplified to a considerable degree before recording. The high fatigue strength is useful when the gage is to operate in a cyclic strain field where the alternating strains exceed 1500 $\mu\epsilon$. In spite of these two advantages, use of the Isoelastic alloy is limited because it is extremely sensitive to temperature changes. When mounted in gage form on a steel specimen, a change in temperature of 1.8°F (1°C) will give an apparent strain indication of 300 to 400 $\mu\epsilon$, as shown in Fig. 6.2. Isoelastic gages can be used in dynamic applications only when the temperature is stable over the time required for the dynamic measurement.

The Karma alloy has properties that are similar to Advance alloy. Indeed, its fatigue limit is higher than Advance but lower than Isoelastic. In addition, Karma exhibits excellent stability with time and is always used when strain measurements are made over extended periods (weeks or months). Another advantage is that the temperature compensation which can be achieved with Karma is better over a wider range of temperature than the compensation which can be achieved with the Advance alloy. The thermally induced apparent strain with temperature for Advance, Isoelastic, and Karma, shown in Fig. 6.2, clearly indicates more accurate compensation at the temperature extremes for the Karma alloy. Finally, Karma can be used to 500°F (260°C) in static strain measurements whereas Advance is limited to 400°F (204°C). The primary disadvantage of Karma is the difficulty in soldering lead wires to the tabs.

The other alloys, Nichrome V, Armour D, and the platinum-tungsten alloy, are metallurgically more stable and oxidation-resistant at higher temperatures. These alloys are used for special-purpose gages which permit measurements of strain to be made at temperatures in excess of 500°F (260°C).

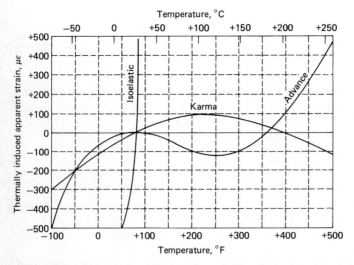

FIGURE 6.2
Thermally induced apparent strain as a function of temperature for three common strain-gage alloys.

6.3 GAGE CONSTRUCTION [6–10]

It is theoretically possible to measure strain with a single length of wire as the sensing element of the strain gage. However, circuit requirements that are usually needed to prevent overloading of the power supply and to minimize heat generated by the gage current place a lower limit of approximately 100 Ω on the gage resistance. As a result, a 100-Ω strain gage fabricated from the finest standard wire is about 4 in (100 mm) long.

The very earliest electrical-resistance strain gages were of the unbonded type, where the conductors were straight wires strung between a movable frame and a fixed frame. This gage was large and required knife edges for mounting, which greatly limited its applicability.

The problem of conductor length and gage mounting was solved in the mid-1930s, when Ruge and Simmons independently developed bonded-wire strain gages. The conductor-length problem was solved by forming the required length of wire into a grid pattern. The attachment problem was solved by bonding the wire grid directly to the specimen with suitable adhesives. Wire gages were produced with both flat-grid and bobbin-type constructions, as illustrated in Fig. 6.3. Bonded-wire strain gages were employed for strain measurements almost exclusively from the mid-1930s to the mid-1950s. They are still used occasionally today when long gage lengths are necessary; but in most instances, they have been replaced by the bonded-foil strain gage.

The first metal-foil strain gages were produced in England in 1952 by Saunders and Roe. With this type of gage, the grid configuration is formed from metal foil by a photoetching process. Since the process is quite versatile, a wide variety of gage sizes and grid shapes can be produced. Typical examples of the variety of gages marketed commercially are shown in Fig. 6.4. The shortest gage length available in a metal-foil gage is 0.008 in (0.20 mm). The longest gage length

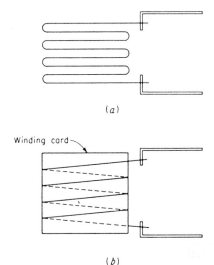

(a)

Winding card

(b)

FIGURE 6.3
(a) Flat-grid and (b) bobbin-type constructions for bonded-wire-type resistance strain gages.

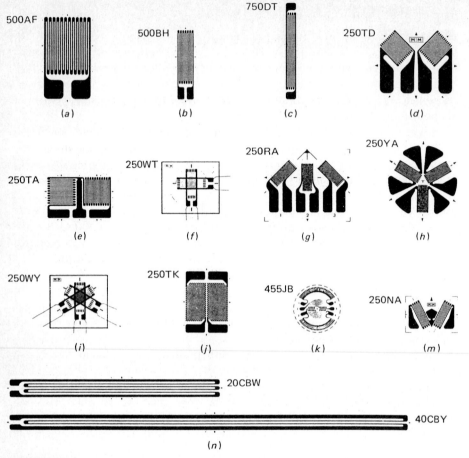

FIGURE 6.4

Configurations of metal-foil strain gages: (a) single-element gage, (b) single-element gage, (c) single-element gage, (d) two-element rosette, (e) two-element rosette, (f) two-element stacked rosette, (g) three-element rosette, (h) three-element rosette, (i) three-element stacked rosette, (j) shear gage, (k) diaphragm gage, (m) stress gage, (n) single-element gage for use on concrete. (*Measurements Group, Inc.*)

is 4.00 in (102 mm). Standard gage resistances are 120 and 350 Ω; however, high gage resistances (500, 1000, and 3000 Ω) are commercially available in select sizes for transducer applications.

Multiple-element gages, shown in Fig. 6.5, are available with 10 gages arranged along a line. These strip gages are usually installed in fillets where high strain gradients occur and it is difficult to locate the point where the strain is a maximum. Two- and three-element rosettes are available in either the in-plane or stacked configuration in a wide range of sizes for use in biaxial stress fields. Two-element rosettes are used when the directions of the principal stresses (or strains) are known. Three-element rosettes are used when the principal directions are not known. Special gage configurations are also available for use in trans-

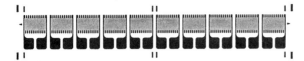

FIGURE 6.5
Linear array of 10 strain gages on a single carrier. (*Measurements Group, Inc.*)

ducers. A typical example is shown in Fig. 6.4*k* for the diaphragm-type pressure transducer.

The etched metal-film grids are very fragile and easy to distort, wrinkle, or tear. For this reason, the metal film is usually bonded to a thin plastic sheet, which serves as a backing or carrier before photoetching. The carrier material also provides electrical insulation between the gage and the component after the gage is mounted. The use of a backing sheet to serve as a carrier for the grid is illustrated in Fig. 6.6, which shows a gage being handled. Markings for the centerline of the gage length and width are also displayed on the carrier, as indicated in Fig. 6.6.

Very thin paper was the first carrier material employed in the production of wire-type gages. However, paper has been replaced with a thin (0.001-in or 0.025-mm) sheet of polyimide, which is a tough and flexible plastic. For transducer applications, where precision and linearity are very important, a very thin high-modulus epoxy is used for the carrier. The epoxy backing is not suitable for general-purpose strain gages since it is brittle and can easily be broken during gage installation. Glass-fiber-reinforced epoxies and/or phenolics are employed as carriers when the strain gage will be exposed to high-level cyclic strains and fatigue

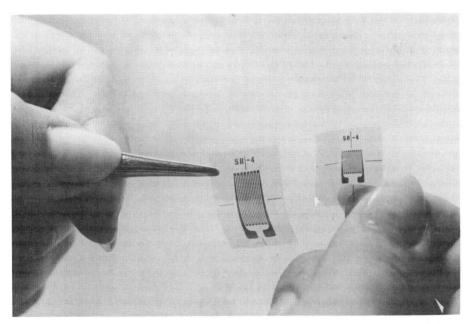

FIGURE 6.6
Backing sheets are necessary for handling fragile foil strain gages. (*BLH Electronics.*)

life of the gage system is important. In this application, the carrier is used to encapsulate the grid, as shown in Fig. 6.7. Glass-reinforced epoxy carriers are also used for moderate temperature applications up to 750°F (400°C). For very high temperature applications, a strippable carrier is used. This carrier is removed during application of the gage, and a ceramic adhesive serves to maintain the grid configuration and to insulate the gage.

Another type of gage, originally developed for high-temperature strain measurement, is the weldable strain gage shown in Fig. 6.8. It consists of a very fine loop of wire which is swaged into a metal case with compacted MgO powder as an insulator. The wire ends are connected to integral lead wires in a metal case. The wire inside the strain tube is reduced in diameter by an etching process; hence, the minimum gage resistance is obtained with a very-small-diameter, high-resistance wire rather than by forming a grid of larger-diameter wire.

Currently, weldable gages are available with resistances which range from 60 to 350 Ω and with lengths which range from 0.375 in (9.5 mm) to 1.09 in (27.7 mm). The gages are suitable for use from cryogenic to elevated temperatures or within the range from −320 to +1200°F (−200 to +650°C). The gages can be welded onto a number of different metals with a capacitor discharge welder and are ready for use immediately. This simplicity in application is a significant advantage over other high-temperature strain-gage systems, which require rather elaborate installation techniques. Indeed, the simplicity of the welding installation is attractive whenever gages must be mounted in the field under adverse conditions regardless of temperature considerations. The weldable gage is extremely rugged and waterproof. A modification of the gage, shown in Fig. 6.8c, can be embedded in concrete to record strains at interior locations in the structure.

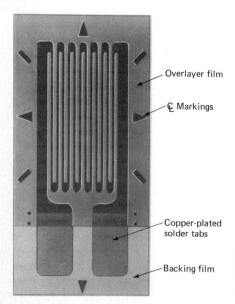

Overlayer film

₵ Markings

Copper-plated
solder tabs

FIGURE 6.7

Backing film

Construction details for a foil strain gage with an encapsulating overlayer and copper-plated solder tabs. (*Measurements Group, Inc.*)

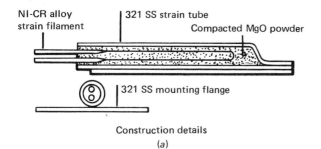

Construction details
(a)

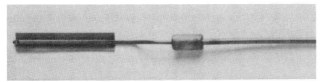

External appearance of a gage
(b)

Gage for embedment in concrete
(c)

FIGURE 6.8
Weldable strain gages: (a) construction details; (b) external appearance of a gage; (c) gage for embedment in concrete. (*Eaton Corp.*)

Of the various gages described, the metal-foil strain gage is the most frequently employed for both general-purpose stress analysis and transducer applications. There are occasional special-purpose applications for which un-bonded-wire, bonded-wire, weldable, or semiconductor gages are more suitable, but these applications are not common.

6.4 STRAIN-GAGE ADHESIVES AND MOUNTING METHODS [11–14]

The bonded type of resistance strain gage of either wire or foil construction is a high-quality precision resistor which must be attached to the specimen with a suitable adhesive. For precise strain measurements, both the correct adhesive and proper mounting procedures must be employed.

The adhesive serves a vital function in the strain-measuring system; it must transmit the strain from the specimen to the gage-sensing element without distortion. It may appear that this role can be easily accomplished if the adhesive is suitably strong; however, the characteristics of the polymeric adhesives used to bond strain gages are such that, as will be shown later, the adhesive can influence gage resistance, apparent gage factor, hysteresis characteristics, resistance to stress relaxation, temperature-induced zero drift, and insulation resistance.

The singularly unimpressive feat of bonding a strain gage to a specimen is perhaps one of the most critical steps in the entire process of measuring strain with a bonded resistance strain gage. The improper use of an adhesive costing a few dollars per test can seriously degrade the validity of an experimental stress analysis which may cost several thousands of dollars.

When mounting a strain gage, it is important to carefully prepare the surface of the component where the gage is to be located. This preparation consists of sanding away any paint or rust to obtain a smooth but not highly polished surface. Next, solvents are employed to remove all traces of oil or grease and the surface is etched with an appropriate acid. Finally, the clean, sanded, degreased, and etched surface is neutralized (treated with a basic solution) to give it the proper chemical affinity for the adhesive.

The gage location is then marked on the specimen and the gage is positioned by using a rigid transparent tape in the manner illustrated in Fig. 6.9. The position

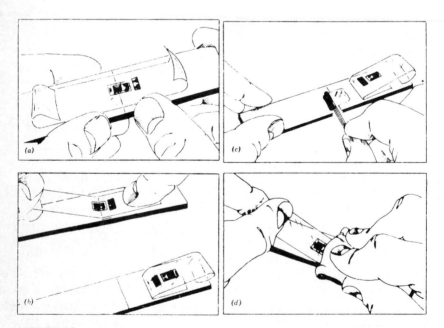

FIGURE 6.9
The tape method of bonding foil strain gages with flexible carriers: (*a*) position gage and overlay with tape; (*b*) peel tape back to expose gage bonding area; (*c*) apply a thin layer of adhesive over the bonding area; (*d*) replace tape in overlay position with a wiping action to clear excess adhesive. (*Measurements Group, Inc.*)

and orientation of the gage are maintained by the tape as the adhesive is applied and as the gage is pressed into place by squeezing out the excess adhesive.

After the gage is installed, the adhesive must be exposed to a proper combination of pressure and temperature for a suitable length of time to ensure a complete cure. This curing process is quite critical since the adhesive will expand because of temperature, experience a volume reduction due to polymerization, exhibit a contraction upon cooling, and on occasion experience postcure shrinkage. Since the adhesive is sufficiently rigid to control deformation of the strain-sensitive element in the gage, residual stresses set up in the adhesive will influence the output of the strain gage. Of particular importance is the postcure shrinkage, which may influence the gage output long after the adhesive is supposedly completely cured. If a long-term strain measurement is attempted with an incompletely cured adhesive, the stability of the gage will be seriously impaired and the accuracy of the measurements compromised.

A wide variety of adhesives are available for bonding strain gages. Factors influencing the selection of a specific adhesive include the carrier material, the operating temperature, the curing temperature, and the maximum strain to be measured. A discussion of the characteristics of several different adhesive systems in common use follows.

6.4.1 Epoxy Cements

Epoxies are a class of thermosetting plastics which, in general, exhibit a higher bond strength and a higher level of strain at failure than other types of adhesives used to mount strain gages. Epoxy systems are usually composed of two constituents, a monomer and a hardening agent. The monomer, or base epoxy, is a light amber fluid which is usually quite viscous. A hardening agent mixed with the monomer will induce polymerization. Amine-type curing agents produce an exothermic reaction which releases sufficient heat to accomplish curing at room temperature or at relatively low curing temperatures. Anhydride-type curing agents require the application of heat to promote polymerization. Temperatures in excess of 250°F (120°C) must be applied for several hours to complete polymerization. In some cases solvents and/or dilutents are added to reduce viscosity. So many epoxies and curing agents are commercially available today that it is impossible to be specific in the coverage of their properties or their behavior. However, the following remarks will be valid and useful for any system of epoxies used in strain-gage applications.

With both types of curing agent, particularly the amine type, the amount of hardener added to the monomer is extremely important. The adhesive curing temperatures and the residual stresses produced during polymerization can be significantly influenced by small deviations from the specified values. For this reason, the quantities of both the monomer and the curing agent should be carefully weighed before they are mixed together.

In general, pure epoxies do not liberate volatiles during cure; therefore, postcure heat cycling is not necessary to evaporate chemical by-products released during polymerization. Solvents which are volatile should not be added to the epoxies to improve their viscosity for general-purpose applications. A filler material with micrometer-sized particles of pure silica can be added in moderate quantities (5 to 10 percent by weight) to improve the bond strength and to reduce the temperature coefficient of expansion of the epoxy.

A modest clamping pressure of 5 to 20 psi (35 to 140 kPa) is recommended for the epoxies during the cure period to ensure as thin a bond line (adhesive layer) as possible. In transducer applications, dilutent-thinned epoxies are frequently specified to reduce viscosity so that extremely thin (less than 200 μin, or 0.005 mm) void-free bond lines can be obtained. Thin bond lines tend to minimize creep, hysteresis, and linearity problems. For these transducer applications, clamping pressures of approximately 50 psi (350 kPa) are recommended.

Many different epoxy systems are available from strain-gage manufacturers in kit form with the components preweighed. These systems are recommended since the epoxies are especially formulated for strain-gage applications. They are easy and convenient to use, they have an adequate pot life, and the time-temperature curve for the curing cycle is specified. The use of hardware-store variety two-tube epoxy systems is discouraged since these systems usually incorporate modifiers or plasticizers to improve the toughness of the adhesive. The modifiers cause large amounts of creep and hysteresis that is undesirable in strain-gage applications.

The best indication of a properly cured adhesive system can be obtained by measuring the resistance between the gage grid and the specimen (resistance through the adhesive layer). A properly cured installation will exhibit a resistance to ground exceeding 10,000 MΩ. Minute traces of either solvent or water in the adhesive will lower the resistance of the adhesive layer and influence gage performance.

26.4.2   Cyanoacrylate Cement

A modified form of a pressure curing adhesive consisting of a methyl-2-cyanoacrylate compound is commonly employed as a strain-gage adhesive. This adhesive system is simple to use, and the strain gage can be employed approximately 10 min after bonding. Chemically, this adhesive is quite unusual in that it requires neither heat nor a catalyst to induce polymerization. Apparently, when this adhesive is spread in a thin film between two components to be bonded, the minute traces of water or other weak bases on the surfaces of the components are sufficient to trigger the polymerization process. A catalyst can be applied to the bonding surfaces to decrease the reaction times, but it is not essential.

In strain-gage applications, a thin film of the adhesive is placed between the gage and the specimen and a gentle pressure is applied for about 1 or 2 min to induce polymerization. Once initiated, the polymerization will continue at room temperature without maintaining the pressure.

The fast room-temperature cure of the cyanoacrylate adhesive makes it ideal for general-purpose strain-gage applications. The performance of this adhesive system, however, will deteriorate markedly with time, moisture absorption, or elevated temperature. It should not be used where extended life of the gage system is important. Coatings such as microcrystalline wax, silicone rubber, polyurethane, etc., can be used to protect the adhesive from moisture in the air and extend the life of an installation.

6.4.3   Polyester Adhesives

Polyesters, like epoxies, are two-component adhesives. The polyesters exhibit a high shear strength and modulus; however, their peel strength is low and they are less resistant to solvents than epoxies. Their primary advantage is the ability to polymerize at a relatively low temperature [40°F (5°C)], which permits gage installations at temperatures only modestly above freezing.

6.4.4   Ceramic Cements

Two different approaches are used in bonding strain gages with ceramic adhesives. The first utilizes a blend of finely ground ceramic powders such as alumina and silica combined with a phosphoric acid. Usually this blend of powders is mixed with a solvent such as isopropyl alcohol and an organic binder to form a liquid mixture which facilitates handling. A precoat of the ceramic cement is applied and fired to form a thin layer of insulation between the gage grid and the component. A second layer of ceramic cement is then applied to bond the gage. In this application, the carrier for the gage is removed, and the grid is totally encased in ceramic.

Since many of the ceramic cements marketed commercially have proprietary compositions, it is often difficult to select the most suitable product. One cement developed by the National Bureau of Standards [recently renamed the *National Institute for Standards and Technology (NIST)*] and described in Table 6.2 is recommended for high-temperature use since it exhibits a very high resistivity at temperatures up to 1800°F (980°C). Gage resistance to ground with this cement will normally exceed 6 MΩ. The ceramic cements are used primarily for high-temperature application or in radiation environments, where organic adhesives cannot be employed.

TABLE 6.2
Composition of NBS-x-142 ceramic cement

Constituent	Parts by weight
Alumina, Al_2O_3	100
Silica, SiO_2	100
Chrome anhydride, CrO_3	2.5
Colloidal silica solution	200
Orthophosphoric acid, H_3PO_4	30

A second method for bonding strain gages with a ceramic material utilizes a flame-spraying Rokide process. A special gun (Fig. 6.10a) is used to apply the ceramic particles to the gage. For this type of application, gages are constructed with a grid fabricated from wire which is mounted on slotted carriers. The carrier, which holds the gage and lead wires in position during attachment, is made from a glass-reinforced Teflon tape. The tape is resistant to the molten flame-sprayed ceramic particles; however, after the gage grid is secured to the component, the tape is removed and the grid is completely encased with flame-sprayed ceramic particles.

The flame-spray gun utilizes an oxyacetylene gas mixture in a combustion chamber to produce very high temperatures. Ceramic material in rod form is fed into the combustion chamber; there the rod decomposes into softened semimelted particles which are forced from the chamber by the burning oxyacetylene gas. The particles impinge on the surface of the component and form a continuous coating. Since the particles (even in their softened state) are somewhat abrasive, foil grids are not normally used because they are subject to damage by erosion due to particle impact.

Most gage installations are made with pure alumina, or with 98 percent alumina with about 2 percent silica. For applications involving temperatures less

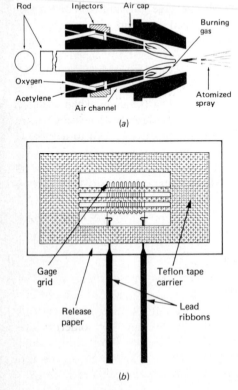

FIGURE 6.10
Flame-spraying ceramic adhesives: (a) particle formation in the spray gun; (b) free-element strain gage with carrier. (*BLH Electronics.*)

than 800°F (425°C), the 98 alumina-2 silica rod is employed since it is easier to melt and apply. The pure alumina rod is intended for higher-temperature applications.

6.4.5 Testing a Strain-Gage Adhesive System

After a strain gage has been bonded to the surface of a specimen, it must be inspected to determine the adequacy of the bond. The inspection procedure attempts to establish whether any voids exist between the gage and the specimen and whether the adhesive is totally cured. Voids can result from bubbles originally in the cement or from the release of volatiles during the curing process. They can be detected by tapping or pressing the gage installation with a soft rubber eraser and observing the effect on a strain indicator. If a strain reading is noted, the gage bond is not satisfactory and voids exist between the sensing element and the specimen.

The stage of the cure of the adhesive is much more difficult to establish. Two procedures are frequently employed which give some indication of the relative completeness of the cure. The first utilizes measurement of the resistance between the gage grid and the specimen as an indication of the stage of the cure. The resistance of the adhesive layer increases as the adhesive cures. Typical gage installations should exhibit a resistance across the adhesive layer of the order of 10,000 MΩ. Resistance values of 100 to 1000 MΩ normally indicate a need for further curing of the adhesive.

The second procedure for determining the completeness of the cure involves subjecting the gage installation to a strain cycle while measuring the resistance change. The change in the zero-load strain reading (zero shift) after a strain cycle or the area enclosed by a strain-resistance change curve is a measure of the degree of cure. Strain-gage installations with completely cured adhesives when cycled to 1000 $\mu\epsilon$ will exhibit zero shifts of less than 2 $\mu\epsilon$ of apparent strain. Should larger values of zero shift be observed, the adhesive should be subjected to a postcure temperature cycle.

In the event that the gages are mounted on a component or a structure which cannot be strain-cycled before testing, a temperature cycle can be substituted for the strain cycle. If the adhesive in a gage installation has been thoroughly cured, no change in the zero strain will be observed when the gage is returned to its original temperature. If the adhesive cure is incomplete, the temperature cycle will result in additional polymerization with associated shrinkage of the adhesive, which causes a shift in the zero reading of the gage.

After the gages have been bonded to the structure or component, it is necessary to attach lead wires so that the change in resistance can be monitored on a suitable instrumentation system. Since the metal-foil strain gages are relatively fragile, care must be exercised in attaching the lead wires to the soldering tabs. Rugged anchor terminals which are bonded to the component are usually employed, as illustrated in Fig. 6.11. A small-diameter wire (32 to 36 gage) is used to connect the gage tab to the anchor terminal. Three lead wires are soldered to

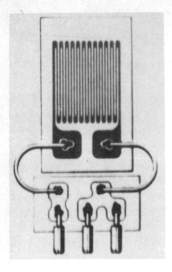

FIGURE 6.11
Use of anchor terminals to connect lead wires to gages.

the anchor terminals. The use of three lead wires to ensure temperature compensation in the Wheatstone-bridge measuring circuit will be discussed in Sec. 7.6. The size of lead wire employed will depend upon the distance between the gage and the instrumentation system. For normal laboratory installations, where lengths rarely exceed 10 to 14 ft (3 to 5 m), wire sizes between 26 and 30 gage are frequently used. Stranded wire is usually preferred to solid wire since it is more flexible and suffers less breakage due to improper stripping or lead-wire movement during the test.

6.5 GAGE SENSITIVITIES AND GAGE FACTOR [15–18]

The strain sensitivity of a single, uniform length of a conductor was previously defined in Chap. 5 as

$$S_A = \frac{dR/R}{\epsilon} \approx \frac{\Delta R/R}{\epsilon} \tag{5.20}$$

where ϵ is a uniform strain along the conductor and in the direction of the axis of the conductor. This sensitivity S_A is a function of the alloy employed to fabricate the conductor and its metallurgical condition. Whenever the conductor is formed into a grid to yield the short gage length often required for measuring strain, the gage exhibits a sensitivity to both axial and transverse strain.

With the older flat-grid wire gages, the transverse sensitivity was due primarily to the end loops in the grid pattern, which placed part of the conductor in the transverse direction. In the foil gages, the end loops are enlarged and desensitized to a large degree. The axial segments of the grid pattern, however, have a large width-to-thickness ratio; thus, some amount of transverse strain will be transmitted through the adhesive and carrier material to the axial segments of the grid pattern to produce a response in addition to the axial-strain response. The magnitude of the transverse strain transmitted to the grid segments depends upon the thickness and elastic modulus of the adhesive, the carrier material, the grid material, and the width-to-thickness ratio of the axial segments of the grid.

The response of a bonded strain gage to a biaxial strain field can be expressed as

$$\frac{\Delta R}{R} = S_a \epsilon_a + S_t \epsilon_t + S_s \gamma_{at} \tag{6.3}$$

where ϵ_a = normal strain along axial direction of gage
$\quad \epsilon_t$ = normal strain along transverse direction of gage
$\quad \gamma_{at}$ = shearing strain
$\quad S_a$ = sensitivity of gage to axial strain
$\quad S_t$ = sensitivity of gage to transverse strain
$\quad S_s$ = sensitivity of gage to shearing strain

In general, the gage sensitivity S_s is small and can be neglected. The magnitude of S_s has yet to be measured. The response of the gage can then be expressed as

$$\frac{\Delta R}{R} = S_a(\epsilon_a + K_t \epsilon_t) \tag{6.4}$$

where $K_t = S_t/S_a$ is defined as the transverse-sensitivity factor for the gage.

Strain-gage manufacturers provide a calibration constant known as the *gage factor* S_g for each gage. The gage factor S_g relates the resistance change to the axial strain as

$$\frac{\Delta R}{R} = S_g \epsilon_a \tag{6.5}$$

The gage factor for each gage type produced from a roll of foil is determined by mounting sample gages from the roll onto a specially designed calibration beam. The beam is then deflected to produce a known strain ϵ_a. The resistance change ΔR is then measured and the gage factor S_g is determined by using Eq. (6.5).

With the beam-in-bending method of calibration, the strain field experienced by the gage is biaxial, with

$$\epsilon_t = -v_0 \epsilon_a \tag{a}$$

where $v_0 = 0.285$ is Poisson's ratio of the beam material. If Eq. (*a*) is substituted into Eq. (6.4), the resistance change in the calibration process is

$$\frac{\Delta R}{R} = S_a \epsilon_a(1 - v_0 K_t) \tag{6.6}$$

Since the resistance changes given by Eqs. (6.5) and (6.6) are identical, the gage factor S_g is related to both S_a and K_t by the expression

$$S_g = S_a(1 - v_0 K_t) \tag{6.7}$$

Typical values of S_g, S_a, and K_t for several different gage configurations are shown in Table 6.3.

It is important to recognize that error will occur in a strain-gage measurement when Eq. (6.5) is employed except for the two special cases where either the stress field is uniaxial or where the transverse-sensitivity factor K_t for the gage is zero. The magnitude of the error can be determined by considering the response of a gage in a general biaxial field with strains ϵ_a and ϵ_t. Substituting Eq. (6.7) into Eq. (6.4) gives

$$\frac{\Delta R}{R} = \frac{S_g \epsilon_a}{1 - v_0 K_t}\left(1 + K_t \frac{\epsilon_t}{\epsilon_a}\right) \tag{b}$$

From Eq. (b), the true value of the strain ϵ_a can be expressed as

$$\epsilon_a = \frac{\Delta R/R}{S_g} \frac{1 - v_0 K_t}{1 + K_t(\epsilon_t/\epsilon_a)} \tag{c}$$

The apparent strain ϵ'_a, which is obtained if only the gage factor is considered, can be determined from Eq. (6.5) as

$$\epsilon'_a = \frac{\Delta R/R}{S_g} \tag{d}$$

TABLE 6.3
Gage factor S_g, axial sensitivity S_a, transverse sensitivity S_t, and transverse-sensitivity factor K_t for several different foil-type strain gages†‡

Gage type	S_g	S_a	S_t	K_t (%)
EA-06-015CK-120	2.13	2.14	0.0385	1.8
EA-06-030TU-120	2.02	2.03	0.0244	1.2
WK-06-030TU-350	1.98	1.98	0.0040	0.2
EA-06-062DY-120	2.03	2.04	0.0286	1.4
WK-06-062DY-350	1.96	1.96	−0.0098	−0.5
EA-06-125RA-120	2.06	2.07	0.0228	1.1
WK-06-125RA-350	1.99	1.98	−0.0297	−1.5
EA-06-250BG-120	2.11	2.11	0.0084	0.4
WA-06-250BG-120	2.10	2.10	−0.0063	−0.3
WK-06-250BG-350	2.05	2.03	−0.0690	−3.4
WK-06-250BF-1000	2.07	2.06	−0.0453	−2.2
EA-06-500AF-120	2.09	2.09	0.0	0
WK-06-500AF-350	2.04	1.99	−0.1831	−9.2
WK-06-500BH-350	2.05	2.01	−0.1347	−6.7
WK-06-500BL-1000	2.06	2.03	−0.0893	−4.4

† Data from Measurements Group, Inc.
‡ Values depend on lot of foil.

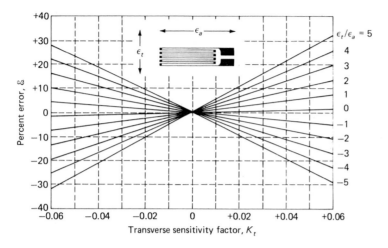

FIGURE 6.12
Error as a function of transverse-sensitivity factor with the biaxial strain ratio as a parameter.

Comparison of Eqs. (c) and (d) shows that

$$\epsilon_a = \epsilon_a' \frac{1 - v_0 K_t}{1 + K_t(\epsilon_t/\epsilon_a)} \tag{6.8}$$

The percent error $\mathscr{E}$ involved in neglecting the transverse sensitivity of the gage is given by

$$\mathscr{E} = \frac{\epsilon_a' - \epsilon_a}{\epsilon_a}(100) \tag{6.9}$$

Substituting Eq. (6.8) into Eq. (6.9) yields

$$\mathscr{E} = \frac{K_t(\epsilon_t/\epsilon_a + v_0)}{1 - v_0 K_t}(100) \tag{6.10}$$

The results of Eq. (6.10), shown graphically in Fig. 6.12, indicate that the error is a function of K_t and the strain biaxiality ratio ϵ_t/ϵ_a. Since the errors can be significant when both K_t and ϵ_t/ϵ_a are large, it is important that corrections be made to account for the transverse sensitivity of the gage. Methods to correct for these errors are presented in Sec. 9.3.

6.6 PERFORMANCE CHARACTERISTICS OF FOIL STRAIN GAGES

Foil strain gages are small precision resistors mounted on a carrier that is bonded to a component part in a typical application. The gage resistance is accurate to ± 0.3 percent, and the gage factor, based on a lot calibration, is certified to ± 0.5 percent. These specifications indicate that foil-type gages provide a means for making precise measurements of strain. The results actually obtained, however, are a function of the installation procedures, the state of strain being measured, and environmental conditions during the test. All these factors affect the performance of a strain-gage system.

6.6.1 Strain-Gage Linearity, Hysteresis, and Zero Shift [19–20]

One measure of the performance of a strain-gage system ("system" here implies gage, adhesive, lead wires, switches, and instrumentation) involves considerations of linearity, hysteresis, and zero shift. If gage output, in terms of measured strain, is plotted as a function of applied strain as the load on the component is cycled, results similar to those shown in Fig. 6.13 will be obtained. A slight deviation from linearity is typically observed, and the unloading curve falls below the loading curve to form a hysteresis loop. Also, when the applied strain is reduced to zero, the gage output indicates a small negative strain, termed *zero shift*. The magnitudes of the deviation from linearity, hysteresis, and zero shift depend on the strain level, the adequacy of the bond, the degree of cold work of the foil material, and the viscoelastic characteristics of the carrier material.

For properly installed gages, deviations from linearity should be approximately 0.1 percent of the maximum strain for polyimide carriers and 0.05 percent for epoxy carriers. First-cycle hysteresis and zero shift depend strongly on the strain range, as shown in Fig. 6.14. Metallurgical changes in the gage alloy produce small but permanent changes in the resistance, which accumulate with the number of cycles. It should be noted that the zero shift per cycle is much larger during the first 5 to 10 cycles. For this reason it is recommended that a strain gage installation be cycled to 125 percent of the maximum test strain for at least 5

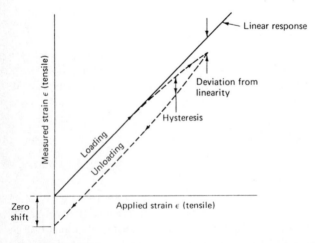

FIGURE 6.13
A typical strain cycle showing nonlinearity, hysteresis, and zero shift (scale exaggerated).

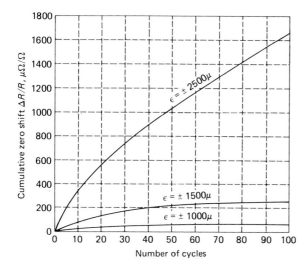

cycles prior to establishing the zero strain readings for all of the gages in the installation.

6.6.2 Temperature Compensation [21–22]

In many test programs, the strain-gage installation is subjected to temperature changes during the test period, and careful consideration must be given to determining whether the change in resistance is due to applied strain or temperature change. When the ambient temperature changes, four effects occur which may alter the performance characteristics of the gage:

1. The strain sensitivity S_A of the metal alloy used for the grid changes.
2. The gage grid either elongates or contracts ($\Delta l/l = \alpha \, \Delta T$).
3. The base material upon which the gage is mounted either elongates or contracts ($\Delta l/l = \beta \, \Delta T$).
4. The resistance of the gage changes because of the influence of the temperature coefficient of resistivity of the gage material ($\Delta R/R = \gamma \, \Delta T$).

   The change in the strain sensitivity S_A of Advance, Karma, and other strain-gage alloys with variations in temperature is shown in Fig. 6.15. These data indicate that $\Delta S_A/\Delta T$ equals 0.735 and $-0.975\%/100°C$ for Advance and Karma alloys, respectively. As a consequence, the variations of S_A with temperature are neglected for room-temperature testing, where the temperature fluctuations rarely exceed $\pm 10°C$. In thermal-stress problems, however, larger temperature variations are possible, and the change in S_A should be taken into account by adjusting the gage factor as the temperature changes during the test period.

   The effects of gage-grid elongation, base-material elongation, and increase

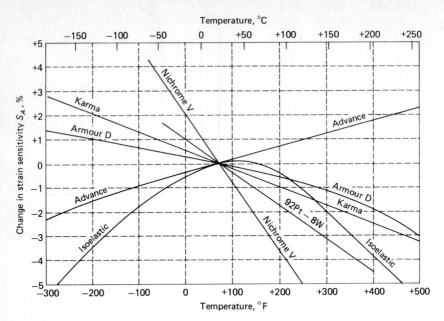

FIGURE 6.15
Change in alloy sensitivity S_A of several strain-gage alloys as a function of temperature.

in gage resistance with increases in temperature combine to produce a temperature-induced change in resistance of the gage $(\Delta R/R)_{\Delta T}$, which can be expressed as

$$\left(\frac{\Delta R}{R}\right)_{\Delta T} = (\beta - \alpha)S_g \Delta T + \gamma \, \Delta T \qquad (6.11)$$

where α = thermal coefficient of expansion of gage material
β = thermal coefficient of expansion of base material
γ = temperature coefficient of resistivity of gage material
S_g = gage factor

If there is a differential expansion between the gage and the base material
due to temperature change (that is, $\alpha \neq \beta$), the gage will be subjected to a
mechanical strain $\epsilon = (\beta - \alpha) \, \Delta T$, which does not occur in the specimen. The gage
reacts to this strain by indicating a change in resistance in the same manner that
it indicates a change for a strain due to the load applied to the specimen.
Unfortunately, it is impossible to separate the apparent strain due to the change
in temperature from the strain due to the applied load. If the gage alloy and the
base material have identical coefficients of expansion, this component of the
thermally induced $(\Delta R/R)_{\Delta T}$ vanishes. The gage may still register a change of
resistance with temperature, however, if the coefficient of resistivity γ is not zero.
This component of $(\Delta R/R)_{\Delta T}$ indicates an apparent strain which does not exist in
the specimen.

Two approaches can be employed to effect temperature compensation in a gage system. The first involves compensation in the gage so that the net effect of the three factors in Eq. (6.11) is canceled out. The second involves compensation for the effects of the temperature change in the signal conditioner required to convert $\Delta R/R$ to voltage output. The second method will be discussed in Chap. 7 when instrument systems are considered in detail.

In producing temperature-compensated gages it is possible to obtain compensation by perfectly matching the coefficients of expansion of the base material and the gage alloy while holding the temperature coefficient of resistivity at zero. Compensation can also be obtained with a mismatch in the coefficients of expansion if the effect of a finite temperature coefficient of resistivity cancels out the effect of the mismatch in temperature coefficients of expansion.

The values of the factors α and γ influencing the temperature response of the strain gage mounted on a specimen with thermal characteristics specified by the value of β are quite sensitive to the composition of the strain-gage alloy, its impurities, and the degree of cold working used in its manufacture. Recently, it has become common practice for strain-gage manufacturers to determine the thermal response characteristics of sample gages from each lot of alloy material which they employ in their production. Because of variations in α and γ between each melt and each roll of foil, it is possible to select foils of Advance and Karma alloys which are suitable for use with almost any type of base material. The gages produced by using this selection technique are known as *selected-melt* or *tempera-ture-compensated gages* and are commercially available with the self-temperature-compensating numbers listed in Table 6.4. Some widely used materials and approximate values for their temperature coefficients of expansion are also listed in Table 6.4.

Unfortunately, these gages are not perfectly compensated over a wide range in temperature because of the nonlinear character of both the expansion coefficients and the resistivity coefficients with temperature. A typical curve showing the apparent strain (temperature-induced) as a function of temperature for a temperature-compensated strain gage fabricated from Advance alloy is shown in Fig. 6.16. These results show that the errors introduced by small changes in temperature in the neighborhood of 75°F (24°C) are quite small with apparent strains of less than 1 $\mu\epsilon/°$F, or $\frac{1}{2}\mu\epsilon/°$C. However, when the change in temperature is large, the apparent strains can become significant, and corrections to account for the thermally induced apparent strains are necessary. These corrections involve measurement of the test temperature at the gage site with a thermocouple and use of a calibration curve similar to the one shown in Fig. 6.16. A calibration curve is provided by the strain-gage manufacturer for each lot of gages produced.

The range of temperature over which an Advance alloy strain gage can be employed is approximately from -20 to 380°F (-30 to 193°C). For temperatures above and below these respective limits, the gage will function; however, very small changes in temperature will produce large apparent strains which can be difficult to account for properly in the analysis of the data.

An extended range of test temperatures is obtained with gages fabricated

TABLE 6.4
Expansion coefficients available in temperature-compensated gages

Specimen material	Coefficient of expansion		Self-temperature-compensating number	
	$10^{-6}/°C$	$10^{-6}/°F$	Advance	Karma
Quartz	0.5	0.3	00	00
Alumina	5.4	3.0	03	03
Zirconium	5.6	3.1		
Glass	9.0	5.0	05	05
Titanium	9.4	5.2		
Cast iron	10.4	5.8	06	06
Steel	11.9	6.6		
Stainless steel	16.7	9.2	09	09
Copper	17.6	9.8		
Bronze	18.2	10.1		
Brass	20.5	11.4	13	13
Aluminum	22.5	12.5		
Magnesium	25.9	14.4	15	15
Polystyrene	72	40	40	
Epoxy resin	90	50	50	
Polymethyl methacrylate	90	50		
Acrylic resin	180	100		

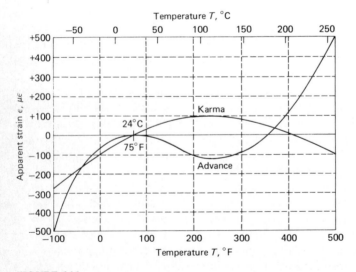

FIGURE 6.16
Apparent strain as a function of temperature for Advance and Karma alloy temperature-compensated strain gages mounted on a specimen having a matching temperature coefficient of expansion.

with Karma alloy. The thermally induced apparent strains as a function of temperature for the Karma alloy are also shown in Fig. 6.16. These data indicate that the strain-temperature curve has a modest slope over the entire range of temperature from -100 to $500°F$ (-73 to $260°C$). Temperature measurements are still required whenever temperature variations are large; however, the small slope of the calibration curve makes it possible to accurately account for the thermally induced apparent strains.

6.6.3 Elongation Limits [23–24]

The maximum strain that can be measured with a foil strain gage depends on the gage length, the foil alloy, the carrier material, and the adhesive. The Advance and Karma alloys with polyimide carriers, used for general-purpose strain gages, can be employed to strain limits of ± 5 and ± 1.5 percent strain, respectively. This strain range is adequate for elastic analyses on metallic and ceramic components, where yield or fracture strains rarely exceed 1 percent; however, these limits can easily be exceeded in plastic analyses, where strains in the postyield range can become large. In these instances, a special postyield gage is normally employed; it is fabricated using a double annealed Advance foil grid with a high-elongation polyimide carrier. Urethane-modified epoxy adhesives are generally used to bond postyield gages to the structure. If proper care is exercised in preparing the surface of the specimen, roughening the back of the gage, formulating a high-elongation plasticized adhesive system, and attaching the lead wires without significant stress raisers, it is possible to approach strain levels of 20 percent before cracks begin to occur in the solder tabs or at the ends of the grid loops.

Special-purpose strain-gage alloys are not applicable for measurements of large strains. The Isoelastic alloy will withstand ± 2 percent strain; however, it undergoes a change of sensitivity at strains larger than 0.75 percent. Armour D and Nichrome V are primarily used for high-temperature measurements and are limited to maximum strain levels of approximately ± 1 percent.

For very large strains, where specimen elongations of 100 percent may be encountered, liquid-metal strain gages can be used. The liquid-metal strain gage is simply a Tygon tube filled with mercury or a gallium-indium-tin alloy, as indicated in Fig. 6.17. When the specimen to which the gage is attached is strained, the volume of the tube cavity remains constant since Poisson's ratio of Tygon is approximately 0.5. Thus the length of the tube increases ($\Delta l = \epsilon l$) while the

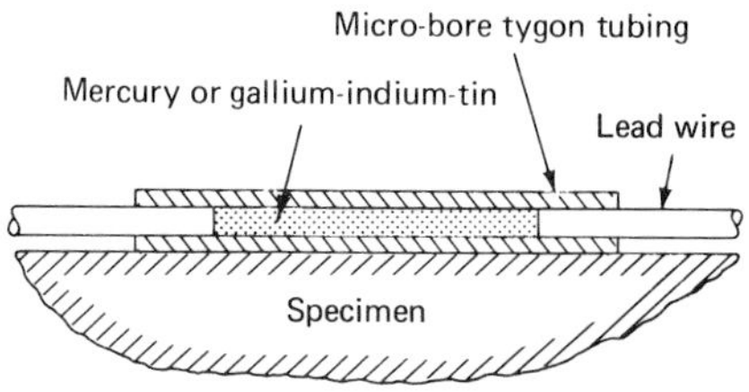

FIGURE 6.17
A liquid-metal electrical-resistance strain gage.

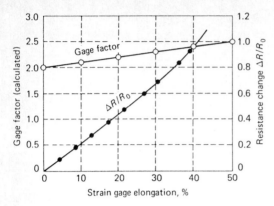

FIGURE 6.18
Resistance change and gage factor as a function of strain for a liquid-metal strain gage. (*From data by Harding.*)

diameter of the tube decreases ($\Delta d = -\nu \epsilon d$). The resistance of such a gage increases with strain, and it can be shown that the gage factor is given by

$$S_g = 2 + \epsilon \qquad (6.12)$$

The resistance of a liquid-metal gage is very small (less than 1 Ω) since the capillary tubes used in their construction have a relatively large inside diameter 0.007 in (0.18 mm). As a consequence, the gages are usually used in series with a large fixed resistor to form a total resistance of 120 Ω so that the gage can be monitored with a conventional Wheatstone bridge. The response of a liquid-metal gage, as shown in Fig. 6.18, is slightly nonlinear with increasing strain due to the increase in gage factor with strain.

6.6.4   Dynamic Response of Strain Gages [25–27]

In dynamic applications of strain gages, the question of their frequency response often arises. This question can be resolved into two parts, namely, the response of the gage in its thickness direction, i.e., how long it takes for an element of the gage to respond to the strain in the specimen beneath it, and the response of the gage due to its length. It is possible to estimate the time required to transmit the strain from the specimen through the adhesive and carrier to the strain-sensing element by considering a gage mounted on a specimen, as shown in Fig. 6.19. A strain wave is propagating through the specimen with velocity c_1. This specimen strain wave induces a shear-strain wave in the adhesive and carrier which propagates with a velocity c_2. The transit time, which is given by $t = h/c_2$, equals 50 ns for typical carrier and adhesive combinations, where $c_2 = 40.000$ in/s (1000 m/s) and $h = 0.002$ in (0.05 mm). The time required for the conductor to respond may exceed this transit time by a factor of 3 to 5; therefore, the response time should be approximately 200 ns. Experiments conducted by Oi and an analysis by Bickle indicate that transit times are approximately 100 ns.

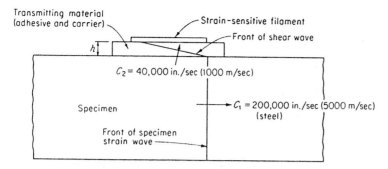

FIGURE 6.19
Dynamic strain transmission between the specimen and the gage.

The rise time in nanoseconds for a strain gage responding to a step pulse is given by

$$t_r = \frac{l_0}{c_1} + 100 \qquad (6.13)$$

where l_0 is the length of the gage and the 100-ns term is added to account for the transmission time through the carrier and adhesive. A typical rise time for a 0.125-in (3.17-mm) gage mounted on a steel bar ($c_1 = 200{,}000$ in/s, or 5000 m/s) is $600 + 100 = 700$ ns. Such short rise times are often neglected when long strain pulses are encountered and the rate of change of strain with time is small. When short, steep-fronted strain pulses are encountered, however, the response time of the gage should be considered since the measured strain pulse can be distorted, as shown in Fig. 6.20. In this instance, the gage is mounted on a specimen which is propagating a strain pulse having an amplitude ϵ_0, a time duration t_0, and a velocity c. The front of the pulse will reach and just pass over a gage of length l_0 in a transit time $t_1 = l_0/c$. In Fig. 6.20 the gage length l_0 is selected so that the transit time t_1 equals $2t_0$. If the gage records average strain over its length, its output will rise linearly to a value of $\epsilon_0/2$ over a time period of t_0. The output will then remain constant at $\epsilon_0/2$ for a period of time equal to t_0 and then decrease linearly to zero over a final time period of t_0. The effect of the gage length in this example was to decrease the amplitude of the output by a factor of 2 and to increase the total time duration of the pulse by a factor of 3. The distortion of the pulse as indicated by the gage will depend upon the ratio t_1/t_0; and as this ratio goes to zero, the distortion vanishes.

It is possible to correct for this distortion by noting that the indicated strain ϵ at time t is given by

$$\bar{\epsilon}(t) = \frac{1}{l_0} \int_{c_1 t - l_0}^{c_1 t} \epsilon(x)\, dx \qquad (6.14)$$

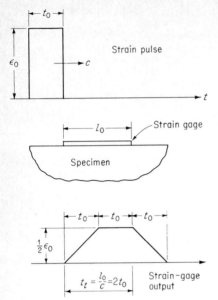

FIGURE 6.20
Dynamic response of a gage to a step-pulse input.

Differentiating gives

$$\frac{d\bar{\epsilon}}{dt} = \frac{1}{l_0}\left\{\int_{c_1t-l_0}^{c_1t} \frac{\partial}{\partial t}[\epsilon(x)]\,dx + \epsilon(a)c_1 - \epsilon(b)c_1\right\} \tag{6.15}$$

where $\epsilon(a)$ and $\epsilon(b)$ are the strains at the two ends of the gage. If the pulse shape remains constant during propagation,

$$\frac{\partial}{\partial t}[\epsilon(x)] = 0$$

and Eq. (6.15) reduces to

$$\frac{d\bar{\epsilon}}{dt} = \frac{c_1}{l_0}[\epsilon(a) - \epsilon(b)] \tag{6.16}$$

Since

$$\epsilon(b, t) = \epsilon\left(a, t - \frac{l_0}{c_1}\right)$$

then

$$\epsilon(a, t) = \frac{l_0}{c_1}\frac{d\bar{\epsilon}}{dt} + \epsilon\left(a, t - \frac{l_0}{c_1}\right) \tag{6.17}$$

Numerical methods used to solve Eq. (6.17) are described in Ref. 26.

6.6.5 Heat Dissipation [28–29]

It is well recognized that temperature variations can significantly influence the output of strain gages, particularly those which are not properly temperature-compensated. The temperature of the gage is of course influenced by ambient-temperature variations and by the power dissipated in the gage when it is connected into a Wheatstone bridge or a potentiometer circuit. The power P is dissipated in the form of heat and the temperature of the gage must increase above the ambient temperature to dissipate the heat. The exact temperature increase required is very difficult to specify since many factors influence the heat balance for the gage. The heat to be dissipated depends upon the voltage applied to the gage and the gage resistance. Thus

$$P = \frac{V^2}{R} = I^2 R \tag{6.18}$$

where P = power, W
I = gage current, A
R = gage resistance, Ω
V = voltage across the gage, V

Factors which govern the heat dissipation include

1. Gage size, w_0 and l_0
2. Grid configuration, spacing and size of conducting elements
3. Carrier, type of polymer and thickness
4. Adhesive, type of polymer and thickness
5. Specimen material, thermal diffusivity
6. Specimen volume in the local area of the gage
7. Type and thickness of overcoat used to waterproof the gage
8. Velocity of the air flowing over the gage installation

A parameter often used to characterize the heat-dissipation characteristics of a strain-gage installation is the power density P_D, which is defined as

$$P_D = \frac{P}{A} \tag{6.19}$$

where P is the power that must be dissipated by the gage and A is the area of the grid of the gage. Power densities that can be tolerated by a gage are strongly related to the specimen which serves as the heat sink because conduction to the sink is much more significant than convection to the air. Recommended values of P_D for different materials and conditions are listed in Table 6.5.

When a Wheatstone bridge with four equal arms is employed, the bridge excitation voltage V_B is related to the power density on the strain gage by

$$V_B^2 = 4AP_D R \tag{6.20}$$

TABLE 6.5
Allowable power densities

Power density P_D		
W/in²	W/mm²	Specimen conditions
5–10	0.008–0.016	Heavy aluminum or copper sections
2–5	0.003–0.008	Heavy steel sections
1–2	0.0015–0.003	Thin steel sections
0.2–0.5	0.0003–0.0008	Fiber glass, glass, ceramics
0.02–0.05	0.00003–0.00008	Unfilled plastics

Allowable bridge voltages for specified grid areas and power densities are shown for 120-Ω gages in Fig. 6.21. Typical grid configurations are identified along the abscissa in the figure to illustrate the effect of gage size on allowable bridge excitation. It should be noted that small gages mounted on a poor heat sink ($P_D < 1$ W/in² or 0.0015 W/mm²) result in lower allowable bridge voltages than those employed in most commercial strain indicators (3 to 5 V). In these instances, it is necessary to use a higher-resistance gage (350 or 1000 Ω in place of 120 Ω) or a gage with a larger grid area.

6.6.6 Stability [30–31]

In certain strain-gage applications it is necessary to record strains over a period of months or years without having the opportunity to unload the specimen and recheck the zero resistance. The duration of the readout period is important and makes this application of strain gages one of the most difficult. All the factors which can influence the behavior of the gage have an opportunity to do so; moreover, there is enough time for the individual contribution to the error from each of the factors to become quite significant. For this reason it is imperative that every precaution be taken in employing the resistance-type gage if meaningful data are to be obtained.

Drift in the zero reading from an electrical-resistance strain-gage installation is due to the effects of moisture or humidity variations on the carrier and the adhesive, the effects of long-term stress relaxation of the adhesive, the carrier, and the strain-gage alloy, and instabilities in the resistors in the inactive arms of the Wheatstone bridge.

Results of an interesting series of stability tests by Freynik are presented in Fig. 6.22. In evaluating a typical general-purpose strain gage with a grid fabricated from Advance alloy and a polyimide carrier, zero shifts of 270 $\mu\epsilon$ were observed after 30 days. Since this installation was carefully waterproofed, the large drifts were attributed to stress relaxation in the polyimide carrier over the period of observation.

Results from a second strain-gage installation with a grid fabricated from

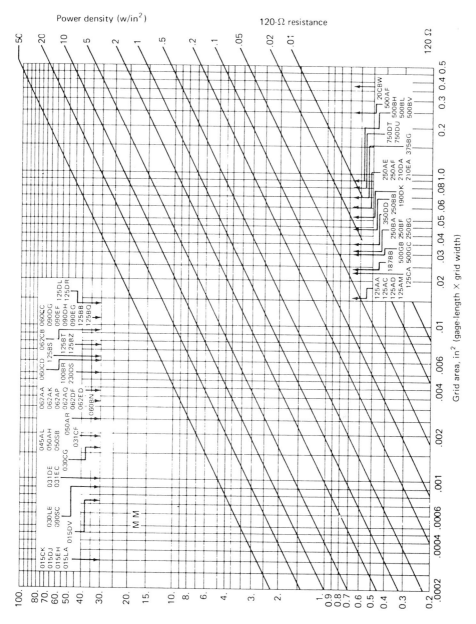

FIGURE 6.21
Allowable bridge voltage as a function of grid area with power density as a parameter for 120-Ω
electrical-resistance strain gages. (*Measurements Group, Inc.*)

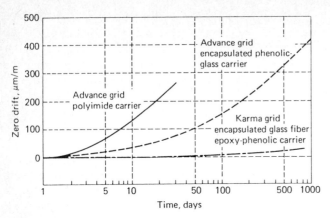

FIGURE 6.22
Zero drift as a function of time for several gage types at 167°F (75°C). (*From data by Freynik.*)

Advance alloy and a carrier from glass-fiber-reinforced phenolic were more satisfactory, with zero drift of approximately 100 $\mu\epsilon$ after 50 days. The presence of the glass fibers essentially eliminated drift due to stress relaxation in the carrier material; the drift measured was attributed to instabilities in the Advance alloy grid at the test temperature of 167°F (75°C). The final and most satisfactory results were obtained with a Karma grid and an encapsulating glass-reinforced epoxy-phenolic carrier. In this case, the zero shift averaged only 30 $\mu\epsilon$ after an observation period of 900 days. In similar tests with this gage installation at room temperature, the zero drift was only -25 $\mu\epsilon$. These results show that electrical-resistance strain gages can be used for long-term measurements provided Karma grids with glass-fiber-reinforced epoxy-phenolic carriers are employed with a well-cured epoxy adhesive system. The gage installation should be waterproofed to minimize the effects of moisture penetration. It is also important to specify hermetically sealed bridge-completion resistors to ensure stability of the bridge over the long observation periods.

6.7 ENVIRONMENTAL EFFECTS

The performance of resistance strain gages is markedly affected by the environment. Moisture, temperature extremes, hydrostatic pressure, nuclear radiation, and cyclic loading produce changes in gage behavior which must be accounted for in the installation of the gage and in the analysis of the data to obtain meaningful results. Each of these parameters is discussed in the following subsections.

6.7.1 Effects of Moisture and Humidity [32–33]

A strain-gage installation can be detrimentally affected by direct contact with water or by the water vapor normally present in the air. The water is absorbed by both the adhesive and the carrier, and the gage performance is affected in several ways. First, the moisture decreases the gage-to-ground resistance. If this value of resistance is reduced sufficiently, the effect is the same as that of placing a shunt resistor across the active gage. The water also degrades the strength and rigidity of the bond and reduces the effectiveness of the adhesive in transmitting the strain from the specimen to the gage. If this loss in adhesive strength or rigidity is sufficient, the gage will not develop its stated calibration factors and measuring errors are introduced.

Plastics also expand when they absorb water and contract when they release it; thus, any change in the moisture concentration in the adhesive will produce strains in the adhesive, which will in turn be transmitted to the strain gage. These moisture-induced adhesive strains will produce a strain-gage response that cannot be separated from the response due to the applied mechanical strain. Finally, the presence of water in the adhesive will cause electrolysis when current passes through the gage. During the electrolysis process, the gage filament will erode and a significant increase in resistance will occur. Again, the strain gage will indicate a tensile strain due to this electrolysis which cannot be differentiated from the applied mechanical strain.

Many methods for waterproofing strain gages have been developed; however, the extent of the measures taken to protect the gage from moisture depends to a large degree on the application and the extent of the gage exposure to water. For normal laboratory work, where the readout time is relatively short, a thin layer of microcrystalline wax or an air-drying polyurethane coating is usually sufficient to protect the gage installation from moisture in the air. For much more severe applications, e.g., prolonged exposure to seawater, it is necessary to build up a seal out of soft wax, synthetic rubber, metal foil, and a final coat of rubber. A cross section of a well-protected gage installation is shown in Fig. 6.23. Care should be exercised in forming the seal at the lead-wire terminal since the seal usually fails at this location. Also, the lead-wire insulation should be rubber, and splices in the cable must be avoided. If water gains entry to the lead wires, it will be transmitted for significant distances by capillary action to the gage installation.

6.7.2 Effects of Hydrostatic Pressure [34–36]

In the stress analysis of pressure vessels and piping systems, strain gages are frequently employed on interior surfaces where they are exposed to a gas or fluid

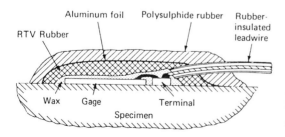

FIGURE 6.23
Waterproofing a gage for severe sea-water exposure.

pressure which acts directly on the sensing element of the gage. Under such conditions, pressure-induced resistance changes occur which must be accounted for in the analysis of the strain-gage data.

Milligan and Brace independently studied this effect of pressure by mounting a gage on a small specimen, placing the specimen in a special high-pressure vessel, and monitoring the strain as the pressure was increased to 140,000 psi (965 MPa). In this type of experiment, the hydrostatic pressure p produces a strain in the specimen which is given by Eq. (2.16) as

$$\epsilon = \frac{-p}{E} - \frac{v}{E}[(-p) + (-p)] = -\frac{1 - 2v}{E}p = K_T p \qquad (6.21)$$

where $K_T = -(1 - 2v)/E$ is often referred to as the *compressibility constant for a material*. The strain gages were monitored during the pressure cycle, and it was observed that the indicated strains were less than the true strains as predicted by Eq. (6.21).

The pressure effect can be characterized by defining the slope of the pressure-strain curve as an apparent compressibility constant for the material. Thus

$$K_1 = -\frac{\Delta\epsilon}{\Delta p} \qquad (6.22)$$

The difference D_p between the observed and predicted effects due to pressure can then be written

$$D_p = \frac{K_T - K_1}{K_T} \qquad (6.23)$$

Experimental results by Milligan, shown in Fig. 6.24, indicate that D_p depends upon the compressibility constant for the specimen material, the curvature of the specimen where the gage is mounted, and the type of strain-sensing alloy used in fabricating the gage. In spite of relatively large values for D_p, however, it was observed that the strain response of the gage remained linear with pressure.

The true strain ϵ_t can be expressed in terms of the indicated strain ϵ_i and a correction term ϵ_{cp} as

$$\epsilon_t = \epsilon_i - \epsilon_{cp} \qquad (6.24)$$

The magnitude of the correction term ϵ_{cp} is expressed in terms of the experimentally determined values of D_p as

$$\epsilon_{cp} = D_p K_T p \qquad (6.25)$$

For a flat steel specimen, $K_T = 133 \times 10^{-10}$ in^2/lb (1.93×10^{-12} m^2/N) and $D_p = 0.3$; hence, the correction ϵ_{cp} is approximately 4 $\mu\epsilon$ for a pressure of 1000 psi (7 MPa). Since this is a very small correction, it is usually possible to neglect pressure effects for pressures less than approximately 3000 psi (20 MPa).

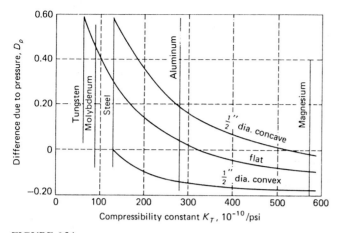

FIGURE 6.24

Difference due to pressure D_p as a function of compressibility constant K_T for Advance foil gages on specimens with different curvatures. [*From data by Milligan* (36).]

For hydrostatic pressure applications, foil strain gages with the thinnest possible carriers should be employed. The gage should be mounted on a smooth surface with a thinned adhesive to obtain the thinnest possible bond line. Bubbles in the adhesive layer cannot be tolerated since the pressure normal to the surface of the gage will force the sensing element into any void beneath the gage and additional erroneous resistance changes will result.

6.7.3 Effects of Nuclear Radiation [37–39]

Several difficult problems are encountered when electrical-resistance strain gages are employed in nuclear-radiation fields. The most serious difficulty involves the change in electrical resistivity of the strain gage and lead wires as a result of the fast-neutron dose. This effect is significant; changes of 2 to 3 percent in $\Delta R/R$ have been observed with a neutron dose of 10^{18} *nvt*. These changes in resistivity produce zero drift with time which can be as large as an apparent strain of 10,000 to 15,000 $\mu\epsilon$. The exact rate of change of resistivity is a function of the strain-gage and lead-wire materials, the state of strain in the gage, and the temperature. Typical changes in resistance with integrated fast-neutron flux are shown in Fig. 6.25. Since the changes in resistivity are a function of the magnitude and sign of the strain, the use of dummy gages to cancel the effects of radiation exposure is not effective. Since the change in electrical resistivity appears to be a linear function of the logarithm of the dose, the most satisfactory solution to neutron-induced zero drift is to employ preexposed strain-gage installations and to reduce test times to a minimum. Frequent unloading and reestablishing the zero resistance of the gage are essential.

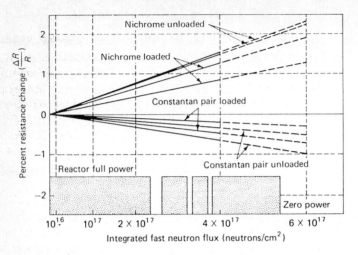

FIGURE 6.25
Percent resistance change as a function of exposure to neutron flux. (*From data by Tallman.*)

The neutrons also produce a change in the sensitivity of the strain-gage alloys. Variations in S_A for Advance alloy, shown in Fig. 6.26, are large and unpredictable when the integrated neutron exposure exceeds 10^{16} *nvt*.

The fast neutrons also produce mechanical effects which are detrimental to strain-gage installations. With exposure to the fast neutrons, the strain-gage alloy exhibits an increase in its yield strength and modulus of elasticity and a decrease in elongation capability. Radiation-induced cross-linking in polymers also destroys the original organic structure of the bond. For this reason, ceramic adhesives are normally employed in any long-term tests where exposure will accumulate.

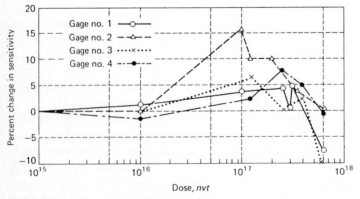

FIGURE 6.26
Percent change in sensitivity of an Advance alloy strain gage with exposure to fast neutrons. (*From data by Tallman.*)

In nuclear-radiation fields with high gamma flux, considerable energy is transferred to the gage and specimen; therefore, temperature changes can be large. For precise measurements, the temperature change must be predicted or determined so that the strain-gage results can be corrected for the effects of temperature.

It is possible to measure strains in strong radiation fields; however, the special precautions outlined in Refs. 37 and 38 must be taken or serious errors will occur.

6.7.4 Effects of High Temperature [40–44]

Resistance-type strain gages can be employed at elevated temperatures for both static and dynamic stress analyses; however, the measurements require many special precautions which depend primarily on the temperature and the time of observation. At elevated temperatures, the resistance R of a strain gage must be considered to be a function of temperature T and time t in addition to strain ϵ. Thus

$$R = f(\epsilon, T, t) \tag{6.26}$$

The resistance change $\Delta R/R$ is then given by

$$\frac{\Delta R}{R} = \frac{1}{R}\frac{\partial f}{\partial \epsilon}\Delta\epsilon + \frac{1}{R}\frac{\partial f}{\partial T}\Delta T + \frac{1}{R}\frac{\partial f}{\partial t}\Delta t \tag{6.27}$$

where $\dfrac{1}{R}\dfrac{\partial f}{\partial \epsilon} = S_g$ = gage sensitivity to strain (gage factor)

$\dfrac{1}{R}\dfrac{\partial f}{\partial T} = S_T$ = gage sensitivity to temperature

$\dfrac{1}{R}\dfrac{\partial f}{\partial t} = S_t$ = gage sensitivity to time

Equation (6.27) can then be expressed in terms of the three sensitivity factors as

$$\frac{\Delta R}{R} = S_g\Delta\epsilon + S_T\Delta T + S_t\Delta t \tag{6.28}$$

In the discussion of performance characteristics of foil strain gages in Secs. 6.5.2 and 6.5.6, it was shown that the sensitivity of the gages to temperature and time was made negligibly small at normal operating temperatures of 0 to 150°F (-18 to 65°C) by proper selection of the strain-gage alloy and carrier materials. As the test temperature increases above this level, however, the performance of the gage deteriorates, and S_T and S_t are not usually negligible.

As the temperature increases, temperature compensation is less effective, and corrections must be made to account for the apparent strain, as shown in Fig.

6.16. Comparisons of these results indicate that the Karma strain-gage alloy is more suitable for the higher-temperature applications than Advance. The Karma gages can be employed to temperatures up to about 500°F (260°C) with temperature-induced apparent strains of less than ± 100 $\mu\epsilon$.

The stability of a strain-gage installation is also affected by temperature; and strain-gage drift becomes a more serious problem as the temperature and the time of observation are increased. Stability is affected by stress relaxation in the adhesive bond and in the carrier material and by metallurgical changes (phase transformations and annealing) in the strain-gage alloy. The upper temperature limit on commercially available Karma gages is controlled by the carrier material. Glass-reinforced epoxy-phenolic carriers are rated at 550°F (288°C); however, Karma gages with this type of carrier drift with time, as shown in Fig. 6.27. If the time of loading and observation is long, corrections must be made for the zero drift. Drift rates will depend upon both the strain level and the temperature. For high-temperature strain analyses, it is recommended that a series of strain-time calibration curves, similar to the one shown in Fig. 6.27, be developed to cover the range of strains and temperatures to be encountered. Zero-drift corrections can then be taken from the appropriate curve.

Changes in the gage factor with temperature are relatively small, as indicated in Fig. 6.15. Also, since the change is linear with temperature, corrections can be made to accurately account for this effect.

The problem of gage stability and apparent strain due to temperature changes is greatly reduced if the period of observation is short. For dynamic analyses, where relatively short strain-time records are used (times usually less than 1 s), the temperature does not have sufficient time to change and temperature effects are therefore small. For this reason, dynamic strain-gage analyses can be made at very high temperatures with good precision if the proper alloy and adhesives are employed.

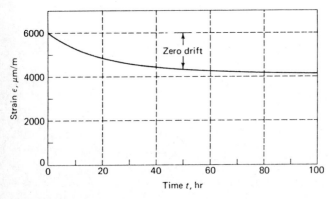

FIGURE 6.27
Typical zero drift as a function of time for a Karma alloy strain gage with a glass-fiber epoxy-phenolic carrier at 560°F (293°C). (*From data by Hayes.*)

Resistance-strain-gage measurements at temperatures higher than 550°F (288°C) require special gages, special techniques for mounting the gages, and special techniques for monitoring the strain-gage signal. At these higher temperatures, polymeric materials can no longer be used for the carrier or the adhesive. The gages must be mounted to the specimen with the ceramic cements described in Sec. 6.4.4. The carrier is either removed entirely or replaced with a thin stainless-steel shim. Strain gages for the very-high-temperature applications are fabricated using materials such as Nichrome V, Armour D, or alloy 479 (See Table 6.1). Of these different materials, the platinum-tungsten alloy is preferred because of its inherent oxidation resistance and metallurgical stability. Performance characteristics of this alloy, shown in Fig. 6.28, indicate that the gage factor drops about 30 percent as the temperature increases from 70 to 1600°F (21 to 871°C). Unfortunately, the temperature coefficient of resistance of this alloy is very high and large apparent strains (40 to 80 $\mu\epsilon/°F$, or 70 to 140 $\mu\epsilon/°C$) are indicated by the gage when the temperature changes. The material is stable, and zero drift is negligible to temperatures of 800°F (427°C). At temperatures above 800°F (427°C), drift will occur, with the rate of drift increasing with temperature.

The effects of drift rate can be minimized by stabilizing the alloy. The stabilization process consists of annealing the alloy at the test temperature for 12 to 16 h. The stabilized drift is relatively low, with rates of 20 $\mu\epsilon/h$ reported at 1400°F (760°C). The effect of apparent strain with temperature is much more difficult to treat and usually requires the use of dual-element or four-element gages, which compensate for the effects of temperature by signal subtraction in the Wheatstone bridge. An example of a four-element (complete-bridge) gage is shown in Fig. 6.29. A reasonable degree of temperature compensation has been achieved ($\pm 300 \ \mu\epsilon$) for temperatures up to 1400°F (760°C). The gage is useful over a wide

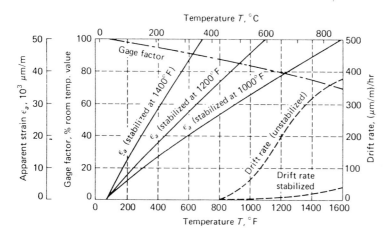

FIGURE 6.28
Performance characteristics of a platinum-tungsten alloy strain gage as a function of temperature. [*From data by Wnuk* (41).]

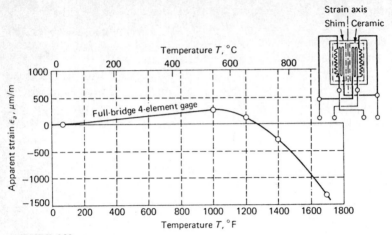

FIGURE 6.29

Apparent strain as a function of temperature for a four-element full-bridge gage. [*From data by Wnuk* (41).]

range of temperatures if the apparent strain is accounted for by correcting the gage output.

6.7.5   High-Temperature Strain Measurements with Capacitive Gages [45–46]

Measurement of static strains at high temperatures is difficult with resistance strain gages. The gage factor changes, the apparent strains are large, zero drift with time is significant, and even the ceramic adhesives which serve as the gage insulation begin to conduct at high temperature. Strain gages based on capacitive sensors are often more suitable than gages using resistive elements as sensors.

The Central Electricity Research Laboratory (CERL) gage, illustrated in Fig. 6.30, utilizes a simple parallel-plate capacitor mounted between two curved-beam elements. The two curved-beam elements are attached at their ends and the assembly is spot-welded to the specimen. As the specimen is strained, the length *l* of the gage changes, the curved beams deflect, and the air gap between the plates

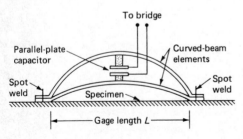

FIGURE 6.30

Schematic illustration of the CERL capacitance strain gage.

changes in proportion to the strain. The curved-beam elements can be fabricated from the same material as the test specimen to provide automatic temperature compensation. Since the dielectric constant of air is nearly constant with temperature, the calibration constant is stable up to 1200°F (650°C).

The capacitance varies nonlinearly with strain changing from about 1.3 to 0.4 pF as the strain increases from 0 to 1 percent. The use of a transformer ratio arm bridge operating at 50 V with a carrier frequency of 1.6 kHz permits 1 $\mu\varepsilon$ resolution of strain.

6.7.6 Effects of Cryogenic Temperatures [47–48]

Strain can be measured with electrical-resistance strain gages to liquid-nitrogen temperatures ($-320°F$, or $-196°C$) and below. Two factors must be carefully accounted for in using strain gages at these temperature extremes. The first effect is the change in gage factor with temperature, illustrated in Fig. 6.31. These results indicate that this correction is small, being only -1.8 and $+3.8$ percent, respectively, for Advance and the selected-melt Karma alloy SK-15 at $-320°F$ ($-196°C$).

The second effect, which is extremely important to account for when measuring strains at cryogenic temperatures, is the very large apparent strains introduced by small changes in temperature. Karma alloys are preferred to Advance alloys because of their better stability at the temperature extremes, as shown in Fig. 6.16. Temperature-induced apparent strains for Karma strain gages mounted on a variety of materials are shown in Fig. 6.32.

Although both Advance and Karma strain gages are temperature-compensated, the compensation is limited to the temperature range from -100 to

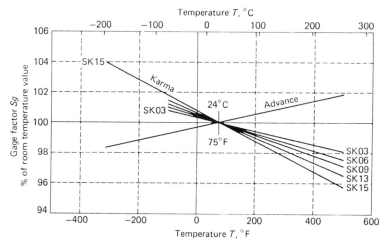

FIGURE 6.31
Changes in gage factor with temperature. [*From data by Telinde* (47).]

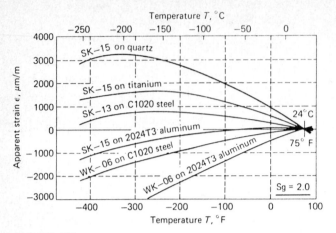

FIGURE 6.32
Apparent strain as a function of temperature for several selected-melt Karma alloys on different materials. [*From data by Telinde* (47).]

500°F (−73 to 260°C). As the test temperature is reduced below −100°F (−73°C), the strain gages become more sensitive to temperature changes and large apparent strains are produced. It is possible to minimize the effects of temperature changes at cryogenic temperatures by selecting compensating gages which are mismatched relative to the component material. For instance, the Karma gage compensated for a material with a temperature coefficient of expansion equal to 13 ppm/°F and 23.4 ppm/°C exhibits a relatively stable response to temperature variations from −100 to −400°F (−73 to −240°C) when mounted on a steel specimen having a temperature coefficient of expansion of 6 ppm/°F (10.8 ppm/°C), as shown in Fig. 6.32.

If temperature sensors are mounted with the strain gages, the measured temperatures can be used with curves similar to those shown in Fig. 6.32 to correct for the apparent strains. It is also possible to use a four-element gage which is wired so that signal cancellation in the Wheatstone bridge compensates for the temperature-induced response of the strain gages.

Cryogenic temperatures are usually obtained in tests by using liquid nitrogen, liquid hydrogen, or liquid helium. All three are excellent insulating materials, and it is not necessary or advisable to use electrical insulation compounds between the gage grid and the cryogenic fluid. The gage should be coated with a silicone grease to provide a heat shield and to eliminate the possibility of boiling of the liquid over the gage grid during the test.

Special consideration must also be given to changes in the mechanical properties of component materials at cryogenic temperatures. The effect of cryogenic temperatures is to increase the elastic modulus from 5 to 20 percent. The higher values of the elastic modulus must be employed in the stress-strain equations when converting the strain data to stress.

6.7.7 Effects of Strain Cycling [49–50]

Strain gages are frequently mounted on components subjected to fatigue or cyclic loading, and the life of the component during which the gage must be monitored can exceed several million cycles. Three factors which must be considered in a fatigue application are zero shift, change in gage factor, and failure of the gage in fatigue.

As the strain gage is subjected to repeated cyclic strain, the gage grid work-hardens and its specific resistance changes. The specific resistance change produces a zero shift. The amount of the zero shift depends on the magnitude of the strain, the number of cycles, the grid alloy, the original state of cold work of the grid alloy, and the type of carrier employed in the gage construction. Typical examples of zero shift as a function of number of strain cycles are shown in Fig. 6.33 for four different gages. The poorest gage for this type of application is the

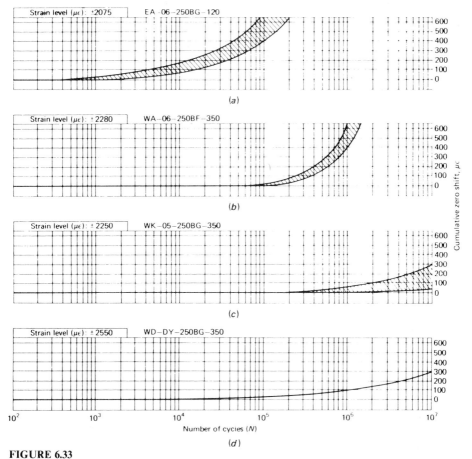

FIGURE 6.33
Zero shift as a function of fatigue exposure. (*a*) Advance, open-face; (*b*) Advance, encapsulated grid; (*c*) Karma, encapsulated grid; and (*d*) isoelastic, encapsulated grid. (*Measurements Group, Inc.*)

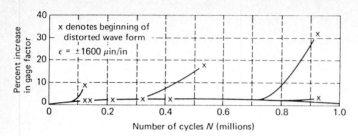

FIGURE 6.34
Increase in gage factor as a function of fatigue exposure for an annealed Advance foil gage.
(*Measurements Group, Inc.*)

open-faced Advance grid (EA-type gage), which begins to exhibit noticeable zero
shift at 10^3 cycles. Encapsulating the grid in a glass-reinforced epoxy-phenolic
resin (WA-type gage) improves the life, and exposures of 10^5 cycles at $\pm 2100\ \mu\epsilon$
occur before zero shift becomes apparent. Further improvement can be achieved
by using Karma (K) or isoelastic (D) alloys fully encapsulated in the glass-
reinforced epoxy-phenolic carrier material with factory-installed lead wires.

The changes in gage factor are quite small due to strain cycling and in general
can be neglected if the zero shift is less than 200 $\mu\epsilon$.

A very important factor to consider when using strain gages under fatigue
conditions is the increase in sensitivity which occurs once a fatigue crack develops
in the grid of the gage. The crack will usually develop in the grid near the lead
wire on the tab. The presence of the crack produces a small increase in the
resistance of the gage, which is monitored as an apparent strain. Thus, the apparent
gage factor has increased with the development and subsequent propagation of
the fatigue crack. Typical results showing this increase in gage factor are shown
in Fig. 6.34. It is important to note that the gage factor increased from 10 to 32
percent for the three gages which failed in fatigue. Also, the error in "calibration"
existed over an appreciable fraction of the life of the gage.

EXERCISES

6.1. The strain sensitivity of most metallic alloys is about 2.0. What portion of this
sensitivity is due to dimension changes in the conductor? What portion is due to
changes in the number of free electrons and their mobility?

6.2. Advance or Constantan alloy exhibits a linear response of $\Delta R/R$ to strain ε for strains
as large as 8 percent. Discuss why this is a remarkably large range of linearity with
respect to strain when the elastic limit of the material may be as low as 145 MPa
and the modulus of elasticity is 152 GPa. Base the discussion on Eq. (5.20) and recall
that Poisson's ratio for the material will increase from 0.3 to 0.5 as the material
transforms from the elastic to the plastic state.

6.3. Describe the advantages and disadvantages of a strain gage fabricated using an
isoelastic alloy.

6.4. Describe the advantages and disadvantages of a strain gage fabricated using a Karma alloy.

6.5. Table 6.1 shows $S_A = 4.1$ for Alloy 479. This sensitivity is higher than that of the isoelastic alloy. Why isn't it used instead of
   (a) The isoelastic alloy
   (b) The Karma alloy
   (c) The Advance alloy

6.6. Why are bonded-wire strain gages still used in selected applications today?

6.7. List four different carrier materials used in strain-gage construction and give the reasons for their use.

6.8. Estimate the shear stress induced in the adhesive layer of a bonded strain gage subjected to 4000 $\mu\varepsilon$. The gage has an l/w ratio of 2 and is fabricated from Advance alloy 30 μm thick. The carrier is polyimide with a thickness of 50 μm.

6.9. Why is postcure shrinkage of an adhesive used to bond a strain gage detrimental?

6.10. Briefly discuss the conditions which would dictate using the following cements in a strain-gage installation:
   (a) Polyester   (b) Epoxy cement
   (c) Cyanoacrylate   (d) Ceramic

6.11. Describe two procedures used to evaluate the completeness of the cure of an adhesive being used to bond a strain gage to a specimen.

6.12. Describe a procedure for determining if bubbles exist under a strain gage.

6.13. If you note a bubble under the center of the grid, will it affect the gage output if the gage is located in a field of:
   (a) Uniform tensile strain?
   (b) Uniform compressive strain?

6.14. Determine the transverse-sensitivity factor for a strain gage fabricated using Constantan alloy with $S_A = 2.11$ if it has been calibrated and found to exhibit a gage factor $S_g = 2.08$.

6.15. The transverse-sensitivity factor K_t for a new grid configuration must be determined. It has been suggested that a simple tension specimen can be used with gages oriented in the axial and transverse directions to make the determination. Develop an expression for K_t in terms of Poisson's ratio v of the specimen material and the ratio of the outputs from the two strain gages.

6.16. A value of $K_t = 0.03$ was obtained for the gage in Exercise 6.15 by assuming Poisson's ratio $v = 0.30$ for the specimen material. The true value of Poisson's ratio was 0.25. Determine the percent error introduced in the determination of K_t.

6.17. Determine the percent error introduced by neglecting transverse sensitivity if a WK-06-062DY-350 type of strain gage is used to measure the longitudinal strain in a steel ($v = 0.30$) beam.

6.18. Determine the percent error introduced by neglecting transverse sensitivity if a WK-06-250BG-350 type of strain gage is used to measure the maximum tensile strain on the surface of a circular steel ($v = 0.30$) shaft subjected to a pure torque.

6.19. Determine the percent error introduced by neglecting transverse sensitivity if an EA-06-125RA-120 type of strain gage is used to measure the maximum tensile strain on the outside surface of a steel ($v = 0.29$) thin-walled cylindrical pressure vessel which is being subjected to an internal gas pressure.

6.20. Specify a test procedure which will determine the zero shift associated with a strain-gage installation. Specify a second test procedure which will minimize this zero shift.

6.21. Determine the change in alloy sensitivity S_A of Constantan as the temperature is increased from -100 to $+200°C$.

6.22. Determine the change in alloy sensitivity S_A of Karma alloy as the temperature is increased from -100 to $+200°C$.

6.23. A strain-gage analysis is to be conducted on a component which will be maintained at $(65 \pm 20°C)$ throughout the test period. Which alloy should be specified for the strain-gage grid? If temperature-compensated gages were specified, what errors would occur due to the temperature fluctuations?

6.24. Repeat Exercise 6.23 by changing $(65 \pm 20°C)$ to $(200 \pm 30°C)$.

6.25. Repeat Exercise 6.23 by changing $(65 \pm 20°C)$ to $(24 \pm 8°C)$.

6.26. Specify the special procedures required for a strain-gage analysis where plastic strains of the order of 15 percent are anticipated.

6.27. For the liquid-metal strain gage illustrated in Fig. 6.17, show that the gage factor S_g is given by $S_g = 2 + \epsilon$.

6.28. A strain wave is propagating in the x direction through a specimen with a characteristic velocity $c_1 = 5000$ m/s. The strain wave may be represented by the equation $\epsilon_1 = A \sin (2\pi/\lambda)(x - c_1 t)$, where A is the amplitude and λ is the wavelength of the wave. If $\lambda = nl_0$, determine the amplitude of strain recorded by a strain gage having a length l_0. Prepare a graph of the output as n varies from 0.1 to 10.

6.29. Determine the allowable bridge voltage for the grid configurations 015LA, 060BN, 125DL, 250BA, 375BG, and 500GH if the 120-Ω gages are mounted on:
(a) An aluminum engine head   (b) A large steel beam
(c) A fiberglass tank   (d) A steel ship hull

6.30. Write a test specification outlining special procedures to be used in measuring strain over a 2-year period on a large steel smokestack subjected to wind loads.

6.31. Write a test specification to be followed by a laboratory technician in waterproofing a strain gage for:
(a) A short-duration test in the laboratory
(b) A long-term test under seawater exposure

6.32. A strain gage is mounted on the inside surface of a large-diameter steel pressure vessel which is subjected to an internal pressure of 150 MPa. Determine the true strain in the hoop direction if the inside diameter is 450 mm and the outside diameter is 800 mm. Compute the error involved in neglecting pressure effects on the output of the strain gages.

6.33. Describe a procedure to minimize the effect of neutron-induced zero drift in a strain-gage installation subjected to intense radiation.

6.34. Describe a procedure for measuring the strain on a mechanical component at 500°F (260°C) if the loading cycle requires:
(a) 1 s   (b) 1 min   (c) 1 h   (d) 1 day   (e) 1 year

6.35. Describe a procedure for measuring the strain on a component at a temperature of 900°F (482°C) if the strain cycle occurs in 1 h.

6.36. Describe a procedure for measuring the strain on a mechanical component at a temperature of 1600°F (872°C) if the strain is transient and a cycle requires only 0.01 s.

6.37. A strain-gage test is to be conducted with a Karma strain gage at $-200°F$ ($-129°F$).

The gage factor is specified as 2.05 at room temperature. What gage factor should be used in reducing the value of $\Delta R/R$ to strain?

6.38. Specify the self-temperature-compensating number for a Karma strain gage to be employed on a stainless-steel tank at liquid-nitrogen temperatures.

6.39. Select a gage to operate under cyclic strain conditions where unloading of the structure is not practical. The strain is

(a) ± 1200 $\mu\epsilon$ for 10^5 cycles

(b) ± 2500 $\mu\epsilon$ for 10^6 cycles

6.40. Will a continuity check on the gage resistance indicate that a fatigue crack has initiated in the grid and that the gage will respond to strain with a supersensitive gage factor? Why?

REFERENCES

1. Thomson, W. (Lord Kelvin): On the Electrodynamic Qualities of Metals, *Phil. Trans. R. Soc.*, vol. 146, pp. 649–751, 1856.
2. Tomlinson, H.: The Influence of Stress and Strain on the Action of Physical Forces, *Phil. Trans. R. Soc.*, vol. 174, pp. 1–172, 1883.
3. Simmons, E. E., Jr.: Material Testing Apparatus, U.S. Patent 2,292,549, Feb. 23, 1940.
4. Kammer, E. W., and T. E. Pardue: Electric Resistance Changes of Fine Wires during Elastic and Plastic Strains, *Proc. SESA*, vol. VII, no. 1, pp. 7–20, 1949.
5. Maslen, K. R.: "Resistance Strain Characteristics of Fine Wires," R. Aircr. Establ. Tech. Note Instr. 127, 1952.
6. Meyer, R. D.: Application of Unbonded-Type Resistance Gages, *Instruments*, vol. 19, no. 3, pp. 136–139, 1946.
7. Campbell, W. R.: "Performance of Wire Strain Gages: I, Calibration Factors in Tension," NACA Tech. Note 954, 1944; "II, Calibration Factors in Compression," NACA Tech. Note 978, 1945.
8. DeForest, A. V.: Characteristics and Aircraft Applications of Wire Wound Resistance Strain Gages, *Instruments*, vol. 15, No. 4, p. 112, 1942.
9. Jones, E., and K. R. Maslen: "The Physical Characteristics of Wire Resistance Strain Gages," R. Aircr. Establ. R and M 2661, November 1948.
10. Maslen, K. R., and I. G. Scott: "Some Characteristics of Foil Strain Gauges," R. Aircr. Establ. Tech. Note Instr. 134, 1953.
11. Wnuk, S. P.: Recent Developments in Flame Sprayed Strain Gages, *Proc. West. Reg. Strain Gage Comm.*, pp. 1–6, 1965.
12. Stein, P. K.: Adhesives: "How They Determine and Limit Strain Gage Performance," in M. Dean and R. D. Douglas (eds.), *Semiconductor and Conventional Strain Gages*, Academic Press, New York, 1962, pp. 45–72.
13. Pitts, J. W., and D. G. Moore: "Development of High Temperature Strain Gages," Natl. Bur. Std. Monogr. 26, 1966.
14. Bodnar, M. J., and W. H. Schroder: Bonding Properties of a Solventless Cyanoacrylate Adhesive, *Mod. Plast.*, vol. 36, no. 1, pp. 142–148, September 1958.
15. Campbell, W. R.: "Performance Tests of Wire Strain Gages: IV, Axial and Transverse Sensitivities," NACA Tech. Note 1042, June 1946.
16. Meier, J. H.: On The Transverse Strain Sensitivity of Foil Gages, *Exp. Mech.*, vol. 1, no. 7, pp. 39–40, 1961.
17. Wu, C. T.: Transverse Sensitivity of Bonded Strain Gages, Exp. Mech., vol. 2, no. 11, pp. 338–344, 1962.
18. "Transverse Sensitivity Errors," Micro-Measurements Tech. Note 509, 1982.
19. Matlock, H., and S. A. Thompson: Creep in Bonded Electrical Strain Gages, *Proc. SESA*, vol. XII, no. 2, pp. 181–188, 1955.

20. Bloss, R. L.: "Characteristics of Resistance Strain Gages," in M. Dean and R. D. Douglas (eds.), op. cit., pp. 123–142.

21. Barker, R. S.: Self-Temperature Compensating SR-4 Strain Gages, *Proc. SESA*, vol. XI, no. 1, pp. 119–128, 1953.

22. Hines, F. F., and L. J. Weymouth: Practical Aspects of Temperature Effects on Resistance Strain Gages, *Strain Gage Readings*, vol. IV, no. 1, 1961.

23. Harding, D.: High Elongation Measurements with Foil and Liquid Metal Strain Gages, *Proc. West. Reg. Strain Gage Comm.*, pp. 23–31, 1965.

24. Stone, J. E., N. H. Madsen, J. L. Milton, W. F. Swinson, and J. L. Turner: "Developments in the Design and Use of Liquid-Metal Strain Gages," in *Strain-Gage and Transducer Techniques*, Soc. Exp. Mech. publication, 1984, pp. 45–55.

25. Stein, P. K.: *Advanced Strain Gage Techniques*, chap. 2, Stein Engineering Services, Phoenix, 1962.

26. Bickle, L. W.: The Response of Strain Gages to Longitudinally Sweeping Strain Pulses, *Exp. Mech.*, vol. 10, no. 8, pp. 333–337, 1970.

27. Oi, K.: Transient Response of Bonded Strain Gages, *Exp. Mech.*, vol. 6, no. 9, pp. 463–469, 1966.

28. Freynik, H. S., Jr.: "Investigation of Current Carrying Capacity of Bonded Resistance Strain Gages," M.S. thesis, Massachusetts Institute of Technology, Cambridge, 1961.

29. "Optimizing Strain Gage Excitation Levels," Micro-Measurements Tech. Note TN-502, 1979.

30. Beyer, F. R., and M. J. Lebow: Long-Time Strain Measurements in Reinforced Concrete, *Proc. SESA*, vol. XI, no. 2, pp. 141–152, 1954.

31. Freynik, H. S., and G. R. Dittbenner: "Strain Gage Stability Measurements for a Year at 75°C in Air," Univ. Calif. Radiat. Lab. Rep. 76039, 1975.

32. Dean, M.: Strain Gage Waterproofing Methods and Installation of Gages in Propeller Strut of U.S.S. Saratoga, *Proc. SESA*, vol. XVI, no. 1, pp. 137–150, 1958.

33. Wells, F. E.: A Rapid Method of Waterproofing Bonded Wire Strain Gages, *Proc. SESA*, vol. XV, no. 2, pp. 107–110, 1958.

34. Milligan, R. V.: The Effects of High Pressure on Foil Strain Gages, *Exp. Mech.*, vol. 4, no. 2, pp. 25–36, 1964.

35. Brace, W. F.: Effect of Pressure on Electrical-Resistance Strain Gages, *Exp. Mech.*, vol. 4, no. 7, pp. 212–216, 1964.

36. Milligan, R. V.: Effects of High Pressure on Foil Strain Gages on Convex and Concave Surfaces, *Exp. Mech.*, vol. 5, no. 2, pp. 59–64, 1965.

37. Anderson, S. D., and R. C. Strahm: Nuclear Radiation Effects on Strain Gages, *Proc. West. Reg. Strain Gage Comm.*, pp. 9–16, 1968.

38. Tallman, C. R.: Nuclear Radiation Effects on Strain Gages, *Proc. West. Reg. Strain Gage Comm.*, pp. 17–25, 1968.

39. Wnuk, S. P.: Progress in High Temperature and Radiation Resistant Strain Gage Development, *Proc. West. Reg. Strain Gage Comm.*, pp. 41–47, 1964.

40. Day, E. E.: Characteristic of Electric Strain Gages at Elevated Temperatures, *Proc. SESA*, vol. IX, no. 1, pp. 141–150, 1951.

41. Wnuk, S. P.: New Strain Gage Developments, *Proc. West. Reg. Strain Gage Comm.*, pp. 1–7, 1964.

42. Hayes, J. K., and G. Roberts: Measurements of Stresses under Elastic-Plastic Strain Conditions at Elevated Temperatures, *Proc. Tech. Comm. Strain Gages*, pp. 20–39, 1969.

43. Denyssen, I. P.: Platinum-Tungsten Foil Strain Gage for High-Temperature Applications. *Proc. West. Reg. Strain Gage Comm.*, pp. 19–23, 1964.

44. Weymouth, L. J.: Strain Measurement in Hostile Environment, *Appl. Mech. Rev.*, vol. 18, no. 1, pp. 1–4, 1965.

45. Noltingk, B. E.: Measuring Static Strains at High Temperatures, *Exp. Mech.*, vol. 15, no. 10, pp. 420–423, 1975.

46. Harting, D. R.: "Evaluation of a Capacitive Strain Measuring System for Use to 1500°F," ISA ASI publication 75251, 1975, pp. 289–297.

47. Telinde, J. C.: Strain Gages in Cryogenic Environment, *Exp. Mech.*, vol. 10, no. 9, pp. 394–400, 1970.

48. Telinde, J. C.: Strain Gages in Cryogenics and Hard Vacuum, *Proc. West. Reg. Strain Gage Comm.*, pp. 45–54, 1968.
49. Dorsey, J.: New Developments and Strain Gage Progress, *Proc. West. Reg. Strain Gage Comm.*, pp. 1–10, 1965.
50. "Fatigue Characteristics of Micro-Measurement Strain Gages," Micro-Measurements Tech. Note 130-2, 1974.

CHAPTER

7

STRAIN-GAGE CIRCUITS

7.1 INTRODUCTION

An electrical-resistance strain gage will change in resistance due to applied strain according to Eq. (6.5), which indicates

$$\frac{\Delta R}{R} = S_g \epsilon_{xx} \qquad (6.5)$$

where the gage axis coincides with the x axis and $\epsilon_{yy} = -0.285\epsilon_{xx}$. In order to apply the electrical-resistance strain gage in any experimental stress analysis, the quantity $\Delta R/R$ must be measured and converted to the strain which produced the resistance change. Two electrical circuits, the potentiometer circuit and the Wheatstone bridge circuit, are commonly employed to convert the value of $\Delta R/R$ to a voltage signal (denoted here as ΔE) which can be measured with a recording instrument.

In this chapter the basic theory for the two circuits is presented, and the circuit sensitivities and ranges are derived. In addition, temperature compensation, signal addition, and loading effects are discussed in detail. Also covered are the constant-current circuits recommended for gages that exhibit large changes in $\Delta R/R$. Methods of calibrating both the potentiometer and the Wheatstone bridge circuits are covered, and the effects of lead wires on noise, calibration, and temperature compensation are discussed.

Insofar as possible, this discussion has been kept fundamental without reference to commercially available circuits. The material presented here is applicable to commercial instruments since their design is based on the fundamental principles covered in this chapter.

The methods of measuring the circuit output voltages ΔE are discussed in Chap 8.

7.2 THE POTENTIOMETER AND ITS APPLICATION TO STRAIN MEASUREMENT [1]

The potentiometer circuit, which is often employed in dynamic strain-gage applications to convert the gage output $\Delta R/R$ to a voltage signal ΔE, is shown in Fig. 7.1. For fixed-value resistors R_1 and R_2 in the circuit, the open-circuit output voltage E can be expressed as

$$E = \frac{R_1}{R_1 + R_2} V = \frac{1}{1 + r} V \tag{7.1}$$

where V is the input voltage and $r = R_2/R_1$ is the resistance ratio for the circuit. If incremental changes ΔR_1 and ΔR_2 occur in the value of the resistors R_1 and R_2, the change ΔE of the output voltage E can be computed by using Eq. (7.1) as follows:

$$E + \Delta E = \frac{R_1 + \Delta R_1}{R_1 + \Delta R_1 + R_2 + \Delta R_2} V \tag{a}$$

Solving Eq. (a) for ΔE gives

$$\Delta E = \left(\frac{R_1 + \Delta R_1}{R_1 + \Delta R_1 + R_2 + \Delta R_2} - \frac{R_1}{R_1 + R_2} \right) V \tag{b}$$

which can be expressed in the following form by introducing $r = R_2/R_1$:

$$\Delta E = \frac{[r/(1 + r)^2](\Delta R_1/R_1 - \Delta R_2/R_2)}{1 + [1/(1 + r)][\Delta R_1/R_1 + r(\Delta R_2/R_2)]} V \tag{7.2}$$

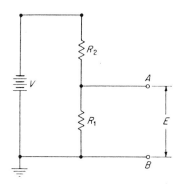

FIGURE 7.1
The potentiometer circuit.

Examination of Eq. (7.2) shows that the voltage signal ΔE from the potentiometer circuit is a nonlinear function of $\Delta R_1/R_1$ and $\Delta R_2/R_2$. To inspect the nonlinear aspects of this circuit further, it is possible to rewrite Eq. (7.2) in the form

$$\Delta E = \frac{r}{(1+r)^2}\left(\frac{\Delta R_1}{R_2} - \frac{\Delta R_2}{R_2}\right)(1-\eta)V \tag{7.3a}$$

where the nonlinear term η is expressed as

$$\eta = 1 - \frac{1}{1 + [1/(1+r)][\Delta R_1/R_1 + r(\Delta R_2/R_2)]} \tag{7.3b}$$

Equations (7.3) are the basic relationships which govern the behavior of the potentiometer circuit, and as such they can be used to establish the applicability of this circuit for strain-gage measurements.

7.2.1   Range and Sensitivity of the Potentiometer Circuit

The range of the potentiometer circuit is effectively established by the nonlinear term η, as expressed in Eq. (7.3b). If a strain gage of resistance R_g is used in the position of R_1, and if a fixed-ballast resistor R_b is employed in the position of R_2, then

$$R_1 = R_g \qquad \Delta R_1 = \Delta R_g \qquad R_2 = R_b \qquad \Delta R_2 = 0$$

and Eq. (7.3b) becomes

$$\eta = 1 - \frac{1}{1 + [1/(1+r)](\Delta R_g/R_g)} \tag{a}$$

By substituting Eq. (6.5) into Eq. (a), the nonlinear term is given in terms of strain ϵ, the gage factor S_g, and the resistance ratio $r = R_b/R_g$ as

$$\eta = 1 - \frac{1}{1 + [1/(1+r)]S_g\epsilon} \tag{b}$$

This expression can be evaluated by means of a series expansion where

$$\eta = 1 - \frac{1}{1+x} = x - x^2 + x^3 - x^4 + \cdots$$

and

$$x = \frac{S_g\epsilon}{1+r} \tag{7.4}$$

The range of the potentiometer circuit depends upon the error which can be tolerated due to the nonlinearities introduced by the circuit. The nonlinear term η depends on S_g, r, and the strain ϵ. For example, if $r = 9$ and $S_g = 2$, then $x = \epsilon/5$

and it is easy to show that η will be 2 percent when $\epsilon = 10$ percent. Increasing the value of r decreases the error due to circuit nonlinearities. The range of the potentiometer circuit, based on an allowable error of 2 percent, varies from 10 to 2 percent strain as r varies from a value of 9 to 2. Obviously the range of the potentiometer circuit is sufficient for determining elastic strains in metallic components. However, elastic-strain determinations in plastic materials and plastic-strain determinations in metallic materials may exceed linearity limits and require corrections of the output signal.

The sensitivity of the potentiometer circuit is defined as the ratio of the output voltage divided by the strain:

$$S_c = \frac{\Delta E}{\epsilon} \tag{7.5}$$

It is apparent that the relative merit of a circuit can be judged by the magnitude of the circuit sensitivity S_c. In particular for dynamic applications, where relatively small strains are being measured, it is quite important that S_c be made as large as possible to reduce the degree of amplification necessary to read the output signal.

The sensitivity of the potentiometer circuit can be established by substituting Eq. (7.3a) into Eq. (7.5) to obtain

$$S_c = \frac{r}{(1+r)^2} \left(\frac{\Delta R_1}{R_1} - \frac{\Delta R_2}{R_2} \right) \frac{V}{\epsilon} \tag{7.6}$$

where $\eta \ll 1$.

If a strain gage is placed in the circuit in place of R_1 and a ballast resistor is employed for R_2, then

$$R_g = R_1 \qquad \Delta R_g = \Delta R_1 \qquad R_b = R_2 \qquad \Delta R_2 = 0$$

and by Eqs. (6.5) and (7.6) the circuit sensitivity becomes

$$S_c = \frac{r}{(1+r)^2} S_g V \tag{c}$$

This equation indicates that the circuit sensitivity is controlled by the resistance ratio r and the circuit voltage V. However, since V can be varied, the limiting factor becomes the power which can be dissipated by the gage. This power must be held within the limits set forth in Sec. 6.6.5. To account for the influence of the power P_g dissipated by the gage, the relationship between P_g and V given by Eq. (6.18) is applied to the potentiometer circuit to give

$$V = I_g R_g (1+r) = \sqrt{P_g R_g}(1+r) \tag{7.7}$$

Substituting Eq. (7.7) into Eq. (c) yields

$$S_c = \frac{r}{1+r} S_g \sqrt{P_g R_g} \tag{7.8}$$

This equation shows that the circuit sensitivity is controlled by two independent parameters, $r/(1 + r)$ and $S_g\sqrt{P_g R_g}$. The first parameter $r/(1 + r)$ is related to the circuit and is dictated by the selection of R_b. It is clear that as r becomes very large, $r/(1 + r)$ approaches 1 and maximum circuit efficiency is obtained. However, for very high values of r, the voltage required to drive the potentiometer becomes excessive. Values of r of about 9 are commonly used, which gives a circuit efficiency of approximately 90 percent. The circuit sensitivity is strongly influenced by the second parameter $S_g\sqrt{P_g R_g}$, which is determined entirely by the selection of the strain gage and is totally independent of the elements used in the design of the potentiometer circuit. In fact, if very high circuit sensitivities are required, much can be gained by careful gage selection, since $S_g\sqrt{P_g R_g}$ can be varied over wide limits (i.e., three orders of magnitude) while circuit efficiency can be varied over a much more limited range (0.5 to 1).

The output voltage from any circuit containing a strain gage is inherently low, and as a consequence S_c is quite often as low as 5 or 10 $\mu V/\mu\epsilon$ of strain. Exercise 7.3 shows typical values of S_c as well as the voltage required to drive a potentiometer circuit with a circuit efficiency of 83 percent.

7.2.2   Temperature Compensation and Signal Addition in the Potentiometer Circuit

It is possible to effect a degree of temperature compensation in the strain-measuring system by employing two strain gages in the potentiometer circuit. The gage used in place of R_1 is known as the *active strain gage* and is used to measure the strain on a specimen. The gage used to replace R_2 is known as the *dummy gage* and is mounted on a small block of material identical to that of the specimen and is exposed to the same thermal environment as the active gage. In the active gage the total change in resistance will be due to strain ΔR_ϵ and to changes in temperature ΔR_T. Hence,

$$\frac{\Delta R_1}{R_1} = \frac{\Delta R_\epsilon}{R_1} + \frac{\Delta R_T}{R_1} \tag{a}$$

However, in the dummy gage the change in resistance is due only to a change in temperature and

$$\frac{\Delta R_2}{R_2} = \frac{\Delta R_T}{R_2} \tag{b}$$

Substituting Eqs. (*a*) and (*b*) into Eq. (7.3*a*) gives

$$\Delta E = \frac{r}{(1 + r)^2}\left(\frac{\Delta R_\epsilon}{R_1} + \frac{\Delta R_T}{R_1} - \frac{\Delta R_T}{R_2}\right)(1 - \eta)V \tag{7.9}$$

Since $R_1 = R_2 = R_g$, it is evident that the ΔR_T terms cancel out of Eq. (7.9) and the output signal ΔE is due to the ΔR_ϵ term alone.

This type of temperature compensation is effective if the dummy and the active gages are from the same lot of gages, if the materials upon which each gage is mounted are identical, and if the temperature changes due to ambient conditions and gage currents on both gages are equal. If any one of these three conditions is violated, complete temperature compensation cannot be achieved.

A loss in circuit efficiency, as determined by $r/(1 + r)$, is the price which must be paid for temperature compensation. Since $R_1 = R_2 = R_g$ is a requirement for temperature compensation, $r = 1$, and the circuit is fixed at 50 percent. If circuit sensitivity is extremely important, temperature compensation by this means should not be specified, since temperature-compensated gages (discussed in Sec. 6.6.2) can be employed to accomplish the same purpose.

Resistance changes from two active strain gages placed in positions R_1 and R_2 can be used to increase the output signal in the potentiometer circuit in certain cases. As an example, consider the application shown in Fig. 7.2, where two gages are mounted on a beam in bending. The gages mounted on the two surfaces of the beam will exhibit resistance changes which can be expressed as

$$\frac{\Delta R_1}{R_1} = S_g f(P) \qquad \text{and} \qquad \frac{\Delta R_2}{R_2} = -S_g f(P) \tag{c}$$

where

$$f(P) = \frac{6Pl_1}{bh^2 E} = \epsilon \tag{d}$$

The signal output from the potentiometer circuit containing these two active gages can be computed from Eqs. (7.3) as

$$\Delta E = \frac{r}{(1 + r)^2} (2S_g \epsilon) V \tag{e}$$

which can be reduced to

$$\Delta E = \frac{r}{1 + r} 2S_g \sqrt{P_g R_g} \epsilon = S_g \sqrt{P_g R_g} \epsilon \tag{7.10}$$

where $r = 1$ since $R_1 = R_2 = R_g$. If one active gage had been employed, the relation for ΔE would have been

$$\Delta E = \frac{r}{1 + r} S_g \sqrt{P_g R_g} \epsilon \tag{f}$$

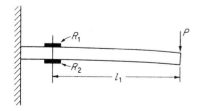

FIGURE 7.2
Strain gages mounted on the top and bottom surfaces of a cantilever beam.

Comparison of Eq. (7.10) and Eq. (f) indicates that the output signal has been doubled, based on a value of $r = 1$. However, if r is increased to, say 9 in Eq. (f), the advantage of using two gages to increase voltage output is almost completely nullified.

Multiple-gage circuits are more important in transducer applications of strain gages, where the signals due to unwanted components of load are canceled in the circuit (see Sec. 7.8).

7.2.3 Potentiometer Output

The output of a potentiometer circuit is measured across terminals A and B, as indicated in Fig. 7.1, to give a total voltage $E + \Delta E$. This output, which corresponds to a strain varying sinusoidally with time, is shown in Fig. 7.3. Normally the voltage E is between 5 and 20 V, and ΔE is measured in terms of microvolts. The voltage related to the strain is ΔE, and this quantity must be measured independently of the large superimposed dc voltage E. Unfortunately, most voltage-measuring instruments do not have sufficient range and sensitivity to measure ΔE when it is superimposed on E. Thus, the potentiometer circuit is usually inadequate for static strain-gage applications where both E and ΔE are constant with respect to time. The potentiometer circuit is quite useful, however, in dynamic applications where ΔE varies with time and E is a dc voltage. In these applications it is possible to block the dc voltage by using a suitable filter while passing the voltage ΔE without significant distortion. The filter serves to reduce the required range of the measuring instrument from $E + \Delta E$ to ΔE and permits the use of highly sensitive but limited-range measuring instruments. A simple RC filter used with a potentiometer circuit is illustrated in Fig. 7.4.

If R_M is large in comparison with $R_1 R_2/(R_1 + R_2)$, the voltage E' measured

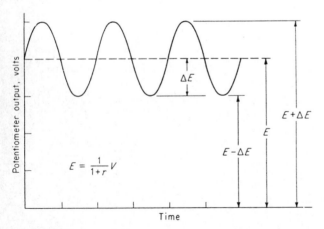

FIGURE 7.3
Potentiometer output as a function of time for a sinusoidally varying strain.

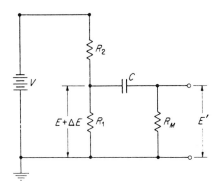

FIGURE 7.4
Filter for blocking out the dc component of the potentiometer-circuit output.

across the resistance of the measuring instrument R_M is

$$E' = \frac{R_M}{\sqrt{R_M^2 + (1/\omega C)^2}} (E + \Delta E) \tag{7.11}$$

where C is the capacitance used in the filter and ω is the angular frequency of the voltage $E + \Delta E$. If one considers that the voltage $E + \Delta E$ is made up of many frequencies varying from zero (direct current) to several kHz, it is possible to show that $E'/(E + \Delta E)$ goes to 0.99 as $\omega R_M C \to 5$. The student is referred to Exercise 7.6, where the validity of this statement is established.

Since the dc component of the potentiometer output voltage $E + \Delta E$ is constant with time (that is, $\omega = 0$), this component of the signal is entirely blocked by the filter. If $R_M = 1$ MΩ and $C = 0.1$ μF, the RC constant for the filter is 10^{-1}, and the frequency components associated with the ΔE voltage will pass through the filter with negligible distortion provided they exceed 30 rad/s or about 5 Hz. This filter arrangement can be employed to adapt the potentiometer circuit to dynamic strain-gage applications whenever the dynamic strains are composed of frequency components which exceed about 5 Hz.

7.2.4 Summary of the Potentiometer Circuit

The equations which govern the behavior of the potentiometer circuit as it is applied to dynamic strain measurement are summarized below:

$$E = \frac{1}{1 + r} V \tag{7.1}$$

$$\Delta E = \frac{r}{(1 + r)^2} \left(\frac{\Delta R_1}{R_1} - \frac{\Delta R_2}{R_2} \right)(1 - \eta)V \tag{7.3a}$$

$$\eta = 1 - \frac{1}{1 + [1/(1 + r)][\Delta R_1/R_1 + r(\Delta R_2/R_2)]} \tag{7.3b}$$

$$V = \sqrt{P_g R_g}(1 + r) \qquad (7.7)$$

$$S_c = \frac{r}{1 + r} S_g \sqrt{P_g R_g} \qquad (7.8)$$

$$E' = \frac{R_M}{\sqrt{R_M^2 + (1/\omega C)^2}}(E + \Delta E) \qquad (7.11)$$

The potentiometer circuit is not usually employed to measure the resistance change of a gage due to static strains because of the voltage E, which appears across its output terminals and swamps out the small voltage ΔE which is related to the resistance change. The circuit can, however, be employed to measure the resistance change in the gage due to dynamic strains if a suitable filter is used to block out the dc voltage E while passing the voltage pulse ΔE without distortion.

For most strain-gage applications, the range over which the potentiometer circuit can respond is larger than the strains which are to be measured. However, for more specialized problems where strains of about 10 percent are to be measured, corrections must be made to account for nonlinearities introduced by the circuit.

The sensitivity of a potentiometer measuring system is controlled by the circuit efficiency $r/(1 + r)$ and the gage selection which dictates the value of $S_g\sqrt{P_g R_g}$. Of the two factors, the gage selection is more important because of its greater variability (three orders of magnitude). The circuit efficiency usually is maintained at about 90 percent by selecting $R_b = 9R_g$ and providing sufficiently high voltage V to drive the allowable current through the gage.

The potentiometer circuit can be used to add or subtract signals from a multiple-strain-gage installation. However, the gain in output voltage by the addition of two strain-gage signals is usually not sufficient to warrant the use of two gages. For transducer applications, multiple-gage circuits should be considered to eliminate response from components (of, say, the load) which are not sought in the measurement.

Temperature compensation can be introduced into the circuit by employing both a dummy and an active gage. However, the circuit efficiency is reduced to 50 percent in this circuit arrangement. If low-magnitude strains are to be measured, it is usually more advisable to employ temperature-compensated gages and seek higher circuit efficiencies by using a fixed-value ballast resistor in place of the dummy gage required for temperature compensation within the circuit.

The output of the potentiometer circuit ΔE is directly proportional to the supply voltage V. Thus, it is imperative that the voltage supply provide a stable voltage V over the period of readout. Batteries serve as excellent power supplies for the potentiometer circuit, provided their rate of decay is small over the period of readout (this period of readout is usually quite short in dynamic applications).

Finally, the potentiometer circuit can be grounded together with the amplifier. This feature represents a real advantage when the signal-to-noise ratio is low and it is important to reduce the electronic noise level.

7.3 THE WHEATSTONE BRIDGE [2]

The Wheatstone bridge is a second circuit that can be employed to determine the change in resistance which a gage undergoes when it is subjected to a strain. Unlike the potentiometer, the Wheatstone bridge can be used to determine both dynamic and static strain-gage readings. The bridge may be used as a direct-readout device, where the output voltage ΔE is measured and related to strain. Also, the bridge may be used as a null-balance system, where the output voltage ΔE is adjusted to a zero value by adjusting the resistive balance of the bridge. In either of these modes of operation, the bridge can be effectively employed in a wide variety of strain-gage applications.

To show the principle of operation of the Wheatstone bridge as a direct-readout device (where ΔE is measured to determine the strain), consider the circuit shown in Fig. 7.5. The voltage drop across R_1 is denoted as V_{AB} and is given as

$$V_{AB} = \frac{R_1}{R_1 + R_2} V \qquad (a)$$

and, similarly, the voltage drop across R_4 is denoted as V_{AD} and is given by

$$V_{CD} = \frac{R_4}{R_3 + R_4} V \qquad (b)$$

The output voltage E from the bridge is equivalent to V_{BD}, which is

$$E = V_{BD} = V_{AB} - V_{AD} \qquad (c)$$

Substituting Eqs. (a) and (b) into Eq. (c) and simplifying gives

$$E = \frac{R_1 R_3 - R_2 R_4}{(R_1 + R_2)(R_3 + R_4)} V \qquad (7.12)$$

The voltage E will go to zero and the bridge will be considered in balance when

$$R_1 R_3 = R_2 R_4 \qquad (7.13)$$

It is this zeroing feature which permits the Wheatstone bridge to be employed for static strain measurements. The bridge is initially balanced ($E = 0$) before strains

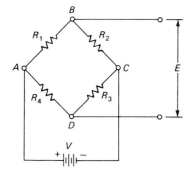

FIGURE 7.5
The Wheatstone bridge circuit.

are applied to the gages in the bridge. Then the strain-induced voltage ΔE can be measured relative to a zero voltage for both static and dynamic applications.

Consider an initially balanced bridge with $R_1 R_3 = R_2 R_4$ so that $E = 0$ and then change each value of resistance R_1, R_2, R_3, and R_4 by an incremental amount ΔR_1, ΔR_2, ΔR_3, and ΔR_4. The voltage output ΔE of the bridge can be obtained from Eq. (7.12), which becomes

$$\Delta E = V \frac{\begin{vmatrix} R_1 + \Delta R_1 & R_2 + \Delta R_2 \\ R_4 + \Delta R_4 & R_3 + \Delta R_3 \end{vmatrix}}{\begin{vmatrix} R_1 + \Delta R_1 + R_2 + \Delta R_2 & 0 \\ 0 & R_3 + \Delta R_3 + R_4 + \Delta R_4 \end{vmatrix}} = V \frac{A}{B} \qquad (d)$$

where A is the determinant in the numerator and B is the determinant in the denominator. By expanding each of these determinants, neglecting second-order terms, and noting that $R_1 R_3 = R_2 R_4$, it is possible to show that

$$A = R_1 R_3 \left(\frac{\Delta R_1}{R_1} - \frac{\Delta R_2}{R_2} + \frac{\Delta R_3}{R_3} - \frac{\Delta R_4}{R_4} \right) \qquad (e)$$

and

$$B = \frac{R_1 R_3 (R_1 + R_2)^2}{R_1 R_2} \qquad (f)$$

Combining Eqs. (d), (e), and (f) yields

$$\Delta E = V \frac{R_1 R_2}{(R_1 + R_2)^2} \left(\frac{\Delta R_1}{R_1} - \frac{\Delta R_2}{R_2} + \frac{\Delta R_3}{R_3} - \frac{\Delta R_4}{R_4} \right) \qquad (7.14)$$

By letting $R_2 / R_1 = r$, it is possible to rewrite Eq. (7.14) as

$$\Delta E = V \frac{r}{(1 + r)^2} \left(\frac{\Delta R_1}{R_1} - \frac{\Delta R_2}{R_2} + \frac{\Delta R_3}{R_3} - \frac{\Delta R_4}{R_4} \right) \qquad (7.15)$$

In reality, Eqs. (7.14) and (7.15) both carry a nonlinear term $(1 - \eta)$, as defined in Exercise 7.8. However, the influence of the nonlinear term is quite small and can be neglected, provided the strains being measured are less than 5 percent. Equation (7.15) thus represents the basic equation which governs the behavior of the Wheatstone bridge in strain measurement.

7.3.1 Wheatstone Bridge Sensitivity

The sensitivity of the Wheatstone bridge must be considered from two points of view: (1) with a fixed voltage applied to the bridge regardless of gage current (a condition which exists in most commercially available instrumentation) and (2) with a variable voltage whose upper limit is determined by the power dissipated by the particular arm of the bridge which contains the strain gage. By recalling the definition for the circuit sensitivity given in Eq. (7.5), and using the basic bridge relationship given in Eq. (7.15), it is clear that

$$S_c = \frac{\Delta E}{\epsilon} = \frac{V}{\epsilon} \frac{r}{(1+r)^2} \left(\frac{\Delta R_1}{R_1} - \frac{\Delta R_2}{R_2} + \frac{\Delta R_3}{R_3} - \frac{\Delta R_4}{R_4} \right) \qquad (a)$$

If a multiple-gage circuit is considered with n gages (where $n = 1, 2, 3,$ or 4) whose outputs sum when placed in the bridge circuit, it is possible to write

$$\sum_{m=1}^{m=n} \frac{\Delta R_m}{R_m} = n \frac{\Delta R}{R} = nS_g \epsilon \qquad (b)$$

Substituting Eq. (b) into Eq. (a) gives the circuit sensitivity as

$$S_c = V \frac{r}{(1+r)^2} nS_g \qquad (7.16)$$

This sensitivity equation is applicable in those cases where the bridge voltage V is fixed and independent of the bridge resistance. The equation shows that the sensitivity of the bridge depends upon the number n of active arms employed, the gage factor S_g, the input voltage V, and the ratio of the resistances $r = R_2/R_1$. A graph of r versus $r/(1+r)^2$ (the circuit efficiency) in Fig. 7.6 shows that maximum circuit sensitivity occurs when $r = 1$. With four active arms in this bridge, a circuit sensitivity of $S_g V$ can be achieved, whereas with one active gage and $r = 1$ a circuit sensitivity of only $S_g V/4$ can be obtained.

When the bridge supply voltage V is selected to drive the gages in the bridge so that they dissipate the maximum allowable power, a different sensitivity equation must be employed. Since the gage current is a limiting factor in this approach, the number of gages used in the bridge and their relative position are important. To show this fact, consider the following four cases, illustrated in Fig. 7.7, which represent the most common bridge arrangements.

CASE 1. This bridge arrangement consists of a single active gage in position R_1 and is employed for many dynamic and some static strain measurements where

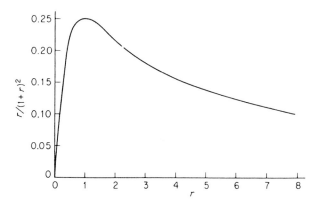

FIGURE 7.6
Circuit efficiency as a function of r for a fixed-voltage bridge.

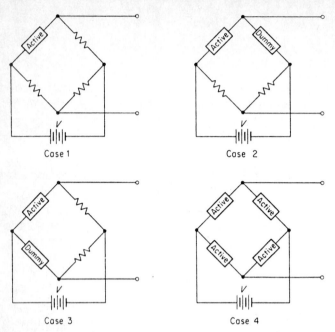

FIGURE 7.7
Four common bridge arrangements.

temperature compensation in the circuit is not critical. The value of R_1 is, of course, equal to R_g, but the value of the other three resistors may be selected to maximize circuit sensitivity, provided the initial balance condition $R_1 R_3 = R_2 R_4$ is maintained. The power dissipated by the gage can be determined from

$$V = I_g(R_1 + R_2) = I_g R_g(1 + r) = (1 + r)\sqrt{P_g R_g} \qquad (c)$$

By combining Eqs. (7.16) and (c) and recalling that $r = R_2/R_1$, it is possible to obtain the circuit sensitivity in the following form:

$$S_c = \frac{r}{1 + r} S_g \sqrt{P_g R_g} \qquad (7.17)$$

Here it is evident that the circuit sensitivity of the bridge is due to two factors, namely, the circuit efficiency, which is given by $r/(1 + r)$, and the gage selection, represented by the term $S_g\sqrt{P_g R_g}$. For this bridge arrangement, r should be selected as high as possible to increase circuit efficiency but not high enough to increase the supply voltage beyond reasonable limits. A value of $r = 9$ gives a circuit efficiency of 90 percent and, with a 120-Ω gage dissipating 0.15 W, requires a supply voltage of 42.4 V.

CASE 2. This bridge arrangement employs one active gage in arm R_1 and one dummy gage in arm R_2 which is utilized for temperature compensation (see

Exercise 7.12). The value of the gage current which passes through both gages (note $R_1 = R_2 = R_g$) is given by

$$V = 2I_g R_g \qquad (d)$$

Substituting Eq. (d) into Eq. (7.16) with $n = 1$ and $r = 1$ gives S_c as

$$S_c = \tfrac{1}{2} S_g \sqrt{P_g R_g} \qquad (7.18)$$

In this instance the circuit efficiency is fixed at a value of 0.5 since the condition $R_1 = R_2 = R_g$ requires that $r = 1$. Thus, it is clear that the placement of a dummy gage in position R_2 to effect temperature compensation reduces circuit efficiency to 50 percent. The gage selection, of course, maintains its importance in this arrangement since the term $S_g \sqrt{P_g R_g}$ again appears in the circuit-sensitivity equation. In fact, this term will appear in the same form in all four bridge arrangements covered in this section. It also appeared in the circuit sensitivity of the potentiometer circuit in Eq. (7.8).

CASE 3. The bridge arrangement in this case incorporates an active gage in the R_1 position and a dummy gage in the R_4 position. Fixed resistors of any value are placed in positions R_2 and R_3. With these gage positions, the bridge is temperature-compensated since the temperature-introduced resistance changes are canceled out in the Wheatstone bridge circuit. The current through both the active gage and the dummy gage is given by

$$V = I_g(R_1 + R_2) = (1 + r)\sqrt{P_g R_g} \qquad (e)$$

Substituting Eq. (e) into Eq. (7.16) and recalling for this case that $n = 1$ gives the circuit sensitivity as

$$S_c = \frac{r}{1 + r} S_g \sqrt{P_g R_g} \qquad (7.19)$$

The circuit sensitivity for this bridge arrangement (case 3) is identical to that which can be achieved with the bridge circuit shown in case 1. Thus, temperature compensation can be obtained without any loss in circuit sensitivity only if the dummy gage is placed in position R_4.

CASE 4. In this bridge arrangement, four active gages are placed in the bridge, with one gage in each of the four arms. If the gages are placed on, say, a beam in bending, as shown in Fig. 7.8, the signals from each of the four gages will add and the value of n given in Eq. (7.16) will be equal to 4. The power dissipated by each of the four gages is given by

$$V = 2I_g R_g = \sqrt{P_g R_g} \qquad (f)$$

Since the resistance is the same for all four gages, $r = 1$. The circuit sensitivity for this bridge arrangement is obtained by substituting Eq. (f) into Eq. (7.16):

$$S_c = 2I_g R_g S_g = 2S_g \sqrt{P_g R_g} \qquad (7.20)$$

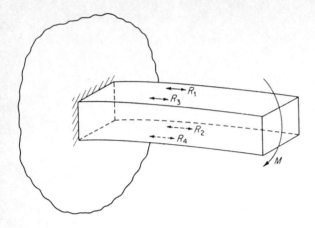

FIGURE 7.8
Positions of four gages employed on a beam in bending to give a bridge factor $n = 4$.

A four-active-arm bridge is slightly more than twice as sensitive as the single-active-arm bridges (case 1 or case 3). Also, this bridge arrangement is temperature-compensated. The four gages employed are a relatively high price to pay for this increased sensitivity. The two-active-arm bridge presented in Exercise 7.13 exhibits a sensitivity which approaches that given in Eq. (7.20).

Examination of these four bridge arrangements plus the fifth arrangement presented in Exercise 7.13 shows that the circuit sensitivity can be varied between 0.5 and 2 times $S_g\sqrt{P_gR_g}$. Temperature compensation for a single active gage in position R_1 can be effected without loss in sensitivity by placing the dummy gage in position R_4. If the dummy gage is placed in position R_2, the circuit sensitivity is reduced by a factor of 2. Circuit sensitivities can be improved by the use of multiple-active-arm bridges, as illustrated in cases 4 and 5 (see Exercise 7.13). However, the cost of installing additional gages for a given strain measurement is rarely warranted except for transducer applications. For experimental stress analyses, single-active-arm bridges are normally employed, and the signal from the bridge is usually amplified from 10 to 1000 times before readings are taken.

7.3.2 Commercial Strain Indicators—Null Balance

One of the most commonly employed bridge arrangements for static strain measurements is the reference bridge, which is used in several commercial instruments. A schematic illustration of the reference bridge arrangement is presented in Fig. 7.9, where two bridges are used together to provide a null-balance system.

In this reference-bridge arrangement the bridge on the left-hand side is employed to contain the strain gage or gages, and the bridge on the right-hand side contains either fixed or variable resistors. When gages are placed in the left-hand bridge, the variable resistance in the right-hand bridge is adjusted to obtain initial balance. Strains producing resistance changes in the left-hand bridge cause an unbalance between the two bridges and an associated meter reading.

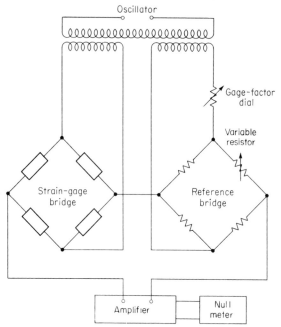

FIGURE 7.9
Schematic illustration of the reference bridge arrangement.

This meter reading is nulled out by further adjustments of the variable resistor in the right-hand bridge. The advantage of the dual-bridge arrangement is that the left-hand bridge is left completely free for the strain gages and all adjustments for both initial and null balance are performed on the right-hand bridge.

The strain-gage bridge is powered by an oscillator with a 1000-Hz square-wave output of 1.5 V (rms). There is no control over the magnitude of this input voltage; however, for null-balance systems, a fixed-value input voltage is not considered a serious limitation since in the null position the readout is independent of V. The strain indicator will function over a range of gage resistances from 50 to 2000 Ω. When the gage resistance is less than 50 Ω, the oscillator becomes overloaded. When the gage resistance is greater than 2000 Ω, the load on the bridge circuit due to the amplifier becomes excessive. The gage factor can be set to give readings which are directly calibrated in terms of strain provided $1.5 \leq S_g \leq 4.5$. The indicator can be read to ± 2 $\mu\epsilon$ and is accurate to ± 0.1 percent of the reading or 5 $\mu\epsilon$, whichever is greater. The range is $\pm 50,000$ $\mu\epsilon$. The unit, shown in Fig. 7.10, is small, lightweight, and portable. It is simple to operate and is adequate for almost all static strain-gage applications.

7.3.3 Commercial Strain Indicators—Direct Reading

Recent advances in voltage regulators, stable amplifiers, and analog-to-digital converters have permitted the development of a direct-reading Wheatstone bridge

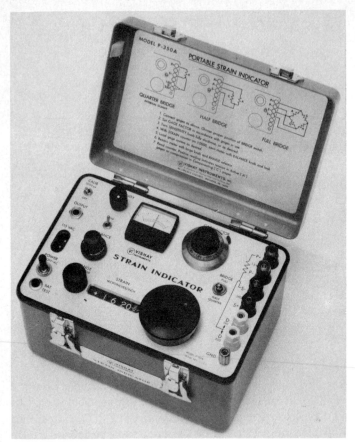

FIGURE 7.10
Model P-350A strain indicator. (*Measurements Group, Inc.*)

for strain-gage applications. A schematic diagram of the circuit elements employed in a direct-reading strain indicator is presented in Fig. 7.11. The bridge is powered by a battery supply equipped with a voltage regulator that applies a fixed 2.000 Vdc to the P^+ and P^- terminals. The P^- terminal is grounded with the negative of the battery supply. The bridge is equipped with fixed resistors and may be used in one-quarter-, one-half-, or full-bridge configurations. The one-quarter- or one-half-bridge arrangement accommodates either 120- or 350-Ω strain gages. Shunt resistors are provided which produce a calibration indication of 5000 $\mu\epsilon$ on the display if the gage factor is set at 2.000.

The output of the bridge is the input of an instrument amplifier with a gain of 40. A potentiometer adjustment is provided to initially balance the amplifier. A potentiometer is also provided to balance the bridge; however, it should be noted that the bridge output is nulled by adjusting a voltage on the instrument amplifier and not by adjusting a resistance in one arm of the bridge. The analog

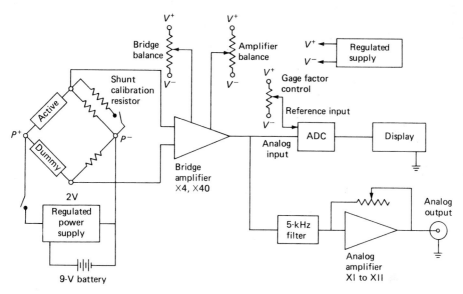

FIGURE 7.11
Schematic block diagram illustrating the key features of a direct-reading strain indicator.

output from the amplifier is converted into digital format and displayed on a $4\frac{1}{2}$-digit liquid-crystal display (LCD). When the amplifier gain is set at 40, a count of 1 in the least-significant digit corresponds to 1 $\mu\epsilon$. With a gain setting of 4 a count of 1 in the least-significant digit corresponds to 10 $\mu\epsilon$.

The gage factor is adjusted through a potentiometer which controls the reference voltage of the analog-to-digital converter. The $4\frac{1}{2}$-digit display is used in monitoring the adjustment of the reference voltage, and in this manner the indicator can be adjusted for a specific gage factor from 0.500 to 9.900 with a resolution of 0.001.

The output from the bridge amplifier is also supplied to a low-pass filter (5 kHz) and then to an analog amplifier. The analog amplifier has a variable gain which can be adjusted from $\times 1$ to $\times 11$. The amplified analog output is provided on a BNC connector for an oscilloscope lead. It should be noted that the bandwidth of the analog output is limited to 0 to 4 kHz and the signal is down 3 db at 4 kHz.

A commercial direct-reading strain indicator, shown in Fig. 7.12, is small ($9 \times 6 \times 6$ in, or $230 \times 150 \times 150$ mm), light, and portable. The battery supply (six alkaline D cells) provides 250 h of operation before replacement is necessary. The accuracy of the model P3500 indicator shown in Fig. 7.12 is 0.05 percent of the reading ± 3 $\mu\epsilon$.

Both the null balance and the direct-reading strain indicators are excellent instruments for static strain determinations. They are simple to operate and provide a long, reliable service life with minimum maintenance. The primary

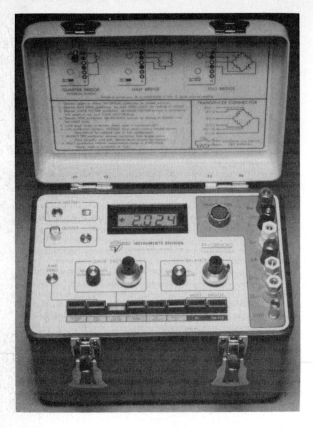

FIGURE 7.12
Model P3500 direct-reading strain
indicator. (*Measurements Group,
Inc.*)

advantage of the direct-reading instrument is in the speed of operation. Readings
can be taken as rapidly as the switch box can be activated and the indication on
the display recorded (about 10 s).

7.3.4 Summary of the Wheatstone Bridge Circuit

The equations which govern the behavior of the Wheatstone bridge circuit, under
initial balance conditions, for both static and dynamic strain measurements are
summarized below:

$$\Delta E = V \frac{R_1 R_2}{(R_1 + R_2)^2} \left(\frac{\Delta R_1}{R_1} - \frac{\Delta R_2}{R_2} + \frac{\Delta R_3}{R_3} - \frac{\Delta R_4}{R_4} \right) \tag{7.14}$$

$$\Delta E = V \frac{r}{(1 + r)^2} \left(\frac{\Delta R_1}{R_1} - \frac{\Delta R_2}{R_2} + \frac{\Delta R_3}{R_3} - \frac{\Delta R_4}{R_4} \right) \tag{7.15}$$

$$S_c = V \frac{r}{(1 + r)^2} n S_g \qquad (7.16)$$

$$S_c = \frac{r}{1 + r} S_g \sqrt{P_g R_g} \qquad (7.19)$$

The Wheatstone bridge circuit can be employed for both static and dynamic strain measurements since it can be initially balanced to yield a zero output voltage. The potentiometer circuit, on the other hand, is suitable only for dynamic measurements. Comparison of Eqs. (7.8) and (7.19) shows that the potentiometer circuit and the Wheatstone bridge circuit have the same circuit sensitivity.

The output voltage ΔE from the Wheatstone bridge is nonlinear with respect to resistance change ΔR, and the measuring instrument used to detect the output voltage can produce loading effects which are similar to those produced with the potentiometer circuit. No emphasis was placed on nonlinearity or on loading effects since they are not usually troublesome in the typical experimental stress analysis. Nonlinear effects can become significant when semiconductor gages are used to measure relatively large strains because of the large ΔR's involved. In these applications, the constant-current circuits described in Sec. 7.4 should be used. Loading effects should never be a problem since high-quality, low-cost measuring instruments with high input impedances are readily available.

The Wheatstone bridge can be used to add or subtract signals from multiple-strain-gage installations. A four-active-arm bridge is slightly more than twice as sensitive as an optimum single-active-arm bridge. For transducer applications, the gain in sensitivity as well as cancellation of gage response from components of load which are not being measured are important factors; therefore, the Wheatstone bridge is used almost exclusively for transducers using strain gages as sensors.

Temperature compensation can be employed without loss of sensitivity provided a separate dummy gage or another active gage is employed in arm R_4 of the bridge. Grounding of the Wheatstone bridge can be accomplished only at point C (see Fig. 7.5); consequently, noise can become more of a problem with the Wheatstone bridge than with the potentiometer circuit. For further details on the treatment of noise in strain-gage circuits, see Sec. 7.7.

7.4 CONSTANT-CURRENT CIRCUITS [3, 4]

The potentiometer and Wheatstone bridge circuits described in Secs. 7.2 and 7.3 were each driven with a voltage source which ideally provides a constant voltage even with changes in the resistance of the circuit. These voltage-driven circuits exhibit nonlinear outputs whenever $\Delta R/R$ is large [see Eq. (7.3b)]. This nonlinear behavior limits their applicability. It is possible to replace the constant-voltage source with a constant-current source, and it can be shown that improvements in both linearity and sensitivity result.

Constant-current power supplies with sufficient regulation for strain-gage applications are relatively new and have been made possible by advances in solid-state electronics. Basically, the constant-current power supply is a high-impedance (1- to 10-MΩ) device which changes output voltage with changing resistive load to maintain constant current.

7.4.1 Constant-Current Potentiometer Circuit

Consider the constant-current potentiometer circuit shown in Fig. 7.13a. When a very high impedance meter is placed across resistance R_1, the measured output voltage E is

$$E = IR_1 \tag{7.21}$$

When resistances R_1 and R_2 change by ΔR_1 and ΔR_2, the output voltage becomes

$$E + \Delta E = I(R_1 + \Delta R_1) \tag{a}$$

Thus from Eqs. (7.21) and (a)

$$\Delta E = (E + \Delta E) - E = I\,\Delta R_1 = IR_1 \frac{\Delta R_1}{R_1} \tag{7.22}$$

It should be noted that ΔR_2 does not affect the signal output. Indeed, even R_2 is not involved in the output voltage, and hence it can be eliminated to give the very simple potentiometer circuit shown in Fig. 7.13b.

Substituting Eq. (6.5) into Eq. (7.22) yields

$$\Delta E = IR_g S_g \epsilon \tag{7.23}$$

Equations (7.22) and (7.23) show that the output signal ΔE is linear with respect to resistance change ΔR and strain ϵ. It is this feature of the constant-current potentiometer circuit which makes it more suitable for measuring large changes in R than the constant-voltage potentiometer circuit.

The circuit sensitivity $S_c = \Delta E/\epsilon$ reduces to

$$S_c = IR_g S_g \tag{7.24}$$

If the constant-current source is adjustable, so that the current I can be increased

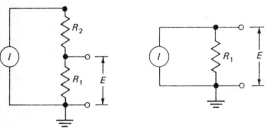

(a)   (b)

FIGURE 7.13
Constant-current potentiometer circuits.

to the power-dissipation limit of the strain gage, then $I = I_g$ and Eq. (7.24) can be rewritten as

$$S_c = \sqrt{P_g R_g S_g} \tag{7.25}$$

Thus, the circuit sensitivity is totally dependent on the strain-gage parameters P_g, R_g, and S_g and is totally independent of circuit parameters except for the capability to adjust the current source. Comparison of Eqs. (8.8) and (7.25) shows that the sensitivities differ by the $r/(1 + r)$ multiplier for the constant-voltage potentiometer; thus, S_c will always be higher for the constant-current potentiometer.

It was noted in Eq. (7.22) that R_2 and ΔR_2 did not affect the signal output of the constant-current potentiometer. This fact indicates that temperature compensation by signal cancellation in the strain-gage circuit or signal addition cannot be performed. However, it is possible to maintain the advantages of high sensitivity and perfect linearity of this circuit and to obtain the capability of signal addition or subtraction by using the double constant-current potentiometer circuit described next.

7.4.2 Double Constant-Current Potentiometer Circuit

Consider the double constant-current potentiometer circuit shown in Fig. 7.14. In this circuit, two constant-current generators I_1 and I_2 are employed, and the output voltage is measured with a very high impedance meter between points A and B. The voltage at points A and B will be

$$V_A = I_1 R_1 \qquad V_B = I_2 R_2$$

and the output voltage E is

$$E = V_A - V_B = I_1 R_1 - I_2 R_2 \tag{7.26}$$

The circuit can be balanced to give a null output ($E = 0$) initially if the current from each supply can be adjusted so that

$$I_1 R_1 = I_2 R_2 \tag{7.27}$$

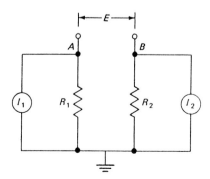

FIGURE 7.14
Double constant-current potentiometer circuit.

If resistances R_1 and R_2 change by ΔR_1 and ΔR_2, Eq. (7.26) yields

$$E + \Delta E = I_1(R_1 + \Delta R_1) - I_2(R_2 + \Delta R_2) \qquad (a)$$

For a circuit which is initially balanced, $E = 0$ and Eq. (a) reduces to

$$\Delta E = I_1 \Delta R_1 - I_2 \Delta R_2 = I_1 R_1 \frac{\Delta R_1}{R_1} - I_2 R_2 \frac{\Delta R_2}{R_2} \qquad (7.28)$$

Examination of Eq. (7.28) shows that the output voltage ΔE is linear with respect to resistance change ΔR; thus, the circuit can be used where large values of $\Delta R/R$ are experienced. Also, since the circuit can be initially balanced, the voltage change ΔE is easy to measure and the circuit is useful for both static and dynamic measurements.

Application of the double constant-current potentiometer circuit to strain-gage measurements usually involves one of the four cases illustrated in Fig. 7.15. The analysis of Case 3 to determine the voltage-strain relation and the circuit sensitivity is covered in Exercise 7.18.

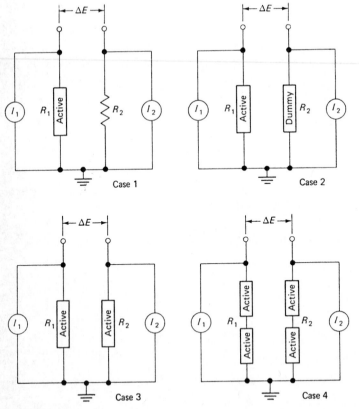

FIGURE 7.15
Four common double potentiometer circuits.

The double constant-current potentiometer circuit has the following advantages:

1. The output voltage ΔE is perfectly linear with respect to $\Delta R/R$, and, as such, the circuit is ideal for use where large values of $\Delta R/R$ may be encountered.
2. The sensitivity of the double constant-current potentiometer circuit is equal to or better than that of the Wheatstone bridge circuit (constant voltage supply) when one or two active gages are used. The sensitivity is twice that of the normal Wheatstone bridge when four active gages are employed.
3. Grounding is possible to eliminate electrical noise.
4. The circuit is the ultimate in simplicity. It can easily be used for either static or dynamic measurements if the current supplies can be adjusted to achieve initial balance and to drive the gages at their maximum power capabilities.

The disadvantage of the circuit pertains to the ripple requirement for the constant-current power supplies, which is severe. However, with battery-driven constant-current supplies, ripple can be limited to 20 ppm for carefully designed circuits. Also, two constant-current sources are required, which increases the costs for an otherwise extremely simple circuit.

7.4.3 Constant-Current Wheatstone Bridge Circuits

Since Wheatstone bridges have been used to measure resistance changes in strain gages since the discovery of the basic phenomenon by Lord Kelvin in 1856, it is logical to consider a bridge driven by a constant-current supply, as shown in Fig. 7.16. The current I delivered by the supply divides at point A of the bridge into currents I_1 and I_2, where $I = I_1 + I_2$. The voltage drop between points A and B of the bridge is

$$V_{AB} = I_1 R_1 \tag{a}$$

and the voltage drop between points A and D is

$$V_{AD} = I_2 R_4 \tag{b}$$

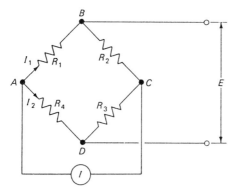

FIGURE 7.16
Wheatstone bridge with a constant-current supply.

Thus, the output voltage E from the bridge can be expressed as

$$E = V_{BD} = V_{AB} - V_{AD} = I_1 R_1 - I_2 R_4 \qquad (7.29)$$

For the bridge to be in balance ($E = 0$) under no-load conditions,

$$I_1 R_1 = I_2 R_4 \qquad (7.30)$$

Consider next the voltage V_{AC} and note that

$$V_{AC} = I_1(R_1 + R_2) = I_2(R_3 + R_4) \qquad (c)$$

from which

$$I_1 = \frac{R_3 + R_4}{R_1 + R_2} I_2 \qquad (d)$$

Since

$$I = I_2 + I_2 \qquad (e)$$

Eq. (d) can be substituted into Eq. (e) to obtain

$$I_1 = \frac{R_3 + R_4}{R_1 + R_2 + R_3 + R_4} I$$

$$I_2 = \frac{R_1 + R_2}{R_1 + R_2 + R_3 + R_4} I \qquad (f)$$

Substituting Eqs. (f) into Eq. (7.29) yields

$$E = \frac{I}{R_1 + R_2 + R_3 + R_4}(R_1 R_3 - R_2 R_4) \qquad (7.31)$$

From Eq. (7.31) it is evident that the balance condition ($E = 0$) for the constant-current Wheatstone bridge is the same as that for the constant-voltage Wheatstone bridge, namely,

$$R_1 R_3 = R_2 R_4 \qquad (7.13)$$

If resistances R_1, R_2, R_3, and R_4 change by amounts of ΔR_1, ΔR_2, ΔR_3, and ΔR_4, the voltage $E + \Delta E$ measured with a high-impedance meter is

$$E + \Delta E = \frac{I}{\sum R + \sum \Delta R}[(R_1 - \Delta R_1)(R_3 + \Delta R_3) - (R_2 + \Delta R_2)(R_4 + \Delta R_4)] \quad (g)$$

where $\sum R = R_1 + R_2 + R_3 + R_4$
$\sum \Delta R = \Delta R_1 + \Delta R_2 + \Delta R_2 + \Delta R_4$

Expanding Eq. (g) and simplifying after assuming initial balance ($E = 0$) gives

$$\Delta E = \frac{IR_1 R_3}{\sum R + \sum \Delta R}\left(\frac{\Delta R_1}{R_1} - \frac{\Delta R_2}{R_2} + \frac{\Delta R_3}{R_3} - \frac{\Delta R_4}{R_4} + \frac{\Delta R_1}{R_1}\frac{\Delta R_3}{R_3} - \frac{\Delta R_2}{R_2}\frac{\Delta R_4}{R_4}\right) \quad (7.32)$$

Inspection of Eq. (7.32) shows that the output signal ΔE is nonlinear with respect to ΔR because of the term $\sum \Delta R$ in the denominator and because of the second-order terms $(\Delta R_1/R_1)(\Delta R_3/R_3)$ and $(\Delta R_2/R_2)(\Delta R_4/R_4)$ within the bracketed quantity. The nonlinearity of the constant-current Wheatstone bridge, however, is less than that with the constant-voltage bridge. Indeed, if the constant-current Wheatstone bridge is properly designed, the nonlinear terms can be made insignificant even for the large $\Delta R/R$'s.

The nonlinear effects in a typical situation can be evaluated by considering the constant-current Wheatstone bridge shown in Fig. 7.17. A single active gage is employed in arm R_1 and a temperature-compensating dummy gage is employed in arm R_4. Fixed resistors are employed in arms R_2 and R_3. Thus

$$R_1 = R_4 = R_g \qquad R_2 = R_3 = rR_g \qquad \Delta R_2 = \Delta R_3 = 0$$

Under stable thermal environments, $\Delta R_1 = \Delta R_g$ and $\Delta R_4 = 0$, and Eq. (7.32) reduces to

$$\Delta E = \frac{IR_g r}{2(1 + r) + \Delta R_g/R_g} \frac{\Delta R_g}{R_g} \tag{7.33}$$

Again it is evident that Eq. (7.33) is nonlinear due to the presence of the term $\Delta R_g/R_g$ in the denominator. To determine the degree of the nonlinearity let

$$\frac{IR_g r}{2(1 + r) + \Delta R_g/R_g} \frac{\Delta R_g}{R_g} = \frac{IR_g r}{2(1 + r)} \frac{\Delta R_g}{R_g}(1 - \eta)$$

when η, the nonlinear term, is

$$\eta = \frac{\Delta R_g/R_g}{2(1 + r) + \Delta R_g/R_g} = \frac{S_g \epsilon}{2(1 + r) + S_g \epsilon} \tag{7.34}$$

Inspection of Eq. (7.34) shows that the nonlinear term η can be minimized by increasing r (making the fixed resistors R_2 and R_3, say, 9 times the value of R_g). In this case, the nonlinear term η will depend on the gage factor and on the magnitude of the strain. Consider, for example, a semiconductor strain gage with

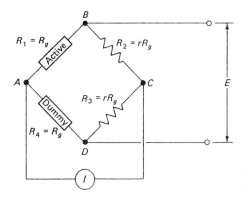

FIGURE 7.17
Constant-current Wheatstone bridge designed to minimize nonlinear effects.

$S_g = 100$; then η will be less than 1 percent for strains less than 2000 $\mu\varepsilon$. Since strains from 1000 to 2000 $\mu\varepsilon$ represent the upper limit for semiconductor strain gages, they can be used with the constant-current Wheatstone bridge if it is properly designed ($r \geq 9$).

7.5   CALIBRATING STRAIN-GAGE CIRCUITS [5]

A strain-measuring system usually consists of a strain gage, a power supply (either constant voltage or constant current), circuit-completion resistors, an amplifier, and a recording instrument of some type. A schematic illustration of a typical system is shown in Fig. 7.18. It is possible to calibrate each component of the system and determine the voltage-strain relationship from the equations developed in Sec. 7.3. However, this procedure is time-consuming and subject to calibration errors associated with each of the components involved in the system. A more precise and direct procedure is to obtain a single calibration for the complete system so that readings from the recording instrument can be directly related to the strains which produced them.

Direct system calibration can be achieved by shunting a fixed resistor R_c across one arm (say R_2) of the Wheatstone bridge, as shown in Fig. 7.18. If the bridge is initially balanced and the switch is closed to place R_c in parallel with R_2, the effective resistance of this arm of the bridge is

$$R_{2e} = \frac{R_2 R_c}{R_2 + R_c} \tag{a}$$

Because of the shunt resistance R_c, the ratio of the change in resistance to the original resistance in arm R_2 of the bridge is

$$\frac{\Delta R_2}{R_2} = \frac{R_{2e} - R_2}{R_2} \tag{b}$$

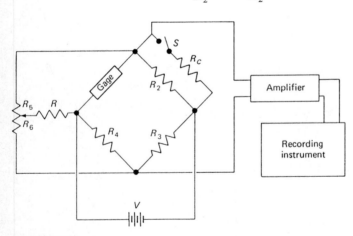

FIGURE 7.18
Typical strain-recording system.

Combining Eqs. (*a*) and (*b*) gives

$$\frac{\Delta R_2}{R_2} = -\frac{R_2}{R_2 + R_c} \tag{c}$$

Substituting Eq. (*c*) into Eq. (7.14) gives the output voltage of the bridge produced by R_c. Thus

$$\Delta E = \frac{R_1 R_2}{(R_1 + R_2)^2} \frac{R_2}{R_2 + R_c} V \tag{d}$$

Note also that a single active gage in position R_1 of the bridge would produce an output due to a strain ϵ of

$$\Delta E = \frac{R_1 R_2}{(R_1 + R_2)^2} (S_g \epsilon) V \tag{e}$$

Equating Eqs. (*d*) and (*e*) gives

$$\epsilon_c = \frac{R_2}{S_g (R_2 + R_c)} \tag{7.35}$$

where ϵ_c is the calibration strain which would produce the same voltage output from the bridge as the calibration resistor R_c.

For example, consider a bridge with $R_2 = R_g = 350\ \Omega$ and $S_g = 2.05$. If $R_c = 100\ \Omega$, the calibration strain $\epsilon_c = 1700\ \mu\epsilon$. If the recording instrument is operated during the period of time when the switch S is closed, an instrument deflection $d_c = 1700\ \mu\epsilon$ will be recorded, as shown in Fig. 7.19. The switch can then be opened and the load-induced strain can be recorded in the normal manner to give a strain-time pulse similar to the one illustrated in Fig.7.19. The peak strain associated with this strain-time pulse produces an instrument deflection d_p which can be numerically evaluated as

$$\epsilon = \frac{d_p}{d_c} \epsilon_c \tag{7.36}$$

This method of shunt calibration is accurate and easy to employ. It provides a means of calibrating the complete system regardless of the number of components in the system. The calibration strain produces a reading on the recording instrument; all other readings are linearly related to this calibration value.

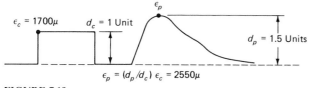

FIGURE 7.19
Calibrated strain-time trace.

7.6 EFFECTS OF LEAD WIRES, SWITCHES, AND SLIP RINGS

The resistance change for a metallic-foil strain gage is quite small ($0.7 \ m\Omega/\mu\epsilon$ for a 350-Ω gage). As a consequence, anything that produces a resistance change within the Wheatstone bridge is extremely important. The components within a Wheatstone bridge almost always include lead wires, soldered joints, terminals, and binding posts. Frequently, switches and slip rings are also included. The effects of each of these components on the output of the Wheatstone bridge circuit are discussed in the following sections.

7.6.1 Effect of Lead Wires [6]

Consider first a two-lead-wire system, illustrated in Fig. 7.20, where a single active gage is positioned on a test structure at a location remote from the bridge and recording system. If the length of the two-lead-wire system is long, three detrimental effects occur: signal attenuation, loss of balancing capability, and loss of temperature compensation.

To show that signal attenuation may occur note that

$$R_1 = R_g + 2R_L \qquad (a)$$

where R_L is the resistance of a single lead wire. Note also that

$$\frac{\Delta R_1}{R_1} = \frac{\Delta R_g}{R_g + 2R_L} = \frac{\Delta R_g/R_g}{1 + 2R_L/R_g} \qquad (b)$$

Equation (b) may be expressed in terms of a signal loss factor $\mathscr{L}$. Thus

$$\frac{\Delta R_1}{R_1} = \frac{\Delta R_g}{R_g}(1 - \mathscr{L}) \qquad (c)$$

From Eqs. (b) and (c), the signal loss factor $\mathscr{L}$ for the two-lead-wire system

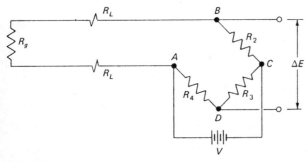

FIGURE 7.20
Two-lead-wire system.

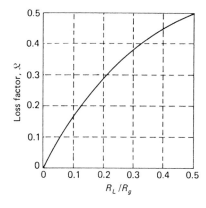

FIGURE 7.21
Loss factor as a function of the ratio of lead wire to gage resistance for a two-lead-wire system.

can be expressed as

$$\mathscr{L} = \frac{2R_L/R_g}{1 + 2R_L/R_g} \approx \frac{2R_L}{R_g} \qquad \text{if } 2R_L/R_g \le 1 \qquad (7.37)$$

The signal loss factor $\mathscr{L}$ is presented as a function of the ratio of lead resistance to gage resistance in Fig. 7.21. This plot clearly shows that $\mathscr{L}$ increases rapidly as R_L becomes a significant fraction of R_g. In order to limit lead-wire losses to less than 2 percent, $R_L/R_g \le 0.01$. If test conditions dictate long leads, then large-gage wire must be employed to limit R_L. It is also advantageous to use 350- or 1000-Ω gages in place of 120-Ω gages. The resistance of a 100-ft (30.5-m) length of lead wire as a function of wire gage size is listed in Table 7.1 for copper wire.

The second detrimental effect of the two-lead-wire system is loss of the ability to initially balance the bridge. For the bridge shown in Fig. 7.20,

$$R_1 = R_g + 2R_L \qquad R_2 = R_3 = rR_g \qquad \text{(fixed resistors)}$$

$$R_4 = R_g \qquad \Delta R_2 = \Delta R_3 = 0$$

TABLE 7.1
Resistance of solid-conductor copper wire in Ω per 100 ft (30.5 m)

Gage size	Resistance	Gage size	Resistance
12	0.159	28	6.490
14	0.253	30	10.310
16	0.402	32	16.41
18	0.639	34	26.09
20	1.015	36	41.48
22	1.614	38	65.96
24	2.567	40	104.90
26	4.081		

With the addition of $2R_L$ in arm R_1 of the bridge, it is obvious that the initial balance condition $R_1 R_3 = R_2 R_4$ is not satisfied. Of course, a parallel-balance resistor similar to the one shown in Fig. 7.18 is available in most commercial bridges to obtain initial balance; however, if $R_L/R_g > 0.02$, the range of the balance potentiometer is exceeded and initial balance of the bridge cannot be achieved.

The third detrimental effect of the two-lead-wire system is that temperature compensation of the measuring circuit is lost. First, consider a temperature-compensating dummy gage with relatively short leads in arm R_4 of the bridge shown in Fig. 7.20. The output voltage due to resistance changes in arms R_1 and R_4, as obtained from Eq. (7.15), is

$$\Delta E = V \frac{r}{(1 + r)^2} \left(\frac{\Delta R_1}{R_1} - \frac{\Delta R_4}{R_4} \right) \tag{d}$$

If the gages are subjected to a temperature difference ΔT at the same time that the active gage is subjected to a strain ϵ, then Eq. (d) becomes

$$\Delta E = V \frac{r}{(1 + r)^2} \left[\left(\frac{\Delta R_g}{R_g + 2R_L} \right)_\epsilon + \left(\frac{\Delta R_g}{R_g + 2R_L} \right)_{\Delta T} + \left(\frac{2\Delta R_L}{R_g + 2R_L} \right)_{\Delta T} - \left(\frac{\Delta R_g}{R_g} \right)_{\Delta T} \right] \tag{7.38}$$

This relation shows that temperature compensation is not achieved in the Wheatstone bridge since the second and fourth terms in the bracketed quantity are not equal. Also, the lead wires can suffer significant resistance changes due to temperature; therefore, the third term in the bracketed quantity can produce significant errors in the measurement of strain with the two-lead-wire system.

The detrimental effects of long lead wires can be minimized by employing the three-lead-wire system shown in Fig. 7.22. In this circuit, both the active and dummy gages are placed at the remote location. One of the three wires is used to transfer terminal A of the bridge to the remote location. It is not considered a lead wire since it is not within either arm R_1 or arm R_4 of the bridge. The active and dummy gages each have one long lead wire with resistance R_L and one short lead wire with negligible resistance (see Fig. 7.22). With the three-lead-wire system,

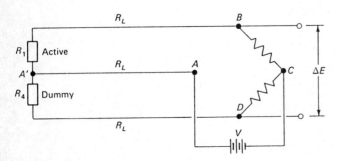

FIGURE 7.22
Three-lead-wire system.

the signal loss factor $\mathscr{L}$ is reduced to

$$\mathscr{L} = \frac{R_L/R_g}{1 + R_L/R_g} \approx \frac{R_L}{R_g} \qquad \text{if } \frac{R_L}{R_g} \leq 1 \tag{7.39}$$

The bridge retains its initial balance capability since the resistance of both arms R_1 and R_4 is increased by R_L. With the three-lead-wire system, Eq. (7.38) becomes

$$\Delta E = V \frac{r}{(1 + r)^2} \left[\left(\frac{\Delta R_g}{R_g + R_l} \right)_c + \left(\frac{\Delta R_g}{R_g + R_L} \right)_{\Delta T} + \left(\frac{\Delta R_L}{R_g + R_L} \right)_{\Delta T} \right.$$
$$\left. - \left(\frac{\Delta R_g}{R_g + R_L} \right)_{\Delta T} - \left(\frac{\Delta R_L}{R_g + R_L} \right)_{\Delta T} \right]$$

Temperature compensation is achieved since all the temperature-related terms in the bracketed quantity cancel.

7.6.2 Effect of Switches [7]

In most strain-gage applications, many strain gages are installed and are monitored several times during the test. When the number of gages is large, it is not economically feasible to employ a separate recording instrument for each gage. Instead, a single recording instrument is used, and the gages are switched in and out of this instrument. Two different methods of switching are commonly used in multiple-gage installations.

The first method, illustrated in Fig. 7.23, involves switching one side of each active gage, in turn, into arm R_1 of the bridge. The other side of each of the active gages is connected to terminal A of the bridge by a common lead wire. A single dummy gage or fixed resistor is used in arm R_4 of the bridge. With this arrangement, the switch is located within arm R_1 of the bridge; therefore, an extremely-high-quality switch with negligible resistance (less than 1 mΩ) must be

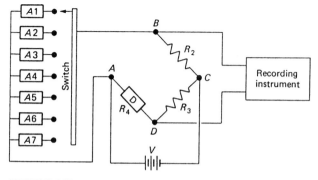

FIGURE 7.23
Switching active gages in arm R_1 of the Wheatstone bridge.

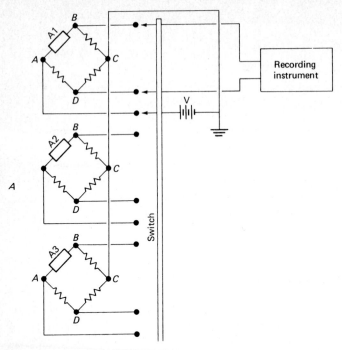

FIGURE 7.24
Switching the complete bridge.

used. If the switching resistance is not negligible, the switch resistance adds to ΔR_g to produce an error in the strain measurement. The quality of a switch can be checked quite easily, since excessive switch resistance produces variations in the zero strain readings.

The second method, illustrated in Fig. 7.24, involves switching the complete bridge. In this method, a three-pole switch is employed in the leads between the bridge and the power supply and the recording instrument. Since the switch is not located in the arms of the bridge, switching resistance is not as important. Switching the complete bridge is more expensive, however, since a separate dummy gage and two bridge-completion resistors are required for each active gage.

7.6.3 Effect of Slip Rings [8]

Strain gages are frequently used on rotating machinery, where it is impossible to use ordinary lead wires to connect the active gages to the recording instrument. Slip rings are employed in these applications to provide lead-wire connections. The rings are mounted on a shaft which is attached to the end of the rotating member so that the rings rotate with the member. The shell of the slip-ring assembly is stationary and usually carries several brushes for each slip ring. Lead wires from the strain-gage bridge, which rotates with the member, are connected to the slip rings. Lead wires from the power supply and recording instrument, which are stationary, are connected through the brushes to the appropriate slip ring. Depending on the design of the slip-ring assembly, satisfactory operation at rotary speeds to 24,000 rpm can be achieved.

Wear debris collecting on the slip rings and brush jump tend to generate electrical noise in this type of strain-gage system. The use of multiple brushes wired in parallel helps to minimize the noise; however, the resistance changes between the rings and the brushes are usually so large that slip rings are not recommended for use within an arm of the bridge. Instead, a complete bridge should be assembled on the rotating member, as shown in Fig. 7.25, and the slip rings should be used to connect the bridge to the power supply and recording instrument. In this way, the effects of resistance changes due to the slip rings are minimized.

7.7 ELECTRICAL NOISE [9, 10]

The voltages ΔE from strain gage circuits are quite small [usually less than $10 \, \mu V/\mu\epsilon$]. As a consequence, electrical noise is an important consideration in strain-gage circuit design. Electrical noise in strain-gage circuits is produced by the magnetic fields generated when currents flow through supply wires located in close proximity to the strain-gage lead wires, as shown in Fig. 7.26. When an alternating current flows in the supply wire, a time-varying magnetic field is produced which cuts both wires of the signal circuit and induces a voltage in the signal loop. The induced voltage is proportional to the current I and the area enclosed by the signal loop but inversely proportional to the distance from the supply wire to the signal circuit.

Since the distances d_1 and d_2 in Fig. 7.26 are not equal, the difference in magnetic fields at the two signal leads induces a noise voltage E_N which is superimposed on the strain-gage signal voltage ΔE. In certain instances, where the magnetic fields are large, the noise voltage becomes significant and makes separation of the true strain-gage signal from the noise signal difficult.

There are three procedures which should normally be employed to reduce noise to a minimum.

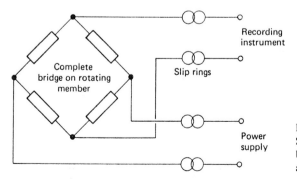

FIGURE 7.25
Slip rings connecting a complete bridge to the recording instrument and power supply.

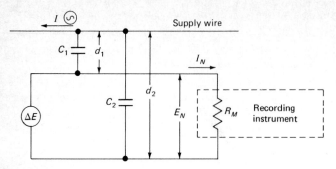

FIGURE 7.26
Generation of electrical noise.

PROCEDURE 1. All lead wires should be tightly twisted together to minimize the area in the signal loop and make the distances d_1 and d_2 equal. In this way, the noise voltage is minimized.

PROCEDURE 2. Shielded cables should be used for the strain-gage leads and the shields should be connected only to the signal ground, as indicated in Fig. 7.27. If the shield is connected to both the signal ground and the system ground, as shown in Fig. 7.28, a ground loop is formed. Since two different grounds are seldom at the same absolute voltage, a noise signal can be generated by the potential difference which exists between the two grounding points. A second ground loop, from the signal source through the cable shield to the amplifier, also occurs with the grounding method shown in Fig. 7.28. Alternating currents in the shield, due to this second ground loop, are coupled to the signal pair through the distributed capacitance in the signal cable (see Fig. 7.28). Either of these ground loops is capable of generating a noise signal 100 times larger than the strain-gage signal.

In most recording instruments, the third conductor in the power cord is used to provide the system ground. Since this ground is connected to the enclosure for

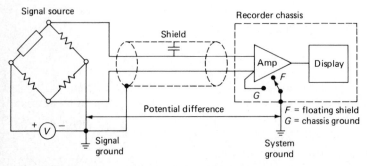

FIGURE 7.27
Proper method of grounding a shielded cable.

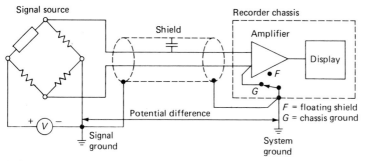

FIGURE 7.28
Incorrect method of grounding a shielded cable, resulting in a ground loop.

the instrument, care should be exercised to insulate the enclosure from any other building ground. In most modern recording instruments, the amplifier can be operated in either a floating mode or a grounded mode, as shown in Figs. 7.27 and 7.28, respectively. In the floating mode, the amplifier is insulated from the system ground; and, with the signal source also isolated from the system ground, the arrangement is correct for minimizing the noise signal. The proper point of attachment for the signal ground on both the potentiometer circuit and the Wheatstone bridge circuit is at the negative terminal of the power supply. The power supply itself should be floated relative to the system ground to avoid a ground loop at the supply.

PROCEDURE 3. The third way to eliminate noise is by common-mode rejection. Here the lead wires are arranged so that any noise signals will be equal in magnitude and phase on both the lead wires. If a differential amplifier is employed in the recording system, the noise signals are rejected and the strain signals are amplified. Unfortunately, the common-mode rejection of the very best differential amplifiers is not perfect, and a very small portion of the common-mode voltage is transmitted by the amplifier. The common-mode rejection for good low-level data amplifiers is about 10^5:1 at 60 Hz; thus, it is evident that significant noise suppression can be achieved in this manner.

7.8   TRANSDUCER APPLICATIONS [11]

Since the electrical-resistance strain gage is such a remarkable measuring device— small, lightweight, linear, precise, and inexpensive—it is used as the sensor in a wide variety of transducers. In a transducer such as a load cell, the unknown quantity, load, is measured by sensing the strain developed in a mechanical member. Because the load is linearly related to the strain, as long as the mechanical member remains elastic, the load cell can be calibrated so that the output signal is proportional to load.

Transducers of many different types and models are commercially available. Included are load cells, torque meters, pressure gages, displacement gages, and accelerometers. In addition, many different special-purpose transducers are custom-designed, with strain gages as the sensing device, to measure other quantities.

7.8.1 Load Cells

The design of strain-gage transducers for general-purpose measurement is relatively easy if accuracy is limited to ± 2 percent. Consider, for instance, a tensile load cell fabricated from a tension specimen, as shown in Fig. 7.29. To convert the simple tension bar into a load cell, four strain gages are mounted on the central region of the bar with two opposite gages in the axial direction and two opposite gages in the transverse direction. When a load P is applied to the tension member, the axial and transverse strains produced are

$$\epsilon_a = \frac{P}{AE} \qquad \epsilon_t = -\frac{vP}{AE} \tag{7.40}$$

where A = cross-sectional area of tension member
$\quad E$ = modulus of elasticity of the tension specimen
$\quad v$ = Poisson's ratio of the tension specimen

If the four gages are positioned in the Wheatstone bridge as shown in Fig. 7.29, the ratio of output voltage to supply voltage $\Delta E/V$ is given by Eq. (7.15) as

$$\frac{\Delta E}{V} = \frac{1}{4}\left(\frac{\Delta R_1}{R_1} - \frac{\Delta R_2}{R_2} + \frac{\Delta R_3}{R_3} - \frac{\Delta R_4}{R_4}\right) \tag{7.41}$$

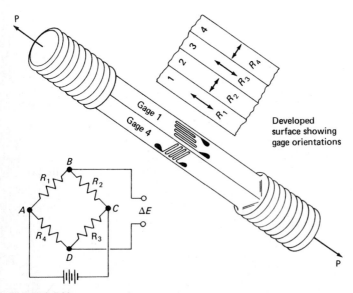

FIGURE 7.29
Strain gages mounted on a simple tension specimen to produce a load cell.

The changes in resistance of the four gages on the tension member are obtained from Eqs. (6.5) and (7.40) as

$$\frac{\Delta R_1}{R_1} = \frac{\Delta R_3}{R_3} = S_g \epsilon_a = \frac{S_g P}{AE}$$

$$\frac{\Delta R_2}{R_2} = \frac{\Delta R_4}{R_4} = S_g \epsilon_t = -\frac{v S_g P}{AE}$$

Substituting into Eq. (7.41) gives

$$\frac{\Delta E}{V} = \frac{S_g P}{2AE}(1 + v) \approx \frac{P}{AE}(1 + v) \tag{7.42}$$

when $S_g \approx 2.00$.

From Eq. (7.42), it is evident that the output signal $\Delta E/V$ is linearly related to the load P. The magnitude of $\Delta E/V$ will depend upon the design of the tension member, i.e., its cross-sectional area A and the material constants E and v. In most commercial load cells $\Delta E/V$ varies between 0.001 and 0.003. Steel with $E = 30 \times 10^6$ psi (207 GPa) and $v = 0.30$ is usually used to fabricate the tension member. The range of the load cell P_R (the maximum load) is then

$$P_R = A \frac{\Delta E}{V} \frac{E}{1 + v} \tag{7.43}$$

The upper limit on the output signal $\Delta E/V$ is determined by the strength of the tensile member and the fatigue limit of the strain gages. The maximum stress in the tension member is obtained from Eq. (7.43) as

$$\sigma = \frac{P}{A} = \frac{\Delta E}{V} \frac{E}{1 + v} \tag{7.44}$$

With steel tension members and $\Delta E/V = 0.003$, the stress $\sigma = 69,000$ psi (476 MPa) developed in the tension member is well within the fatigue limit of heat-treated alloy steel such as 4340. However, the axial strain $\epsilon_a = 2300 \ \mu\epsilon$ is near the fatigue limit of most strain gages.

Placement of the strain gages on the four sides of the tension member, as shown in Fig. 7.29, provides a load cell which is essentially independent of either bending or torsional loads. Consider a bending moment M applied to the tension member either by a transverse load or by an eccentrically applied axial load. The moment M may have any direction relative to the axes of symmetry of the cross section, as shown in Fig. 7.30. The components M_1 and M_2 of the moment M will produce resistance changes in the gages as follows:

$$\left.\frac{\Delta R_2}{R_2}\right|_{M1} = -\left.\frac{\Delta R_4}{R_4}\right|_{M1} \quad \text{and} \quad \left.\frac{\Delta R_1}{R_1}\right|_{M1} = \left.\frac{\Delta R_3}{R_3}\right|_{M1} = 0$$

$$\left.\frac{\Delta R_3}{R_3}\right|_{M2} = -\left.\frac{\Delta R_1}{R_1}\right|_{M2} \quad \text{and} \quad \left.\frac{\Delta R_2}{R_2}\right|_{M2} = \left.\frac{\Delta R_4}{R_4}\right|_{M2} = 0$$

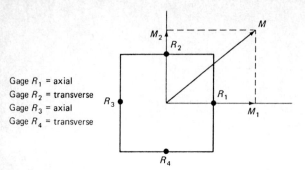

Gage R_1 = axial
Gage R_2 = transverse
Gage R_3 = axial
Gage R_4 = transverse

FIGURE 7.30
Resolution of the moment M into components M_1 and M_2.

Substitution of these relations into Eq. (7.41) shows that the effects of bending moments applied to the load cell are canceled in the Wheatstone bridge since $\Delta E/V$ vanishes for both M_1 and M_2.

Consider next the tension member subjected to a torque T, as shown in Fig. 7.31. The state of stress in the tension member for this form of loading has been shown to be

$$\tau_{max} = \frac{4.81T}{a^3} \qquad \sigma_{xx} = \sigma_{yy} = \sigma_{zz} = 0$$

Thus

$$\epsilon_{xx} = \epsilon_{yy} = \epsilon_{zz} = 0 \qquad (a)$$

When Eqs. (a) are substituted into Eq. (6.5),

$$\frac{\Delta R_1}{R_1} = \frac{\Delta R_3}{R_3} = S_g \epsilon_{xx} = 0$$

$$\frac{\Delta R_2}{R_2} = \frac{\Delta R_4}{R_4} = S_g \epsilon_{xx} = 0 \qquad (b)$$

Substitution of Eqs. (b) into Eq. (7.41) indicates that the output of the tensile load cell is independent of the applied torque since $\Delta E/V$ is again equal to

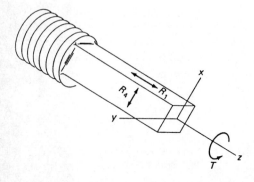

FIGURE 7.31
Torque T applied to the tension member of an axial-load cell.

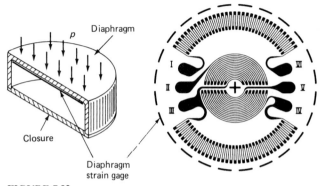

FIGURE 7.32
Diaphragm-type pressure transducer.

zero. Temperature compensation is also achieved with the four active strain gages in the bridge.

7.8.2 Diaphragm Pressure Transducers [12–14]

A second type of transducer which utilizes a strain gage as the sensing element is the diaphragm type of pressure transducer. Here a special-purpose strain gage is mounted on one side of the diaphragm while the other side is exposed to the pressure, as shown in Fig. 7.32. The diaphragm pressure transducer is small, easy to fabricate, and inexpensive, and has a relatively high natural frequency.

The special-purpose diaphragm strain gage, shown in Fig. 7.33, has been designed to maximize the output voltage from the transducer. The strain distribution in the diaphragm is given by

$$\epsilon_{rr} = \frac{3p(1 - v^2)}{8Et^2}(R_o^2 - 3r^2)$$

$$\epsilon_{\theta\theta} = \frac{3p(1 - v^2)}{8Et^2}(R_o^2 - r^2)$$

(7.45)

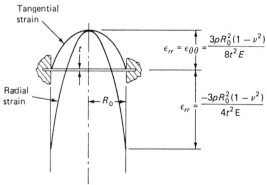

Tangential strain

Radial strain

$$\epsilon_{rr} = \epsilon_{\theta\theta} = \frac{3pR_0^2(1 - v^2)}{8t^2E}$$

$$\epsilon_{rr} = \frac{-3pR_0^2(1 - v^2)}{4t^2E}$$

FIGURE 7.33
Distribution of strain over a diaphragm.

where p = pressure
  t = thickness of diaphragm
  R_o = outside radius of diaphragm
  r = position parameter

Examination of this strain distribution indicates that the circumferential strain $\epsilon_{\theta\theta}$ is always positive and assumes its maximum value at $r = 0$. The radial strain ϵ_{rr} is positive in some regions but negative in others and assumes its maximum negative value at $r = R_o$. Both these strain distributions are shown in Fig. 7.33. The special-purpose diaphragm strain gage has been designed to take advantage of this distribution. Circumferential grids are employed in the central region of the diaphragm, where $\epsilon_{\theta\theta}$ is maximum. Similarly, radial grids are employed near the edge of the diaphragm, where ϵ_{rr} is maximum. It should also be noted that the circumferential and radial grids are each divided into two parts so that the special-purpose gage actually consists of four separate strain sensors. Terminals are provided which permit the individual gages to be connected into a bridge with the circumferential elements in arms R_1 and R_3 and the radial elements in arms R_2 and R_4.

If the strains are averaged over the areas of the circumferential and radial grids, and if the average values of $\Delta R/R$ obtained are substituted into Eq. (7.41), the signal output can be approximated by

$$\frac{\Delta E}{V} = 0.82 \frac{pR_o^2(1 - v^2)}{t^2 E} \tag{7.46}$$

Special-purpose diaphragm strain gages are commercially available in four sizes ranging from 0.182 to 0.455 in (4.62 to 11.56 mm) in diameter with resistances of 120, 350, and 1000 Ω.

Under the action of the pressure, the diaphragm deflects and changes from a flat circular plate to a segment of a large-radius shell. As a consequence, the strain in the diaphragm is nonlinear with respect to the applied pressure. Acceptable linearity can be maintained by limiting the deflection of the diaphragm. The center deflection w_c of the diaphragm can be expressed as

$$w_c = \frac{3pR_o^4(1 - v^2)}{16t^3 E} \tag{7.47}$$

If $w_c \leq t/4$, $\Delta E/V$ will be linear to within 0.3 percent over the pressure range of the transducer.

Frequently, diaphragm pressure transducers are employed to measure pressure transients. In these dynamic applications, the natural frequency of the diaphragm should be considerably higher than the highest frequency present in the pressure pulse. Depending upon the degree of damping incorporated in the design of the transducer, an undamped natural frequency of 5 to 10 times greater than the highest applied frequency should be sufficient to avoid resonance effects.

The natural frequency f_n of the diaphragm can be expressed as

$$\omega_n = 2\pi f = \frac{10.21t}{R_o^2} \sqrt{\frac{gE}{12(1 - v^2)\gamma}} \tag{7.48}$$

where γ is the density of the diaphragm material and g is the gravitational constant. If the thickness of the diaphragm cannot be determined accurately, the natural frequency can be determined experimentally by tapping the transducer at the center of the diaphragm to excite the fundamental mode and recording the vibratory response on an oscilloscope. The peak-to-peak period is the reciprocal of the natural frequency.

EXERCISES

7.1. Verify Eqs. (7.3a) and (7.3b) from Eq. (7.2).

7.2. Determine the magnitude of the nonlinear term as a function of strain for a potentiometer circuit ($r = 5$) with a single active gage ($S_g = 2.00$). Show a graph of the results obtained, then discuss the error due to circuit-induced nonlinearities.

7.3. Select the following gages from a strain-gage catalog:

(a) The gage having the smallest gage length

(b) The gage having the largest gage length

(c) A gage having an area of approximately 38 mm², a resistance of 350 Ω, and a gage factor of about 2

(d) A gage having an area of approximately 19 mm², a resistance of 1000 Ω, and a gage factor of about 2

(e) A gage having an area of approximately 19 mm², a resistance of 350 Ω, and a gage factor of about 3.5

If the allowable power density is 0.78 mW/mm² for each of these gages, determine S_c, V, and R_b for a potentiometer circuit with $r = 5$. Discuss the results.

7.4. For the best and worst cases in Exercise 7.3, construct a graph of output voltage ΔE as a function of strain ϵ. For strain levels associated with the yield strength of low-carbon steels ($\sigma_{ys} = 250$ MPa), determine the magnitude of ΔE. Is this a high- or low-level signal?

7.5. The sawtooth voltage pulse shown in Fig. E7.5 is the input to the filter. Resolve this voltage pulse into its Fourier components and consider the pulse distortion as the first five components pass through the filter. Discuss the results obtained.

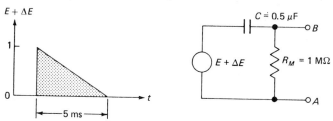

FIGURE E7.5

7.6. How could the simple filter shown in Fig. E7.5 be changed to improve the fidelity of the signal?

7.7. The triangular voltage pulse shown in Fig. E7.7 is the input to the filter. Resolve this voltage pulse into its Fourier components and consider the pulse distortion as the first five components pass through the filter. Discuss the results obtained and indicate how this filter design can be improved.

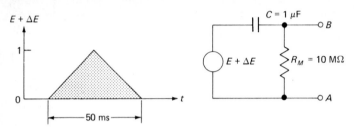

FIGURE E7.7

7.8. Verify Eq. (7.14) and the preceding Eqs. (*e*) and (*f*). Show that:
(a) If the second-order terms are neglected, Eq. (7.14) is linear in terms of $\Delta R/R$.
(b) If the second-order terms are retained, Eq. (7.14) is nonlinear in terms of $\Delta R/R$ and

$$\Delta E = V \frac{R_1 R_2}{(R_1 + R_2)^2} \left(\frac{\Delta R_1}{R_1} - \frac{\Delta R_2}{R_2} + \frac{\Delta R_3}{R_3} - \frac{\Delta R_4}{R_4} \right)(1 - \eta)$$

where

$$\eta = \frac{1}{1 + \dfrac{r + 1}{\Delta R_1/R_1 + \Delta R_4/R_4 + r(\Delta R_2/R_2 + \Delta R_3/R_3)}} \qquad \text{and} \qquad r = \frac{R_2}{R_1}$$

7.9. Three strain gages are placed in series in arm R_1 of a fixed-voltage Wheatstone bridge, as shown in Fig. E7.9. If the strain experienced by all gages is the same, determine the increase in circuit sensitivity over that obtained with a single gage.

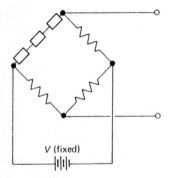

V (fixed)

FIGURE E7.9

7.10. If the voltage can be varied in the Wheatstone bridge of Fig. E7.9, does the insertion of the additional gages improve the sensitivity? If $r = 4$, determine the percent change in sensitivity for the three gages versus a single gage. Compute the required voltage V for the three series-connected gages and for a single gage if $R_g = 350\ \Omega$ and $P_g = 0.02$ W.

7.11. Compare the circuit sensitivities of the three bridges shown in Fig. E7.11. Compute the required voltages for bridges (b) and (c).

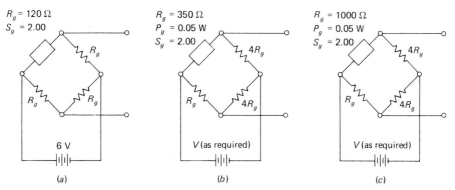

$R_g = 120\ \Omega$ $S_g = 2.00$	$R_g = 350\ \Omega$ $P_g = 0.05$ W $S_g = 2.00$	$R_g = 1000\ \Omega$ $P_g = 0.05$ W $S_g = 2.00$

6 V   V (as required)   V (as required)

(a)   (b)   (c)

FIGURE E7.11

7.12. Show that the bridge arrangements represented in Cases 2 to 4 of Fig. 7.7 are all temperature-compensating. Follow the procedure outlined in Sec. 7.2.2.

7.13. Determine the circuit sensitivity for the bridge arrangement shown in Fig. E7.13. Compare this sensitivity with that given by Eq. (7.20) for Case 4 of Fig. 7.7. Discuss the results obtained.

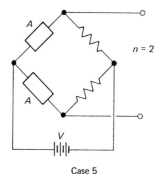

$n = 2$

Case 5   **FIGURE E7.13**

7.14. Describe the differences in the two most recent designs of commercial strain indicators shown in Figs. 7.10 and 7.12.

7.15. For the constant-current potentiometer circuit shown in Fig. 7.13b, show that

$$\Delta E = R_1 I \frac{\Delta I}{I} + R_1 I \frac{\Delta R_1}{R_1}$$

where ΔI is the ripple current from an imperfect constant-current power supply.

7.16. For a semiconductor strain gage employed as R_1 in Fig. 7.13b, specify the ripple requirement $\Delta I/I$ to limit the ripple signal to an equivalent strain of 1 μm/m. Assume $S_g = 100$.

7.17. Rederive Eq. (7.28) after permitting I_1 and I_2 to change by ΔI_1 and ΔI_2 to obtain

$$\Delta E = I_1 R_1 \left(\frac{\Delta I_1}{I_1} + \frac{\Delta R_1}{R_1} \right) - I_2 R_2 \left(\frac{\Delta I_2}{I_2} + \frac{\Delta R_2}{R_2} \right)$$

Will the terms $I_1 R_1 (\Delta I_1/I_1)$ and $I_2 R_2 (\Delta I_2/I_2)$ tend to cancel? Why?

7.18. Determine ΔE if two semiconductor gages ($S_g = 100$) are employed in the double-potentiometer circuit shown under case 3 of Fig. 7.15. Both active gages have $R_1 = R_2 = R_g = 500 \ \Omega$. Assume $I_g = 10$ mA and $\epsilon = +1000 \ \mu\epsilon$ on R_1 and $-1000 \ \mu\epsilon$ on R_2.

7.19. Beginning with Eq. (7.33), verify Eq. (7.34).

7.20. Prepare a graph showing the calibration strain ϵ_c as a function of R_c for $R_2 = 120$, 350, 500, and 1000 Ω. Assume $S_g = 2.00$.

7.21. Could the calibration resistance be placed across arm R_1 of the bridge which contains the active gage? When would this procedure be recommended?

7.22. Determine the signal loss factor for a two-lead-wire system if 18-gage wire is employed and the recording instrument and strain gage are separated by 200 m. Determine the signal loss factor if a three-lead-wire system is employed.

7.23. Determine the signal loss factor for a three-lead-wire system as a function of lead length ranging from 1 to 1000 m for wire sizes
(a) 12   (b) 18   (c) 22   (d) 26

7.24. For the two-lead-wire system of Exercise 7.22, determine the apparent strain introduced by an average temperature change of 18°F (10°C) over the length of the lead wires. Assume $S_g = 2.00$.

7.25. Specify a switch which would be adequate to use in the switching arrangement shown in Fig. 7.23.

7.26. Specify a switch which would be adequate to use in the switching arrangement shown in Fig. 7.24.

7.27. Briefly describe electrical noise and indicate why it occurs.

7.28. Outline three procedures which should be followed to minimize electrical noise in a strain-gage recording system.

7.29. Describe signal ground, system ground, and building ground. Describe a ground loop. How are ground loops avoided? How many grounds are usually employed in a recording system? How is a ground loop avoided if more than one ground is employed?

7.30. Design a torque cell to measure torques which range from 0 to 100 kN · m. Specify the size and shape of the mechanical member, the gage layout, the circuit design, and the voltage required. The torque cell must be insensitive to axial loads and bending moments.

7.31. Prove that the torque cell designed in Exercise 7.30 is insensitive to axial loads and bending moments.

7.32. Design a load cell to measure a 100-MN axial tensile load. Specify the size and shape of the mechanical member, the gage layout, the circuit design, and the voltage required.

7.33. A very compact compression-load cell can be designed by using the mechanical member shown in Fig. E7.33. If the maximum load is 1000 kN and the ratio $\Delta E/V = 0.002$, specify the size of the mechanical member, the gage layout, the circuit design, and the voltage required.

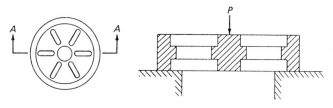

FIGURE E7.33

7.34. Design a displacement transducer having a cantilever beam as the mechanical member. The transducer must be capable of measuring the displacement to an accuracy of ± 0.004 mm over a range of 5 mm. Specify the size of the mechanical member, the gage layout, the circuit design, and the voltage required.

7.35. Design a strain extensometer with a strain range of 20 percent, a gage length of 100 mm, and a ratio $\Delta E/V = 0.003$. Use the thin semicircular ring segment shown in Fig. E7.35 as the mechanical member.

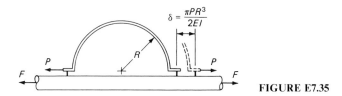

$$\delta = \frac{\pi P R^3}{2EI}$$

FIGURE E7.35

7.36. Design a line of diaphragm-type pressure transducers to measure pressures of 100, 200, 500, 1000, 2000, 5000, and 10,000 kPa. If the diameter of the transducer line is 15 mm and the diaphragms are fabricated from stainless steel, specify the thickness t for each pressure if $\Delta E/V = 0.003$. Determine the displacement at the center of the diaphragm at each maximum pressure. Determine the natural frequency of the diaphragm for each case.

7.37. Four electrical-resistance strain gages have been mounted on a torque wrench and wired into a fixed-voltage Wheatstone bridge, as shown in Fig. E7.37. The designer

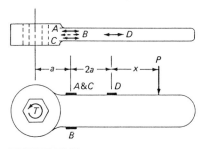
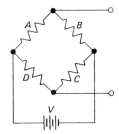

FIGURE E7.37

of the system claims that the output signal from the bridge is proportional to the torque applied by the wrench irrespective of the point of application of the load P so long as it remains to the right of the gages. Analyze the system and prove or disprove the claim.

7.38. Design a load-measuring transducer having a cantilever beam as the mechanical member and four electrical-resistance strain gages as the sensing elements. Position the gages on the beam and connect them into a fixed-voltage Wheatstone bridge to achieve maximum sensitivity. The output from the system must be proportional to the load P, which can be positioned anywhere on the right half of the beam, as shown in Fig. E7.38.

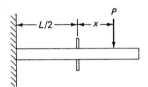

FIGURE E7.38

7.39. Design a scale to be used to measure the weight of first class mail. The scale is a subsystem in a larger automatic system used to place the correct postage on a large volume of mail each day.

7.40. Design a probe to be used to measure the viscosity of oil flowing in a pipe with a diameter of 200 mm. The range of the absolute viscosity is from 1 to 10^4 MPa·s.

REFERENCES

1. Geldmacher, R. C.: Ballast Circuit Design, *Proc. SESA*, vol. XII, no. 1, pp. 27–32, 1954.
2. Wheatstone, C.: An Account of Several New Instruments and Processes for Determining the Constants of a Voltaic Circuit, *Phil. Trans. R. Soc.*, vol. 133, pp. 303ff., 1843.
3. Stein, P. K.: "Strain Gage Circuits for Semiconductor Gages," in M. Dean and R. D. Douglas (eds.), *Semiconductor and Conventional Strain Gages*, Academic Press, New York, 1962, pp. 273–282.
4. Frank, E.: "Strain Indicator for Semiconductor Gages," in M. Dean and R. D. Douglas (eds.), ibid., pp. 283–306.
5. Haakana, C. H.: Shunt Bridge Balancing in Strain Gage Indicators, *Electronics*, vol. 32, no. 30, pp. 50–51, 1959.
6. Murray, W. M., and P. K. Stein: *Strain Gage Techniques*, 1960. Copyright by authors; lectures and laboratory exercises by authors at the Massachusetts Institute of Technology, Cambridge.
7. Warshawsky, L.: "A Multiple Bridge for Elimination of Contact-Resistance Errors in Resistance Strain Gage Measurements," NACA Tech. Note 1031, 1946.
8. Perry, C. C., and H. R. Lissner: *The Strain Gage Primer*, 2d ed., McGraw-Hill, New York, 1962, pp. 200–217.
9. "Elimination of Noise in Low-Level Circuits," report of Brush Instrument Division of Clevite Corporation, Cleveland, OH.
10. "Noise Control in Strain Gage Measurements," Measurements Group, Inc., Tech. Note 501, 1980.
11. "Strain Gage Based Transducers," Measurements Group, Inc., 1988.

12. "Design Considerations for Diaphragm Pressure Transducers," Measurements Group, Inc., Tech. Note 105, 1982.
13. Timoshenko, S.: "Strength of Materials," part II, *Advanced Theory and Problems*, 3d ed., Van Nostrand, Princeton, N.J., 1956, pp. 96–97.
14. Timoshenko, S.: *Vibration Problems in Engineering*, 3d ed., Van Nostrand, Princeton, N.J., 1955, pp. 449–451.

CHAPTER
8

RECORDING
INSTRUMENTS

8.1 INTRODUCTION [1, 2]

In strain-gage applications, the signal generated by the strain gage through the use of a potentiometer, Wheatstone bridge, or constant-current circuit must be measured accurately to determine the strain. This chapter deals with instruments commonly employed by the experimental stress analyst to measure the small signal voltages produced by the gages. In selecting the type of voltage-measuring instrument, several criteria are considered which include:

1. Availability of instruments
2. Number of gages involved and location of the test
3. Type of analysis required
4. Number and frequency of reports
5. Frequency content of the strain signals

Most laboratories have a limited inventory of instruments and a more limited budget to acquire modern equipment capable of recording small signals from strain gages with fidelity. While the limited selection of available instruments may prolong the test or experiment and require manual interaction and long, tedious calculation, the instrument must not introduce excessive error. If the equipment is malfunctioning, out of calibration, or inadequate in frequency response, the experiment must be delayed or canceled. It is better to report nothing than to report misleading results which are in error.

Given a wide variety of instruments, the selection of the appropriate instrument will depend upon the number of gages required in the experimental stress analysis. If the experiment is static and involves only a small number of gages (say 10 or less), a simple strain indicator and a switch box with manual recording is sufficient. The use of a data logger probably is not warranted if less than 100 strain readings are to be made. If the experiment is conducted in the field where the instruments must be portable, battery power and simplicity are essential.

The type of analysis to be performed on the data affects the choice of the instrument. If one is to compute static stresses from a two-element rosette, a hand calculator is sufficient and the strain readings do not need to be digitized, handled in real time, or stored in a digitized format. On the other hand, if a strain-time record for a fracture specimen is to be analyzed to give the stress intensity factor for a propagating crack, it is essential that the data be digitized with a sampling period of 100 to 200 ns, and that the thousands of data samples be stored in real time and processed on a digital computer. The instrument should match the test requirements. Overmatch is costly in terms of time, accuracy, and skill requirements. Undermatch is usually disastrous since fidelity of signal may be lost or the data may be incomplete or in error.

The number of reports which must be processed by the laboratory is a consideration. If the laboratory must issue a large number of reports, then instruments which convert the voltage from the bridge into report-ready graphs are needed even if the experiment is simple and involves only a few strain gages. In this instance, a data logger with a computer interface and a friendly software package with excellent graphics output is preferred.

The frequency content of the strain signal is a critical consideration. If the strain signal is static, the time available for recording is relatively long and null-balance methods, potentiometer recorders, and slow A/D converters can be employed to record the signal output. In recent years, high-quality, low-cost integrating digital voltmeters have become available and have led to the development of static data-logging systems. With such systems, the output signals from several gages (systems accommodating 100 or more gages are common) are measured with the integrating digital voltmeter and automatically recorded in digital form on magnetic tape or disks for processing with relatively simple computer programs.

If the strain-gage signal is dynamic, the time available to measure the output voltage is limited and recording is much more difficult and expensive. Also, automatic processing of dynamic data requires digitizing with high-speed ADCs. In treating dynamic signals, the frequency of the strain signal is an important consideration in selecting the most appropriate recording system. For very low frequencies (0 to 3 Hz), quasistatic recording instruments such as the integrating digital voltmeter, the potentiometer recorder, and the xy recorder can be used very effectively. For intermediate frequencies (0 to 10,000 Hz), the oscillograph with either a pen (0 to 100 Hz) or a light-writing galvanometer (0 to 10,000 Hz) is the most common recording method. For the high-frequency range (0 to

20,000 Hz), an FM (frequency-modulated) instrument tape recorder can be used to monitor the signal. While this method of recording is expensive, it has an advantage in that the analog signal on the tape can be digitized and processed automatically. Finally, for very high frequencies (greater than 20,000 Hz), either an analog or digital oscilloscope is recommended. If the number of gages is small, this is a very effective recording system.

Each of these recording methods is described in detail in the following sections. In most instances, the description is qualitative with available equipment shown and capabilities, as they pertain to a given recording system, listed. However, coverage of the galvanometer and the oscillograph is detailed since the galvanometer is a low-input-impedance recording device which must be properly matched to the bridge to give the correct frequency response and the maximum sensitivity.

Telemetry is covered in the final section. While telemetry is not a recording method, it is closely aligned with recording since frequency modulation, multiplexing, and demodulation are techniques common to recording on magnetic tape and to automatic data processing. The description presented covers a relatively simple single-bridge, short-range transmission of data. A brief discussion of multichannel long-range transmission is also provided.

8.2 STATIC RECORDING AND DATA LOGGING

8.2.1 Manual Strain Indicators

The manual strain indicator, either null-balance or direct reading, previously described in Secs. 7.3.2 and 7.3.3, is the most common recording method used for static strain signals. These strain indicators, which are commercially available from several firms at modest cost, are accurate and simple to use in quarter-, half-, or full-bridge arrangements. The indicator can be used for multiple-strain-gage installations by adding a switch or a switch and balance circuit. Since the gage factor can be dialed directly into the indicator, the strain indicator is calibrated to read out directly in terms of strain. Readout requires approximately 10 s per gage for the direct reading indicator and 30 s per gage for the null balance indicator. Strain readings can be obtained from an installation involving 100 gages in approximately 20 min and 1 h, respectively. Normally, the output from each gage is recorded manually on a data sheet. The data are usually processed by hand or with a simple programmable calculator.

Some of the direct-reading indicators can also be used to log data automatically since the output from the digital voltmeter can be printed on paper or recorded in digital form on either punched tape or magnetic tape. The primary advantage of utilizing the more sophisticated (and more costly) direct-reading indicators is that data are obtained in a form which can be automatically processed. Data recorded in digital form on a magnetic tape can be processed by a computer to yield stresses, graphs of stress distributions, etc., or to provide failure or yield predictions.

8.2.2 Automatic Data-Acquisition Systems

Automatic data-acquisition systems should be considered by laboratories where relatively large strain-gage installations are employed on a regular basis. Automatic data-acquisition systems suitable for use with large numbers of strain gages are expensive ($20,000 to $30,000), are much more complex to operate, and require more maintenance than the simpler manual systems. However, if the volume of strain-gage readings is sufficiently large, the additional capital expenditure can be justified by the reduction in time required to test and analyze the data and by the immediate availability of test results in a report-ready format.

Automatic data-acquisition systems usually involve many optional components and can essentially be custom-designed. In each system, however, there are always six basic subsystems which include the controller, the signal conditioner, the multiplexer amplifier, the analog-to-digital converter, the storage or memory unit, and the readout devices. A schematic illustration of the elements of a data-acquisition system is presented in Fig. 8.1.

To illustrate these six subsystems, consider the commercial data-acquisition system shown in Fig. 8.2. The controller is a microprocessor that serves as the interface between the operator and the data-acquisition system. Directions to the controller are entered by the operator with the front panel key pads. A liquid crystal display (LCD) provides a readout showing the system operating parameters

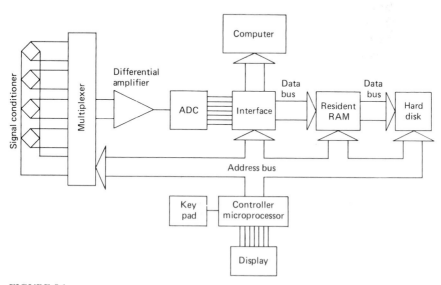

FIGURE 8.1
Schematic diagram showing the key elements in a digital data-acquisition system.

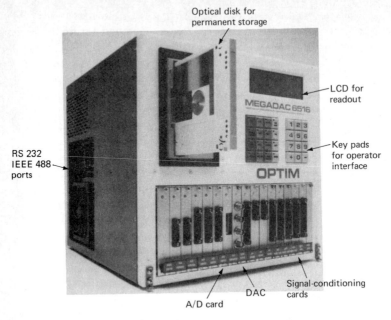

Optical disk for
permanent storage

LCD for
readout

MEGADAC 6516

Key pads
for operator
interface

RS 232
IEEE 488
ports

OPTIM

Signal-conditioning
cards

DAC

A/D card

FIGURE 8.2
A 128-channel—250,000 samples/s—portable data-acquisition system. (*Optim Electronics Corp.*)

and select readings of the strains. The controller is programmed with parameters which affect the data flow such as the sampling rate, the sequence of channels to be monitored, strain levels to trigger and to stop recording, time limits for the same purpose, and activation limits for supervisory alarms.

The controller also serves to direct the flow of data collected in the random-access memory (RAM). Depending on programming, the data stored in the RAM buffer can be held for one cycle, erased, or transferred to a permanent storage media such as a hard disk, or on high-capacity systems, an optical disk.

The signal conditioner consists of the power supply, the Wheatstone bridges, and the terminals used to connect a large number of gages in turn to the multiplexer. Usually, several bridges are contained on a plug-in circuit board, which can be modified by adding or deleting fixed resistors to provide for quarter-, half-, or full-bridge arrangements. A single power source is often used to power a number of individual bridges (from 10 to 100, depending upon the design of the system). Usually, the power supply is a highly regulated constant-current supply which can be adjusted to provide about 4 mA to the bridge.

The multiplexer portion of the signal conditioner-scanner subassembly consists of two parts: (1) A bank of switches serves to switch the two output leads and the cable shield from the bridge to the differential amplifier. In modern systems, solid-state switching devices (junction field-effect transistors) are employed. (2) The multiplexer also contains the circuits which control the switching sequence as programmed in the controller. The low-voltage signal from the bridge is switched through the multiplexer to the differential amplifier. The signal is amplified to ± 10 V full scale before conversion to a digital signal.

The amplified analog signal is converted into a digital signal by using an analog-to-digital converter. Two different ADCs are in common usage—the dual-slope integrating type or the successive approximation type. The dual-slope integrating ADC is relatively slow, usually less than 1000 samples/s and it is often employed on the lower-performance data-logging systems. The successive approximation ADCs are much faster and are used in more advanced acquisition systems which operate at 20,000 to 250,000 samples/s. Since the ADC is the most important component in digital acquisition and digital processing of data, the theory of operation of the ADC and the important aspects of digital codes are covered in detail in Sec. 8.3.

The data from the ADC is output from the interface unit (see Fig. 8.1) on a parallel wired data bus. Usually it is stored temporarily in the random-access memory on a first-in first-out basis. The data can be processed in real time on a host computer, or in less critical experiments the RAM can be downloaded onto a permanent storage media such as a hard disk, or in very-high-capacity systems onto an optical disk. The address bus carries instructions and addresses from the controller and provides for the flow of the data and the organization of the memory arrays.

The disks provide the input data, in digital format, to an off-line computer where the data is processed. Software is usually available from the suppliers of the data-acquisition systems to assist in the organization of data files and the transfer of data files to spreadsheets for subsequent processing. Data manipulation and the graphics output are usually dependent upon the capabilities of the commercially available spreadsheets such as Lotus 1-2-3. Graphical output of processed data in a report-ready format is a significant advantage of the digital data-acquisition systems.

Prior to processing the data, it is usually advisable to select the more important gages and to monitor the output from these gages. The examination of select output from specified gages is accomplished by displaying the voltages or voltages converted to strain on the LCD. For high-performance systems, with high-sampling-rate ADCs, the digital signals are taken out of the RAM and reconverted to analog signals with digital-to-analog converters (DACs). These analog signals are then displayed on oscilloscopes to show transient behavior of the strain-time records.

Rapid advances in developing higher-performance and lower-cost electronic components such as ADCs, RAMs, multiplexers, and storage devices have markedly improved the capability of digital data-acquisition systems. A decade ago these systems were very slow, 10 to 20 samples/s, and considered as suitable for recording only static events. Today, with systems capable of 250,000 samples/s, a digital data-acquisition system can process several channels of dynamic signals and has the capability of an oscillograph.

8.2.3   PC-Based Data-Acquisition Systems

In recent years relatively low-cost circuit boards have become available which contain many of the elements found in higher-cost, more elaborate data-acquisition systems. One typical board, illustrated in Fig. 8.3, is identified as a *multisensor interface unit*. This card supports several different sensors such as strain gages, resistance temperature detectors (RTDs), thermocouples, and thermistors. The sensor support includes sensor excitation, linearization, cold reference compensation, and conversion of the output to engineering units.

These data cards typically contain four sections. The first performs the signal conditioning for the sensors, multiplexes to the appropriate sensor, and amplifies the signal with a programmable gain. The second section performs A/D conversion. Typical cards utilize 14-bit successive approximation type ADCs with conversion times of about 2 ms. The third section incorporates a microcomputer with on-board memory that is used in processing data by performing tasks such as linearization, reference junction compensation, and engineering unit conversions. The microprocessor also provides the logic to scan the sensors, adjust the amplifier gain, and transfer the data to the standard bus registers. The standard bus interface incorporates drivers and receivers to facilitate communication with a personal computer- or PC-type host computer. The input/output (I/O) ports vary from card to card, but a typical configuration utilizing serial communication has three ports: one port to transfer data and instructions to the card, and two ports, one to transfer data and the other to indicate status to the host computer.

The card is programmed from the PC host computer and the digitized data is transferred on the standard bus from the card to the memory of the PC. At this point all further processing and preparation of graphics are performed on the host computer.

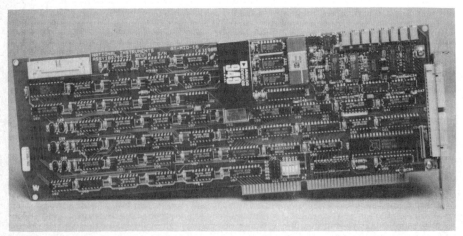

FIGURE 8.3
Multisensor interface board. (*National Instruments.*)

8.3 DIGITAL PROCESSING OF ANALOG SIGNALS

During the past two decades enormous progress has been made in developing digital instrumentation. Indeed, the combination of analog instrumentation with digital processing accomplished through analog-to-digital conversion has given new dimensions to both engineering analysis and process control because of the great savings of time associated with digital processing.

The analog side of an instrument system, which includes the bridges, power supply, multiplexer, and differential amplifier, interfaces with the digital side of the system with an analog-to-digital converter. The ADC takes the analog signal, a voltage $E(t)$ as an input, and converts it at discrete times into an equivalent digital code. Once digitized, the signal, now in the format of a digital code, can be stored in memory, transmitted at high speed over a data bus, processed in a digital computer, and displayed in pixel format on a cathode ray tube, liquid crystal display, or plasma panel.

The binary code is the most common digital code used in the conversion process. Resolution and resolution error based on the code count are introduced as the analog signal is converted. Resolution, an important concept in digital conversion, is described in terms of the binary code in Sec. 8.3.1.

8.3.1 Binary Code and Resolution

Digital systems employ logic gates which act as switches that are either on (1) or off (0). Several gages are arranged to provide a digital word with a length of n bits. Consider a digital word consisting of a 4-bit array, say 1011. The 1 at the extreme left is the most-significant bit (MSB) and the 1 at the extreme right is the least-significant bit (LSB). With a binary code, the least-significant bit has a weight 2^0, the next bit has a weight 2^1, the next 2^2, and the most-significant bit has a weight 2^{n-1}. When all 4 bits are zero (i.e., 0000), the equivalent count is 0. When all 4 bits are 1 (i.e., 1111), the equivalent count is $2^3 + 2^2 + 2^1 + 2^0 = 8 + 4 + 2 + 1 = 15$. A complete listing of the count possible with a 4-bit binary code is presented in Table 8.1.

It is evident from Table 8.1 that a 4-bit binary word permits a count of 2^4 or 16, arranged from 0 to 15. The total count C is determined from:

$$C = (2^n - 1) \tag{8.1}$$

where n is the number of bits in the digital word.

The least-significant bit represents the resolution R of a digital count containing n bits, which can be written as

$$R = \frac{2^0}{2^n - 1} = \frac{1}{2^n - 1} = \frac{1}{C} \tag{8.2}$$

This result indicates that the resolution which can be achieved with logic gates arranged to yield an 8-bit digital word (a byte) is $\frac{1}{255}$ or 0.392 percent of full scale.

TABLE 8.1
Count for a 4-bit binary code

MSB*	Bit 2	Bit 3	LSB*	Count
0	0	0	0	0
0	0	0	1	1
0	0	1	0	2
0	0	1	1	3
0	1	0	0	4
0	1	0	1	5
0	1	1	0	6
0	1	1	1	7
1	0	0	0	8
1	0	0	1	9
1	0	1	0	10
1	0	1	1	11
1	1	0	0	12
1	1	0	1	13
1	1	1	0	14
1	1	1	1	15

* MSB and LSB are the most- and least-significant bits, respectively.

Resolution is an important concept in digital instrumentation since it defines the number of bits required for a specified error in a measurement, or for the conversion of an analog voltage into a digital count representing that voltage. Values for C, R, and resolution error $\mathscr{E}$ as a function of n are presented in Table 8.2. Clearly an 8-bit digital word that provides resolution to within ± 1 count out of a total or full-scale count of 256 will limit the resolution error to 0.39 percent.

8.3.2 Conversion Processes

Analog voltages are converted to digital code with an analog-to-digital converter. The ADC is a one-way device converting only from analog to digital. To convert from a digital code to an analog voltage requires a digital-to-analog converter. The DAC is also a one-way device. As the DAC and ADC are the key functional elements in analog/digital instrument systems, they will be described in detail.

First, consider a 4-bit DAC where the input is a digital code ranging from 0000 to 1111, as listed in Table 8.1. The digital input is shown as the independent variable plotted along the abscissa of Fig. 8.4 and the analog voltage output is shown along the ordinate ranging from 0 to $\frac{15}{16}$ of full scale. While full scale, 16, is not available as a digital input, it represents the reference quantity to which the analog voltage is normalized. For example, let 10 V be the full-scale voltage; then the digital code 1000 will give, under ideal conditions, an analog voltage of $\frac{8}{16}(10) = 5$ V.

TABLE 8.2
Count C, resolution R, and error $\mathscr{E}_R$
as a function of the number of binary
bits n

n	C	R, ppm	$\mathscr{E}_R$, %
4	15	66666	6.6666
5	31	32258	3.2258
6	63	15873	1.5873
7	127	7874	0.7874
8	255	3922	0.3922
9	511	1957	0.1957
10	1023	978	0.0978
11	2047	489	0.0489
12	4095	244	0.0244
13	8191	122	0.0122
14	16383	61	0.0061
15	32767	31	0.0031
16	65535	15	0.0015

Next, consider an ADC with the analog voltage being the independent variable shown in Fig. 8.5. As all the analog voltages between zero and full scale can exist, they must be quantized by dividing the range of voltage into subranges. If V is the full-scale analog voltage input, the subrange for a 4-bit ADC is $\frac{1}{16}$ or 0.0625 V. All analog voltages within a given subrange are represented by the same digital code. The illustration in Fig. 8.5 shows that the digital code corresponds to the midpoint in each subrange. The quantizing process which replaces a linear

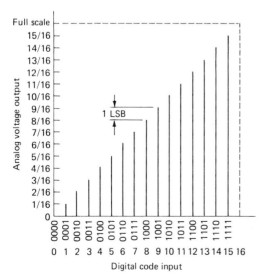

FIGURE 8.4
Relation between digital count and analog voltage for a digital-to-analog converter.

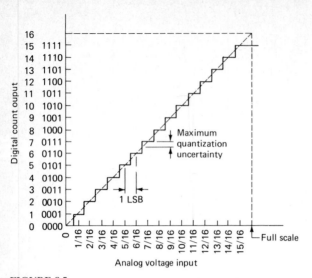

FIGURE 8.5

Relation between analog voltage input and digital count/code output for a 4-bit analog-to-digital converter.

analog function with a staircase digital representation results in a quantization uncertainty of $\pm\frac{1}{2}$ LSB.

8.3.3   A DAC Circuit

Many different circuits are employed in the design of DACs and it is beyond the scope of this book to cover the more sophisticated ones. Instead, a simple circuit will be described which shows the essential features involved in the digital-to-analog conversion process. A 4-bit DAC is illustrated in Fig. 8.6. A voltage reference is connected to a set of precision resistors arranged with binary values through a set of switches. The switches are operated by digital logic with 0 representing open and 1 representing closed. The resistors are binary-weighted, which means that resistance is doubled for each switch or bit so that

$$R_n = 2^n R_f \qquad (8.3)$$

where $R_n =$ the resistance of the nth bit
$\quad\quad R_f =$ the feedback resistance on the operational amplifier

When the switches are closed, a binary-weighted current I_n flows to the summing bus. This current is

$$I_n = \frac{E_R}{R_n} = \frac{E_R}{2^n R_f} \qquad (8.4)$$

The operational amplifier converts the currents to voltage, and furnishes a

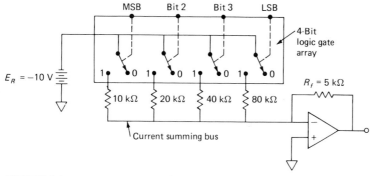

FIGURE 8.6
Schematic diagram of a simple 4-bit DAC.

low-impedance output. The analog output voltage E_o is given by

$$E_o = -R_f \sum_{n=1}^{k} I_n \tag{8.5}$$

where I_n is summed only if the switch n is closed.

Consider as an example the digital code 1011 or 11 with the circuit shown in Fig. 8.6. From Eq. (8.4) it is clear that $I_1 = 1$ mA, $I_2 = 0$, $I_3 = \frac{1}{4}$ mA, and $I_4 = \frac{1}{8}$ mA. Summing these currents and multiplying by $R_f = 5$ kΩ as indicated by Eq. (8.5) gives $E_o = 6.875$ V, which is the same as $\frac{11}{16}$ of the full-scale (reference) voltage.

Commercial DACs are more complex than the schematic shown in Fig. 8.6 because they contain more bits (8, 12, and 16 are common), and have several regulated voltages, integrated circuits (IC) for switching, and on-chip registers. The large number of bits are serviced with a parallel input called a *bus*. The voltage on each conductor in the bus is either high or low and activates each of the switches to provide the digital code as input to the device. The analog output is constant with respect to time as long as the digital code is held at the same value on the input.

8.3.4   ADC Circuits

Conversion of analog signals to digital code is extremely important in any instrument system involving digital processing of the analog output signals from the signal conditioners. Of the many ADC circuits available, three of the most common will be described here. These include the method of successive approximations, integration techniques, and flash conversion types.

CONVERSION BY SUCCESSIVE APPROXIMATIONS. . The method of successive approximations is illustrated in Fig. 8.7, where a bias voltage E_b is shown as an

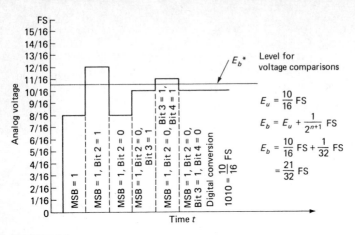

FIGURE 8.7
Successive-approximations method for A/D conversion.

approximation to an unknown analog voltage E_u given by

$$E_b = E_u + \frac{1}{2^{n+1}} \text{FS} \qquad (8.6)$$

The term $(1/2^{n+1})$FS in Eq. (8.6) is added to the unknown voltage to place the DAC output in the center of an output quantum, as illustrated in Fig. 8.4. In the case illustrated in Fig. 8.7, a 4-bit DAC is employed to convert a fixed analog input voltage $E_u = \frac{10}{16}$FS. The bias voltage shown in Fig. 8.7 is

$$E_b = \frac{10}{16} \text{FS} + \frac{1}{2^5} \text{FS} = \frac{21}{32} \text{FS}$$

The bias voltage E_b is compared to a sequence of precise voltages generated by an input controlled DAC. The digital input to the DAC is successively adjusted until the known output E_o from the DAC compares closely with E_b. The accuracy of the approximation will be defined after the conversion process is described.

At the start of the conversion process, the input to the DAC is set at 1000 (i.e., the most-significant bit is on and the others are off). The analog voltage output $E_o = \frac{8}{16}$FS, and a voltage comparison shows $E_o < E_b$. As this first approximation underestimates E_b, the MSB is locked at 1 and bit 2 is turned on to give a digital input of 1100 to the DAC. The output from this second approximation is $E_o = \frac{12}{16}$FS, as shown in Fig. 8.7. The voltage comparison shows $E_o > E_b$, an overestimate which turns bit 2 off and locks it at zero. The third approximation involves turning bit 3 on to give a digital input of 1010 and $E_o = \frac{10}{16}$FS. Since the voltage comparison shows $E_o < E_b$, bit 3 is locked on. The fourth and final approximation turns bit 4 on to give the digital input of 1011 and $E_o = \frac{11}{16}$FS. Since $E_o > E_u$, bit 4 is turned off to give a final result of 1010 at the input to the DAC after four approximations. It should be noted that the successive-approximations method is analogous to weighing an unknown mass on a balance using a set of binary weights.

The input of the DAC, in this case 1010, is transferred to the output register of the ADC. In this simple example the conversion process was exact since the unknown voltage E_u was selected at a 4-bit binary value $\frac{10}{16}$FS. However, an uncertainty in the conversion will occur when the unknown voltage differs from a binary value. The uncertainty, illustrated in Fig. 8.4, referenced to full scale is

$$\frac{\Delta E}{\text{FS}} = \frac{1}{2^{n+1}} \tag{8.7}$$

where n is the number of bits used in the approximation.

Equation (8.7) shows that analog-to-digital conversion results in a maximum uncertainty that is equivalent to $\frac{1}{2}$LSB.

In this example, the analog input voltage did not change during the conversion process. In fact, accurate conversion cannot be accomplished when the input voltage changes with time. To avoid problems associated with fluctuating voltages, the ADC utilizes an input device that samples and holds the signal constant for the period required for conversion. The complexity of the process increases because it is necessary to sample, hold, convert, and release from hold when the conversion is complete. The time required for conversion depends on the number of bits and the speed of the transistors used in switching. In 1990, high-performance 8- and 12-bit successive-approximation type ADCs typically required 0.25 and 1 μs (settling time), respectively, to convert an analog signal.

INTEGRATION METHOD FOR A/D CONVERSION. Analog-to-digital conversion by integration is based on counting clock pulses. A typical circuit for the dual-slope ADC is shown in Fig. 8.8a. At the start of a conversion, the unknown input voltage E_u is applied together with a reference voltage E_R to a summing amplifier which gives an output voltage

$$E_a = -\tfrac{1}{2}(E_u + E_R) \tag{a}$$

This voltage is imposed on an integrator which integrates E_a with respect to time to obtain

$$E_i = \frac{(E_u + E_R)t^*}{2RC} \tag{8.8}$$

where t^* is a fixed time of integration.

Upon completion of the integration at time t^*, a switch on the input of the integrator is activated which disconnects the summing amplifier and connects the reference voltage E_R to the integrator, as shown in Fig. 8.8b. The output voltage of the integrator then decreases with a slope of

$$\frac{\Delta E_i}{\Delta t} = -\frac{E_R}{RC} \tag{8.9}$$

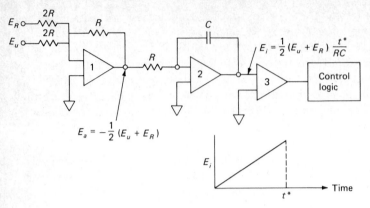

$$E_a = -\frac{1}{2}(E_u + E_R)$$

$$E_i = \frac{1}{2}(E_u + E_R)\frac{t^*}{RC}$$

(a) Charging ramp

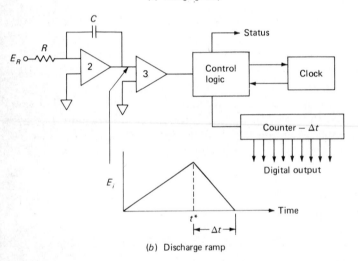

(b) Discharge ramp

FIGURE 8.8
(a) and (b) Dual-slope integration method for A/D conversion. ① Summing amplifier; ② integrator; ③ comparator.

The comparator monitors the output voltage E_i and issues a signal to the control logic when E_i goes to zero. This zero crossing occurs when

$$\frac{(E_u + E_R)t^*}{2RC} = \frac{E_R \Delta t}{RC} \tag{b}$$

Reducing Eq. (b) yields

$$\frac{\Delta t}{t^*} = \frac{1}{2}\left(\frac{E_u}{E_R} + 1\right) \tag{8.10}$$

It is clear from Eq. (8.10) that $\Delta t/t^*$, a proportional count, is related to E_u/E_R. If a counter is initiated by the switch on the integrator, then the counter will give a binary number representing the unknown voltage E_u.

The integration method for analog-to-digital conversion has several advantages. It is extremely accurate, as its output is independent of R, C, and clock frequency because these quantities affect the up and down ramps in the same way. The influence of noise on the unknown is markedly attenuated since the high-frequency noise signal is averaged toward zero during the integration period t^*. The primary disadvantage of the integration method is the speed of conversion, which is less than $\frac{1}{2}t^*$ conversions per second. To eliminate 60-Hz noise $t^* \geq 16.67$ ms and the rate of conversions must be less than 30 per second. This conversion rate is too slow for large multipurpose data-acquisition systems, but it is satisfactory for digital voltmeters and smaller data-logging systems.

PARALLEL OR FLASH A/D CONVERSION. Parallel or flash analog-to-digital conversion is the fastest but most expensive method for designing ADCs. The flash converter, illustrated in Fig. 8.9, employs $2^n - 1$ analog comparators arranged in parallel. Each comparator is connected to the unknown voltage E_u. The reference voltage is applied to a resistance ladder so that the reference voltage applied to a given comparator is 1 LSB higher than the reference voltage applied to the lower adjacent comparator.

When the analog signal is presented to the comparator bank, all the comparators with $E_{Rn}^* < E_u$ will go high and those with $E_{Rn}^* > E_u$ will stay low. As these are latching-type comparators, they hold high or low until they are downloaded to a system of digital logic gates which converts the parallel comparator word into a binary coded word.

The illustration shown in Fig. 8.9 is deceptively simple, as an 8-bit parallel ADC contains $2^8 - 1 = 255$ latching comparators and resistances and about 1000

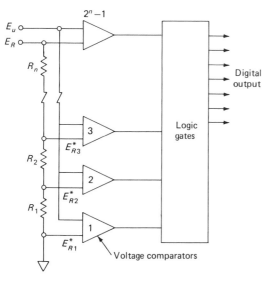

FIGURE 8.9
Schematic diagram for a parallel or flash ADC.

logic gates to convert the output to binary code. Also, the accuracy is improved by placing a track-and-hold device on the input so that the input voltage does not change over the period required to operate the comparators.

Flash-type ADCs are employed on high-speed digital acquisition systems such as waveform recorders or digital oscilloscopes. Conversion in these applications is made at 10 to 100 MHz, which gives sampling times of 100 to 10 ns.

8.4 DYNAMIC RECORDING AT VERY LOW FREQUENCIES

The methods used to record dynamic signals from strain-gage circuits depend upon the frequencies of the components which make up the transient strain pulse. The coverage of dynamic recording methods in this chapter is divided into four parts which deal with frequencies classified as very low (0 to 10 Hz) (Sec. 8.4), intermediate (0 to 5000 Hz)(Sec. 8.5), high (0 to 25,000 Hz)(Sec. 8.6), and very high (0 to over 100,000 Hz)(Sec. 8.7). The division into four frequency ranges was made to correspond with typical frequency limits on four different types of recording instruments.

For very low frequencies, instruments such as potentiometer (or strip-chart) recorders and xy recorders, which employ servomotors together with feedback control and null-balance positioning, can be used to measure the output voltage from the strain-gage bridge. The operating principle used in this type of instrument is illustrated in Fig. 8.10.

Potentiometer recorders are multirange instruments which can be used to measure voltages from $1\ \mu V$ to 100 V. The sensitivity is achieved by the stable and high-gain amplification provided by the servo drive system. Since no current flows in the potentiometer circuit in the balanced condition, the voltage measurements are independent of the length of the lead wires. The response of the potentiometer recorder is inherently slow because of the inertia

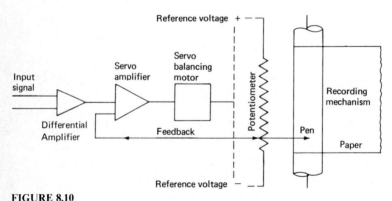

FIGURE 8.10
Schematic diagram of a servomotor-driven null-balanced (potentiometer) circuit for voltage measurement.

effects of the balancing motor. Most commercial instruments require between 0.2 and 5 s for the pointer to traverse the full scale of the indicator. Because of this low-frequency response, the potentiometer cannot be employed in strain-gage applications where the strain signal has frequency components greater than about 1 Hz. However, since the stability of the amplifier and servo system is excellent, accurate measurements of bridge voltages can be made over extended periods of time. The chart speeds can be varied over a wide range [1 in/h to 2 in/s (25 mm/h to 50 mm/s)] to provide an extremely versatile recording system for monitoring strain versus time for slowly changing loads on structures. The cost is relatively low in comparison with other dynamic recording instruments.

Another instrument which utilizes servo-driven pen motors is the xy recorder, which is very similar to the strip-chart recorder except that it simultaneously records one voltage along the x axis and another along the y axis. The two potentiometers (slide wires) are mounted along the traveling paths of the arm and pen carriage. The xy recorder is also a multirange device which can record voltages from about 5 μV to 100 V, depending upon the range selected. The time response of an xy recorder is determined by the slewing speed of the pen and the distance through which the pen must travel to record the voltage inputs. Slewing speeds of 20 in/s (500 mm/s) are common, and high-speed units are available with slewing speeds of 30 in/s (750 mm/s). The xy recorder is quite useful in monitoring load versus strain and is often employed in material-testing applications to record stress-strain curves. In this application, the output from a load cell is used to drive the y (stress) axis, and the output from an extensometer is used to drive the x (strain) axis.

In recent years the classic design of the potentiometer recorder (either strip chart or xy) has been modified. Today's recorders combine both analog and digital technologies to provide an instrument which is capable of real-time recording and digital plotting. A schematic diagram of the combination recorder-plotter is presented in Fig. 8.11.

The analog signals, two channels shown in Fig. 8.11, are amplified and then alternately switched into an ADC for conversion into digital format. The digital

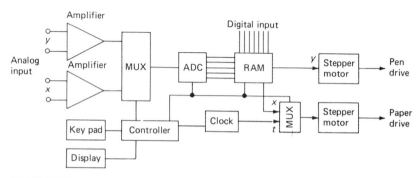

FIGURE 8.11
Schematic diagram of a combined analog recorder and digital plotter.

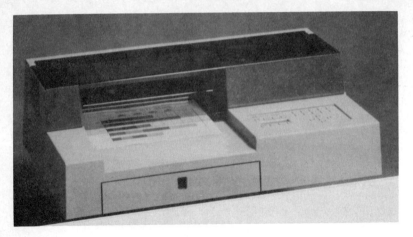

FIGURE 8.12
A combined analog recorder and digital plotter. (*Hewlett-Packard.*)

signals are stored in a RAM-type memory buffer. The digital signals are then transformed into a train of pulses which drives a stepper motor. The pen mechanism used in plotting is driven with this y stepper motor. When the system is used to record in real time, a clock is used to drive the second stepper motor. This motor in turn drives a pair of rolls which moves the paper in controlled movements through the recorder to give the time axis. The recorder is also used in the x-y mode by using the x analog input to drive the second stepper motor.

In addition to service in recording analog signals (y, t) or (x, y), the combination recorder-plotter can be used as a hard copier for other digital systems. For example, the output from a commercial spreadsheet processed on a personal computer can be downloaded into the RAM buffer and plotted out.

The capabilities of the combined recorder-plotters such as the one shown in Fig. 8.12 are impressive. The HP 7090A unit has a bandwidth from 0 (dc) to 3 kHz corresponding to a 33.3-kHz sampling rate. Three analog channels are sampled simultaneously with a 12-bit ADC. The amplifiers have a sensitivity of 5 mV full-scale and programmed calibration for direct readout in the appropriate engineering units. The RAM memories for each channel store 1000 words each 12 bits long. The controller with a keypad provides a user interface to facilitate annotation, data manipulation, and storage and retrieval of the data.

8.5 DYNAMIC RECORDING AT INTERMEDIATE FREQUENCIES

8.5.1 The Oscillograph

Oscillographs employing galvanometers as the recording device are the most widely used method for recording intermediate-frequency strain signals. There are four different types of oscillographs: (1) the pen-writing type, where the galvano-meter drives a pen or hot stylus; (2) the light-writing type, where the galvanometer drives a mirror and a beam of light is used to write on a photosensitive paper; (3) the cathode-ray-tube type, where the light from the tube is collected with fiber optics and focused on a light-sensitive paper; and (4) a digitizing oscillograph where the digitized signal activates an 8-dot/mm thermal head which writes on a thermally sensitive paper. The pen-writing galvanometer can be used for fre-quencies up to about 60 Hz, and the hot-stylus type can record frequencies up to about 100 Hz. The inertia of the pen arm limits the frequency response of this type of instrument. For these lower frequencies, the pen-type oscillograph is preferred over the light-writing oscillograph since the recordings on chart paper by either a pen or a hot stylus are of higher quality, more permanent, and less expensive than comparable recordings made on photosensitive paper with a light-writing oscillograph.

When the dynamic strain signal contains frequency components from 0 to 5 kHz, the cathode-ray-tube fiber-optic oscillograph is generally used as a record-ing method. This recording instrument is a combination of three technologies, namely, the oscillograph, the oscilloscope, and fiber optics. In this hybrid recorder, the galvanometers have been replaced with a special-purpose fiber-optic cathode ray tube. The fiber-optic cathode ray tube is unorthodox in that it provides only a time sweep with no provision for a y deflection of the electron stream.

Each input channel in this oscillograph contains a voltage-to-time converter that produces a pulse with a time duration proportional to the amplitude of the input signal. These pulses are used to produce a bright point of light on the CRT (a single point for each channel) whose duration corresponds to the input voltage from the bridge. The CRT is swept at a fixed frequency of 50 kHz to give a sampling rate of 20 μs/point. A dotted trace on the oscillograph record is avoided by using a memory circuit to interpolate between sampling points and produce a continuous trace.

The electron beam impinges on the fluorescent screen on the inside face of the CRT. The light emitted by the phosphorus, upon impact of the electrons, is collected by an array of approximately 10×10^6 glass fibers that have been fused together to form a 5×200 mm head that transmits the light from the CRT to the recording paper. A photosensitive paper is driven past the fiber-optic trans-mitter at speeds which can vary from 2.5 to 3000 mm/s. The result of this interesting mix of technologies is an instrument which is easy to use and provides the capability of displaying up to 18 channels of data on a photosensitive paper.

The frequency response of the fiber-optic cathode ray oscillograph is de-termined by the sweep rate of the CRT. With a definition of 10 points to describe a waveform, the sweep rate of 50 kHz gives a frequency response of 5 kHz. A modern fiber-optic cathode ray oscillograph is shown in Fig. 8.13.

For somewhat higher frequencies (up to 25 kHz), the light-writing oscillo-graph using high-gain amplifiers and fluid-damped galvanometers is usually employed. The galvanometer drives a mirror, which reflects light through a lens system onto a light-sensitive paper as shown in Fig. 8.14. The deflection of the

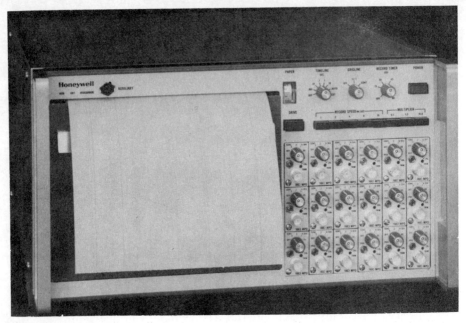

FIGURE 8.13
A fiber-optic cathode ray oscillograph. (*Honeywell Test Instruments Division.*)

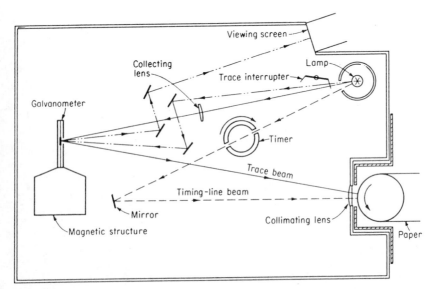

FIGURE 8.14
Schematic illustration of the operating principle of a light-writing oscillograph.

galvanometer is amplified optically and provides a y displacement on the record proportional to the strain magnitude. The paper speed is controlled by the motor and gear train driving the paper-feed mechanism and can be adjusted to give various time scales on the abscissa of the record. The frequency responses of this type of oscillograph will vary, depending upon the galvanometer employed; however, frequencies from 0 to approximately 25 kHz can be recorded if the proper galvanometer and paper speeds are employed.

Since the galvanometer represents the heart of the oscillograph and can be changed to vary the instrument's sensitivity and frequency response, it is important that the experimenter thoroughly understand its principle of operation. The essential elements of an electromagnetic galvanometer, which consists of a filament suspension system, a rotating coil, a mirror, and a pair of pole pieces, are shown in Fig. 8.15.

A current I, which passes through the filament suspension and the coil, produces a pair of equal and opposite forces F, which act to rotate the coil through an angle θ, as illustrated in Fig. 8.15. The magnitude of these current-induced forces is given by

$$F = nBIl \qquad (a)$$

where n = number of turns on the coil
 B = strength of the magnetic field
 l = length of the coil

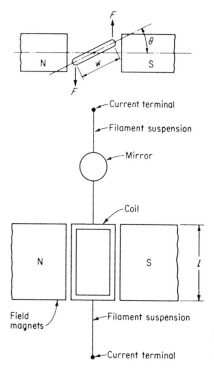

FIGURE 8.15
Schematic illustration showing the key elements of an electromagnetic galvanometer.

These coil forces produce a torque T_1 that is dependent on the angle of rotation θ and the coil width w as follows:

$$T_1 = Fw \cos \theta \qquad (b)$$

By combining Eqs. (a) and (b), the coil torque T_1 can be expressed in terms of the electrical and mechanical parameters of the galvanometer as

$$T_1 = nBIA \cos \theta \qquad (8.11)$$

where $A = wl$ is the area of the coil.

The rotation of the coil is resisted by the torsional constraint of the filament suspension system. The opposing torque T_2 can be expressed as

$$T_2 = G\theta \qquad (8.12)$$

Under static conditions, that is, where current I is independent of time, the two torques T_1 and T_2 must be equal to ensure rotational equilibrium. Hence

$$nBIA \cos \theta = G\theta \qquad (c)$$

Solving Eq. (c) for the rotational deflection θ of the galvanometer gives

$$\frac{\theta}{\cos \theta} = \frac{nBA}{G} I = CI \qquad (8.13)$$

where $C = nBA/G$ is the galvanometer constant.

The rotational deflection θ as expressed by Eq. (8.13) is a nonlinear function of the galvanometer current I due to the $\cos \theta$ term. This nonlinear behavior can be mitigated if the angular deflection θ is kept sufficiently small to ensure that

$$\frac{\theta}{\cos \theta} \approx \theta \qquad (d)$$

In order to increase the sensitivity of the galvanometer to accurately measure very small currents from strain gage bridges, it is necessary to increase the galvanometer constant C to as large a value as possible. This increase in sensitivity can be accomplished by increasing the number of turns in the coil and the coil area; however, this approach increases the inertia of the coil and reduces the frequency response of the galvanometer in dynamic applications. The galvanometer constant C can also be increased by increasing the field strength B. Unfortunately, the materials from which the permanent magnets are constructed limit the value of B that can be achieved. Finally, the sensitivity of the galvanometer can be enhanced by reducing the torsional spring constant G of the filament suspension system. This is accomplished by reducing the diameter of the wire in the suspension system, but this results in a delicate and fragile galvanometer.

In spite of these difficulties, sensitive and reasonably rugged galvanometers are commercially available for use in oscillographs. A sectioned view showing construction details of a commercially available unit is presented in Fig. 8.16. In dynamic strain-gage applications, the sensitivity of the galvanometer is usually

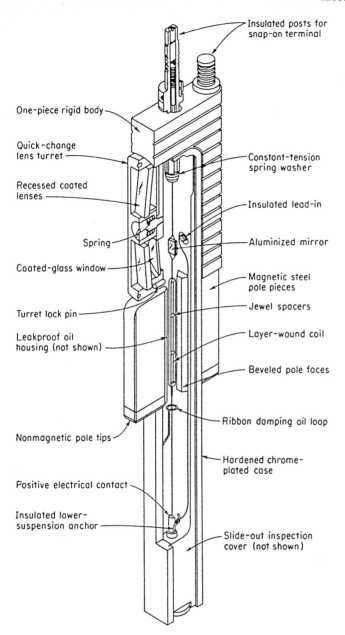

Insulated posts for snap-on terminal

One-piece rigid body

Quick-change lens turret

Constant-tension spring washer

Recessed coated lenses

Insulated lead-in

Spring

Aluminized mirror

Coated-glass window

Magnetic steel pole pieces

Jewel spacers

Turret lock pin

Layer-wound coil

Leakproof oil housing (not shown)

Beveled pole faces

Ribbon damping oil loop

Nonmagnetic pole tips

Hardened chrome-plated case

Positive electrical contact

Insulated lower-suspension anchor

Slide-out inspection cover (not shown)

FIGURE 8.16
Construction details of a galvanometer.

sacrificed to some degree to provide adequate frequency response. The dynamic response of a galvanometer will be covered in more detail in Sec. 8.5.2.

8.5.2   Transient Response of Galvanometers

Since galvanometers are largely employed in dynamic strain-gage applications, it is important to consider their response to a step pulse of current, which can be applied by closing the switch in the circuit shown in Fig. 8.17. The equation of motion which describes the angular deflection of the coil as a function of time after closing the switch is given by

$$T_1 - T_2 - T_3 = J \frac{d^2\theta}{dt^2} \tag{a}$$

where T_1 = the applied torque [Eq. (8.11)]
  T_2 = the restraining torque due to the suspension system [Eq. (8.12)]
  $T_3 = D_0 \, d\theta/dt$ is the restraining torque due to fluid damping
  D_0 = the fluid-damping constant

Substituting the three torques T_1, T_2, and T_3 in Eq. (a) gives

$$J \frac{d^2\theta}{dt^2} + D_0 \frac{d\theta}{dt} + G\theta = nBIA \cos \theta \tag{b}$$

By considering only small values of θ, where $\cos \theta \approx 1$, and using Eq. (8.13) it is possible to rewrite Eq. (b) as

$$J \frac{d^2\theta}{dt^2} + D_* \frac{d\theta}{dt} + G\theta = GCI_s \tag{8.14}$$

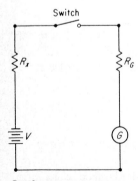

R_G = Galvanometer resistance
R_x = External resistance

FIGURE 8.17
Circuit for applying a step pulse of current to a galvanometer.

where I_s is the steady-state current in the coil and

$$D_* = D_0 + \frac{G^2 C^2}{R_x + R_G} \tag{8.14a}$$

is the combined damping constant.

The complementary solution of the differential equation given in Eq. (8.14) depends upon the constants J, D, and G. Three cases which arise must be treated individually, as summarized below:

Case 1, overdamped: $\left(\dfrac{D_*}{2J}\right)^2 > \dfrac{G}{J}$

Case 2, critically damped: $\left(\dfrac{D_*}{2J}\right)^2 = \dfrac{G}{J}$

Case 3, underdamped: $\left(\dfrac{D_*}{2J}\right)^2 < \dfrac{G}{J}$

The solutions of Eq. (8.14) for these three cases are:

Case 1, overdamped: $\dfrac{\theta}{\theta_s} = 1 - e^{-\alpha t}\left(\dfrac{\alpha}{\beta}\sinh \beta t + \cosh \beta t\right)$ $\tag{8.15}$

where $\alpha = D_*/2J$
$\beta^2 = (D_*/2J)^2 - G/J$
$\theta_s = $ steady-state deflection of galvanometer

In this, the overdamped case, the response of the galvanometer is sluggish, as indicated in Fig. 8.18, where the time required for the galvanometer to assume its steady-state position is quite long. In most applications, the overdamped galvanometer is to be avoided since the time required for a stable galvanometer reading is unduly long.

Case 2, critically damped: $\dfrac{\theta}{\theta_s} = 1 - e^{-\alpha t}(1 + \alpha t)$ $\tag{8.16}$

When the galvanometer is critically damped, $\alpha^2 = G/J$ and the response of the galvanometer is improved in comparison with the overdamped case. The galvanometer deflection θ approaches the steady-state deflection θ_s but never exceeds this value. This characteristic response of the galvanometer is also shown in Fig. 8.18.

Case 3, underdamped: $\dfrac{\theta}{\theta_s} = 1 - e^{-\alpha t}\left(\dfrac{\alpha}{\beta}\sin \beta t + \cos \beta t\right)$ $\tag{8.17}$

Underdamped galvanometers respond vigorously to a transient signal and initially overshoot the steady-state deflection then oscillate about θ_s for some time. Figure 8.18 also shows a plot of θ/θ_s for the underdamped case, which illustrates the character of the oscillation of the galvanometer about θ_s.

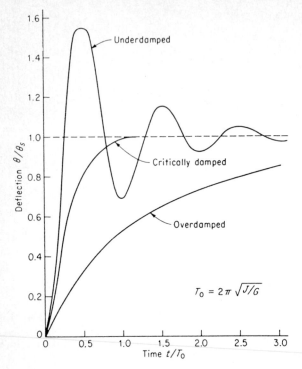

$$T_0 = 2\pi \sqrt{J/G}$$

FIGURE 8.18
Response of overdamped, critically damped, and underdamped galvanometers to an applied step pulse.

The degree of damping d

$$d = \frac{\alpha}{\sqrt{G/J}} \tag{8.18}$$

which is desirable to reduce the time of response of the galvanometer, depends upon the accuracy required in the measurement. This fact is illustrated in Fig. 8.19, where accuracy limits of ± 10 percent are placed on the value of θ/θ_s. The dimensionless time required for a critically damped galvanometer ($d = 1$) to record θ totally within this accuracy band is defined at point A. For an underdamped galvanometer with $d = 0.2$ and 0.5, the times required for the galvanometer to reach and remain in the ± 10 percent limit bands is given by points B and C, respectively. The relative positions of points A, B, and C will, of course, depend upon the accuracy requirements which establish the bandwidth. The locations of these points are important because they establish the response time of the galvanometer.

In most applications galvanometers are employed in a slightly underdamped condition, where the first overshoot and subsequent oscillations about θ_s do not exceed the bandwidth imposed by the accuracy limits. The degree of damping is normally selected so that the overshoot makes a tangent at its peak with the upper accuracy limit. The response time is therefore minimized by the selection of this particular degree of damping.

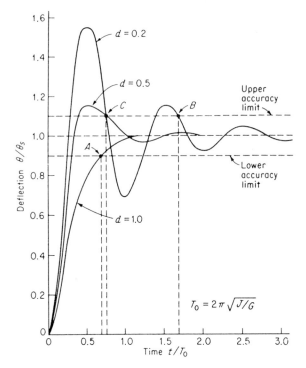

FIGURE 8.19
Accuracy limits superimposed on the response curves of a galvanometer.

8.5.3   Response of a Galvanometer to a Sinusoidal Signal

When a galvanometer is used to measure a sinusoidal signal which is periodic and of constant magnitude, the initial transient response of the galvanometer is of little importance. Instead, it is more important to determine how well the galvanometer deflection θ is following the input signal. The response of the galvanometer to this type of signal can be determined by considering its equation of motion. Thus

$$J \frac{d^2\theta}{dt^2} + D_0 \frac{d\theta}{dt} + G\theta = GCI_s \sin \omega t \tag{8.19}$$

where the instantaneous current in the galvanometer coil is given by

$$I = \frac{V}{R_x + R_G} \sin \omega t - \frac{GC}{R_x + R_G} \frac{d\theta}{dt} = I_s \sin \omega t - \frac{GC}{R_x + R_G} \frac{d\theta}{dt}$$

where $GC(d\theta/dt)$ is the back electromotive force on the coils.

This differential equation is identical to that given by Eq. (8.14) except for the term on the right-hand side. As such, the complementary solution is also the same; however, in this measurement, the solution sought is the steady-state response of the galvanometer, which is obtained from the particular solution of

Eq. (8.19). The particular solution for Eq. (8.19) leads to

$$\theta_s = \frac{GCI_s}{(\omega D_0)^2 + (G - \omega^2 J)^2} [(G - \omega^2 J) \sin \omega t - \omega D_0 \cos \omega t] \qquad (8.20)$$

A more concise form for Eq. (8.20) is given by

$$\frac{\theta_s}{CI_s} = \frac{1/2d}{(\omega/\omega_0)\sqrt{1 + (1/2d)^2(\omega_0/\omega - \omega/\omega_0)^2}} \sin(\omega t - \phi) \qquad (8.21)$$

where

$$\phi = \tan^{-1} \frac{1}{(1/2d)(\omega_0/\omega - \omega/\omega_0)} \qquad (8.22)$$

$$\omega_0 = \sqrt{G/J} \text{ is the free angular frequency} \qquad (8.23a)$$

$$T_0 = 2\pi/\omega_0 \text{ is the free period} \qquad (8.23b)$$

The angle ϕ represented in Eq. (8.22) is a phase angle, which indicates that the galvanometer deflection is following the signal with some phase angle ϕ. Provided this angle ϕ remains constant, it is not of serious concern since it represents a shift on the time scale of the oscillograph record. The phase angle ϕ is dependent upon the galvanometer characteristics (as defined by ω_0) and the degree of damping d. The results of Eq. (8.22) are shown in graphical form in Fig. 8.20, where ϕ is plotted as a function of the frequency ratio ω/ω_0 for different values of d.

In recording any complex waveform, the pulse will not be distorted if the phase angle increases linearly with frequency. This lack of distortion is due to the fact that with a linear phase angle the second harmonic will be shifted twice as

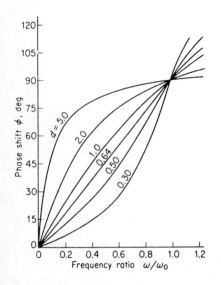

FIGURE 8.20
Phase shift as a function of frequency ratio.

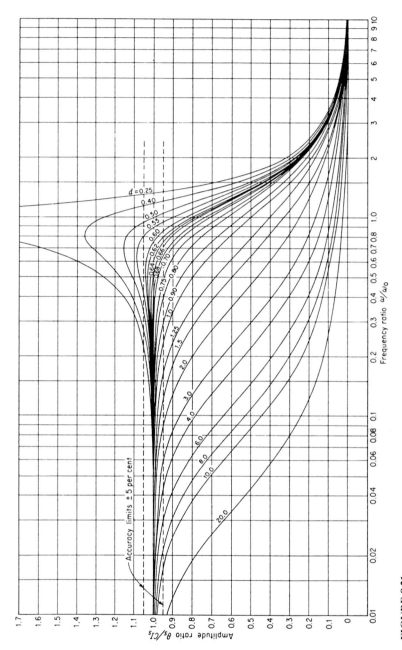

FIGURE 8.21
Amplitude ratio as a function of frequency ratio.

many degrees as the fundamental harmonic, etc. The implication here is that all the harmonics will experience the same time lag in seconds as the fundamental, and no distortion of the strain-time record will occur.

The frequency response of the galvanometer is described in Eq. (8.21), where the amplitude ratio θ_s/CI_s is shown to be a function of the frequency ratio ω/ω_0 and the degree of damping d. The results of this equation are shown in Fig. 8.21, where the amplitude ratio θ_s/CI_s is given as a function of ω/ω_0.

The curves shown in Fig. 8.21 represent the theoretical deflection of a galvanometer to a sinusoidal current which is constant in magnitude (I_s) but of variable frequency (ω). The frequency response of the galvanometer is dependent upon the accuracy limit and the degree of damping associated with the circuit. In Fig. 8.21, accuracy limits of ± 5 percent have been superimposed upon the response curves over the entire frequency range considered. For this accuracy requirement, it is clear that the degree of damping $d = 0.59$ maximizes the frequency response of the galvanometer; hence the response of the galvanometer will be flat, i.e., will be within the ± 5 percent band, for frequencies ω from 0 to 85 percent of ω_0. For other degrees of damping, the frequency response is reduced appreciably. For instance, with $d = 1.0$ the response of the galvanometer will lie within the ± 5 percent accuracy limits for frequencies from 0 to 22 percent of ω_0. As the accuracy limits are decreased, the range of the frequency response of the galvanometer also decreases. Accuracy requirements of ± 2 percent achieved with $d = 0.64$ can be realized over a frequency range from 0 to 67 percent of ω_0. This frequency response of $0.67\omega_0$ compares with $0.85\omega_0$ for ± 5 percent accuracy requirements.

8.6   DYNAMIC RECORDING AT HIGH FREQUENCIES

Magnetic-tape analog-data recording systems of the type illustrated in Fig. 8.22 offer an excellent method of recording and storing dynamic strain-gage signals. Multichannel recording of signals, ranging in frequency from 0 to 1 MHz, can be accomplished with this type of equipment. Data recorded and stored on magnetic tape are usually played back and displayed on an oscillograph. By varying the tape speed during playback, the time base can be extended or compressed. Detailed data analysis can be accomplished by utilizing an A/D converter and processing the digital data directly on a digital oscilloscope or a computer. The storage feature of magnetic tape is also important since it permits repeated examination of the data at any convenient time after completion of the test. Information stored on magnetic tape can be reliably retrieved any number of times; therefore, it can be subjected to many different types of analyses as the need arises.

In a magnetic-tape recorder, the tape ($\frac{1}{2}$ to 1 in wide) is driven at a constant speed by a servoed dc capstan motor over either the record or reproduce heads, as shown in Fig. 8.23. The speed of the capstan motor is monitored with a photocell (see Fig. 8.23) and compared with the frequency from a crystal oscillator to provide the feedback signal in a closed-loop servo system designed

FIGURE 8.22
Magnetic-tape recorder/reproducer. (*Honeywell Test Instruments Division.*)

to maintain constant tape speeds. Tape speeds have been standardized at $\frac{15}{16}$, $1\frac{7}{8}$, $3\frac{3}{4}$, $7\frac{1}{2}$, 15, 30, 60, 120, and 240 in/s.

The signal written on the tape by the record magnetic head assemblies is in the form of variations in the level of magnetism imposed on the magnetic coating of the tape. The reproduce heads convert these variations in magnetism back into electrical signals. Multichannel recorders employ four head assemblies which are precisely positioned on a single baseplate to ensure alignment with the tape. Two of the heads are record heads and the other two are reproduce heads, as shown

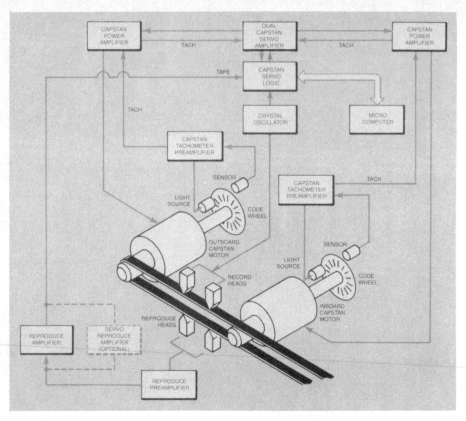

FIGURE 8.23
DC capstan magnetic-tape drive. (*Honeywell Test Instruments Division.*)

in Fig. 8.23. As the tape passes the first head assembly, odd-channel data are recorded; it then passes the second head assembly where the even-channel data are recorded. This procedure minimizes interchannel cross talk by maximizing the spacing between individual heads in the stacked-head assembly. Use of two recording-head stacks permits seven channels of recordings on $\frac{1}{2}$-in tape and 14 or 28 channels of recording on 1-in tape.

Most instrument tape systems can be used in either direct-recording or frequency-modulation modes. With direct recording, the intensity of magnetization on the tape is proportional to the instantaneous amplitude of the input signal. Direct recording is easy to accomplish with simple, moderately priced electronics. Extremely high frequencies (4 MHz) can be recorded. In playback, however, a signal is generated in the reproduce head only in response to changes in the magnetic flux on the tape; consequently, the direct-recording process cannot reproduce the dc components of the input signal. The direct-recording method is also subject to significant signal variations caused by randomly distributed surface inhomogeneities in the tape. Because of these irregularities, direct recording is usually limited to audio recording where the human ear, on playback, can average the amplitude errors, or to recordings where the signal frequency and not the signal amplitude is of primary importance.

Practically all strain-gage signals are written on magnetic tape with the FM method of recording. With FM recording, a carrier oscillator is frequency-modulated by the input signal. The oscillator has a center frequency which corresponds to a zero input signal. Deviations from the center frequency are proportional to the input signal. The polarity of the input signal determines the direction of the deviation. FM recording preserves the dc component in the signal and is much more accurate than direct recording. The price paid for these improvements is a reduction in frequency response, illustrated in Table 8.3, which shows bandwidth for direct and FM recording with three different interrange instrumentation group (IRIG) bands.

A detailed discussion of an instrument-quality analog tape recording system is beyond the scope of this text. The systems are complex and require considerable care in specifying the tape system and the associated electronics. A typical system adapted for strain-gage-data recording includes bridge-conditioning units, an input-signal coupler, a control panel, direct-record data amplifiers, FM-record data amplifiers, an FM frequency plug-in for each tape speed, direct-reproduce data amplifiers, FM-reproduce data amplifiers, the tape system, and related visual recorders. Complete system prices range from $70,000 to $100,000 depending on the IRIG band specified and the associated accessory equipment included. Even for the most affluent laboratory, the analog instrument tape recorder represents a large investment which should be carefully specified to meet the needs of the laboratory while minimizing the cost of the system.

TABLE 8.3
Bandwidth and signal-to-noise ratios for frequency-modulated and direct tape recording

IRIG band	Bandwidth, kHz	Signal-to-noise ratio, dB
Frequency-modulated		
±40 FM Int	dc–80	48
±40 FM WB-I	dc–160	48
±30 FM WB-II	dc–1000	36
Direct recording:		
1.2-MHz system	0.6–1200	34
4.0-MHz system	0.8–4000	23

Note: For 240 in/s tape speed, 28-track IB tape.

8.7 DYNAMIC RECORDING AT VERY HIGH FREQUENCIES

8.7.1 Analog Oscilloscopes

Very-high-frequency recording can be accomplished with cathode ray oscilloscopes which have bandwidths up to 1 GHz. The cathode ray oscilloscope is, in effect, a voltmeter which can be employed to measure transient voltage signals. The essential element in the oscilloscope is the cathode ray tube, illustrated in Fig. 8.24. The cathode ray tube is a highly evacuated tube in which electrons are produced by heating a cathode. These electrons are then collected, accelerated, and focused on a fluorescent screen on the face of the tube. The impinging stream of electrons forms a bright point of light on the fluorescent screen. Voltages applied to the horizontal and vertical deflection plates in the cathode ray tube will deflect the stream of electrons and move the bright point of light on the face of the tube. It is this ability to deflect the stream of electrons which permits the cathode ray tube to be employed as a dynamic voltmeter with an inertialess indicating system.

In recent years, advances in cathode-ray-tube technology have provided storage-type tubes, which retain and display the image of an electrical waveform on the tube face after the waveform ceases to exist. The image retention may be from a few seconds to several hours. The stored display can be instantaneously erased to ready the tube for display of the next waveform. Since the storage tubes utilize a bistable phosphor, they can be operated in either the storage or conventional (nonstorage) mode.

Storage oscilloscopes provide clear visual displays of slowly changing phenomena that would appear as a moving dot on a conventional tube. The storage tube also gives visual displays of rapidly changing nonrepetitive waveforms and allows for on-scope evaluation of the recorded data. This feature reduces the number of photographs which must be taken to give strain-time traces. The unwanted displays are erased and only the important displays are photographed.

For dynamic strain-gage applications, the cathode ray oscilloscope is an ideal voltage-measuring instrument. The input impedance of the instrument is

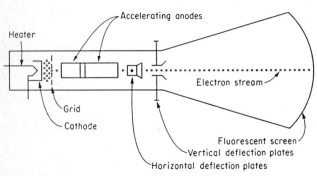

FIGURE 8.24
Basic elements of a cathode ray tube.

where f_{bw} = the bandwidth, MHz
  t_r = the signal rise time, ns

An additional requirement is the sample rate, which determines the time interval between data points. For example, a digital oscilloscope with a 4-MHz sample rate can sample, hold, convert, and store a data point in 250 ns.

Second, the number of bits of resolution used in the A/D conversion determines the accuracy relative to full scale. Resolutions of 6, 7, 8, 10, and 12 bits are used in commercially available digital oscilloscopes, which produce resolution errors ranging from 1.59 to 0.024 percent (see Table 8.2). Higher sampling rates are available with lower-resolution instruments and a tradeoff between sampling rate and resolution is necessary in selecting dynamic digital recorders.

Next, the size of the memory is important because memory length controls the length of signal which can be recorded. Recall that memory width is the same as the number of bits provided by the ADC. For example a digital oscilloscope with 8-bit resolution contains an 8K word memory and operates each of two channels at, say, 10 MHz. The memory size is 8192 words, with each word containing 8 bits for a total memory of 65,536 bits. The memory is divided into two equal segments of 4096 words with one segment for channel A and the second for channel B. At the maximum sampling rate of 10 samples/μs the recording period is $4096/10 = 409.67$ μs when both channels are in use. Reducing the sampling rate to 20 samples/s extends the recording period to 204.8 s. Clearly the range of time which can be covered with digital oscilloscopes with variable sampling rates is large (500,000/1 is common).

A final feature of importance is whether the A/D conversion method is designed for single-shot pulse measurements or measurements of repetitive periodic signals. Repetitive signals are easier to measure as sampling can be repeated on the second and subsequent waveforms to give instruments with apparent sampling rates which are higher than the real sampling rates. For pulse measurements, the signal occurs once and only once and repetitive measurements cannot be used to extend the sampling rate.

In operation, a trigger initiates the A/D conversion at a prescribed sampling rate with the words going to storage until the buffer is full. The buffer address for each data word is proportional to the time at which the data were taken. After a sweep, when the buffer is full, the data are transferred from buffer memory to mainframe memory where they are processed and displayed as a voltage-time trace on the CRT.

The fact that the input signal has been stored offers many advantages for data display or data processing. The data are displayed on the CRT in a repetitive manner so that traces of one-shot, transient events appear stationary. The trace can also be manipulated by expanding either the horizontal or vertical scales or both. This expansion feature permits a small region of the record to be enlarged and examined in detail. Readout of the data from the trace is also much easier and more accurate with digital oscilloscopes. A pair of marker lines called *cursors* (one vertical and the other horizontal) can be positioned anywhere on the screen.

The procedure is to position the vertical line at a time on the trace when a reading of the voltage is needed. The horizontal marker (or cross hair) automatically positions itself on the trace. The coordinates of the cross hair, intersection with the trace, are presented as a numerical display on the screen.

Modern digital oscilloscopes are usually equipped with a powerful microprocessor which gives several different on-board analysis features that often include:

1. Pulse characterization—rise time, fall time, baseline and top line width, overshoot, period, frequency, and duty cycle
2. Frequency analysis—power, phase, and magnitude spectrum
3. Math package—add, subtract, multiply, integrate, and differentiate
4. Counter—average frequency and event crossings
5. Display control—x zoom, x position, y gain, y offset
6. Plotting display
7. Mass storage to floppy disk, hard disk, or bubble devices

If additional processing is required, the data can be downloaded to a host computer for final data analysis.

Digital oscilloscopes are relatively new; early models were introduced only in 1972. Initially, performance of digital oscilloscopes was limited due to the relatively low-bandwidth capability (10 kHz or less); however, improvements in A/D converters, microprocessors, and high-speed DRAM memory chips have greatly enhanced speed of conversion, improved resolution, and expanded the amount of data which can be stored. Except for very-high-speed transient signals with very short rise times, the digital oscilloscope is superior in every respect to the analog oscilloscope. Costs are decreasing as performance improves in digital oscilloscopes, and the digital methods of recording are rapidly replacing the analog methods.

8.7.3   Waveform Recorders

Waveform recorders are similar to digital oscilloscopes in that they incorporate a high-speed A/D converter and store transient pulses or high-frequency waveforms in high-speed memory (RAM). They differ from digital oscilloscopes in that they must be used with an auxiliary display device such as an oscilloscope, oscillograph, or the monitor associated with a computer terminal to display either the original or the processed output.

Waveform recorders usually have superior resolution and longer memories than digital oscilloscopes with comparable sampling rates. In a sense they can be considered as an extremely high-performance tape recorder where the tape is replaced by silicon memory. A typical two-channel waveform recorder incorporates 16K words of memory with 10 bits per word. At a maximum sampling rate of 20M samples/s (corresponding to a minimum sampling time of 50 ns), with the memory divided equally between the two channels, the minimum recording time

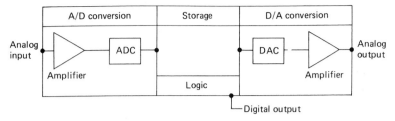

FIGURE 8.26
Block diagram of a waveform recorder and generator.

is $8192/20 = 409.6 \ \mu s$. A total of 8192 data points are recorded and stored at 50-ns intervals for each channel. At the recorder's lowest sampling rate of 20 samples/s, the recording time is 409.6 s or 6.83 min.

The waveform recorder has three main components, as illustrated in Fig. 8.26. At the input side, the analog signal is digitized with an independent ADC for each channel. The digitized output goes to the second component, a memory buffer (silicon RAM), where it is stored. The digital signal can be downloaded over a data bus to a host computer for processing and subsequent display. Alternatively, the digital signal can be reconverted to an analog signal and replayed repeatedly into a display device such as an analog oscilloscope.

8.8 DIGITAL CONVERSION RATES AND FREQUENCY RESPONSE

The conversion rate of the ADC controls the frequency of the unknown analog signal which can be measured. If the ADC is to resolve the analog signal into n bits to within 1 LSB, then the rate of change of the unknown signal is

$$\frac{dE_u}{dt} = \frac{E_u^*}{2^n T} \tag{8.25}$$

where E_u^* = the full-scale voltage of the ADC
T = the conversion period

Considering the unknown voltage as a sinusoid with

$$E_u = \tfrac{1}{2} E_u^* \sin 2\pi f t \tag{8.26}$$

differentiating Eq. (8.26) with respect to time yields

$$\frac{dE_u}{dt} = E_u^* \pi f \cos 2\pi f t \tag{8.27}$$

Substituting Eq. (8.27) into Eq. (8.25) yields

$$f = \frac{1}{2^n T \pi} \tag{8.28}$$

where $\cos 2\pi f t = 1$ and f is the maximum frequency that can be processed by the ADC.

For example, a 10-bit ADC converting at 20 samples/s can monitor a frequency $f = 0.0062$ Hz which corresponds to a voltage rate of change of 0.390 V/s in a 20-V span (± 10 V) for the ADC.

8.8.1 Aliasing

Digital data-acquisition systems depend upon an ADC which converts an analog signal to a digital signal at a specified sampling rate. This sampling rate is extremely important in dynamic measurements where high-frequency analog signals are being processed. Sampling theory indicates that sampling rates must exceed at least twice the frequency of a periodic analog signal.

If the frequency of the analog signal exceeds twice the sampling rate, the sampling process is inadequate and the output will give low-frequency signals, called *aliases*, that differ from the true analog signal. To illustrate aliasing, consider a sinusoidal analog signal with a frequency f which is the input to an ADC with a sampling rate of f_s. The output from the ADC is recorded as the difference frequency or aliasing frequency f_a given by

$$f_a = f_s - f \tag{8.29}$$

where f_a = the aliasing frequency
f_s = the sampling rate
f = the frequency of a sinusoidal analog signal

The results of Eq. (8.29) are illustrated in Fig. 8.27 for the example where $f_s = \frac{3}{2}f$. Note that the recorded digital signal exhibits the difference frequency $f_a = f/2$.

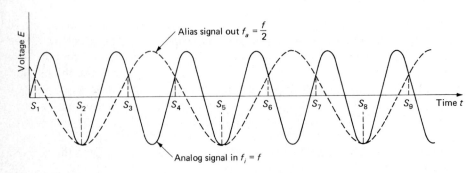

FIGURE 8.27
Effect of aliasing when a sinusoid is sampled at a frequency $f_s = 3/2f$.

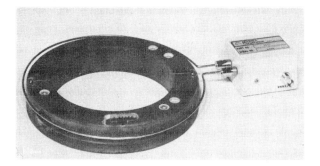

(*a*) Rotating Collar and Stationary Loop Antenna

(*b*) Readout and Display Unit

(*c*) Torsion-Measuring System on a Shaft

FIGURE 8.28
Torsion-measuring system utilizing telemetry for data transmission. (*Acurex Corporation.*)

8.9 TELEMETRY SYSTEMS

Test conditions sometimes preclude the use of lead wires, e.g., on rotating machinery, measurements in space, or measurements at great distances. In these instances, it is often necessary to transmit the signal from the measuring point to the recording site by telemetry. In relatively simple telemetry systems, the output from a single bridge is used to modulate a radio signal. The strain gage, bridge, power supply, and radio transmitter are located at the measuring point, and the receiver and recorder are at the recording site. The distance of transmission is a function of the power of the transmitter. For low-power transmission, where the range of operation is limited to a few feet, licensing is not required and the carrier frequency used for transmission is not limited.

An example of a telemetry system used to convert an operating shaft into a torsional load cell is illustrated in Fig. 8.28. This system incorporates a split collar which fits over the shaft and houses the dc power supply for the strain-gage bridge, the modulator, and the voltage-controlled oscillator (VCO). The signal from the bridge is used to pulse-width-modulate a constant-amplitude 5-kHz square wave (where the width of the output pulse is directly proportional to the voltage output from the bridge). This square wave serves to vary the frequency of the VCO, which is centered at 10.7 MHz. The VCO signal is transmitted by a rotating antenna mounted on the split collar and is received by a stationary loop antenna which encircles the shaft. The signal is demodulated, filtered, and amplified before being displayed. The transmitting unit is completely self-contained. The power to drive the bridge, modulator, and VCO is obtained by inductively coupling a 160-kHz signal to the power supply through the stationary loop antenna.

For longer-range multiple-transducer applications, the systems become more complex. First, licensing is required since the frequencies available for transmission are limited and crowded. Telemetry transmissions have been restricted to the bands from 1435 to 1535 and 2200 to 2300 MHz in the United States. Next, as the number of transducers increases at the measuring site, it becomes less practical to provide a separate transmitter for each; therefore, most of the long-range multiple-transducer systems combine the signals from a number of transducers into a single signal for transmission by one radio transmitter. The process of combining the different signals is called *multiplexing*. Each individual signal in the group retains its original information by using either frequency-division or time-division multiplexing. At the radio receiver, the individual signals are separated from the composite signal and recorded on individual recording instruments.

EXERCISES

8.1. Prepare a list of equipment and supplies to be taken into the field to perform a strain-gage analysis of a truck suspension system as the truck is operated over a rough road several miles long.

8.2. Prepare a list of equipment and supplies to be taken into the field to perform a strain-gage analysis of a pipeline in a rugged and remote site. Indeed the site is so remote that the nearest road is 3 miles from the test location and the site is 200 miles from your laboratory.

8.3. List the advantages and disadvantages of employing manually operated strain indicators and switch boxes to acquire strain data.

8.4. List the advantages and disadvantages of employing an automatic data-acquisition system to record strain data.

8.5. Four strain gages ($S_g = 2.05$) are mounted on a steel tension specimen having a cross-sectional area of 6.0 mm² to produce a load cell. If two of the gages are mounted in the axial direction and the other two in the transverse direction, what gage factor should be dialed into a manual null-balance strain indicator so that it will give a direct reading of the load P applied to the tension load cell?

8.6. Two 350-Ω strain gages are mounted in a Wheatstone bridge, one to serve as an active gage and the other to serve as a dummy gage. The gage factor for both gages is $S_g = 2.04$, and a strain of 1500 $\mu\epsilon$ is anticipated in the test. Describe the balance and calibration procedure to be followed so that the digital readout provided with a DVM that measures the bridge output voltage will be direct in terms of strain in units of $\mu\epsilon$.

8.7. A pressure vessel is being subjected to a series of hydrostatic tests, where the pressure changes from zero to maximum in 30 min on each run, and 200 strain gages have been installed on the vessel. Describe in complete detail how an automatic data-processing system would be employed to monitor, process, and analyze the data from each test run.

8.8. Assume that a 100-kN load cell with an output of 3 mV/V and a 2.5-mm strain extensometer with a 50-mm gage length and an output of 3 mV/V are used to monitor the stress and strain in a standard mild-steel tension specimen. Select appropriate input voltages for each transducer and select the amplifier setting on an xy recorder so that the stress-strain curve can be recorded directly on a 180×250-mm sheet of graph paper.

8.9. What are the six basic subsystems found in any automatic data-acquisition system?

8.10. Write a one-page engineering brief explaining the theory of operation of a modern data-acquisition system. The brief is intended for (a) your manager and (b) your laboratory technician.

8.11. If a digital data-acquisition system capable of 250×10^3 samples/s is used to monitor a single dynamic signal, what is the maximum frequency allowed for this signal?

8.12. Describe the standard bus interface used with PC-based data-acquisition systems.

8.13. Prepare a listing, similar to that shown in Table 8.1, showing the switch positions (0 or 1) and count for a 6-bit binary code.

8.14. Prepare a graph showing resolution R and count C as a function of the number of bits n. Let n range from 4 to 16.

8.15. For an 8-bit ADC, prepare a graph showing the quantization uncertainty and determine the quantization error.

8.16. Determine the currents flowing in the circuit shown in Fig. 8.6 for a digital code 0110. Also, find the voltage E_o and compare E_o to the reference voltage E_R.

8.17. Draw a circuit, similar to that shown in Fig. 8.6, for an 8-bit DAC. Show all resistance values and voltages.

8.18. Prepare an illustration, similar to that shown in Fig. 8.7, if the ADC has 4 bits and the unknown voltage is $\frac{7}{16}$FS.

8.19. Prepare an illustration, similar to that shown in Fig. 8.7, if the ADC has 6 bits and the unknown voltage is $\frac{55}{64}$FS.

8.20. Write a one-page engineering brief describing the theory of operation of a dual-slope A/D converter. The brief is to be read and understood by (a) your manager and (b) your laboratory technician.

8.21. Prepare a circuit diagram showing the voltage comparators used in a 3-bit flash ADC. Determine the resistances in the resistance ladder. For a (FS) reference voltage of 1 V and an unknown voltage of 0.65 V, identify the comparators which go high and those which remain low. Also give the binary output from the ADC.

8.22. Why do responsible engineers still purchase strip-chart recorders when a variety of digital recording systems are currently available for the same purpose?

8.23. Describe the differences between the last generation and the new generation recorder/plotter.

8.24. Solve Eq. (8.14) for the overdamped, critically damped, and underdamped conditions and verify Eqs. (8.15) to (8.17).

8.25. For a ± 3 percent accuracy band on a step-pulse input, determine the minimum response time of a galvanometer characterized by constants α, J, and G. Also, find the degree of damping d associated with this minimum response time.

8.26. Determine response time (to a step input) as a function of accuracy if an optimum degree of damping of the galvanometer is employed in the measurement. Show a graph of response time normalized with respect to the period of the galvanometer for percent accuracies ranging from 0 to 5 percent.

8.27. A galvanometer with an undamped frequency ω_0 is to be used to measure a sinusoidal current to an accuracy of

(a) ± 1 percent   (b) ± 2 percent
(c) ± 5 percent   (d) ± 10 percent

In each case, determine the frequency response and the degree of damping required to maximize this frequency response.

8.28. Verify Eq. (8.20) by solving the differential equation (8.19).

8.29. For the Wheatstone bridge with the low-impedance measuring system shown in Fig. E8.29, prove that

(a) $R_e = R_1 \left(\dfrac{r}{1+r} \right) \left(\dfrac{1+q}{q} \right)$ where $q = R_2/R_3$

(b) $I_G = \dfrac{q}{1+q} \, nS_g \epsilon \sqrt{P_g/R_g} \left(\dfrac{R_e}{R_e + R_M} \right)$

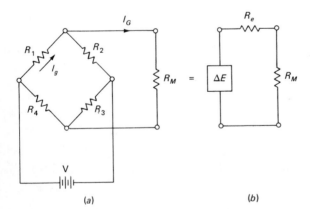

FIGURE E8.29

(a)   (b)

8.30. If the sensitivity of the oscillograph is defined by $S_\theta = \theta/\varepsilon$, show that S_θ for the circuit of Fig. E8.29 is

$$S_\theta = \frac{q}{1+q} \, nS_g C \sqrt{\frac{P_q}{R_g}} \left(\frac{R_e}{R_e + R_M} \right)$$

8.31. For quasistatic strain-gage applications where the results of Exercise 8.30 are valid, verify the sensitivity factors $S_\theta/(S_g C \sqrt{P_g/R_q})$ for the bridge arrangements shown in Fig. E8.31.

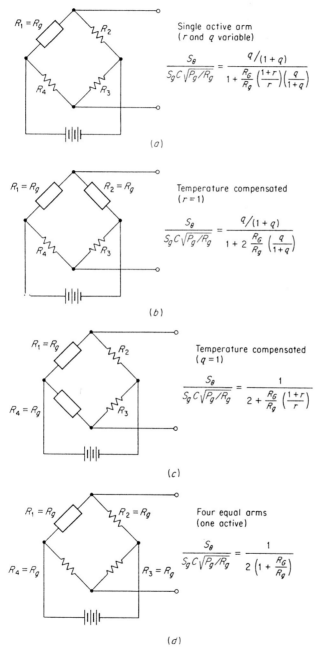

Single active arm
(r and q variable)

$$\frac{S_\theta}{S_g C \sqrt{P_g/R_g}} = \frac{q/(1+q)}{1 + \frac{R_G}{R_g}\left(\frac{1+r}{r}\right)\left(\frac{q}{1+q}\right)}$$

(a)

Temperature compensated
($r=1$)

$$\frac{S_\theta}{S_g C \sqrt{P_g/R_g}} = \frac{q/(1+q)}{1 + 2\frac{R_G}{R_g}\left(\frac{q}{1+q}\right)}$$

(b)

Temperature compensated
($q=1$)

$$\frac{S_\theta}{S_g C \sqrt{P_g/R_g}} = \frac{1}{2 + \frac{R_G}{R_g}\left(\frac{1+r}{r}\right)}$$

(c)

Four equal arms
(one active)

$$\frac{S_\theta}{S_g C \sqrt{P_g/R_g}} = \frac{1}{2\left(1 + \frac{R_G}{R_g}\right)}$$

(d)

FIGURE E8.31

8.32. Prepare a graph showing sensitivity factor $S_\theta/(S_g C \sqrt{P_g/R_g})$ as a function of R_G/R_g for the four bridge arrangements shown in Fig. E8.31. From these results determine
(a) The resistance ration R_G/R_g for maximum sensitivity
(b) The best bridge arrangement for temperature compensation
(c) The sensitivity of the four-equal-arm bridge

8.33. For a static or quasistatic strain-gage application, where the external resistance of the galvanometer circuit is not critical, determine the sensitivity factor $S_\theta/(S_g C \sqrt{P_g/R_g})$ for the bridge arrangements shown in Fig. E8.33. Compare these results with the results shown in Fig. E8.31 and draw conclusions regarding the use of multiple-gage installations to improve sensitivity.

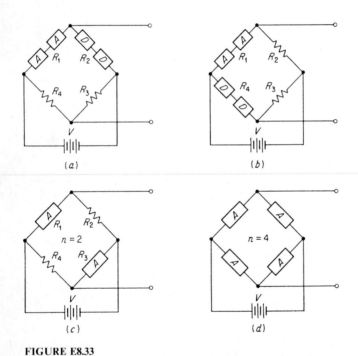

FIGURE E8.33

8.34. A single active strain gage ($R_1 = R_g = 120\ \Omega$) and a dummy gage are employed in a four-equal-arm bridge; that is, $r = q = 1$. An M40-120A galvanometer is used to measure the bridge output. For this combination of bridge and galvanometer, determine the system sensitivity $S_\theta/(S_g \sqrt{P_g/R_g})$. If the strain-gage characteristics are $S_g = 2.06$ and $P_g = 0.05\ \mathrm{W}$, determine the galvanometer deflection for a strain of $600\ \mu\epsilon$. The characteristics of a "line" of galvanometers are shown in Table 8.4.

8.35. A single active gage ($R_1 = R_g = 350\ \Omega$) and a dummy gage are employed in a four-equal-arm bridge; that is, $r = q = 1$. An M40-350A galvanometer is used to measure the bridge output. For this combination of bridge and galvanometer, determine the system sensitivity $S_\theta/(S_g \sqrt{P_g/R_g})$ and compare the results with those obtained in Exercise 8.34.

TABLE 8.4
Honeywell series "M" miniature galvanometer characteristics

Galvo type no.	Nominal undamped natural frequency Hz	Flat (±5%) frequency response Hz	Required external damping resistence R_x ohms	Nominal coil resistence R_G ohms	Current (±5%) I_g	Current (±5%) $\frac{1}{I_g}$	Galvanometer Voltage V_g	Galvanometer Voltage $\frac{1}{V_g}$	Circuit Voltage V_c mV/cm	Circuit Voltage $\frac{1}{V_c}$ cm/mV	Max. safe current mA	Max. deflection DC cm	Max. deflection AC peak to peak cm	balance std mm/g	balance precision mm/g
					Electromagnetically-damped types										
					μA/cm	cm/μA	mV/cm	cm/mV	mV/cm	cm/mV					
*M40-120A	40	0- 24	120	30.0	3.15	.317	.094	10.6	.472	2.12	10	10.0	20.0	.89	.46
*M40-350A	40	0- 24	350	66.0	1.61	.621	.106	9.43	.673	1.49	10	10.0	20.0	.89	.46
*M100-120A	100	0- 60	120	52.2	3.94	.254	.206	4.85	.677	1.48	10	10.0	20.0	.41	.21
*M100-350	100	0- 60	350	75.0	2.48	.403	.187	5.35	1.06	.943	10	10.0	20.0	.56	.28
*M200-120	200	0- 120	120	62.0	10.0	.100	.622	1.61	1.82	.546	10	10.0	20.0	.41	.21
*M200-350	200	0- 180	350	62.0	10.0	.100	.622	1.61	4.13	.242	10	10.0	20.0	.41	.21
*M400-120	400	0- 240	120	125.0	30.3	.033	3.79	.264	7.44	.134	10	10.0	20.0	.41	.21
*M400-350	400	0- 360	350	125.0	30.3	.033	3.79	.264	14.4	.069	10	10.0	20.0	.41	.21
"M600-350	600	0- 540	350	320.0	51.2	.020	16.4	.061	34.3	.029	15	10.0	20.0	.41	
					Fluid-damped types										
					mA/cm	cm/mA	V/cm	cm/V			mA	cm	cm	mm/g	mm/g
*M1000	1000	0- 600	150	39.0	1.04	.961	.041	24.4			70	10.0	20.0	.21	
*M1650	1650	0- 1000	100	26.8	3.66	.273	.098	10.2			70	10.0	20.0	.21	
*M3300	3300	0- 2000	100	32.0	7.87	.127	.252	3.97			70	5.0	14.5	.21	
*M5000	5000	0- 3000	100	39.5	12.3	.081	.484	2.07			70	3.05	8.9	.21	
*M8000	8000	0- 4800	100	35.0	15.7	.064	.551	1.81			70	2.0	5.8	.21	
*M10000	10000	0- 6000 ± 12%	100	35.0	15.7	.064	.551	1.81			70	2.0	5.8	.21	
*M13000	13000	0-13000 ± 8%	100	71.6	32.1	.031	2.30	.435			55	1.0	2.8	.21	
*M25K	22000	0-25000 ± 15%	100	49.0	49.2(±15%)	.020	2.41	.415			75	1.0	2.8	.21	

8.36. Determine the system sensitivity $S_\theta/(S_g\sqrt{P_g/R_g})$ if an M400-120 galvanometer is used in place of the M40-120A galvanometer in Exercise 8.34. Compare the two systems on the basis of sensitivity and frequency response.

8.37. Determine the system sensitivity $S_\theta/(S_g\sqrt{P_g/R_g})$ if an M400-350 galvanometer is used in place of the M40-350A galvanometer in Exercise 8.35. Compare the two systems on the basis of sensitivity and frequency response.

8.38. Determine the system sensitivity $S_\theta/(S_g\sqrt{P_g/R_g})$ if an M40-120A galvanometer must be used in place of the M40-350A galvanometer in Exercise 8.35. Compare with the results of Exercise 8.35 and discuss.

8.39. Determine the system sensitivity $S_\theta/(S_g\sqrt{P_g/R_g})$ if an M40-350A galvanometer must be used in place of the M40-120A galvanometer in Exercise 8.34. Compare with the results of Exercise 8.34 and discuss.

8.40. Briefly describe the difference between direct (AM) and frequency modulation (FM) recording and give the reasons for not using AM recording with instrument-quality magnetic tape.

8.41. Briefly discuss the advantages and disadvantages of magnetic-tape recording systems relative to the oscillograph recording method.

8.42. Consider a dynamic application where only two or three strain gages are to be monitored as a function of time. Discuss the relative merits of oscillograph, oscilloscope, and magnetic-tape recording.

8.43. If an oscilloscope has a bandwidth of 1 MHz, determine the minimum rise time of a dynamic signal which can be recorded without significant distortion.

8.44. A 12-bit ADC is used in a digital oscilloscope. If the oscilloscope samples at 100 ns/sample, determine the size of RAM memory required for a single channel with a 1000-μs recording interval.

8.45. Verify Eq. (8.28).

8.46. A 12-bit ADC operates over a span of ± 10 V with a rate of 100 samples/s. Determine the maximum frequency which can be converted.

8.47. Prepare a graph similar to that shown in Fig. 8.27 if $f_s = \frac{4}{3}f$. Show the alias signal and determine its frequency.

REFERENCES

1. Dally, J. W., W. F. Riley, and K. G. McConnell: *Instrumentation for Engineering Measurements*, John Wiley, New York, 1984.
2. McConnell, K. G., and W. F. Riley: "Strain Gage Instrumentation and Data Analysis," in A. A. Kobayashi (ed.), *Handbook on Experimental Mechanics*, Prentice-Hall, Englewood Cliffs, N.J., pp. 79–116, 1987.

CHAPTER
9

STRAIN-ANALYSIS METHODS

9.1 INTRODUCTION

Electrical-resistance strain gages are normally employed on the free surface of a specimen to establish the stress at a particular point on this surface. In general it is necessary to measure three strains at a point to completely define either the stress or the strain field. In terms of principal strains, it is necessary to measure ϵ_1, ϵ_2, and the direction of ϵ_1 relative to the x axis as given by the principal angle ϕ. Conversion of the strains into stresses requires, in addition, a knowledge of the elastic constants E and v of the specimen material.

In certain special cases the state of stress can be established with a single strain gage. Consider first a uniaxial state of stress where $\sigma_{yy} = \tau_{xy} = 0$ and the direction of σ_{xx} is known. In this case a single-element strain gage is mounted with its axis coincident with the x axis. The stress σ_{xx} is given by

$$\sigma_{xx} = E\epsilon_{xx} \qquad (9.1)$$

Next, consider an isotropic state of stress where $\sigma_{xx} = \sigma_{yy} = \sigma_1 = \sigma_2$ and $\tau_{xy} = 0$. In this case a strain gage may be mounted with any orientation ϕ since all directions are principal, and the magnitude of the stresses can be established from

$$\sigma_{xx} = \sigma_{yy} = \sigma_1 = \sigma_2 = \frac{E}{1 - v}\epsilon_\theta \qquad (9.2)$$

where ϵ_θ is the strain measured in any direction in the isotropic stress field.

When less is known beforehand regarding the state of stress in the specimen, it is necessary to employ multiple-element strain gages to establish the magnitude of the stress field. If the specimen being investigated has an axis of symmetry, or if a brittle-coating analysis has been conducted to establish the principal-stress directions, this knowledge can be used to reduce the number of gage elements required from three to two. A two-element rectangular rosette similar to those illustrated in Fig. 6.4d and e is mounted on the specimen with its axes coincident with the principal directions. The two principal strains ϵ_1 and ϵ_2 obtained from the gages can be employed to give the principal stresses σ_1 and σ_2:

$$\sigma_1 = \frac{E}{1 - v^2}(\epsilon_1 + v\epsilon_2)$$

$$\sigma_2 = \frac{E}{1 - v^2}(\epsilon_2 + v\epsilon_1)$$

(9.3)

These relations give the complete state of stress since the principal directions are known a priori. The stresses on any plane can be established by employing Eqs. (1.16) with the results obtained from Eqs. (9.3).

In the most general case, no knowledge of the stress field or its directions is available before the experimental analysis is conducted. Three-element rosettes are required in these instances to completely establish the stress field. To show that three strain measurements are sufficient, consider three strain gages aligned along axes A, B, and C, as shown in Fig. 9.1.

From Eqs. (2.18) it is evident that

$$\epsilon_A = \epsilon_{xx} \cos^2 \theta_A + \epsilon_{yy} \sin^2 \theta_A + \gamma_{xy} \sin \theta_A \cos \theta_A$$

$$\epsilon_B = \epsilon_{xx} \cos^2 \theta_B + \epsilon_{yy} \sin^2 \theta_B + \gamma_{xy} \sin \theta_B \cos \theta_B$$

$$\epsilon_C = \epsilon_{xx} \cos^2 \theta_C + \epsilon_{yy} \sin^2 \theta_C + \gamma_{xy} \sin \theta_C \cos \theta_C$$

(9.4)

The cartesian components of strain ϵ_{xx}, ϵ_{yy}, and γ_{xy} can be determined from a simultaneous solution of Eqs. (9.4). The principal strains and the principal directions can then be established by employing Eqs. (2.7), (1.12), and (1.14). The

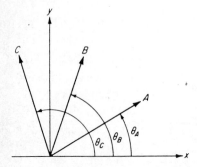

FIGURE 9.1
Three gage elements placed at arbitrary angles relative to the x and y axes.

results are

$$\epsilon_1 = \tfrac{1}{2}(\epsilon_{xx} + \epsilon_{yy}) + \tfrac{1}{2}\sqrt{(\epsilon_{xx} - \epsilon_{yy})^2 + \gamma_{xy}^2}$$

$$\epsilon_2 = \tfrac{1}{2}(\epsilon_{xx} + \epsilon_{yy}) - \tfrac{1}{2}\sqrt{(\epsilon_{xx} - \epsilon_{yy})^2 + \gamma_{xy}^2} \qquad (9.5)$$

$$\tan 2\phi = \frac{\gamma_{xy}}{\epsilon_{xx} - \epsilon_{yy}}$$

where ϕ is the angle between the principal axis (σ_1) and the x axis. The principal stresses can then be computed from the principal strains by utilizing Eqs. (9.3).

In actual practice, three-element rosettes with defined angles (that is, θ_A, θ_B, and θ_C fixed at specified values) are employed to provide sufficient data to completely define the stress field. These rosettes are described by the fixed angles as the rectangular rosette, the delta rosette, and the tee-delta rosette. Examples of commercially available three-element rosettes are presented in Fig. 6.4g and h. Also shown in this figure (Fig. 6.4m) is a stress gage which can be employed to give the stress in an arbitrary direction. The three-element rosette gage will be discussed in detail in Sec. 9.2.

9.2 THE THREE-ELEMENT RECTANGULAR ROSETTE

The three-element rectangular rosette employs gages placed at the 0°, 45°, and 90° positions, as indicated in Fig. 9.2. For this particular rosette it is clear from Eqs. (9.4) that

$$\epsilon_A = \epsilon_{xx} \qquad \epsilon_B = \tfrac{1}{2}(\epsilon_{xx} + \epsilon_{yy} + \gamma_{xy}) \qquad \epsilon_C = \epsilon_{yy}$$

and that $\qquad\qquad\qquad\qquad\qquad\qquad\qquad\qquad\qquad\qquad\qquad$ (9.6)

$$\gamma_{xy} = 2\epsilon_B - \epsilon_A - \epsilon_C$$

Thus, by measuring the strains ϵ_A, ϵ_B, and ϵ_C, the cartesian components of strain ϵ_{xx}, ϵ_{yy}, and γ_{xy} can be quickly and simply established through the use of Eqs.

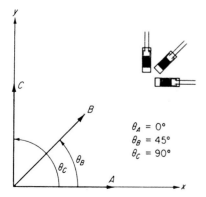

$$\theta_A = 0°$$
$$\theta_B = 45°$$
$$\theta_C = 90°$$

FIGURE 9.2
Gage positions in a three-element rectangular rosette.

(9.6). Next, by utilizing Eqs. (9.5), the principal strains ϵ_1 and ϵ_2 can be established as

$$\epsilon_1 = \tfrac{1}{2}(\epsilon_A + \epsilon_C) + \tfrac{1}{2}\sqrt{(\epsilon_A - \epsilon_C)^2 + (2\epsilon_B - \epsilon_A - \epsilon_C)^2}$$
$$\epsilon_2 = \tfrac{1}{2}(\epsilon_A + \epsilon_C) - \tfrac{1}{2}\sqrt{(\epsilon_A - \epsilon_C)^2 + (2\epsilon_B - \epsilon_A - \epsilon_C)^2}$$

(9.7a)

and the principal angle ϕ is given by

$$\tan 2\phi = \frac{2\epsilon_B - \epsilon_A - \epsilon_C}{\epsilon_A - \epsilon_C}$$

(9.7b)

The solution of Eq. (9.7b) gives two values for the angle ϕ, namely, ϕ_1, which refers to the angle between the x axis and the axis of the maximum principal strain ϵ_1, and ϕ_2, which is the angle between the x axis and the axis of the minimum principal strain ϵ_2. These angles are illustrated in the Mohr's strain circle shown in Fig. 9.3. It is possible to show (see Exercise 9.5) that the principal axes can be identified by applying the following rules:

$$
\begin{aligned}
0 < \phi_1 < 90° \qquad &\text{when } \epsilon_B > \tfrac{1}{2}(\epsilon_A + \epsilon_C) \\
-90° < \phi_1 < 0 \qquad &\text{when } \epsilon_B < \tfrac{1}{2}(\epsilon_A + \epsilon_C) \\
\phi_1 = 0 \qquad &\text{when } \epsilon_A > \epsilon_C \text{ and } \epsilon_A = \epsilon_1 \\
\phi_1 = \pm 90° \qquad &\text{when } \epsilon_A < \epsilon_C \text{ and } \epsilon_A = \epsilon_2
\end{aligned}
$$

(9.8)

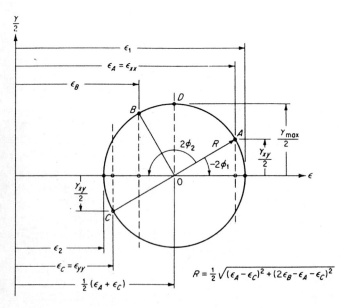

FIGURE 9.3
Graphical solution for the principal strains and their directions from a rectangular rosette.

Finally, the principal stresses occurring in the component can be established by employing Eqs. (9.7) together with Eqs. (9.3) to obtain

$$\sigma_1 = E\left[\frac{\epsilon_A + \epsilon_C}{2(1-v)} + \frac{1}{2(1+v)}\sqrt{(\epsilon_A - \epsilon_C)^2 + (2\epsilon_B - \epsilon_A - \epsilon_C)^2}\right]$$

$$\sigma_2 = E\left[\frac{\epsilon_A + \epsilon_C}{2(1-v)} - \frac{1}{2(1+v)}\sqrt{(\epsilon_A - \epsilon_C)^2 + (2\epsilon_B - \epsilon_A - \epsilon_C^2}\right]$$

(9.9)

The use of Eqs. (9.6) to (9.9) permits a determination of the cartesian components of strain, the principal strains and their directions, and the principal stresses by a totally analytical approach. However, it is also possible to determine these quantities with a graphical approach, as illustrated in Fig. 9.3. A Mohr's strain circle is initiated by laying out the ϵ (abscissa) and the $\frac{1}{2}\gamma$ (ordinate) axes. The three strains ϵ_A, ϵ_B, and ϵ_C are then plotted as points on the abscissa. Vertical lines are then drawn through these three points. The shearing strain γ_{xy} is computed from Eqs. (9.6), and $\frac{1}{2}\gamma_{xy}$ is plotted positive downward or negative upward along the vertical line drawn through ϵ_A to establish point A. This shearing strain may also be plotted as positive upward or negative downward along the vertical line through ϵ_C to establish point C. The diameter of the circle is then determined by drawing a line between A and C which intersects the abscissa and defines the center of the circle at a distance $\frac{1}{2}(\epsilon_A + \epsilon_C)$ from the origin. A circle is then drawn from this center passing through points A and C. The circle will intersect the vertical line drawn through ϵ_B, and this point of intersection is labeled as B. A straight line through the center of the circle and point B should be a perpendicular bisector of the diameter AC. The values of the principal strains ϵ_1 and ϵ_2 are given by the intersections of the circle with the abscissa. The principal angle $2\phi_1$ is given by the angle $AO\epsilon_1$ and is negative if point A lies above the ϵ axis. The principal angle $2\phi_2$ is given by the angle $AO\epsilon_2$ and is positive if point A lies above the ϵ axis. The maximum shearing strain is established by a vertical line drawn through the center of the circle to give point D at the intersection. The projection of point D onto the $\frac{1}{2}\gamma$ axis determines the value of $\frac{1}{2}\gamma_{max}$. The principal stresses can be determined directly from the principal strains by employing Eqs. (9.3).

This graphical approach for reducing the data obtained from a rectangular rosette is quite applicable when the data to be reduced are limited to a few gages. However, if large amounts of data must be reduced, the analytical approach is normally preferred since it requires less time.

Special strain-gage computers and nomographs are available for use when data from a very large number of rosettes must be reduced. Also, the analytical approach for obtaining the principal strains and stresses and their directions can be programmed for a digital computer. With a digital instrumentation system, where the strain-gage output is available in digital format, the digital computer can be employed quite effectively. Exercises are provided at the end of the chapter which pertain to computer programs for determining the principal stresses and their directions for several different rosettes.

9.3   CORRECTIONS FOR TRANSVERSE STRAIN EFFECTS [6–9]

In Sec. 6.5 it was noted that foil-type resistance strain gages exhibit a sensitivity S_t to transverse strains. Reference to Fig. 6.12 shows that in certain instances this transverse sensitivity can lead to large errors, and it is important to correct the data to eliminate this effect. Two different procedures for correcting data have been developed.

The first procedure requires a priori knowledge of the ratio ϵ_t/ϵ_a of the strain field. The correction factor is evident in Eq. (6.8), where

$$\epsilon_a = \epsilon_a' \frac{1 - v_0 K_t}{1 + K_t \epsilon_t/\epsilon_a} \tag{6.8}$$

The term ϵ_a' is the apparent strain, and the correction factor CF is given by

$$\text{CF} = \frac{1 - v_0 K_t}{1 + K_t \epsilon_t/\epsilon_a} \tag{9.10}$$

It is possible to correct the strain gage for this transverse sensitivity by adjusting its gage factor. The corrected gage factor S_g^* which should be used with the strain indicator is

$$S_g^* = S_g \frac{1 + K_t \epsilon_t/\epsilon_a}{1 - v_0 K_t} \tag{9.11}$$

Correction for the cross-sensitivity effect when the strain field is unknown is more involved and requires the experimental determination of strain in both the x and y directions. If ϵ_{xx}' and ϵ_{yy}' are the apparent strains recorded in the x and y directions, respectively, then from Eq. (6.8) it is evident that

$$\epsilon_{xx}' = \frac{1}{1 - v_0 K_t} (\epsilon_{xx} + K_t \epsilon_{yy})$$

$$\epsilon_{yy}' = \frac{1}{1 - v_0 K_t} (\epsilon_{yy} + K_t \epsilon_{xx}) \tag{9.12}$$

where the unprimed quantities ϵ_{xx} and ϵ_{yy} are the true strains. Solving Eqs. (9.12) for ϵ_{xx} and ϵ_{yy} gives

$$\epsilon_{xx} = \frac{1 - v_0 K_t}{1 - K_t^2} (\epsilon_{xx}' - K_t \epsilon_{yy}')$$

$$\epsilon_{yy} = \frac{1 - v_0 K_t}{1 - K_t^2} (\epsilon_{yy}' - K_t \epsilon_{xx}') \tag{9.13}$$

Equations (9.13) give the true strains ϵ_{xx} and ϵ_{yy} in terms of the apparent strains ϵ_{xx}' and ϵ_{yy}'. Correction equations for transverse strains in two- and three-element rosettes are given in Ref. 9.

9.4 THE STRESS GAGE [10–12]

The transverse sensitivity which was shown in Sec. 9.3 to result in errors in strain measurements can be employed to produce a special-purpose transducer known as a *stress gage*. The stress gage looks very much like a strain gage (see Fig. 6.4*m*) except that its grid is designed to give a select value of K_t so that the output $\Delta R/R$ is proportional to the stress along the axis of the gage. The stress gage serves a very useful purpose when a stress determination in a particular direction is the ultimate objective of the analysis, for it can be obtained with a stress gage rather than a three-element rosette.

The principle of a stress gage is shown in the following derivation. The output $\Delta R/R$ of any gage is expressed by Eq. (6.4) as

$$\frac{\Delta R}{R} = S_a(\epsilon_a + K_t\epsilon_t) \tag{6.4}$$

The relationship between stress and strain for plane stress is given by Eqs. (2.19) as

$$\epsilon_a = \frac{1}{E}(\sigma_a - v\sigma_t)$$
$$\epsilon_t = \frac{1}{E}(\sigma_t - v\sigma_a) \tag{2.19}$$

Substituting Eqs. (2.19) into Eq. (6.4) yields

$$\frac{\Delta R}{R} = \frac{\sigma_a S_a}{E}(1 - vK_t) + \frac{\sigma_t S_a}{E}(K_t - v) \tag{9.14}$$

Examination of Eq. (9.14) indicates that the output of the gage $\Delta R/R$ will be independent of σ_t if $K_t = v$. It can also be shown that the axial sensitivity S_a of a gage is related to the alloy sensitivity S_A by the expression

$$S_a = \frac{S_A}{1 + K_t} \tag{9.15}$$

Substituting Eq. (9.15) into Eq. (9.14) and letting $K_t = v$ leads to

$$\sigma_a = \frac{E}{S_A(1 - v)}\frac{\Delta R}{R} \tag{9.16}$$

Since the factor $E/S_A(1 - v)$ is a constant for a given gage alloy and specimen material, the gage output in terms of $\Delta R/R$ is linearly proportional to stress.

In practice the stress gage is made with a V-type grid configuration, as shown in Fig. 6.4*m*. Further analysis of the stress gage is necessary to understand its operation in a strain field which is unknown and in which the strain gage is placed in an arbitrary direction. Consider the placement of the gage, as shown in Fig. 9.4, along an arbitrary *x* axis which is at some unknown angle ϕ with the principal axis corresponding to σ_1. The grid elements are at a known angle θ relative to the *x* axis.

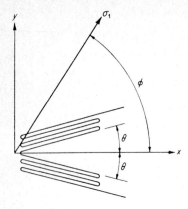

FIGURE 9.4
The stress gage relative to the x axis and the principal axis associated with σ_1.

The strain along the top grid element is given by a modified form of Eq. (9.4) as

$$\epsilon_{\phi-\theta} = \tfrac{1}{2}(\epsilon_1 + \epsilon_2) + \tfrac{1}{2}(\epsilon_1 - \epsilon_2)\cos 2(\phi - \theta) \tag{a}$$

The strain along the lower grid element is

$$\epsilon_{\phi+\theta} = \tfrac{1}{2}(\epsilon_1 + \epsilon_2) + \tfrac{1}{2}(\epsilon_1 - \epsilon_2)\cos 2(\phi + \theta) \tag{b}$$

Summing Eqs. (a) and (b) and expanding the cosine terms yields

$$\epsilon_{\phi-\theta} + \epsilon_{\phi+\theta} = (\epsilon_1 + \epsilon_2) + (\epsilon_1 - \epsilon_2)\cos 2\phi \cos 2\theta \tag{c}$$

Note from the Mohr's strain circles presented earlier in this chapter that

$$\epsilon_{xx} + \epsilon_{yy} = \epsilon_1 + \epsilon_2 \tag{d}$$

$$\epsilon_{xx} - \epsilon_{yy} = (\epsilon_1 - \epsilon_2)\cos 2\phi \tag{e}$$

Substituting Eqs. (d) and (e) into Eq. (c) gives

$$\begin{aligned}
\epsilon_{\phi-\theta} + \epsilon_{\phi+\theta} &= (\epsilon_{xx} + \epsilon_{yy}) + (\epsilon_{xx} - \epsilon_{yy})\cos 2\theta \\
&= 2(\epsilon_{xx}\cos^2\theta + \epsilon_{yy}\sin^2\theta) \\
&= 2\cos^2\theta(\epsilon_{xx} + \epsilon_{yy}\tan^2\theta)
\end{aligned} \tag{f}$$

If the gage is manufactured so that θ is equal to $\tan^{-1}\sqrt{v}$, then

$$\tan^2\theta = v \qquad \cos^2\theta = \frac{1}{1+v}$$

and Eq. (f) becomes

$$\epsilon_{\phi-\theta} + \epsilon_{\phi+\theta} = \frac{2}{1+v}(\epsilon_{xx} + v\epsilon_{yy}) \tag{g}$$

Substituting Eq. (*g*) into Eqs. (2.20) yields

$$\sigma_{xx} = \frac{E}{2(1-v)}(\epsilon_{\phi-\theta} + \epsilon_{\phi+\theta}) \tag{9.17}$$

where $\frac{1}{2}(\epsilon_{\phi-\theta} + \epsilon_{\phi+\theta})$ is the average strain indicated by the two elements of the gage and is equal to $(\Delta R/R)/S_g$.

The gage reading will give $\frac{1}{2}(\epsilon_{\phi-\theta} + \epsilon_{\phi+\theta})$, and it is only necessary to multiply this reading by $E/(1-v)$ to obtain σ_{xx}. The stress gage will thus give σ_{xx} directly with a single gage. However, it does not give any data regarding σ_{yy} or the principal angle ϕ. Moreover, σ_{xx} may not be the most important stress since it may differ appreciably from σ_1. If the directions of the principal stresses are known, the stress gage may be used more effectively by choosing the *x* axis to coincide with the principal axis corresponding to σ_1 so that $\sigma_{xx} = \sigma_1$. In fact, when the principal directions are known, a conventional single-element strain gage can be employed as a stress gage.

This adaption is possible if the gage is located along a line which makes an angle θ with respect to the principal axis, as shown in Fig. 9.5. In this case, the strains will be symmetrical about the principal axis; hence it is clear that

$$\epsilon_{\phi-\theta} = \epsilon_{\phi+\theta} = \epsilon_\theta$$

and Eq. (9.17) reduces to

$$\sigma_1 = \frac{E}{1-v}\epsilon_\theta \tag{9.18}$$

The value ϵ_θ is recorded on the strain gage and converted to σ_1 directly by multiplying by $E/(1-v)$. This procedure reduces the number of gages necessary if only the value of σ_1 is to be determined. The saving of a gage is of particular importance in dynamic strain measurements when the instrumentation required becomes complex and the number of available channels of recording equipment is limited.

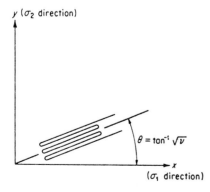

FIGURE 9.5
A single-element strain gage employed as a stress gage when the principal directions are known.

9.5 PLANE-SHEAR OR TORQUE GAGE [13]

Consider two strain gages A and B oriented at angles $\theta_A = -\theta_B$ with respect to the x axis, as shown in Fig. 9.6. The strains along the gage axes are given by a modified form of Eqs. (2.18) as

$$\epsilon_A = \frac{\epsilon_{xx} + \epsilon_{yy}}{2} + \frac{\epsilon_{xx} - \epsilon_{yy}}{2} \cos 2\theta_A + \frac{\gamma_{xy}}{2} \sin 2\theta_A$$

$$\epsilon_B = \frac{\epsilon_{xx} + \epsilon_{yy}}{2} + \frac{\epsilon_{xx} - \epsilon_{yy}}{2} \cos 2\theta_B + \frac{\gamma_{xy}}{2} \sin 2\theta_B$$

(9.19)

From Eqs. (10.19), the shear strain γ_{xy} is

$$\gamma_{xy} = \frac{2(\epsilon_A - \epsilon_B) - (\epsilon_{xx} - \epsilon_{yy})(\cos 2\theta_A - \cos 2\theta_B)}{\sin 2\theta_A - \sin 2\theta_B}$$

(a)

Since gages A and B are oriented such that $\theta_A = -\theta_B$, then

$$\cos 2\theta_A = \cos(-2\theta_B) = \cos 2\theta_B$$

(b)

because the cosine is an even function. As a result, Eq. (a) reduces to

$$\gamma_{xy} = \frac{\epsilon_A - \epsilon_B}{\sin 2\theta_A - \sin 2\theta_B} = \frac{\epsilon_A - \epsilon_B}{2 \sin 2\theta_A}$$

(9.20)

because the sine is an odd function. Thus, the shearing strain γ_{xy} is proportional to the difference between the normal strains experienced by gages A and B when they are oriented with respect to the x axis, as shown in Fig. 9.6. The angle $\theta_A = -\theta_B$ can be arbitrary; however, for the angle $\theta_A = \pi/4$, Eq. (9.20) reduces simply to

$$\gamma_{xy} = \frac{\epsilon_A - \epsilon_B}{2}$$

(9.21)

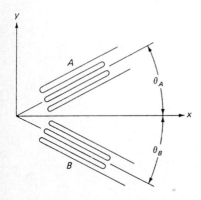

FIGURE 9.6
Positions of gages A and B for measuring γ_{xy}.

Equation (9.21) indicates that the shearing strain γ_{xy} can be measured with a two-element rectangular rosette by orienting the gages at 45 and $-45°$ with respect to the x axis and connecting one gage into arm R_1 and the other into arm R_4 of a bridge. The subtraction $\epsilon_A - \epsilon_B$ will be performed automatically in the bridge, and the output will give $2\gamma_{xy}$ directly.

Both two- and four-element shear gages are marketed commercially for this measurement. The four-element gages provide a complete four-arm bridge with twice the output of the two-element gages. Typical shear gages are illustrated in Fig. 9.7.

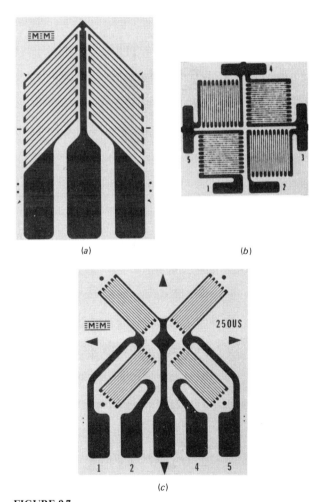

(a)   (b)

(c)

FIGURE 9.7
Two- and four-element shear strain gages: (a) two-element gage; (b) four-element gage; (c) four-element gage. (*Measurements Group, Inc.*)

9.6   THE STRESS INTENSITY FACTOR GAGE K_I [14]

Consider a two-dimensional body with a single-ended through crack as shown in Fig. 4.7. The stability of this crack is determined by the opening-mode stress intensity factor K_I. If the specimen is fabricated from a brittle material, the crack will be initiated when

$$K_I > K_{Ic} \tag{4.2}$$

It is possible to determine K_I as a function of loading on a structure by placing one or more strain gages near the crack tip. To show an effective approach to this measurement, consider a series representation, using three terms, of Eqs. (4.49). The three-term representation of the strain field is

$$E\epsilon_{xx} = A_0 r^{-1/2} \cos \frac{\theta}{2} \left[(1 - v) - (1 + v) \sin \frac{\theta}{2} \sin \frac{3\theta}{2} \right] + 2B_0$$

$$+ A_1 r^{1/2} \cos \frac{\theta}{2} \left[(1 - v) + (1 + v) \sin^2 \frac{\theta}{2} \right]$$

$$E\epsilon_{yy} = A_0 r^{-1/2} \cos \frac{\theta}{2} \left[(1 - v) + (1 + v) \sin \frac{\theta}{2} \sin \frac{3\theta}{2} \right] - 2vB_0 \tag{9.22}$$

$$+ A_1 r^{1/2} \cos \frac{\theta}{2} \left[(1 - v) - (1 + v) \sin^2 \frac{\theta}{2} \right]$$

$$2\mu\gamma_{xy} = A_0 r^{-1/2} \left[\sin \theta \cos \frac{3\theta}{2} \right] - A_1 r^{1/2} \left[\sin \theta \cos \frac{\theta}{2} \right]$$

where A_0, B_0, and A_1 are unknown coefficients which depend on the geometry of the specimen and the loading. Recall that A_0 and K_I are related by

$$A_0 = \frac{K_I}{\sqrt{2\pi}} \tag{4.35}$$

Equations (9.22) could be used to determine the unknown coefficients A_0, B_0, and A_1 if three or more strain gages were placed at appropriate positions in the near-field region. However, the number of gages required for the determination of A_0 or K_I can be reduced to one by considering a gage oriented at an angle α and positioned along the Px' axis as shown in Fig. 9.8. For the rotated coordinates shown in Fig. 9.8, the strain $\epsilon_{x'x'}$ is obtained from Eqs. (9.22) and Eqs. (2.18) as

$$2\mu\epsilon_{x'x'} = A_0 r^{-1/2} \left[k \cos \frac{\theta}{2} - \tfrac{1}{2} \sin \theta \sin \frac{3\theta}{2} \cos 2\alpha \right.$$

$$\left. + \tfrac{1}{2} \sin \theta \cos \frac{3\theta}{2} \sin 2\alpha \right] + B_0(k + \cos 2\alpha)$$

$$+ A_1 r^{1/2} \cos \frac{\theta}{2} \left[k + \sin^2 \frac{\theta}{2} \cos 2\alpha - \tfrac{1}{2} \sin \theta \sin 2\alpha \right] \tag{9.23}$$

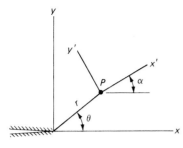

FIGURE 9.8
Definition of a rotated coordinate system positioned at point P.

where

$$k = \frac{1 - v}{1 + v} \tag{9.24}$$

The coefficient of the B_0 term is eliminated by selecting the angle α as

$$\cos 2\alpha = -k = -\frac{1 - v}{1 + v} \tag{9.25}$$

Next, the coefficient of A_1 vanishes if the angle θ is selected as

$$\tan \frac{\theta}{2} = -\cot 2\alpha \tag{9.26}$$

By the proper placement of a single strain gage with angles α and θ determined to satisfy Eqs. (9.25) and (9.26), the strain $\epsilon_{x'x'}$ is related directly to the stress intensity factor K_1 by:

$$2\mu\epsilon_{x'x'} = \frac{K_1}{\sqrt{2\pi r}} \left[k \cos \frac{\theta}{2} - \tfrac{1}{2} \sin \theta \sin \frac{3\theta}{2} \cos 2\alpha \right.$$

$$\left. + \tfrac{1}{2} \sin \theta \cos \frac{2\theta}{2} \sin 2\alpha \right] \tag{9.27}$$

The choice of the angles α and θ depends only on Poisson's ratio, as indicated in Table 9.1.

TABLE 9.1
Angles α and θ as a function of
Poisson's ratio v

v	θ, deg	α, deg
0.250	73.74	63.43
0.300	65.16	61.29
0.333	60.00	60.00
0.400	50.76	57.69
0.500	38.97	54.74

Consider, for example, an aluminum plate ($v = \frac{1}{3}$) with a through crack. In this case $\alpha = \theta = 60°$ (see Table 9.1) and a single strain gage is placed at any point located by r in the near-field region along a $60°$ radial line drawn from the crack tip. For this example with $v = \frac{1}{3}$ and $\alpha = \theta = 60°$, Eq. (9.27) reduces to

$$K_1 = E\sqrt{\tfrac{8}{3}\pi r}\,\epsilon_g \tag{9.28}$$

where $\epsilon_g = \epsilon_{x'x'}$ is the strain indicated by the single gage.

9.6.1 Strain-Gradient Errors

The use of strain gages to measure stress intensity factors was delayed for nearly 30 years because of an unfounded concern about error introduced by the strain-gradient effect. To explore the magnitude of the error due to strain gradients, consider a single-element gage positioned in the near-field region with $v = \frac{1}{3}$, $k = \frac{1}{2}$, and $\alpha = \theta = 60°$. The strain along this radial line is determined from Eq. (9.27) as

$$\epsilon_{x'x'} = qr^{-1/2} \tag{9.29}$$

where

$$q = \frac{K_1}{E\sqrt{\tfrac{8}{3}\pi}} \tag{9.30}$$

The gage senses the strain $\epsilon_{x'x'}$ and the gage signal represents the average strain over the gage length, which is given by

$$(\epsilon_{x'x'})_{\text{avg}} = \frac{q}{r_o - r_i}\int_{r_i}^{r_o} r^{-1/2}\,dr \tag{9.31}$$

where r_o and r_i are positions defining the location of the gage grid as shown in Fig. 9.9. Integrating Eq. (9.31) yields

$$(\epsilon_{x'x'})_{\text{avg}} = \frac{2q}{\sqrt{r_o} + \sqrt{r_i}} \tag{a}$$

The strain-gage response $(\epsilon_{x'x'})_{\text{avg}}$ corresponds to the true strain $\epsilon_{x'x'}$ at a specific point r_t along the gage length. By equating Eq. (a) with Eq. (9.29), it is clear that r_t is

$$r_t = \tfrac{1}{4}[\sqrt{r_o} + \sqrt{r_i}]^2 \tag{9.32}$$

Note that the position of the center of the gage r_c is

$$r_c = \tfrac{1}{2}(r_o + r_i) \tag{b}$$

Defining Δr as the distance between the center of the gage and the true strain location r_t gives

$$\Delta r = r_c - r_t = \tfrac{1}{4}(r_o - 2\sqrt{r_o}\sqrt{r_i} + r_i) \tag{c}$$

Next, write the gage length L as

$$L = r_o - r_i \tag{d}$$

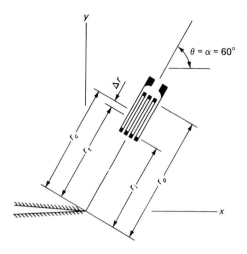

FIGURE 9.9
Definition of radii associated with strain-gage placement near a crack tip.

Finally, combine Eqs. (*b*), (*c*), and (*d*) to obtain

$$\frac{\Delta r}{r_c} = \frac{1}{2}\left\{1 - \left[1 - \left(\frac{L}{2r_c}\right)^2\right]^{1/2}\right\}$$

(9.33)

where $r_c > L/2$ to avoid placing any portion of the gage over the crack tip. The results of Eq. (9.33) are shown in Fig. 9.10. The strain-gradient effect is a maximum with $\Delta r/r_c = 0.5$ when the gage is placed as close as possible to the crack tip with

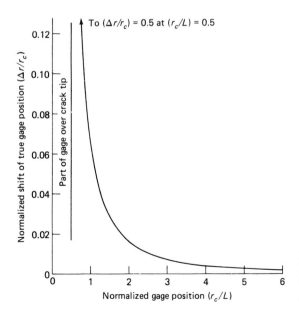

FIGURE 9.10
Normalized shift of true gage position $\Delta r/r_c$ as a function of normalized gage position r_c/L.

$r_c = L/2$. This is a serious error; however, placement at this location should be avoided in any event since it locates the gage in region 1 (see Fig. 4.12) where the plane-stress analysis of the strain field is not valid. For gages placed in the near-field region, r_c/L will probably exceed 2 and $\Delta r/r_c < 0.016$. If a correction is required to account for Δr, the radius r_t is determined from Eq. (9.32) and this value is substituted into Eq. (9.28) to determine K_1. In many applications the difference between r_c and r_t is negligible. For example, consider a gage with $L = 1$ mm positioned at $r_c = 5L$ and note from Fig. 9.10 that $\Delta r/r_c = 0.0025$. The correction $\Delta r = 0.0125$ mm is much less than the accuracy that can be achieved in measuring r_c.

9.7 DETERMINING MIXED-MODE STRESS INTENSITY FACTORS [15]

Frequently the loading on a body containing a crack produces a mixed-mode state of stress. The mixed mode indicates that the opening mode and the shearing mode occur together. The strain field in the region near the crack tip, for a mixed-mode state of stress, is given by superposition of Eqs. (4.49) and (4.53). Consider again a strain gage positioned at point P with an orientation α, as shown in Fig. 9.8. The strain in the x' direction may be written by using Eqs. (4.49), (4.53), and the strain transformation equation Eq. (2.18) to give

$$
\begin{aligned}
E\epsilon_{x'x'} = A_0 r^{-1/2} \Bigg\{ &\cos\frac{\theta}{2}\left[(1-v)-(1+v)\sin\frac{\theta}{2}\sin\frac{3\theta}{2}\right]\cos^2\alpha \\
&+ \cos\frac{\theta}{2}\left[(1-v)+(1+v)\sin\frac{\theta}{2}\sin\frac{3\theta}{2}\right]\sin^2\alpha \\
&+ (1+v)\sin\theta\cos\frac{3\theta}{2}\sin\alpha\cos\alpha \Bigg\} + 2B_0\left(\cos^2\alpha - v\sin^2\alpha\right) \\
&+ A_1 r^{1/2}\Bigg\{\cos\frac{\theta}{2}\left[(1-v)+(1+v)\sin^2\frac{\theta}{2}\right]\cos^2\alpha \\
&+ \cos\frac{\theta}{2}\left[(1-v)-(1+v)\sin^2\frac{\theta}{2}\right]\sin^2\alpha \\
&- (1+v)\sin\theta\cos\frac{\theta}{2}\sin\alpha\cos\alpha\Bigg\} \\
&+ 2B_1 r\left[\cos\theta\cos^2\alpha - v\cos\theta\sin^2\alpha - 2(1+v)\sin\theta\sin\alpha\cos\alpha\right] \\
&+ C_0 r^{-1/2}\left\{\sin\frac{\theta}{2}\left[(1+v)\cos\frac{\theta}{2}\cos\frac{3\theta}{2}+2v\right]\sin^2\alpha\right\}
\end{aligned}
$$

$$- \sin \frac{\theta}{2} \left[(1 + v) \cos \frac{\theta}{2} \cos \frac{3\theta}{2} + 2 \right] \cos^2 \alpha$$

$$+ 2(1 + v) \cos \frac{\theta}{2} \left[1 - \sin \frac{\theta}{2} \sin \frac{3\theta}{2} \right] \sin \alpha \cos \alpha \bigg\}$$

$$+ C_1 r^{1/2} \bigg\{ \sin \frac{\theta}{2} \left[(1 + v) \cos^2 \frac{\theta}{2} + 2 \right] \cos^2 \alpha$$

$$- \sin \frac{\theta}{2} \left[(1 + v) \cos^2 \frac{\theta}{2} + 2v \right] \sin^2 \alpha$$

$$+ 2(1 + v) \cos \frac{\theta}{2} \left[\sin^2 \frac{\theta}{2} + 1 \right] \sin \alpha \cos \alpha \bigg\}$$

$$+ 2D_1 r[\sin \theta \, (\cos^2 \alpha - v \sin^2 \alpha)] \tag{9.34}$$

where an eight-term series is used in Eq. (9.34) to represent the mixed-mode strain field.

Again, if $\cos 2\alpha = -k$, the coefficients of the B_0 and D_1 terms vanish. Also, if $\tan (\theta/2) = -\cot 2\alpha$, the coefficient of the A_1 term vanishes. The angles α and θ which permit these simplifications are exactly the same as those listed in Table 9.1. The D_0 term is not required. The reduction of Eq. (9.34) to a four-term series containing A_0, B_1, C_0, and C_1 is given as an exercise.

To show an application of strain gages to the independent measurement of K_I and K_{II} for a mixed-mode state of stress, consider an aluminum specimen with $v = \frac{1}{3}$ and $\alpha = \theta = \pm 60°$. For strain gages deployed along the radial line defined by $\theta = +60°$, Eq. (9.34) reduces to

$$E\epsilon_{x'x'+} = \frac{\sqrt{3}}{2} A_0 r^{-1/2} - 2B_1 r + \tfrac{1}{2} C_0 r^{-1/2} + C_1 r^{1/2} \tag{9.35}$$

and for strain gages deployed along the radial line defined by $\theta = -60°$,

$$E\epsilon_{x'x'-} = \frac{\sqrt{3}}{2} A_0 r^{-1/2} - 2B_1 r - \tfrac{1}{2} C_0 r^{-1/2} - C_1 r^{1/2} \tag{9.36}$$

If Eqs. (9.35) and (9.36) are added:

$$E(\epsilon_{x'x'+} + \epsilon_{x'x'-}) = \sqrt{3} A_0 r^{-1/2} - 4B_1 r \tag{9.37}$$

The result of Eq. (9.37) shows that the sum of the two strains measured at the same position r along the $+60°$ and $-60°$ radial lines gives a quantity that is independent of the shearing mode. If $(\epsilon_{x'x'+} + \epsilon_{x'x'-})$ is determined at two or more positions of r, then the unknown coefficient can be calculated. With two values of $(\epsilon_{x'x'+} + \epsilon_{x'x'-})$, deterministic methods of computation are employed. However, if more than two values of $(\epsilon_{x'x'+} + \epsilon_{x'x'-})$ are measured, then overdeterministic methods of solution of the equations are useful.

Subtracting Eq. (9.36) from Eq. (9.35) yields

$$E(\epsilon_{x'x'+} - \epsilon_{x'x'-}) = C_0 r^{-1/2} + 2C_1 r^{1/2} \tag{9.38}$$

This result shows that the difference of two strains measured at the same position r along the $+60°$ and $-60°$ radial lines gives a quantity that is independent of the opening mode. Solution of Eq. (9.38) by either deterministic or overdeterministic methods yields C_0, which is related to the shearing-mode stress intensity factor by the equation

$$K_{\mathrm{II}} = \sqrt{2\pi} C_0 \tag{9.39}$$

9.8 OVERDETERMINISTIC METHODS OF STRAIN ANALYSIS [16]

In fracture mechanics one knows the form of the solution for the stresses or strains in the near-field region. This solution is given by series relations such as those given in Eqs. (4.48) and (4.49). As the size of the near-field region is enlarged, the number of terms in the series representation must be increased to accurately describe the field quantities. The increase in the number of terms in the series increases the number of unknown coefficients A_n and B_m, and eventually the use of deterministic methods becomes inappropriate. With a large number of unknown coefficients, small errors in the measurement of strain, $\epsilon_{x'x'}$, gage position (r, θ), and gage orientation α combine to produce large errors in the determination of A_n and B_m. To avoid these errors, additional strain measurements are made so that the amount of data available exceeds the number of unknowns by a factor of say 3 or 4. With this additional data, overdeterministic solutions for the unknown coefficients A_n and B_m can be employed and the results are improved by averaging in a least-squares sense.

To illustrate the overdeterministic method, consider that n strain gages are deployed in a radial pattern ($\alpha = \theta$) in the near-field region with an arbitrary selection of r and θ. These gages record a radial strain ϵ_{rr}, which is described by

$$
\begin{aligned}
2\mu\epsilon_{rr} = &\ A_0 r^{-1/2}\left(k \cos \frac{\theta}{2} - \tfrac{1}{2} \sin \theta \sin \frac{3\theta}{2} \cos 2\theta + \sin^2 \theta \cos \theta \cos \frac{3\theta}{2}\right) \\
&+ B_0(k + \cos 2\theta) + A_1 r^{1/2} \cos \frac{\theta}{2}\left(k + \sin^2 \frac{\theta}{2} \cos 2\theta - \sin^2 \theta \cos \theta\right) \\
&+ B_1 r \cos \theta(k + 1 - 6 \sin^2 \theta) \\
&+ A_2 r^{3/2}\left(k \cos \frac{3\theta}{2} - \tfrac{3}{2} \sin \frac{\theta}{2} \cos 2\theta - 3 \sin^2 \theta \cos \theta \cos \frac{\theta}{2}\right) \\
&+ B_2 \frac{2}{1+v} r^2[1 - (3 + 2v) \sin^2 \theta \cos^2 \theta + v \sin^2 \theta(1 + \sin^2 \theta)] \tag{9.40}
\end{aligned}
$$

For n strain readings for a given load, Eq. (9.40) is used repeatedly to form a system of equations in terms of the unknown coefficients A_0, A_1, A_2, B_0, B_1, and B_2. Thus

$$2\mu\epsilon_{rr1} = A_0 r_1^{-1/2} g_{01} + B_0 r_1^0 h_{01} + A_1 r_1^{1/2} g_{11} + B_1 r_1^1 h_{11} + A_2 r_1^{3/2} g_{21} + B_2 r_1^2 h_{21}$$

$$2\mu\epsilon_{rr2} = A_0 r_2^{-1/2} g_{02} + B_0 r_2^0 h_{02} + A_1 r_2^{1/2} g_{12} + B_1 r_2^1 h_{12} + A_2 r_2^{3/2} g_{22} + B_2 r_2^2 h_{22}$$

$$\cdots\cdots\cdots\cdots\cdots\cdots\cdots\cdots\cdots\cdots\cdots\cdots\cdots\cdots\cdots\cdots$$

$$\cdots\cdots\cdots\cdots\cdots\cdots\cdots\cdots\cdots\cdots\cdots\cdots\cdots\cdots\cdots\cdots$$

$$\cdots\cdots\cdots\cdots\cdots\cdots\cdots\cdots\cdots\cdots\cdots\cdots\cdots\cdots\cdots\cdots$$

$$2\mu\epsilon_{rrn} = A_0 r_n^{-1/2} g_{0n} + B_0 r_n^0 h_{0n} + A_1 r_n^{1/2} g_{1n} + B_1 r_n^1 h_{1n} + A_2 r_n^{3/2} g_{2n} + B_2 r_n^2 h_{2n}$$

$$(9.41)$$

where g and h are the functions of k and θ given in Eq. (9.40). It is more convenient to write Eqs. (9.41) in matrix form as

$$\{\epsilon\} = [fg]\{AB\} \tag{9.42}$$

where $\{\epsilon\}$ = a column matrix containing n data points
   $[fg]$ = a $6 \times n$ matrix containing the known coefficients
   $\{AB\}$ = a column matrix containing the six unknown coefficients

Equation (9.42) is overdeterministic with six unknown coefficients. A least-squares solution for these six unknowns requires only two matrix manipulations. First, multiply both sides by the transpose $[fg]^T$

$$[fg]^T\{\epsilon\} = [fg]^T\{AB\} \tag{a}$$

to obtain a 6×6 system of linear equations. Second, solve for $\{AB\}$ by noting that

$$\{AB\} = [[fg]^T[fg]]^{-1}[fg]\{\epsilon\} \tag{9.43}$$

where $[[fg]^T[fg]]^{-1}$ is the inverse matrix.

   The matrix solutions for the unknown coefficients A_n, B_m are easy to program. The method provides an effective technique for using statistical procedures (least squares) to minimize the errors in the determination of A_0 or K_I.

9.9   RESIDUAL STRESS   DETERMINATION
[17–21]

9.9.1   Uniaxial Residual Stresses

Residual stresses are self-equilibrated, and often occur in a component or structure without the application of any loading. Residual stresses are usually the result of fabrication processes or assembly processes, although they can be produced if the structure is loaded in the plastic range and then unloaded.

   Residual stresses are difficult to measure since they are independent of the load and they are imposed on the component before strain gages can be installed. To sense the presence of residual stresses with strain gages, it is necessary to unload (relieve) the residual stresses after the strain gage is mounted at some point P on the component. Residual stresses are relieved by cutting away selected regions of the component, and in this process the component is destroyed. For this reason, the method of residual stress determination with strain gages is classified as destructive.

There are several different methods for relieving residual stresses but the hole drilling technique is the simplest and the most widely used. To show the application of the hole drilling technique, consider a thin plate subjected to a uniaxial residual stress σ_0 which is in the y direction and uniformly distributed through the thickness h of the plate and over the region of measurement. The state of stress in polar coordinates prior to drilling the hole at any point P (r, θ) is obtained from Eq. (1.11) as

$$\sigma_{rr}^l = \frac{\sigma_0}{2}(1 - \cos 2\theta)$$

$$\sigma_{\theta\theta}^l = \frac{\sigma_0}{2}(1 + \cos 2\theta) \tag{9.44}$$

$$\tau_{r\theta}^l = \frac{\sigma_0}{2}\sin 2\theta$$

A hole of radius a is now drilled completely through the thin plate, as shown in Fig. 9.11. Drilling the hole markedly changes the stress distribution in the plate since a cylinder of stressed material of radius r_0 has been removed. To determine the stress distribution in the postdrilled state, consider the stress state in a uniaxially loaded infinite plate with a circular hole obtained by Kirsch and discussed in Chap. 3. The stresses are given by Eqs. (3.46) as

$$\sigma_{rr} = \frac{\sigma_0}{2}\left\{\left(1 - \frac{a^2}{r^2}\right)\left[1 + \left(\frac{3a^2}{r^2} - 1\right)\cos 2\theta\right]\right\}$$

$$\sigma_{\theta\theta} = \frac{\sigma_0}{2}\left[\left(1 + \frac{a^2}{r^2}\right) + \left(1 + \frac{3a^4}{r^4}\right)\cos 2\theta\right] \tag{3.46}$$

$$\tau_{r\theta} = \frac{\sigma_0}{2}\left[\left(1 + \frac{3a^2}{r^2}\right)\left(1 - \frac{a^2}{r^2}\right)\sin 2\theta\right]$$

where $r > a$.

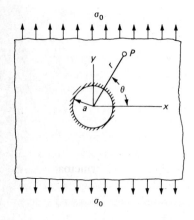

FIGURE 9.11
Coordinates of point P which is located in a region of a plate adjacent to a through hole of radius a.

Subtracting the initial stresses [Eqs. (9.44)] from the stresses after drilling [Eqs. (3.46)] gives

$$\sigma_{rr}^{II} = -\frac{\sigma_0 a^2}{2r^2}\left[1 + \left(\frac{3a^2}{r_2} - 4\right)\cos 2\theta\right]$$

$$\sigma_{\theta\theta}^{II} = \frac{\sigma_0 a^2}{2r^2}\left(1 + \frac{3a^2}{r^2}\cos 2\theta\right) \tag{9.45}$$

$$\tau_{r\theta}^{II} = -\frac{\sigma_0 a^2}{2r^2}\left[\left(\frac{3a^2}{r^2} - 2\right)\sin 2\theta\right]$$

Substituting Eqs. (9.45) into Eqs. (2.19) gives

$$\epsilon_{rr} = -\frac{\sigma_0 a^2(1 + v)}{2Er^2}\left[1 + \frac{3a^2}{r^2}\cos 2\theta - \frac{4\cos 2\theta}{1 + v}\right]$$

$$\epsilon_{\theta\theta} = \frac{\sigma_0 a^2(1 + v)}{2Er^2}\left[1 + \frac{3a^2}{r^2}\cos 2\theta - \frac{4v\cos 2\theta}{1 + v}\right] \tag{9.46}$$

These relations give the strain distribution in the region about the hole which is due to relief of the uniaxial residual stress σ_0. It is possible to simplify this analysis by considering the strain along the x axis where $\theta = 0°$, $\epsilon_{\theta\theta} = \epsilon_{yy}$, and $\epsilon_{rr} = \epsilon_{xx}$:

$$\frac{2E\epsilon_{rr}}{\sigma_0(1 + v)} = -\left(\frac{a}{r}\right)^2\left[1 + 3\left(\frac{a}{r}\right)^2 - \frac{4}{1 + v}\right]$$

$$\frac{2E\epsilon_{\theta\theta}}{\sigma_0(1 + v)} = \left(\frac{a}{r}\right)^2\left[1 + 3\left(\frac{a}{r}\right)^2 - \frac{4v}{1 + v}\right] \tag{9.47}$$

The normalized strains $3E\epsilon/2\sigma_0$, corresponding to Eqs. (9.47) with $v = \frac{1}{3}$, are shown as a function of position r/a in Fig. 9.12. Inspection of these curves shows that both strains exhibit sharp gradients with respect to r at locations close to the edge of the hole. This implies that precise positioning of a strain gage relative to the hole is necessary for gage locations with $1 < r/a < 1.5$. If a single gage is to be used to determine σ_0, it should be positioned at $r/a = 1.75$ and oriented along the x axis to measure ϵ_{rr}. With this placement of the gage, the residual stress for a material with $v = \frac{1}{3}$ is given by

$$\sigma_0 = 4.502E\epsilon_g \tag{9.48}$$

The placement of the gage at $r/a = 1.75$ maximizes the signal due to ϵ_{rr} in the region where strain gradients are minimized.

This analysis of uniaxial residual stress illustrates the approach used to measure σ_0 in the simplest possible case. However, residual stresses are usually biaxial and it is necessary to determine the magnitudes of the principal residual stresses σ_1^R and σ_2^R as well as the principal directions. The strain-gage method used for the determination of biaxial residual stresses is covered in the following section.

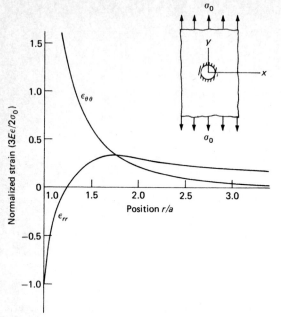

FIGURE 9.12
Distribution of strains along the x axis due to residual stress relief by hole drilling ($\nu = \frac{1}{3}$).

9.9.2 Biaxial Residual Stresses

Consider a biaxial state of residual stress described with σ_1^R and σ_2^R as illustrated in Fig. 9.13. Drilling a through hole of radius a in the plate relieves the residual stresses and produces a change in the strain in the local region near the hole. The change in strain can be determined by using Eqs. (9.46) and the principle of superposition. Before applying the principle of superposition, rewrite the ϵ_{rr} part of Eqs. (9.46) in the following concise form:

$$\epsilon_{rr} = \sigma_1^R[C_1 - C_2 \cos 2\theta_2] \tag{9.49}$$

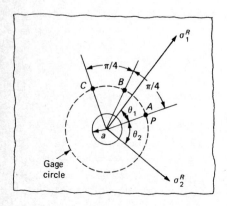

FIGURE 9.13
A biaxial state of residual stress σ_1^R and σ_2^R uniformly distributed through the thickness of the plate.

where

$$C_1 = -\frac{1+v}{2E}\left(\frac{a}{r}\right)^2$$

$$C_2 = -\frac{1+v}{2E}\left(\frac{a}{r}\right)^2\left[-3\left(\frac{a}{r}\right)^2 + \frac{4}{1+v}\right]$$

and θ_2 is the angle between the principal direction σ_2 and the point P. Note in writing Eq. (9.49) that σ_1^R was used in place of σ_0. In a similar manner, the strain ϵ_{rr} due to relief of σ_2^R can be written as

$$\epsilon_{rr} = \sigma_2^R[C_1 - C_2 \cos 2\theta_1] \qquad (a)$$

Since

$$\cos 2\theta_1 = \cos 2\left(\theta_2 + \frac{\pi}{2}\right) = -\cos 2\theta_2 \qquad (b)$$

it is clear that

$$\epsilon_{rr} = \sigma_2^R[C_1 + C_2 \cos 2\theta_2] \qquad (9.50)$$

Superimposing the radial strains given by Eqs. (9.49) and (9.50) gives

$$\epsilon_{rr} = C_1(\sigma_1^R + \sigma_2^R) + C_2(\sigma_2^R - \sigma_1^R) \cos 2\theta_2 \qquad (9.51)$$

At this point in the analysis we have the basic relation for ϵ_{rr} that is necessary to solve for the unknowns σ_1^R and σ_2^R, and the angle defining the orientation of this orthogonal pair of principal stresses. A three-gage rectangular rosette with the gages deployed as shown in Fig. 9.14 is used to measure three strains ϵ_A, ϵ_B, and ϵ_C. For the strains measured at points A, B, and C in Fig. 9.13, Eq. (9.51) yields

$$\epsilon_A = C_1(\sigma_1^R + \sigma_2^R) + C_2(\sigma_2^R - \sigma_1^R) \cos 2\theta_2$$

$$\epsilon_B = C_1(\sigma_1^R + \sigma_2^R) + C_2(\sigma_2^R - \sigma_1^R) \cos 2\left(\theta_2 + \frac{\pi}{4}\right) \qquad (9.52)$$

$$\epsilon_C = C_1(\sigma_1^R + \sigma_2^R) + C_2(\sigma_2^R - \sigma_1^R) \cos 2\left(\theta_2 + \frac{\pi}{2}\right)$$

Solving Eqs. (9.52) for σ_1^R and σ_2^R gives

$$\sigma_1^R = \frac{\epsilon_A + \epsilon_C}{4C_1} + \frac{\sqrt{2}}{4C_2}\sqrt{(\epsilon_A - \epsilon_B)^2 + (\epsilon_B - \epsilon_C)^2}$$

$$\sigma_2^R = \frac{\epsilon_A + \epsilon_C}{4C_1} - \frac{\sqrt{2}}{4C_2}\sqrt{(\epsilon_A - \epsilon_B)^2 + (\epsilon_B - \epsilon_C)^2} \qquad (9.53)$$

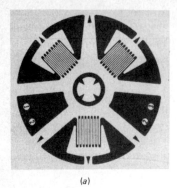

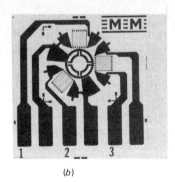

(a)

(b)

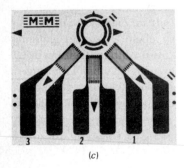

(c)

FIGURE 9.14
Rectangular rosettes used with the hole-drilling technique for residual stress determination: (a) original design by Rendler and Vigness; (b) more rugged version of the original design with encapsulation; (c) design with gage elements on one side of the hole to facilitate application near weld lines. (*Measurements Group, Inc.*).

and

$$\tan 2\theta = \frac{\epsilon_A - 2\epsilon_B + \epsilon_C}{\epsilon_C - \epsilon_A}$$

where θ is the angle from gage A to the nearer of the two principal axes. If $\epsilon_C > \epsilon_A$, θ refers to σ_1 but if $\epsilon_C < \epsilon_A$, θ refers to σ_2.

The rosettes used in residual stress investigations are illustrated in Fig. 9.14. The strain gages are all positioned about a circle of radius r_0. The hole position and the hole radius a are located with patterns etched in foil. All the gages and the hole centering aids are carried on a single film carrier. For the rosettes illustrated in Fig. 9.14b and c, a covering film is used to encapsulate the installation. The covering film protects the fragile gage grids from the chips produced as the hole is drilled.

EXERCISES

9.1. Suppose a state of pure shear stress occurs (say, on a circular shaft under pure torsion). Show how a single-element strain gage can be employed to determine the principal stresses σ_1 and σ_2.

9.2. In a thin-walled cylinder loaded with internal pressure $(\sigma_h = 2\sigma_a)$, show how a single-element strain gage in a hoop direction can be used to establish the hoop and axial stresses σ_h and σ_a.

9.3. A two-element rectangular rosette is being used to determine the two principal stresses at the point shown in Fig. E9.3. If $\epsilon_1 = 860\ \mu\epsilon$ and $\epsilon_2 = -390\ \mu\epsilon$, determine σ_1 and σ_2. Use $E = 207$ GPa and $v = 0.30$.

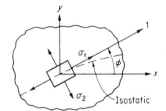

FIGURE E9.3

9.4. If strain data from the two-element rectangular rosette shown in Fig. E9.3 yields $\sigma_1 = 180$ MPa and $\sigma_2 = 60$ MPa, find σ_{xx}, σ_{yy}, and τ_{xy} when ϕ is (a) 30°, (b) 45°, (c) 60°, and (d) 90°.

9.5. Prove the validity of Eqs. (9.8) by using the Mohr's strain diagram presented in Fig. 9.3.

9.6. Derive an expression for the maximum shear stress $\tau_{\max}$ at a point in terms of the strains obtained from a three-element rectangular rosette by using Eqs. (9.9).

9.7. The following observations are made with a rectangular rosette mounted on a steel $(E = 207$ GPa and $v = 0.30)$ specimen. Determine the principal strains, the principal stresses, and the principal angles ϕ_1 and ϕ_2.

Case number	ϵ_A, $\mu\epsilon$	ϵ_B, ϵ	ϵ_C, ϵ
1	1000	− 500	0
2	1800	600	− 400
3	− 1000	400	400
4	1600	− 200	− 1800
5	− 400	0	400

9.8. Write a computer program to determine the principal strains ϵ_1 and ϵ_2, the principal angle ϕ, and the principal stresses σ_1 and σ_2 using the strains ϵ_A, ϵ_B, and ϵ_C measured with a three-element rectangular rosette.

9.9. For the three-element delta rosette shown in Fig. E9.9, determine the equation for the principal strains ϵ_1 and ϵ_2 in terms of ϵ_A, ϵ_B, and ϵ_C.

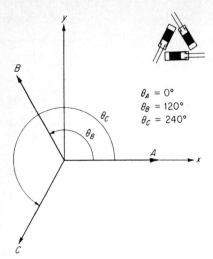

$\theta_A = 0°$
$\theta_B = 120°$
$\theta_C = 240°$

FIGURE E9.9

9.10. For the three-element delta rosette shown in Fig. E9.9, determine the equation for the principal angle ϕ in terms of ϵ_A, ϵ_B, and ϵ_C.

9.11. Construct a Mohr's circle for strains using the strains ϵ_A, ϵ_B, and ϵ_C measured with a three-element delta rosette. Show that the radius of the circle is

$$R = \{[\epsilon_A - \tfrac{1}{3}(\epsilon_A + \epsilon_B + \epsilon_C)]^2 + \tfrac{1}{3}(\epsilon_C - \epsilon_B)^2\}^{1/2}.$$

9.12. Using the result of Exercise 9.11 show that the principal angles can be identified by applying the following rules:

$$0° < \phi_1 < 90° \qquad \text{when } \epsilon_C > \epsilon_B$$
$$-90° < \phi_1 < 0° \qquad \text{when } \epsilon_C < \epsilon_B$$
$$\phi_1 = 0 \qquad \text{when } \epsilon_B = \epsilon_C \text{ and } \epsilon_A > \epsilon_B = \epsilon_C$$
$$\phi_1 = \pm 90° \qquad \text{when } \epsilon_B = \epsilon_C \text{ and } \epsilon_A < \epsilon_B = \epsilon_C$$

9.13. Using the results of Exercise 9.9, derive the equations for the principal stresses σ_1 and σ_2 in terms of the strains ϵ_A, ϵ_B, and ϵ_C measured with a three-element delta rosette.

9.14. The following observations were made with a three-element delta rosette mounted on a steel ($E = 207$ GPa and $v = 0.30$) specimen. Determine the principal strains ϵ_1 and ϵ_2, the principal stresses σ_1 and σ_2, and the principal angles ϕ_1 and ϕ_2.

Case number	$\epsilon_A, \mu\epsilon$	$\epsilon_B, \mu\epsilon$	$\epsilon_C, \mu\epsilon$
1	800	−400	400
2	1600	800	0
3	−1200	600	800
4	1400	0	−1400
5	−600	200	800

9.15. Verify the expressions listed in Table 9.2 for the rectangular four-element rosette.

TABLE 9.2
A summary of the equations used to determine principal strains, principal stresses, and their directions from four types of rosettes

Type of rosette	Gage arrangement	Principal strain and principal stress	Principal angle	Identification $0 < \phi_1 < 90°$
Three-element, rectangular	(diagram: gages A at $45°$, B at $45°$, C; x–y axes)	$\epsilon_{1,2} = \dfrac{\epsilon_A + \epsilon_C}{2} \pm \dfrac{1}{2}\sqrt{(\epsilon_A - \epsilon_C)^2 + (2\epsilon_B - \epsilon_A - \epsilon_C)^2}$ $\sigma_{1,2} = \dfrac{E}{2}\left[\dfrac{\epsilon_A + \epsilon_C}{1-\nu} \pm \dfrac{1}{1+\nu}\sqrt{(\epsilon_A - \epsilon_C)^2 + (2\epsilon_B - \epsilon_A - \epsilon_C)^2}\right]$	$\tan 2\phi_1$ $= \dfrac{2\epsilon_B - \epsilon_A - \epsilon_C}{\epsilon_A - \epsilon_C}$	$\epsilon_B > \dfrac{\epsilon_A + \epsilon_C}{2}$
Delta	(diagram: gages A, B, C at $120°$)	$\epsilon_{1,2} = \dfrac{\epsilon_A + \epsilon_B + \epsilon_C}{3} \pm \dfrac{\sqrt{2}}{3}\sqrt{(\epsilon_A - \epsilon_B)^2 + (\epsilon_B - \epsilon_C)^2 + (\epsilon_C - \epsilon_A)^2}$ $\sigma_{1,2} = \dfrac{E}{3}\left[\dfrac{\epsilon_A + \epsilon_B + \epsilon_C}{(1-\nu)} \pm \dfrac{\sqrt{2}}{(1+\nu)}\sqrt{(\epsilon_A - \epsilon_B)^2 + (\epsilon_B - \epsilon_C)^2 + (\epsilon_C - \epsilon_A)^2}\right]$	$\tan 2\phi_1$ $= \dfrac{\sqrt{3}(\epsilon_C - \epsilon_B)}{2\epsilon_A - (\epsilon_B + \epsilon_C)}$	$\epsilon_C > \epsilon_B$
Four-element, rectangular	(diagram: gages C, D, B, A at $45°$)	$\epsilon_{1,2} = \dfrac{\epsilon_A + \epsilon_B + \epsilon_C + \epsilon_D}{4} \pm \dfrac{1}{2}\sqrt{(\epsilon_A - \epsilon_C)^2 + (\epsilon_B - \epsilon_D)^2}$ $\sigma_{1,2} = \dfrac{E}{2}\left[\dfrac{\epsilon_A + \epsilon_B + \epsilon_C + \epsilon_D}{2(1-\nu)} \pm \dfrac{1}{1+\nu}\sqrt{(\epsilon_A - \epsilon_C)^2 + (\epsilon_B - \epsilon_D)^2}\right]$	$\tan 2\phi_1 = \dfrac{\epsilon_B - \epsilon_D}{\epsilon_A - \epsilon_C}$	$\epsilon_B > \epsilon_D$
Tee-delta	(diagram: gages B, D, A, C; $30°$, $90°$, $120°$)	$\epsilon_{1,2} = \dfrac{\epsilon_A + \epsilon_D}{2} \pm \dfrac{1}{2}\sqrt{(\epsilon_A - \epsilon_D)^2 + \dfrac{4}{3}(\epsilon_C - \epsilon_B)^2}$ $\sigma_{1,2} = \dfrac{E}{2}\left[\dfrac{\epsilon_A + \epsilon_D}{1-\nu} \pm \dfrac{1}{1+\nu}\sqrt{(\epsilon_A - \epsilon_D)^2 + \dfrac{4}{3}(\epsilon_C - \epsilon_B)^2}\right]$	$\tan 2\phi_1$ $= \dfrac{2(\epsilon_C - \epsilon_B)}{\sqrt{3}(\epsilon_A - \epsilon_D)}$	$\epsilon_C > \epsilon_B$

9.16. Verify the expressions listed in Table 9.2 for the tee-delta rosette.

9.17. Determine the error due to cross sensitivity when a WK-06-125RA-350 type of strain gage is used on aluminum ($E = 70$ GPa and $v = 0.33$) to measure (a) a state of hydrostatic compression, (b) a state of pure shear, and (c) Poisson's ratio.

9.18. Solve Exercise 9.17 if the gage is used on steel ($E = 207$ GPa and $v = 0.30$).

9.19. The following apparent strain data were obtained with two-element rectangular rosettes:

Rosette number	ϵ'_{xx}, μϵ	ϵ'_{yy}, μϵ
1	1200	600
2	−400	1400
3	2400	800
4	1200	−600

Determine the true strains ϵ_{xx} and ϵ_{yy} if $K_t = 0.02$. In each case, determine the error which would have occurred if the cross sensitivity of the gage had been neglected.

9.20. Solve Exercise 9.19 if $K_t = 0.015$.

9.21. Solve Exercise 9.19 if $K_t = -0.015$.

9.22. Determine the included angle between elements in a V-type stress gage designed for use on (a) glass ($v = 0.25$), (b) steel ($v = 0.30$), and (c) aluminum ($v = 0.33$).

9.23. Determine $\Delta R/R$ for the circular-arc gage element shown in Fig. E9.23. Neglect transverse-sensitivity effects.

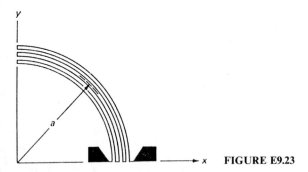

FIGURE E9.23

9.24. Show that two circular-arc gage elements (see Fig. E9.23) located in adjacent quadrants and positioned in arms R_1 and R_4 of a Wheatstone bridge have an output $\Delta E/V$ proportional to $2\gamma_{xy}/\pi$.

9.25. Determine $\Delta E/V$ for a shear gage consisting of four circular-arc elements arranged in all four quadrants and connected into the four arms of a Wheatstone bridge.

9.26. Show how a V-type stress gage designed to measure σ_1 on steel can be used to measure the shear strain γ_{xy} on any material.

9.27. Verify Eq. (9.23) beginning with Eqs. (9.22).

9.28. Determine the orientation angle α and the position angle θ for a single gage used to determine the stress intensity factor K_1 on a material with v equal to (a) 0.20, (b) 0.35, and (c) 0.45.

9.29. Equation (9.28) gives a relation for K_1 in terms of strain for a single strain gage mounted on aluminum. Develop a similar relation for a gage mounted on steel with $v = 0.300$. The angles θ and α are according to the specifications shown in Table 9.1.

9.30. Beginning with Eq. (9.29) verify Eq. (9.33).

9.31. If a single strain gage with a gage length of 2 mm is placed a distance $r_c = 6$ mm from the crack tip, determine the distance Δr between the center of the gage and the true strain location r_t.

9.32. Repeat Exercise 9.31 if the gage length is reduced to 1 mm.

9.33. Repeat Exercise 9.31 if the gage is placed at a position $r_c = 9$ mm from the crack tip.

9.34. A strain gage with a gage length of 1 mm is mounted on aluminum with $\theta = \alpha = 60°$ at a distance of 6 mm from the crack tip. Determine the stress intensity factor K_1 if the strain is 900 $\mu\epsilon$.

9.35. Verify Eq. (9.34).

9.36. If $\cos 2\alpha = -k$ and if $\tan(\theta/2) = -\cot 2\alpha$, show that Eq. (9.34) reduces to a four-term series containing A_0, B_1, C_0, and C_1.

9.37. Verify Eq. (9.37), beginning with the results of Exercise 9.36 subjected to $v = \frac{1}{3}$ and $\theta = \alpha = 60°$.

9.38. Verify Eq. (9.38).

9.39. Verify Eq. (9.40).

9.40. Write a program in either FORTRAN or BASIC languages which implements the theory described in Sec. 9.8. The program should accept as input ϵ_{rr}, r, θ, v, and k and give as output the unknown coefficients A_0, A_1, A_2, B_0, B_1, and B_2.

9.41. Write a program to determine the effectiveness of the solution provided by the results of Exercise 9.40. Explain the strategy used as the basis for this program.

9.42. An aluminum panel with a uniaxial residual stress $\sigma_0 = 100$ MPa in the y direction was investigated using the hole-drilling technique. If a strain gage is placed at a location 12 mm from the center of the hole and oriented in the x direction, find the strain developed as an 8-mm-diameter hole is drilled through the panel.

9.43. If the hole in Exercise 9.42 is enlarged from a diameter of 8 mm to diameters of 12, 16, and then 20 mm, determine the strain developed at the gage as the hole is enlarged. Does this exercise suggest an overdeterministic method for determining residual stresses?

9.44. Verify Eq. (9.51).

9.45. Beginning with Eq. (9.51) verify Eqs. (9.53).

9.46. A residual stress rosette like the one shown in Fig. 9.14 is designed with $r/a = 3$. This type of rosette was mounted on three steel panels and strain measurements were taken for each plate after drilling the center hole. Determine the biaxial residual stresses in the three plates.

Plate	ϵ_A, $\mu\epsilon$	ϵ_B, $\mu\epsilon$	ϵ_C, $\mu\epsilon$
1	700	−700	0
2	300	600	900
3	500	500	500

REFERENCES

1. Ades, C. S.: Reduction of Strain Rosettes in the Plastic Range, *Exp. Mech.*, vol. 2, no. 11, pp. 345–349, 1962.
2. Bossart, K. J., and G. A. Brewer: A Graphical Method of Rosette Analysis, *Proc. SESA*, vol. IV, no. 1, pp. 1–8, 1946.
3. McClintock, F. A.: On Determining Principal Strains from Strain Rosettes with Arbitrary Angles, *Proc. SESA*, vol. IX, no. 1, pp. 209–210, 1951.
4. Murray, W. M.: Some Simplifications in Rosette Analysis, *Proc. SESA*, vol. XV, no. 2, pp. 39–52, 1958.
5. Stein, P. K.: A Simplified Method of Obtaining Principal Stress Information from Strain Gage Rosettes, *Proc. SESA*, vol. XV, no. 2, pp. 21–38, 1958.
6. Meier, J. H.: On the Transverse-Strain Sensitivity of Foil Gages, *Exp. Mech.*, vol. 1, no. 7, pp. 39–40, 1961.
7. Meyer, M. L.: A Simple Estimate for the Effect of Cross Sensitivity on Evaluated Strain-Gage Measurements, *Exp. Mech.*, vol. 7, no. 11, pp. 476–480, 1967.
8. Wu, C. T.: Transverse Sensitivity of Bonded Strain Gages, *Exp. Mech.*, vol. 2, no. 11, pp. 338–344, 1962.
9. "Errors Due to Transverse Sensitivity in Strain Gages," Micro-Measurements publication TN-509, 1982.
10. Williams, S. B.: Geometry in the Design of Stress Measuring Circuits, *Proc. SESA*, vol. XVII, no. 2, pp. 161–178, 1960.
11. Hines, F. F.: The Stress-Strain Gage, *Proc. 1st Int. Congr. Exp. Mech. 1963*, pp. 237–253.
12. Lissner, H. R., and C. C. Perry: Conventional Wire Strain Gage Used as a Principal Stress Gage, *Proc. SESA*, vol. XIII, no. 1, pp. 25–32, 1955.
13. Perry, C. C.: Plane-Shear Measurement with Strain Gages, *Exp. Mech.*, vol. 9, no. 1, pp. 19N–22N, 1969.
14. Dally, J. W., and R. J. Sanford: Strain Gage Methods for Measuring the Opening Mode Stress Intensity Factor, *Exp. Mech.*, vol. 27, no. 4, pp. 381–388, 1987.
15. Dally, J. W., and J. R. Berger: Strain Gage Method for Determining K_I and K_{II} in a Mixed Mode Field, *Proc. 1986 Spring Conf. SEM*, June 1986.
16. Berger, J. R., and J. W. Dally: An Overdeterministic Approach for Measuring the Opening Mode Stress Intensity Factor Using Strain Gages, *Exp. Mech.*, vol. 28, no. 2, pp. 142–145, 1988.
17. "Determining Residual Stresses by the Hole-Drilling Strain-Gage Method," ASTM Standard E837–81, Philadelphia, 1981.
18. Rendler, J. J., and I. Vigness: Hole-Drilling Strain-Gage Method of Measuring Residual Stresses, *Proc. SESA*, vol. XXIII, no. 2, pp. 577–586, 1966.
19. Bynum, J. E.: Modifications to the Hole-Drilling Technique of Measuring Residual Stresses for Improved Accuracy and Reproducibility, *Exp. Mech.*, vol. 21, no. 1, pp. 21–33, 1981.
20. Niku-Lari, A., J. Lu, and J. F. Flavenot: Measurement of Residual Stress Distribution by the Incremental Hole-Drilling Method, *Exp. Mech.*, vol. 25, no. 2, pp. 175–185, 1985.
21. "Measurement of Residual Stresses by the Hole Drilling Strain Gage Method," Measurements Group, Inc., TN-503-1, 1988.

PART III

OPTICAL METHODS OF STRESS ANALYSIS

CHAPTER
10

BASIC OPTICS

10.1 THE NATURE OF LIGHT

The phenomenon of light has attracted attention from the earliest historical times. The ancient Greeks considered light to be an emission of small particles by a luminous body which entered the eye and returned to the body. Empedocles (484–424 B.C.) suggested that light takes time to travel from one point to another; however, Aristotle (384–322 B.C.) later rejected this idea as being too much to assume. The ideas of Aristotle concerning the nature of light persisted for approximately 2000 years.

In the seventeenth century, considerable effort was devoted to a study of the optical effects associated with thin films, lenses, and prisms. Huygens (1629–1695) and Hooke (1635–1703) attempted to explain some of these effects with a wave theory. In the wave theory, a hypothetical substance of zero mass, called the *ether*, was assumed to occupy all space. Initially, light propagation was assumed to be a longitudinal vibratory disturbance moving through the ether. The idea of secondary wavelets, in which each point on a wavefront can be regarded as a new source of waves, was proposed by Huygens to explain refraction. Huygens' concept of secondary wavelets is widely used today to explain, in a simple way, other optical effects such as diffraction and interference. At about the same time, Newton (1642–1727) proposed his corpuscular theory, in which light is visualized as a stream of small but swift particles emanating from shining bodies. The theory was able to explain most of the optical effects observed at the time.

A revival of interest in the wave theory of light began with the work of Young (1773–1829), who demonstrated that the presence of a refracted ray at an interface between two materials was to be expected from a wave theory while the corpuscular theory of Newton could explain the effect only with difficulty. His two-pinhole experiment, which demonstrated the interference of light, together with the work of Fresnel (1788–1827) on polarized light, which required transverse rather than longitudinal vibrations, firmly established the transverse ether wave theory of light.

The next major step in the evolution of the theory of light was due to Maxwell (1831–1879). His electromagnetic theory predicts the presence of two vector fields in light waves, an electric field and a magnetic field. Since these fields can propagate through space unsupported by any known matter, the need for the hypothetical ether of the previous wave theory was eliminated. The electromagnetic wave theory also unites light with all the other invisible entities of the electromagnetic spectrum, e.g., cosmic rays, gamma rays, x-rays, ultraviolet rays, infrared rays, microwaves, radio waves, and electric-power-transmission waves. The wide range of wave-lengths and frequencies available for study in the electromagnetic spectrum has led to rapid development of additional theory and understanding. For most of the effects to be described in later sections of this text, the wave properties of light are important and the particle characteristics of individual photons have little application. For this reason, simple wave theory will be used in most of the discussions which follow.

10.2   WAVE THEORY OF LIGHT [1]

Electromagnetic radiation is predicted by Maxwell's theory to be a transverse wave motion which propagates with an extremely high velocity. Associated with the wave are oscillating electric and magnetic fields which can be described with electric and magnetic vectors **E** and **H**. These vectors are in phase, perpendicular to each other, and at right angles to the direction of propagation. A simple representation of the electric and magnetic vectors associated with an electro-magnetic wave at a given instant of time is illustrated in Fig. 10.1. For simplicity and convenience of representation, the wave has been given sinusoidal form.

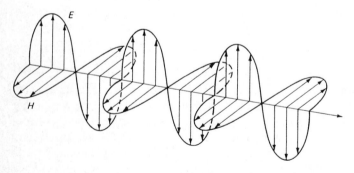

FIGURE 10.1
The electric and magnetic vectors associated with a plane electromagnetic wave.

All types of electromagnetic radiation propagate with the same velocity in free space, approximately 3×10^8 m/s (186,000 mi/s). The parameters used to differentiate between the various radiations are wavelength and frequency. These two quantities are related to the velocity by the relationship

$$\lambda f = c \qquad (10.1)$$

where λ = wavelength
  f = frequency
  c = velocity of propagation

The electromagnetic spectrum has no upper or lower limits. The radiations commonly observed have been classified in the broad general categories shown in Fig. 10.2.

Light is usually defined as radiation that can affect the human eye. From Fig. 10.2 it is evident that the visible range of the spectrum is a small band centered about a wavelength of approximately 550 nm. The limits of the visible spectrum are not well defined because the eye ceases to be sensitive at both long and short wavelengths; however, normal vision is usually in the range from 400 to 700 nm. Within this range the eye interprets the wavelengths as the different colors listed in Table 10.1. Light from a source that emits a continuous spectrum with nearly equal energy for every wavelength is interpreted as *white light*. Light of a single wavelength is known as *monochromatic light*.

Electromagnetic waves can be classified as one-, two-, or three-dimensional according to the number of dimensions in which they propagate energy. Light waves which emanate radially from a small source are three-dimensional. Two quantities associated with a propagating wave which will be useful in discussions involving geometrical and physical optics are wavefronts and rays. For a three-dimensional pulse of light emanating from a source, both the electric vector and the magnetic vector exhibit the periodic variation in magnitude shown in Fig. 10.1 along any radial line. The locus of points on different radial lines from the source

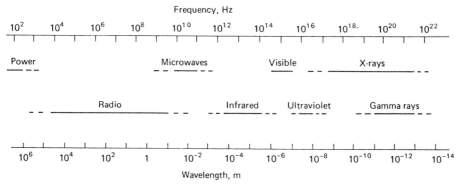

FIGURE 10.2
The electromagnetic spectrum.

TABLE 10.1
The visible spectrum

Wavelength range, nm	Color
400–450	Violet
450–480	Blue
480–510	Blue-green
510–550	Green
550–570	Yellow-green
570–590	Yellow
590–630	Orange
630–700	Red

exhibiting the same disturbance at a given instant of time, e.g., maximum or minimum values, is a surface known as a *wavefront*. The surface moves as the pulse propagates. If the medium is optically homogeneous and isotropic, the direction of propagation is at right angles to the wavefront. A line normal to the wavefront, indicating the direction of propagation of the waves, is called a *ray*. When the waves are propagated out in all directions from a point source, the wavefronts are spheres and the rays are radial lines. At large distances from the source, the spherical wavefronts have very little curvature, and over a limited region they can be treated as plane. Plane wavefronts can also be produced by using a lens or mirror to direct a portion of the light from a point source into a parallel beam.

In ordinary light, which is emitted from, say, a tungsten-filament light bulb, the light vector is not restricted in any sense and may be considered to be composed of a number of arbitrary transverse vibrations. Each of the components has a different wavelength, a different amplitude, a different orientation (plane of vibration), and a different phase. The vector used to represent the light wave can be either the electric vector or the magnetic vector. Both exist simultaneously, as shown in Fig. 10.1, and either can be used to describe the optical effects associated with photoelasticity, moiré, and holography.

10.2.1   The Wave Equation

Since the disturbance producing light can be represented by a transverse wave, it is possible to express the magnitude of the light (electric) vector in terms of the solution of the one-dimensional wave equation:

$$E = f(z - ct) + g(z + ct) \tag{10.2}$$

where E = magnitude of light vector
  z = position along axis of propagation
  t = time
  $f(z - ct)$ = wave propagation in positive z direction
  $g(z + ct)$ = wave propagation in negative z direction

Optical effects of interest in experimental stress analysis can be described with a simple sinusoidal or harmonic waveform. Thus, light propagating in the positive z direction away from the source can be represented by Eq. (10.2) as

$$E = f(z - ct) = \frac{K}{z} \cos \frac{2\pi}{\lambda} (z - ct) \tag{10.3}$$

where K is related to the strength of the source and K/z is an attenuation coefficient associated with the expanding spherical wavefront. At distances far from the source, the attenuation is small over short observation distances, and therefore it is frequently neglected. For plane waves, the attenuation is not a factor since the beam of light maintains a constant cross section. Equation (10.3) can then be written as

$$E = a \cos \frac{2\pi}{\lambda} (z - ct) \tag{10.4}$$

where a is the amplitude of the wave. A graphical representation of the magnitude of the light vector as a function of position along the positive z axis, at two different times, is shown for a plane light wave in Fig. 10.3. The length from peak to peak on the magnitude curve for the light vector is the wavelength λ. The time required for passage of two successive peaks at some fixed value of z is defined as the period T of the wave and is given by

$$T = \frac{\lambda}{c} \tag{10.5}$$

The frequency of the light vector is the number of oscillations per second. Clearly, the frequency is the reciprocal of the period, or

$$f = \frac{1}{T} = \frac{c}{\lambda} \tag{10.6}$$

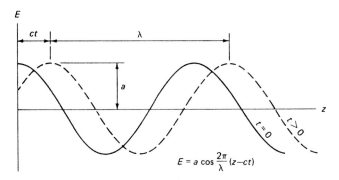

FIGURE 10.3
Magnitude of the light vector as a function of position along the axis of propagation at two different times.

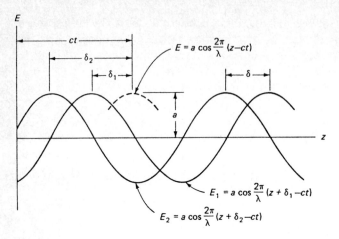

FIGURE 10.4
Magnitude of the light vector as a function of position along the axis of propagation for two waves with different initial phases.

The terms "angular frequency" and "wave number" are frequently used to simplify the argument in a sinusoidal representation of a light wave. The angular frequency ω and the wave number ξ are given by

$$\omega = \frac{2\pi}{T} = 2\pi f \tag{10.7}$$

$$\xi = \frac{2\pi}{\lambda} \tag{10.8}$$

Substituting Eqs. (10.7) and (10.8) into Eq. (10.4) yields

$$E = a \cos(\xi z - \omega t) \tag{a}$$

Two waves having the same wavelength and amplitude but a different phase are shown in Fig. 10.4. The two waves can be expressed by

$$E_1 = a \cos \frac{2\pi}{\lambda}(z + \delta_1 - ct)$$

$$E_2 = a \cos \frac{2\pi}{\lambda}(z + \delta_2 - ct) \tag{10.9}$$

where $\delta_1 =$ initial phase of wave E_1
$\delta_2 =$ initial phase of wave E_2
$\delta = \delta_2 - \delta_2 =$ the linear phase difference between waves

The linear phase difference δ is often referred to as a *retardation* since wave 2 trails wave 1.

The magnitude of the light vector can also be plotted as a function of time at a fixed position along the beam. This representation is useful for many applications since the eye, photographic films, video cameras, and other light-detecting devices are normally located at fixed positions for observations.

10.2.2 Superposition of Waves

In later chapters on photoelasticity and moiré, the phenomena associated with the superposition of two waves having the same frequency but different amplitude and phase will be encountered. At a fixed position z_0 along the light beam, where the observations will be made, the equations for the waves can be expressed as

$$E_1 = a_1 \cos \frac{2\pi}{\lambda} (z_0 + \delta_1 - ct) = a_1 \cos (\phi_1 - \omega t)$$

$$E_2 = a_2 \cos \frac{2\pi}{\lambda} (z_0 + \delta_2 - ct) = a_2 \cos (\phi_2 - \omega t)$$

(10.10)

where ϕ_1 = phase angle associated with wave E_1 at position z_0
ϕ_2 = phase angle associated with wave E_2 at position z_0
a_1 = amplitude of wave E_1
a_2 = amplitude of wave E_2

Consider first the case where the light vectors associated with the two waves oscillate in the same plane. The magnitude of the resulting light vector **E** is the sum

$$\mathbf{E} = E_1 + E_2 \tag{a}$$

Substituting Eqs. (10.10) into Eq. (*a*) yields

$$
\begin{aligned}
\mathbf{E} &= a_1(\cos \omega t \cos \phi_1 + \sin \omega t \sin \phi_1) \\
&\quad + a_2(\cos \omega t \cos \phi_2 + \sin \omega t \sin \phi_2) \\
&= a \cos (\phi - \omega t)
\end{aligned}
\tag{b}
$$

where

$$a^2 = a_1^2 + a_2^2 + 2a_1 a_2 \cos (\phi_2 - \phi_1) \tag{10.11}$$

and

$$\tan \phi = \frac{a_1 \sin \phi_1 + a_2 \sin \phi_2}{a_1 \cos \phi_1 + a_2 \cos \phi_2} \tag{10.12}$$

Equation (*b*) indicates that the resulting wave has the same frequency as the original waves but a different amplitude and a different phase angle. This procedure can easily be extended to the addition of three or more waves.

A special case frequently arises where the amplitudes of the two waves are equal ($a_1 = a_2$). In this case the amplitude of the resulting wave is given by Eq.

(10.11) as

$$a = \sqrt{2a_1^2\left(1 + \cos\frac{2\pi\delta}{\lambda}\right)} = \sqrt{4a_1^2 \cos^2\frac{\pi\delta}{\lambda}} \tag{c}$$

In all applications in stress analysis, the amplitude of the resulting wave is important, but the time variation is not. This results from the fact that the eye and all light-sensing instruments respond to the intensity of light (intensity is proportional to the square of the amplitude) but cannot detect the rapid time variations (for sodium light the frequency is 5.1×10^{14} Hz). For the special case of two waves of equal amplitude, the combined intensity is given by

$$I \sim a^2 = 4a_1^2 \cos^2\frac{\pi\delta}{\lambda} \tag{10.13}$$

Equation (10.13) indicates that the intensity of the light wave resulting from the superposition of two waves of equal amplitude is a function of the linear phase difference δ between the waves. The intensity of the resultant wave assumes its maximum value when $S = n\lambda$, $n = 0, 1, 2, 3, \ldots$. When the linear phase difference is an integral number of wavelengths,

$$I = 4a_1^2 \tag{d}$$

and the intensity of the resultant wave is four times the intensity of one of the individual waves. The intensity of the resultant wave assumes its minimum value when $\delta = [(2n + 1)/2]\lambda$, $n = 0, 1, 2, 3, \ldots$. When the linear phase difference is an odd number of half-wavelengths,

$$I = 0 \tag{e}$$

The modification of intensity by superposition of light waves is referred to as an *interference effect*. The effect represented by Eq. (d) is *constructive interference*. The effect represented by Eq. (e) is *destructive interference*. Interference effects have important applications in photoelasticity, moiré, and holography.

When the electric vector used to describe the light wave is restricted to a single plane, the condition is known as *plane* or *linearly polarized light*. Two other important forms of polarized light arise as a result of the superposition of two linearly polarized light waves having the same frequency but mutually perpendicular planes of vibration, as shown in Fig. 10.5. At a fixed position z_0 along the light beam, the equations for the two waves can be expressed as

$$E_x = a_x \cos\frac{2\pi}{\lambda}(z_0 + \delta_x - ct) = a_x \cos(\phi_x - \omega t)$$

$$E_y = a_y \cos\frac{2\pi}{\lambda}(z_0 + \delta_y - ct) = a_y \cos(\phi_y - \omega t) \tag{10.14}$$

$$\mathbf{E} = a \cos(\phi - \omega t)$$

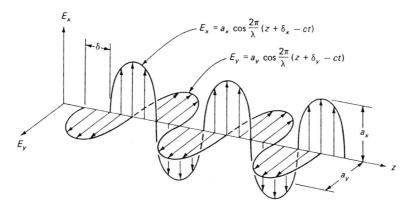

FIGURE 10.5
Two linearly polarized light waves having the same frequency but mutually perpendicular planes of vibration.

where ϕ_x, ϕ_y = phase angles associated with waves in the xz and yz planes
a_x, a_y = the amplitudes of waves in the xz and yz planes

The magnitude of the resulting light vector is given by vector addition as

$$E = \sqrt{E_x^2 + E_y^2} \qquad (f)$$

Considerable insight into the nature of the light resulting from the super-position of two mutually perpendicular waves is provided by a study of the trace of the tip of the resulting electric vector on a plane perpendicular to the axis of propagation at the point z_0. An expression for this trace can be obtained by eliminating time from Eqs. (10.14) to give

$$\frac{E_x^2}{a_x^2} - 2\frac{E_x E_y}{a_x a_y}\cos(\phi_y - \phi_x) + \frac{E_y^2}{a_y^2} = \sin^2(\phi_y - \phi_x) \qquad (g)$$

or since

$$\phi_y - \phi_x = \frac{2\pi}{\lambda}(\delta_y - \delta_x) = \frac{2\pi\delta}{\lambda}$$

$$\frac{E_x^2}{a_x^2} - 2\frac{E_x E_y}{a_x a_y}\cos\frac{2\pi\delta}{\lambda} + \frac{E_y^2}{a_y^2} = \sin^2\frac{2\pi\delta}{\lambda} \qquad (10.15)$$

Equation (10.15) is the equation of an ellipse, and light exhibiting this behavior is known as *elliptically polarized light*. The tips of the electric vectors at different positions along the z axis form an elliptical helix, as shown in Fig. 10.6. During an interval of time t, the helix will translate a distance $z = ct$ in the positive direction. As a result, the electric vector at position z_0 will rotate in a counter-clockwise direction as the translating helix is observed in the positive z direction.

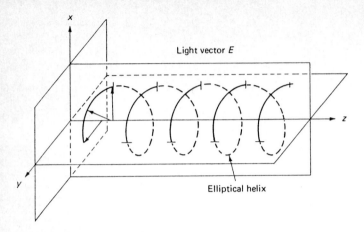

FIGURE 10.6
The elliptical helix formed by the tips of the light vectors along the axis of propagation at a fixed instant of time.

The locus of points representing the trace of the tip of the light vector on the perpendicular plane is the ellipse described by Eq. (10.15) and illustrated in Fig. 10.7.

A special case of elliptically polarized light occurs when the amplitudes of the two waves E_x and E_y are equal and $\delta = [(2n + 1)/4]\lambda$, $n = 0, 1, 2, \ldots$, so that Eq. (10.15) reduces to the equation of a circle:

$$E_x^2 + E_y^2 = a^2 \qquad (h)$$

Light exhibiting this behavior is known as *circularly polarized light*, and the tips of the light vectors form a circular helix along the z axis. For $\delta = \lambda/4$, $5\lambda/4, \ldots$, the helix is a left circular helix, and the light vector rotates counter-clockwise with time when viewed from a distant position along the z axis. For $\delta = 3\lambda/4, 7\lambda/4, \ldots$, the helix is a right circular helix, and the electric vector rotates clockwise.

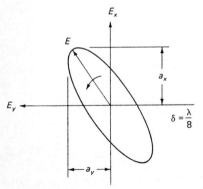

FIGURE 10.7
Trace of the tip of the light vector on the perpendicular plane at position z_0.

A second special case of elliptically polarized light occurs when the linear phase difference δ between the two waves E_x and E_y is an integral number of half-wavelengths ($\delta = n\lambda/2$, $n = 0, 1, 2, \ldots$). For this case, Eq. (10.15) reduces to

$$E_y = \frac{a_y}{a_x} E_x \qquad (i)$$

Equation (i) describes a straight line, and light exhibiting this behavior is known as *plane-* or *linearly polarized light*. The amplitude of the plane wave depends upon the amplitudes of the two original waves since

$$a = \sqrt{a_x^2 + a_y^2} \qquad (k)$$

In this discussion, light is treated as a wave motion without beginning or end. The light emitted by a conventional light source, e.g., a tungsten-filament light bulb, consists of numerous short pulses originating from a large number of different atoms. Each pulse consists of a finite number of oscillations known as a *wave train*. Each wave train is thought to be a few meters long with a duration of approximately 10^{-8} s. Since the light emissions occur in individual atoms which do not act together in a cooperative manner, the wave trains differ from each other in the plane of vibration, frequency, amplitude, and phase. Radiation produced in this manner is referred to as *incoherent light*. Light sources such as the laser, in which the atoms act cooperatively in emitting light, produce *coherent light*, in which the wave trains are monochromatic, in phase, linearly polarized, and extremely intense. For the interference effects discussed previously, coherent wave trains are required.

10.2.3 The Wave Equation in Complex Notation

A convenient way to represent both the amplitude and phase of a light wave, such as the one represented by Eq. (10.4), for calculations involving a number of optical elements is through the use of complex or exponential notation. Recall the Euler identity

$$e^{i\theta} = \cos\theta + i\sin\theta \qquad (a)$$

where $i^2 = -1$. The sinusoidal wave of Eq. (10.4) is represented by the real part of the complex expression with

$$\bar{E} = ae^{i(2\pi/\lambda)(z-ct)} = ae^{i(\phi-\omega t)} \qquad (b)$$

The imaginary part of Eq. (a) could also be used to represent the wave; however, it is normally assumed that the real part of a complex quantity is the one having physical significance.

If the amplitude of the wave is also considered to be a complex quantity, then

$$\bar{a} = a_r + ia_i = ae^{i[(2\pi/\lambda)\delta]} \qquad (c)$$

where

$$a = \sqrt{a_r^2 + a_i^2} \tag{d}$$

and

$$\tan \frac{2\pi}{\lambda} \delta = \frac{a_i}{a_r} \tag{e}$$

A wave with an initial phase δ can be expressed in exponential notation as

$$\bar{E} = \bar{a}e^{i(2\pi/\lambda)(z-ct)} = ae^{i(2\pi/\lambda)(z+\delta-ct)} \tag{10.16}$$

The physical waves previously represented by Eqs. (10.9) are simply the real part of Eq. (10.16) when represented in exponential notation. Superposition of two or more waves having the same frequency but different amplitude and phase is easily performed with the exponential representation. The real and imaginary parts of the amplitudes of the individual waves are added separately in an algebraic manner. The resultant complex amplitude gives the amplitude and phase of a single wave equivalent to the sum of the individual waves. Extensive use will be made of this representation in Chap. 12, where the theory of photoelasticity is discussed.

The real and imaginary parts of a complex quantity such as the amplitude of a wave may also be written

$$a_r = \tfrac{1}{2}(\bar{a} + \bar{a}^*)$$
$$a_i = \tfrac{1}{2}(\bar{a} - \bar{a}^*) \tag{f}$$

where

$$\bar{a}^* = a_r - ia_i \tag{g}$$

is the complex conjugate of the original complex amplitude

$$\bar{a} = a_r + ia_i \tag{h}$$

From Eq. (d).

$$a^2 = a_r^2 + a_i^2 \tag{i}$$

From Eq. (i) and Eq. (f)

$$a^2 = \bar{a}\bar{a}^* \tag{10.17}$$

This representation for the square of the amplitude of a complex quantity will be used in developments of optical methods based on the intensity of light.

10.3 REFLECTION AND REFRACTION [2]

In Sec. 10.2 the electromagnetic wave nature of light was discussed, and wavefronts and rays were defined. The discussions were limited to light propagating in free space. Most optical effects of interest, however, occur as a result of the interaction between a beam of light and some physical material. In free space, light propagates with a velocity c, which is approximately 3×10^8 m/s. In any other medium, the velocity is less than the velocity in free space. The ratio of the velocity in free space to the velocity in a medium depends on the index of refraction n of the medium. The index of refraction for most gases is only slightly greater than unity (for air, $n = 1.0003$). Values for liquids range from 1.3 to 1.5 (for water, $n = 1.33$) and for solids range from 1.4 to 1.8 (for glass, $n = 1.5$). The index of refraction for a material is not constant but varies slightly with wavelength of the light being transmitted. This dependence of index of refraction on wavelength is referred to as *dispersion*.

Since the frequency of a light wave is independent of material, the wavelength is shorter in a material than in free space. For this reason a wave propagating in a material will develop a linear phase shift δ with respect to a similar wave propagating in the free space. The magnitude of the phase shift, in terms of the index of refraction of the material, can be developed as follows. The time required for passage through a material of thickness h is

$$t = \frac{h}{v} \tag{a}$$

where h is the thickness of the material along the path of light propagation and v is the velocity of light in the material. The distance s traveled during the same time by a wave in free space is

$$s = ct = \frac{ch}{v} = nh \tag{b}$$

The distance δ by which the wave in the material trails the wave in free space is given by the difference as

$$\delta = s - h = nh - n = h(n - 1) \tag{10.18}$$

The retardation δ is a positive quantity since the index of refraction of a material is always greater than unity. The relative position of one wave with respect to another can be described by including the retardations in the phase of the appropriate wave equation.

When a beam of light strikes a surface between two transparent materials with different indices of refraction, it is divided into a reflected ray and a refracted ray, as shown in Fig. 10.8. The reflected and refracted rays lie in the plane formed by the incident ray and the normal to the surface, known as the *plane of incidence*. The angle of incidence α, the angle of reflection β, and the angle of refraction γ are related as

For reflection: $$\alpha = \beta \tag{10.19}$$

For refraction: $$\frac{\sin \alpha}{\sin \gamma} = \frac{n_2}{n_1} = n_{21} \tag{10.20}$$

where n_1 = index of refraction of material 1
n_2 = index of refraction of material 2
n_{21} = index of refraction of material 2 with respect to material 1

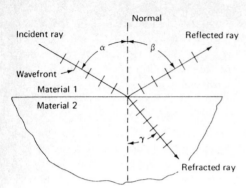

FIGURE 10.8
Reflection and refraction of a plane light wave at an interface between two transparent materials.

If the incident light beam is in the material having the higher index of refraction, n_{21} will be a number less than unity. Under these conditions, some critical angle of incidence α_c is reached for which the angle of refraction is $90°$. For angles of incidence greater than the critical angle, there is no refracted ray and total internal reflection occurs. Total internal reflection cannot occur when the incident beam is in the medium with the lower index of refraction.

The laws of reflection and refraction give the direction of reflected and refracted rays but do not describe the intensity. Intensity relationships, which are derived from Maxwell's equations, indicate that the intensity of a reflected beam depends upon both the angle of incidence and the direction of polarization of the incident beam. Consider a completely unpolarized beam of light falling on a surface between two transparent materials, as shown in Fig. 10.9. The electric vector for each wave train in the beam can be resolved into two components, one perpendicular to the plane of incidence (the perpendicular component) and the other parallel to the plane of incidence (the parallel component). For completely unpolarized incident light, the two components have equal intensity. The intensity of the reflected beam can be expressed as

$$I_r = RI_i \tag{10.21}$$

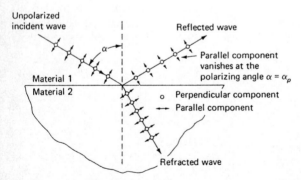

FIGURE 10.9
Reflection and refraction at the polarizing angle.

where I_i = intensity of incident beam
$\quad I_r$ = intensity of reflected beam
$\quad R$ = reflection coefficient

The reflection coefficient is different for the perpendicular and parallel components of the reflected wave. For the perpendicular component:

$$R_{90^\circ} = \frac{\sin^2(\alpha - \gamma)}{\sin^2(\alpha + \gamma)} \tag{10.22}$$

For the parallel component:

$$R_{0^\circ} = \frac{\tan^2(\alpha - \gamma)}{\tan^2(\alpha + \gamma)} \tag{10.23}$$

Reflection coefficients for an air-glass interface ($n_{21} = 1.5$) are shown in Fig. 10.10. These data indicate that there is a particular angle of incidence for which the reflection coefficient for the parallel component is zero. This angle is referred to as the *polarizing angle* α_p. Since the parallel component vanishes when the angle of incidence is equal to the polarizing angle, the beam reflected from the surface is plane-polarized with the plane of vibration perpendicular to the plane of incidence. It is also observed that a plane phase change of $\delta = \lambda/2$ occurs when light is incident from the medium with the lower index of refraction. When light is incident from the medium with the higher index of refraction, no phase change occurs upon reflection.

Metal surfaces exhibit relatively large reflection coefficients, as shown in Fig. 10.11. At oblique incidence, the coefficients for light polarized parallel to the plane of incidence are less than the coefficients for the perpendicular component, and a change of phase also occurs. Unfortunately, the phase change varies with both the angle of incidence and direction of polarization. As a result, plane-polarized light is converted by oblique reflection to elliptically polarized light.

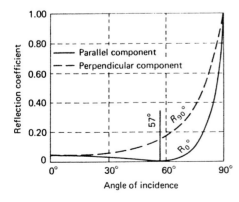

FIGURE 10.10
Reflection coefficients for an air-glass interface ($n_{21} = 1.5$).

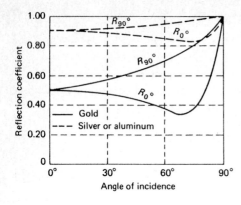

FIGURE 10.11
Reflection coefficients for several air-metal interfaces.

10.4 IMAGE FORMATION BY LENSES AND MIRRORS [2]

In Sec. 10.3, reflection and refraction of a plane light wave at a plane interface between two materials were considered. More complicated situations frequently arise in the optical systems used for experimental stress-analysis work. Since lenses and mirrors are widely used in many of these systems, a brief discussion of the significant features of these elements is provided here for future reference.

10.4.1 Plane Mirrors

An object O placed at a distance u in front of a plane mirror is shown in Fig. 10.12. The light from each point on the object (such as point A) is a spherical wave which reflects from the mirror. When the eye or other light-sensing instrument intercepts the reflected rays, an image I of the object O at a distance v behind the mirror is perceived. In this instance, the image I is a virtual image since light rays do not pass through image points such as A'. From the geometry of Fig. 10.12 it is obvious that the magnitudes of u and v are the same. The image is erect and has the same height as the object. One difference between the object and the image, not apparent from Fig. 10.12, is that left and right are interchanged (an image of a left hand appears as a right hand in the mirror).

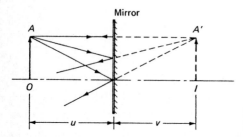

FIGURE 10.12
Image formation by a plane mirror.

10.4.2 Spherical Mirrors

An object O placed at a distance u in front of a concave spherical mirror is shown in Fig. 10.13. The center of curvature of the mirror is located at C, and the focal point is at F. The light from each point on the object reflects as shown in Fig. 10.13 for a typical point A. In this instance, the image I is a real image since light rays pass through the image point A'. From the geometry of Fig. 10.13 it can be shown that if all rays from the object make a small angle with respect to the axis of the mirror, then the distance to the image satisfies the expression

$$\frac{1}{u} + \frac{1}{v} = \frac{2}{r} = \frac{1}{f} \tag{10.24}$$

where r is the radius of curvature. For the case illustrated, both u and v are positive since both the object and the image are real. The distance v will be negative when dealing with a virtual image. With this sign convention, Eq. (10.24) applies to all concave, plane, and convex mirrors. The ratio of the size of the image to the size of the object is known as the *magnification M* and is given by the expression

$$M = -\frac{v}{u} \tag{10.25}$$

where the minus sign is used to indicate an inverted image. Equation (10.25) also applies to all concave, plane, and convex mirrors. For example, when Eqs. (10.24) and (10.25) are applied to the plane mirror of Fig. 10.12,

$$v = -u \quad \text{and} \quad M = 1$$

These results indicate that the image, which is imaginary (virtual), will be erect and identical in size to the object.

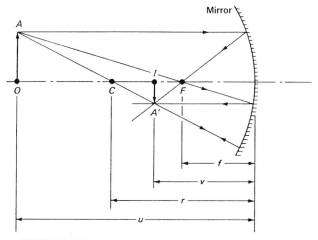

FIGURE 10.13
Image formation by a concave spherical mirror.

10.4.3 Thin Lenses

At least one and often a series combination of lenses are employed in optical equipment to magnify and focus the image of an object on a photographic plate. For this reason, the passage of light through a single convex lens and a pair of convex lenses arranged in series will be examined in detail. The treatment will be limited to instances where the thin-lens approximation can be applied; i.e., the thickness of the lens can be neglected with respect to other distances such as the focal length f of the lens and the object and image distances u and v.

SINGLE-LENS SYSTEM. The optical representation of a single-lens system is shown in Fig. 10.14. The light from each point on the object can be considered as a spherical wave which is reflected and refracted at the air-glass interface, as indicated in Sec. 10.3. With mirrors, the reflected rays are of interest; however, with lenses, the refracted rays produce the desired optical effects. From the geometry of Fig. 10.14 and the small-angle assumption it is clear that the object distance u, the image distance v, and the focal length f of the lens are related by the expression

$$\frac{1}{u} + \frac{1}{v} = \frac{1}{f} \qquad (a)$$

and that the magnification is given by the expression

$$M = -\frac{v}{u} \qquad (b)$$

Equations (a) and (b) are identical to Eqs. (10.24) and (10.25) for mirrors. The image in Fig. 10.14 is a real image; therefore, v is positive even though the image is on the opposite side of the lens from the object. With v positive, Eq. (b) indicates that the magnification is negative (thus the image is inverted, as shown).

The situation illustrated in Fig. 10.14 occurs when the object is located beyond the focal point of the lens. The object may also be placed between the focal point and the lens surface, as illustrated in Fig. 10.15. In this case, the distance u is positive and v is negative since the image formed is virtual. Equation (b) then yields a positive magnification, which indicates that the image is erect (as shown).

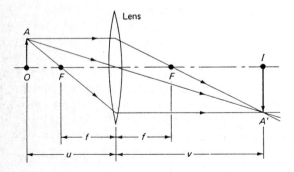

FIGURE 10.14
Image formation by a single convex lens (object outside the focal point).

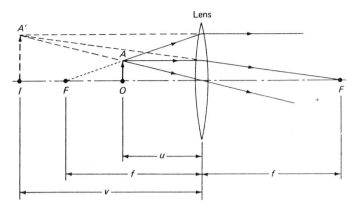

FIGURE 10.15
Image formation by a single convex lens (object inside the focal point).

A similar analysis for a concave lens indicates that Eqs. (10.24) and (10.25) apply so long as v is considered negative for virtual images and positive for real images. The proof is left as an exercise.

If Eqs. (10.24) and (10.25) are combined and solved for u and v in terms of M and f,

$$u = \frac{M - 1}{M} f \quad \text{and} \quad v = (1 - M)f \tag{10.26}$$

The total length L from the model (object) to the focusing plane of a camera (image) in terms of M and f is

$$L = u + v = \left(2 - M - \frac{1}{M}\right)f \tag{10.27}$$

These results show that the distances u, v, and L are directly related to the focal length of the lens. Since the focal length of the camera lens on many optical instruments must be relatively large, the length of an optical bench often exceeds 3 m (10 ft). This length, of course, depends upon the magnification capabilities of the camera, as shown in Table 10.2.

TABLE 10.2
Influence of the magnification M on the lengths u, v, and L in a camera assembly

$-M$	L/f	u/f	v/f
$\frac{1}{4}$	6.25	5.00	1.25
$\frac{1}{2}$	4.50	3.00	1.50
1	4.00	2.00	2.00
2	4.50	1.50	3.00
4	6.25	1.25	5.00

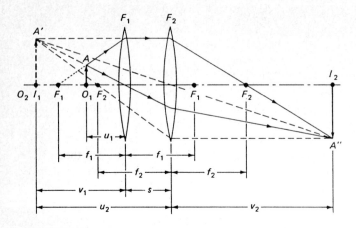

FIGURE 10.16
Image formation by a series combination of two convex lenses.

TWO LENSES IN SERIES. The classical optical representation of a series combination of two convex lenses is shown in Fig. 10.16. This series arrangement of convex lenses is often employed in optical arrangements to decrease their total length. The object is positioned inside the focal point of the long-focal-length field lens F_1. The slowly diverging rays from lens F_1 are then converged by a shorter-focal-length condenser lens F_2. By applying Eq. (10.24) to both lenses, it is clear that:

For the field lens F_1:
$$\frac{1}{u_1} + \frac{1}{v_1} = \frac{1}{f_1} \tag{a}$$

For the condenser lens F_2:
$$\frac{1}{u_2} + \frac{1}{v_2} = \frac{1}{f_2} \tag{b}$$

Since the virtual image from the field lens serves as the object for the condenser lens,

$$u_2 = -v_1 + s \tag{c}$$

The magnification M_c for the combined lens system is

$$M_c = M_1 M_2 = \left(-\frac{v_1}{u_1}\right)\left(-\frac{v_2}{u_2}\right) = \frac{v_1 v_2}{u_1 u_2} \tag{d}$$

By combining Eqs. (a) to (d) and solving for M, u_1, v_2, and s, it is possible to show that

$$M = \frac{f_1 f_2}{(u_1 - f_1)(s - f_2) - u_1 f_1} \tag{10.28}$$

$$s = \frac{f_2(f_1 - u_1) - u_1 f_1 - f_1 f_2/M}{f_1 - u_1} \tag{10.29}$$

$$u_1 = \frac{f_1 f_2 [(1 - M)/M] + s f_1}{s - f_1 - f_2} \qquad (10.30)$$

$$v_2 = M \left[s \left(\frac{u_1}{f_1} - 1 \right) - u_1 \right] \qquad (10.31)$$

These results can be employed to design a polariscope, i.e., to choose f_1 and f_2 to give a range of magnification M within a certain length $L = u_1 + v_2$. Or if the polariscope already exists, where f_1 and f_2 are fixed and u_1 and s can be varied within limits, the equations can be employed to determine M within these limits. It should be noted that M is a function of f_1, f_2, u_1, and s; thus, a given magnification M can be achieved in a number of different ways.

All the relationships for the series lens arrangement, where the object O_1 is placed inside the focal point of lens F_1, are applicable for the case where the object is placed outside the focal point of the lens. This can be illustrated by assuming in Fig. 10.16 that the image I_2 is the object O_1. If the rays are then traced through the lens combination in a reverse direction from the ones shown, the object O_1 in the figure becomes the image I_2. The magnification in the two cases is different. In practice, the position of the object and the location of the viewing screen will depend on the object being studied and the characteristics of the lenses available for the series combination.

The previous discussion for lenses and mirrors was based on the assumption that all light rays from an object were confined to a region near the axis of the element and made small angles with respect to the axis. In systems where the rays are not confined to a region near the axis, all rays from an object point do not focus at exactly the same image point. This effect is known as a *spherical aberration*. Similarly, when an object point is located such that rays from the point make an appreciable angle with respect to the axis of the element, the image tends to become two mutually perpendicular line segments rather than a point image. This defect in optical elements is known as *astigmatism*. Other lens defects which must be considered in the design of optical systems include distortion, coma, curvature of field, and chromatic aberration. The interested reader should consult an optics textbook for a more complete discussion of defects in images produced by mirrors and lenses.

10.5 OPTICAL DIFFRACTION AND INTERFERENCE [1, 2]

Consider light issued by a point source that is being used to illuminate a viewing screen. If an opaque plate with a large hole is placed between the source and the screen, the area illuminated on the screen can be defined by rays drawn from the source to the boundary of the hole and extended to the screen. As the hole in the plate is made smaller and smaller, the area illuminated on the screen decreases until some minimum area of illumination is achieved. Further decreases in the size of the hole cause the area of illumination to increase. This phenomenon, associated

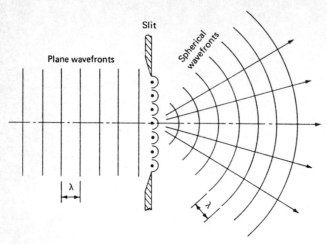

FIGURE 10.17
Diffraction of plane wavefronts at a narrow rectangular slit.

with an apparent bending of the light rays, is known as *diffraction*. Diffraction plays an important role in several methods of experimental stress analysis.

Diffraction of waves at a small aperture is explained by Huygens' principle, which states that "each point on a wavefront may be regarded as a new source of waves." For example, consider a series of plane wavefronts impinging on a rectangular slit as shown in Figure 10.17. When the width of the slit is small (less than 0.10 mm), diffraction becomes important. In accordance with Huygens' principle, the light emerging from the slit can be considered to consist of a series of spherical wavefronts which expand and ultimately illuminate a viewing screen placed in the path of the propagating waves. Equation (10.3) can be used to describe these spherical wavefronts as

$$E = \frac{Kb}{r} \cos \frac{2\pi}{\lambda} (r - ct)$$

The amplitude of the wave Kb/r depends on the strength of the source K, the width of the slit b, and the radial position r of the expanding spherical wavefront.

If the viewing screen is placed at a large distance from the slit, or if a lens is used to focus parallel rays from the slit to a point P on the screen, the intensity distribution I as a function of the angle of inclination θ of the rays is easily calculated. For example, consider the emerging parallel rays from each spherical source shown in Fig. 10.18. The contribution to the total amount of light reaching P from an increment of width ds of the slit above and below the optical axis can be expressed as

$$dE_P(s) = \frac{K\,ds}{r} \cos \frac{2\pi}{\lambda} (r - ct - s \sin \theta) \qquad (a)$$

$$dE_P(-s) = \frac{K\,ds}{r} \cos \frac{2\pi}{\lambda} (r - ct + s \sin \theta) \qquad (b)$$

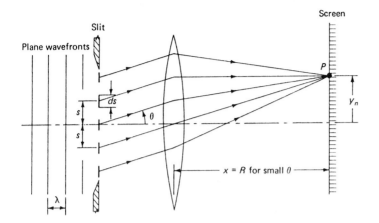

FIGURE 10.18
Light rays from a rectangular slit of width b which contribute to the intensity at P.

The total amount of light reaching P can be determined by summing the contributions from each increment of width of the slit. This is most easily accomplished by adding Eqs. (a) and (b) and integrating over half the width of the slit. Thus

$$E_P = \int_0^{b/2} dE_P(s) + dE_P(-s)$$

$$= \int_0^{b/2} \frac{2K}{r} \left[\cos \frac{2\pi}{\lambda} (r - ct) \cos \frac{2\pi}{\lambda} (s \sin \theta) \right] ds$$

$$= \frac{Kb}{r} \frac{\sin \beta}{\beta} \cos \frac{2\pi}{\lambda} (r - ct) \qquad (10.32)$$

where

$$\beta = \frac{\pi b \sin \theta}{\lambda} \qquad (10.32a)$$

Since the intensity at P is proportional to the square of the amplitude,

$$I = \left(\frac{Kb}{r} \right)^2 \frac{\sin^2 \beta}{\beta^2} \qquad (10.33)$$

For maximum or minimum intensity $dI/d\beta = 0$, and

$$\frac{2 \sin \beta}{\beta^3} (\beta \cos \beta - \sin \beta) = 0$$

Thus either

$$\beta = n\pi \qquad \text{or} \qquad \beta = \tan \beta \qquad (c)$$

The condition $\beta = n\pi$ yields zero intensity except at $\beta = 0$. The roots of $\beta = \tan \beta$ define positions of maximum intensity. The maximums are located at $\beta = 0$, 1.430π, 2.459π, 3.471π, 4.477π, ... If the intensity at $\beta = 0$ ($\theta = 0$) is arbitrarily taken as 1, the intensities of successive maxima are $\frac{1}{21}$, $\frac{1}{61}$, $\frac{1}{120}$, ..., .

The dark bands which form in regions of zero intensity are called *diffraction fringes*. For a rectangular slit, the fringes will be parallel straight lines. For a circular pinhole, the fringes will be a series of concentric circles about the axis of the optical system. If the incident light is parallel and a lens is used to produce a focused image on a screen, the phenomenon is referred to as *Fraunhofer diffraction*. If the incident light comes from a point source at a finite distance from the slit or pinhole, the wavefronts are divergent and the phenomenon is known as *Fresnel diffraction*.

For the narrow slit, the fringe positions can be determined by substituting Eq. (c) into Eq. (10.32a) to give

$$\frac{\pi b \sin \theta}{\lambda} = n\pi$$

For small values θ, $\sin \theta = y_n/R$, where R is the distance between the lens and the screen and y_n is the distance from the axis of the optical system to the diffraction fringe of order n. Thus

$$y_n = \frac{n\lambda R}{b} \qquad n = 1, 2, 3, \ldots \qquad (10.34)$$

The spacing of diffraction fringes is used as the basis for several strain-measurement methods (see Figs. 5.4 and 5.5).

10.6   OPTICAL INSTRUMENTS: THE POLARISCOPE

The polariscope is an optical instrument that utilizes the properties of polarized light in its operation. For experimental stress-analysis work, two types are frequently employed, the plane polariscope and the circular polariscope. The names follow from the type of polarized light used in their operation.

In practice, plane-polarized light is produced with an optical element known as a *plane* or *linear polarizer*. Production of circularly polarized light or the more general elliptically polarized light requires the use of a linear polarizer together with an optical element known as a *wave plate*. A brief discussion of linear polarizers, wave plates, and their series combination follows.

10.6.1   Linear or Plane Polarizers [3–5]

When a light wave strikes a plane polarizer, this optical element resolves the wave into two mutually perpendicular components, as shown in Fig. 10.19. The

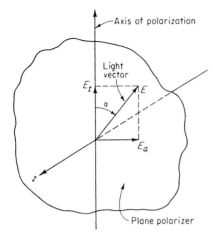

FIGURE 10.19
Absorbing and transmitting characteristics of a plane polarizer.

component parallel to the axis of polarization is transmitted while the component perpendicular to the axis of polarization is absorbed.

For a plane polarizer fixed at z_0, the equation for the light vector is

$$\mathbf{E} = a \cos \frac{2\pi}{\lambda} (z_0 - ct) \qquad (a)$$

Since the initial phase of the wave is not important in the development which follows, Eq. (a) can be reduced to

$$\mathbf{E} = a \cos 2\pi ft = a \cos \omega t \qquad (10.35)$$

by using Eqs. (10.6) and (10.7). The quantity $\omega = 2\pi f$ is the circular frequency of the wave. The absorbed and transmitted components of the light vector are

$$E_a = a \cos \omega t \sin \alpha$$
$$E_t = a \cos \omega t \cos \alpha \qquad (b)$$

where α is the angle between the axis of polarization and the incident light vector.

Polaroid filters are almost always used for producing polarized light in polariscopes. They have the advantage of providing a large field of very well polarized light at a relatively low cost. Most modern polariscopes containing linear polarizers employ Polaroid H sheet,[1] a transparent material with strained and oriented molecules. In the manufacture of H-type Polaroid films, a thin sheet of polyvinyl alcohol is heated, stretched, and immediately bonded to a supporting sheet of cellulose acetate butyrate. The polyvinyl face of the assembly is then stained by a liquid rich in iodine. The amount of iodine diffused into the sheet

---

[1] Manufactured by the Polaroid Corporation, Cambridge, Massachusetts.

determines its quality. Filters are available in five grades, denoted according to their transmittance of light as HN-22, HN-32, HN-35, HN-38, and HN-42. Since the quality of a polarizer is judged by its transmission ratio, grade HN-22 (with the best transmission ratio) is recommended for photoelastic purposes.

10.6.2 Wave Plates [3, 6, 7]

A wave plate is an optical element which has the ability to resolve a light vector into two orthogonal components and to transmit the components with different velocities. Such a material is called *doubly refracting* or *birefringent*. The doubly refracting plate illustrated in Fig. 10.20 has two principal axes labeled 1 and 2. The transmission of light along axis 1 proceeds at velocity c_1 and along axis 2 at velocity c_2. Since c_1 is greater than c_2, axis 1 is called the *fast axis* and axis 2 the *slow axis*.

If this doubly refracting plate is placed in a field of plane-polarized light so that the light vector $\mathbf{E}_t$ makes an angle β with axis 1 (the fast axis), then on entering the plate the light vector is resolved into two components E_{t1} and E_{t2} along axes 1 and 2 with magnitudes E_{t1} and E_{t2} given by

$$E_{t1} = \mathbf{E}_t \cos \beta = a \cos \alpha \cos \omega t \cos \beta = k \cos \omega t \cos \beta$$

$$E_{t2} = \mathbf{E}_t \sin \beta = a \cos \alpha \cos \omega t \sin \beta = k \cos \omega t \sin \beta$$

(a)

where $k = a \cos \alpha$. The light components E_{t1} and E_{t2} travel through the plate with different velocities c_1 and c_2; therefore, the two components emerge from the plate at different times. In other words, one component is retarded in time relative to the other. This retardation produces a relative phase shift between the two components. From Eq. (10.18), the linear phase shifts for components E_{t1} and E_{t2}

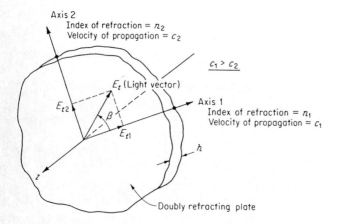

FIGURE 10.20
A plane-polarized light vector entering a doubly refracting plate.

with respect to a wave in air can be expressed as

$$\delta_1 = h(n_1 - n)$$
$$\delta_2 = h(n_2 - n)$$

(b)

where n is the index of refraction of air.

The relative linear phase shift is then computed simply as

$$\delta = \delta_2 - \delta_1 = h(n_2 - n_1)$$ (10.36)

The relative angular phase shift Δ between the two components as they emerge from the plate is given by

$$\Delta = \frac{2\pi}{\lambda} \delta = \frac{2\pi h}{\lambda} (n_2 - n_1)$$ (10.37)

The relative angular phase shift Δ produced by a doubly refracting plate is dependent upon its thickness h, the wavelength of the light λ, and the properties of the plate as described by $n_2 - n_1$. When the doubly refracting plate is designed to give $\Delta = \pi/2$, it is called a *quarter-wave plate*. Doubly refracting plates designed to give angular retardations of π and 2π are known as *half-* and *full-wave plates*, respectively. Upon emergence from a wave plate exhibiting a retardation Δ, the two components of light are described by the equations

$$E'_{t1} = k \cos \beta \cos \omega t$$
$$E'_{t2} = k \sin \beta \cos (\omega t - \Delta)$$

(10.38)

With this representation, only the relative phase shift between components has been considered. The additional phase shift suffered by both components, as a result of passage through the wave-plate material (as opposed to free space), has been neglected since it has no effect on the phenomenon being considered.

The amplitude of the light vector which is equivalent to these two components can be expressed as

$$E'_t = \sqrt{(E'_{t1})^2 + (E'_{t2})^2}$$
$$= k\sqrt{\cos^2 \beta \cos^2 \omega t + \sin^2 \beta \cos^2 (\omega t - \Delta)}$$

(10.39)

The angle that the emerging light vector makes with axis 1 is given by

$$\tan \gamma = \frac{E'_{t2}}{E'_{t1}} = \frac{\cos (\omega t - \Delta)}{\cos \omega t} \tan \beta$$ (10.40)

It is clear that both the amplitude and the rotation of the emerging light vector can be controlled by the wave plate. Controlling factors are the relative phase difference Δ and the orientation angle β. Various combinations of Δ and β and their influence on the type of polarized light produced will be discussed in Sec. 10.6.3.

Wave plates employed in a photoelastic polariscope usually consist of a single plate of quartz or calcite cut parallel to the optic axis, or a sheet of oriented polyvinyl alcohol. In recent years, as the design of the modern polariscope has tended toward a field of relatively large diameter, most wave plates employed have been fabricated from oriented sheets of polyvinyl alcohol. These wave plates, called *retarders*, are manufactured by the Polaroid Corporation by warming and uni-directionally stretching the sheet. Since the oriented polyvinyl alcohol sheet is only about 20 μm thick (for a quarter-wave plate), the commercial retarders are usually laminated between two sheets of cellulose acetate butyrate.

10.6.3   Conditioning of Light by a Series Combination of a Linear Polarizer and a Wave Plate

The magnitude and direction of the light vector emerging from a series combination of a linear polarizer and a wave plate are given by Eqs. (10.39) and (10.40). The light emerging from this combination of optical elements is always polarized; however, the type of polarization may be plane, circular, or elliptical. The factors which control the type of polarized light produced by this combination are the relative phase difference Δ imposed by the wave plate and the orientation angle β. Three well-defined cases exist.

CASE 1: PLANE-POLARIZED LIGHT. If the angle β is set equal to zero and the relative retardation Δ is not restricted in any sense, the magnitude and direction of the emerging light vector are given by Eqs. (10.39) and (10.40) as

$$E'_t = k \cos \omega t \qquad \text{and} \qquad \gamma = 0 \qquad\qquad (a)$$

Since $\gamma = 0$, the light vector is not rotated as it passes through the wave plate and the light remains plane-polarized. The wave plate in this instance does not influence the light except to produce a retardation with respect to a wave in free space which depends on the plate thickness and the index of refraction n_1. Similar results are obtained by letting $\beta = \pi/2$ with

$$E'_t = k \cos (\omega t - \Delta) \qquad \text{and} \qquad \gamma = \frac{\pi}{2} \qquad\qquad (b)$$

CASE 2: CIRCULARLY POLARIZED LIGHT. If a wave plate is selected so that $\Delta = \pi/2$, that is, a quarter-wave plate, and $\beta = \pi/4$, the magnitude and direction of the light vector as it emerges from the plate are given by Eqs. (10.39) and (10.40) as

$$E'_t = \frac{\sqrt{2}}{2} k \sqrt{\cos^2 \omega t + \sin^2 \omega t} = \frac{\sqrt{2}}{2} k \qquad \text{and} \qquad \gamma = \omega t \qquad\qquad (c)$$

The light vector described here has a constant magnitude and the tip of the light vector traces out a circle as it rotates. The vector rotates with a constant angular velocity in a counterclockwise direction when viewed in the positive z direction. Such light is known as *left circularly polarized light*. Right circularly polarized light could be obtained by setting β equal to $3\pi/4$. The light vector would then rotate with a constant angular velocity in the clockwise direction.

CASE 3: ELLIPTICALLY POLARIZED LIGHT. If a quarter-wave plate ($\Delta = \pi/2$) is selected and $\beta \neq n\pi/4$ ($n = 0, 1, 2, 3, \ldots$), then, by Eqs. (10.39) and (10.40), the magnitude and direction of the emerging light vector are

$$E'_t = k\sqrt{\cos^2 \beta \cos^2 \omega t + \sin^2 \beta \sin^2 \omega t}$$

$$\tan \gamma = \tan \beta \tan \omega t \qquad (d)$$

The amplitude of the light vector in this case varies with angular position in such a way that the tip of the light vector traces out an ellipse as it rotates. The shape and orientation of the ellipse and the direction of rotation of the light vector depend on β.

Consider the significance of Eq. (10.37) in the production of circularly polarized light, and note that a quarter-wave plate is required; therefore, the phase difference $\Delta = \pi/2$. It is clear that the thickness h can be determined to give $\Delta = \pi/2$ once the plate material, $n_2 - n_1$, is selected and the wavelength λ of the light is fixed. However, a quarter-wave plate suitable for one wavelength of monochromatic light, i.e., a constant wavelength, will not be suitable for a different wavelength. Also, quarter-wave plates cannot be designed for white light because white light is comprised of different wavelengths.

10.6.4 Arrangement of the Optical Elements in a Polariscope [8–11]

THE PLANE POLARISCOPE. The plane polariscope is the simplest optical system used in photoelasticity; it consists of two linear polarizers and a light source arranged as illustrated in Fig. 10.21a.

The linear polarizer nearest the light source is called the *polarizer*, while the second linear polarizer is known as the *analyzer*. In the plane polariscope the two axes of polarization are always crossed, no light is transmitted through the analyzer, and this optical system produces a dark field. In operation a photoelastic model is inserted between the two crossed elements and viewed through the analyzer. The behavior of the photoelastic model in a plane polariscope will be covered in Sec. 12.4.

THE CIRCULAR POLARISCOPE. As the name implies, the circular polariscope employs circularly polarized light. The photoelastic apparatus contains four optical elements and a light source, as illustrated in Fig. 10.21b.

The first element following the light source is the polarizer. It converts the ordinary light into plane-polarized light. The second element is a quarter-wave plate set at an angle $\beta = \pi/4$ to the plane of polarization. This quarter-wave plate converts the plane-polarized light into circularly polarized light. The second quarter-wave plate is set with its fast axis parallel to the slow axis of the first

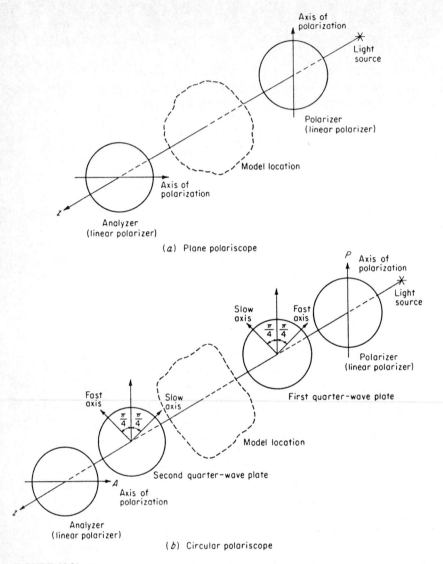

FIGURE 10.21
Arrangement of the optical elements in a plane polariscope and in a circular polariscope.

quarter-wave plate. The purpose of this element is to convert the circularly polarized light into plane-polarized light vibrating in the vertical plane. The last element is the analyzer, with its axis of polarization horizontal, and its purpose is to extinguish the light. This series of optical elements constitutes the standard arrangement for a circular polariscope, and it produces a dark field. Four arrangements of the optical elements in the circular polariscope are possible, depending upon whether the polarizers and quarter-wave plates are crossed or parallel. These optical arrangements are described in Table 10.3.

TABLE 10.3
Four arrangements of the optical elements in a circular polariscope

Arrangement	Quarter-wave plates	Polarizer and analyzer	Field
A[†]	Crossed	Crossed	Dark
B	Crossed	Parallel	Light
C	Parallel	Crossed	Light
D	Parallel	Parallel	Dark

[†] Shown in Fig. 10.21

Arrangements A and B are normally recommended for light- and dark-field use of the polariscope since the error introduced by imperfect quarter-wave plates (i.e., both quarter-wave plates differ from $\pi/2$ by a small amount) is minimized. Since quarter-wave plates are often of poor quality, this fact is important in selecting the optical arrangement.

10.6.5 Construction Details of Diffused-Light and Lens-Type Polariscopes [8–11]

THE DIFFUSED-LIGHT POLARISCOPE. The arrangement of the optical elements discussed previously is not sufficiently complete or detailed for the visualization of a working polariscope. The degree of complexity of a polariscope varies widely with the investigator and ranges from highly complex lens systems with servo-motor drives on the four optical elements to very simple arrangements with no lenses and no provision for rotation of any element.

The diffused-light polariscope described here is one of the simplest and least expensive polariscopes available; yet it can be employed to produce very-high-quality photoelastic results. This polariscope requires only one lens; however, its field can be made very large since its diameter is dependent only upon the size of the available linear polarizers and quarter-wave plates. Diffused-light polariscopes with field diameters up to 450 mm can readily be constructed.

A schematic illustration of the construction details of a diffused-light polariscope is shown in Fig. 10.22.

THE LENS POLARISCOPE. In the earlier days of photoelasticity, Nicol prisms (available only in small diameters) were used almost exclusively as the polarizing elements. Consequently, it was necessary to employ a lens system to expand the field of view so that reasonably sized models could be studied. However, with the advent of high-quality large-diameter sheets, it is no longer necessary to extend the diameter of the field through the use of a multiple-lens system. Instead, lens polariscopes should be employed only where parallel light over the whole field is a necessity. Instances where parallel light is important include applications where precise definition of the entire model boundary is critical and where partial mirrors

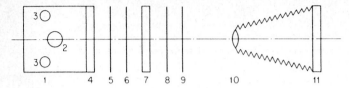

FIGURE 10.22
Design of a circular diffused-light polariscope with both white and monochromatic light sources: (1) light house (flat-white diffusing paint on interior); (2) monochromatic light source, low-pressure sodium street lamp 12 in long, 3 in in diameter, 10,000 lm; (3) white-light source, 300-W tungsten-filament lamp located on side of light house; (4) diffusing plates, flashed opal glass; (5) polarizer, glass-laminated Polaroid; (6) first quarter-wave plate, glass-laminated orientated polyvinyl alcohol; (7) loading frame; (8) second quarter-wave plate, glass-laminated orientated polyvinyl alcohol; (9) analyzer, glass-laminated Polaroid; (10) camera lens, good-quality process lens with about 20- to 24-in focal length; (11) camera.

are to be employed in the photoelastic bench (for fringe sharpening and fringe multiplication).

Several variations of the lens systems are possible. The arrangement shown in Fig. 10.23 is one of the simplest types that can be employed to obtain parallel light. The polarizer, quarter-wave plates, and analyzer should be placed in the parallel beam between the two field lenses to achieve more complete polarization and to avoid problems associated with internal stresses in the field lenses. A point source of light is required for parallel light, and this point source is usually approached by employing a high-intensity mercury lamp with a very short arc. When a photoelastic model is placed in the field, a slight degree of scattering of the light occurs, which disturbs the parallelism of the light. To improve the parallelism of the light, it is appropriate to regard the model as the source of illumination and to control the light as it emerges from the model. The effects of the light scattered by the model can be minimized by placing a diaphragm stop (10) at the focal point of the second field lens. As the diameter of the stop is reduced, the parallelism of the light is improved; however, the intensity of the light striking the camera back is decreased and film exposure times increase.

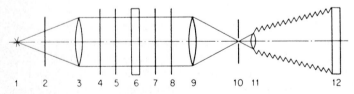

FIGURE 10.23
Construction details of a circular lens-type polariscope. (1) Light source (usually a small mercury arc), (2) color filter, (3) first field lens, (4) polarizer, (5) first quarter-wave plate, (6) loading frame and model, (7) second quarter-wave plate, (8) analyzer, (9) second field lens, (10) diaphragm stop, (11) camera lens, (12) camera back.

10.7 OPTICAL INSTRUMENTS: THE INTERFEROMETER

An interferometer is an optical device which can be used to measure lengths or changes in length with great accuracy by means of interference fringes. The modification of intensity of light by superposition of lightwaves was defined in Sec. 10.2 as an interference effect. The intensity of the wave resulting from the superposition of two waves of equal amplitude was shown by Eq. (10.13) to be a function of the linear phase difference δ between the waves. A fundamental requirement for the existence of well-defined interference fringes is that the light waves producing the fringes have a sharply defined phase difference which remains constant with time. When light beams from two independent sources are super-imposed, interference fringes are not observed since the phase difference between the beams varies in a random way (the beams are incoherent). Two beams from the same source, on the other hand, interfere, since the individual wavetrains in the two beams have the same phase initially (the beams are coherent) and any difference in phase at the point of superposition results solely from differences in optical paths. In this treatment optical-path length is defined as

$$\sum_{i=1}^{i=m} n_i L_i \qquad (a)$$

where L_i is the mechanical-path length in a material having an index of refraction n_i.

The concept of optical-path difference and its effect on the production of interference fringes can be illustrated by considering the reflection and refraction of light rays from a transparent plate having a thickness h, as shown in Fig. 10.24. Consider a plane wavefront associated with the light ray A which strikes the plate at an angle of incidence α. Ray B is reflected from the front surface of the plate and, as discussed in Sec. 10.3, undergoes a phase change of $\lambda/2$. A second ray C is due to refraction at the front surface, reflection at the back surface, and then refraction from the front surface before emerging from the front surface. The

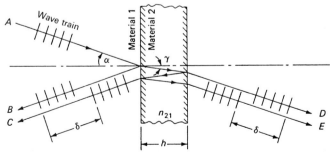

FIGURE 10.24
Reflection and refraction of light from a transparent plate ($n_2 > n_1$).

optical-path difference between rays B and C is

$$\delta = \frac{2h}{\cos \gamma} (n_{21} - \sin \gamma \sin \alpha)$$

But from Eq. (10.20)

$$\sin \alpha = n_{21} \sin \gamma$$

Therefore

$$\delta = \frac{2hn_{21}}{\cos \gamma} (1 - \sin^2 \gamma) = 2hn_{21} \cos \gamma \tag{10.41}$$

Since ray B undergoes a phase change of $\lambda/2$ on reflection, rays B and C will interfere destructively and produce minimum intensity when

$$\delta = m\lambda \qquad m = 0, 1, 2, 3, \ldots$$

If the light beam illuminates an extended area of the plate, and if the thickness of the plate varies slightly with position, the locus of points experiencing the same order of extinction will combine to form an interference fringe. The fringe spacing will represent thickness variations of approximately 7 μin or 180 nm (glass with mercury light and a small angle γ).

Rays emerging from the back surface of the plate can also be used to produce interference effects. In Fig. 10.24, a third ray is refracted at both the front and back surfaces of the plate before emerging as ray D. A fourth ray undergoes two internal reflections before being refracted from the back surface of the plate as ray E. The optical-path difference between rays D and E is identical to the difference between rays B and C as given by Eq. (10.41). Since neither ray D nor E suffers a phase change on reflection, the two rays will interfere destructively when

$$\delta = (2m + 1) \frac{\lambda}{2} \qquad m = 0, 1, 2, 3, \ldots$$

The previous discussion serves to illustrate the principles associated with measurements employing interference effects. For the system illustrated in Fig. 10.24, the optical-path difference δ would be large with $m > 10^3$ and involve high-order interference. It would require extreme coherence and long wave trains such as those provided by a laser for successful visualization of the interference fringe pattern. Other systems which utilize low-order interference ($m = 0, 1, 2, 3$) place less stringent requirements on the light source. One low-order system used in experimental stress-analysis work is the Mach-Zehnder interferometer.

10.7.1 The Mach-Zehnder Interferometer

The essential features of a Mach-Zehnder interferometer are illustrated in Fig. 10.25. The beam from a light source is divided into a reference beam and an active beam with a beam splitter (partial mirror). The beams are recombined after the

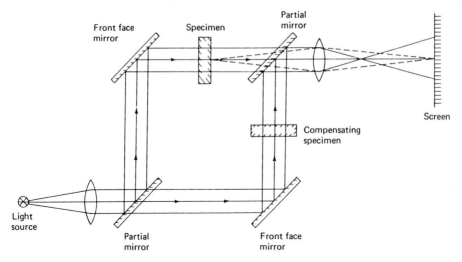

FIGURE 10.25
Light paths through a Mach-Zehnder interferometer.

active beam passes through the specimen of interest by the second partial mirror which is adjusted to bring the optical axes of the two beams in concurrence. A lens is used to focus the recombined beams on a screen where the interference pattern is displayed.

If a laser is used as a light source with the Mach-Zehnder interferometer, it is not necessary to employ a compensation specimen to adjust for the change in the optical path due to the insertion of the model. The coherent light from the laser permits the formation of very-high-order ($m \gg 0$) interference fringes. Low-order interference is obtained by inserting a compensating specimen in the reference beam to adjust for the difference in index of refraction between the specimen material and air.

10.8 OPTICAL INSTRUMENTS: SHADOW CAUSTICS [12–14]

The method of shadow caustics is relatively new, as it was developed in the last 25 years by Manogg, Theocaris, and Kalthoff. The principle of the method, illustrated in Fig. 10.26, is simple in concept. The incident light, which can be ordinary or coherent, is collimated to produce a system of parallel rays. These rays encounter a transparent specimen with its plane oriented so that the light is at normal incidence. When the specimen is subjected to in-plane stresses, a change in thickness $\Delta h(x, y)$ occurs which is given by the expression

$$\Delta h = -\frac{hv}{E}(\sigma_{xx} + \sigma_{yy}) \tag{10.42}$$

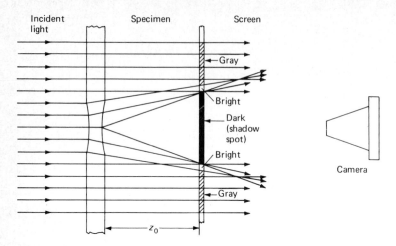

FIGURE 10.26
Optical system used to form the shadow spot.

Since $\sigma_{xx} + \sigma_{yy}$ varies with x and y, the front and back surfaces of the specimen deform as shown in Fig. 10.26. The deformed surfaces cause the light rays to deflect (like a lens), and upon exit from the specimen, the rays are not parallel. If a screen is placed at some location z_0 downstream from the specimen, the light rays produce an interesting optical pattern of gray, dark, and bright regions. The gray regions correspond to locations where the gradient of the sum of the in-plane stresses ($\sigma_{xx} + \sigma_{yy}$) is small and the parallel rays pass through the model without significant deflection. The dark areas give the shadow spot after which the method is named. These dark regions are the result of the deflection of the light rays from this local area on the screen. The light regions are due to the added intensity of the normal rays plus those deflected rays which impinge on this area of the screen. This pattern of gray, bright, and dark regions as it appears on the screen is photographed to provide an image that in effect describes the gradient ∇ of the sum of the in-plane stresses.

The optical system presented in Fig. 10.26 is the transmission arrangement used with transparent specimens. If the specimen is opaque, a similar arrangement can be employed to develop shadow spots if one surface of the specimen is mirrored. With the specimen acting as a mirror, the reflected light produces a pattern of bright, dark, and gray regions. However, the reflected pattern is observed as a virtual image which is formed on the opposite side of the specimen from that shown in Fig. 10.26.

10.8.1 Mapping

The mapping of the specimen plane π_s on the image plane π_i for a transparent model which forms a real image is shown in Fig. 10.27. The light ray deflection

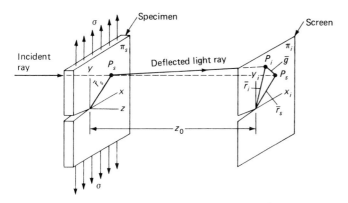

FIGURE 10.27
Coordinate system for developing the mapping equations.

is due to surface curvature which produces a divergent lens effect in the local region near the root of a notch. The point P_s is placed in this local region near the geometric discontinuity. The deflected light ray propagates down the optical bench a distance d_0 and impinges on a screen to form an image at point P_i. The location of point P_i on the image plane π_i is given by the vector $\bar{r}_i$, which is written as

$$\bar{r}_i = \bar{r}_s + \bar{g} \tag{10.43}$$

The movement of the point P_s on the specimen plane to P_i on the image plane is described by the vector $\bar{g}$, which is related to the screen position and the local slope of the specimen surface as

$$\bar{g} = -z_0 \nabla(\Delta s) \tag{10.44}$$

where $\nabla = \bar{i}(\partial/\partial x) + \bar{j}(\partial/\partial y) + \bar{k}(\partial/\partial z)$ is a vector operation and Δs is the change in the optical path length due to stress-induced changes in the thickness of the model and the index of refraction n. For a specimen with a uniform thickness h, the change in optical path length Δs is

$$\Delta s = (n - 1)\Delta h + h\,\Delta n \tag{10.45}$$

where the change in thickness Δh is given by Eq. (10.42) for plane-stress conditions ($\sigma_{zz} = 0$).

The changes in the index of refraction are related to the principal stresses by the Maxwell relations described in detail later in Sec. 12.3. By using the Maxwell equations, Δ_n is related to the stresses by

$$\Delta n_1 = c_1\sigma_1 + c_2\sigma_2$$
$$\Delta n_2 = c_1\sigma_2 + c_2\sigma_1 \tag{10.46}$$

where c_1 and c_2 are material-dependent optical constants. Substituting Eqs.

(10.42) and (10.46) into Eq. (10.45) yields

$$\Delta s_1 = h(a\sigma_1 + b\sigma_2)$$
$$\Delta s_2 = h(a\sigma_2 + b\sigma_1)$$

(10.47)

where

$$a = c_1 - \frac{(n-1)v}{E}$$

$$b = c_2 - \frac{(n-1)v}{E}$$

(10.47a)

It is possible to recast Eqs. (10.47) in terms of the sum and difference of the principal stresses as

$$\Delta s_1 = C_1 h[(\sigma_1 + \sigma_2) + C_2(\sigma_1 - \sigma_2)]$$ (10.48a)

$$\Delta s_2 = C_1 h[(\sigma_1 + \sigma_2) - C_2(\sigma_1 - \sigma_2)]$$ (10.48b)

where

$$C_1 = \frac{c_1 + c_2}{2} - \frac{(n-1)v}{E}$$

(10.49a)

$$C_2 = \frac{c_1 - c_2}{c_1 + c_2 - [2(n-1)v/E]}$$

(10.49b)

The constants c_1, c_2, n, C_1, and C_2 for a number of transparent polymers are presented in Table 10.4.

The fact that Eqs. (10.48) show two different relations for Δs indicates that a double caustic is formed in the image plane. The formation of the second caustic, due to Δs_2, is the result of the optical anisotropic nature of the model material

TABLE 10.4
Constants for transparent materials[†]

| Material | Elastic constants | | | General optical constants | | Shadow optical constants | |
	E, MN/m²	v	n	c_1, μm²/N	c_2, μm²/N	C_1, μm²/N	C_2,
Araldite B	3,660	0.392	1.592	−5.6	−62.0	−97.0	−0.288
CR-39	2,580	0.443	1.504	−16.0	−52.0	−120.0	−0.148
Plate glass	73,900	0.231	1.517	+0.32	−2.5	−2.7	−0.519
Homalite 100	4,820	0.310	1.561	−44.4	−67.2	−92.0	−0.121
PMMA[‡]	3,240	0.350	1.491	−5.0	−57.0	−108.0	0

[†] After Kalthoff, Ref. 14.
[‡] Optically isotropic.

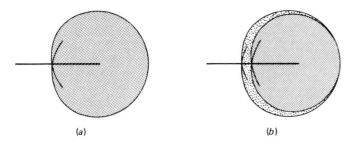

(a)   (b)

FIGURE 10.28
Single and double caustics. (*a*) Optically isotropic material. (*b*) Optically anisotropic (birefringent)
material.

with $\Delta n_1 \neq \Delta n_2$. When the model material is optically isotropic, $\Delta n_1 = \Delta n_2$ and
it is clear from Eqs. (10.46) and (10.49) that

$$c_1 = c_2 = c$$

$$C_1 = c - \frac{(n-1)v}{E} \qquad\qquad (10.50)$$

$$C_2 = 0$$

For optically isotropic materials it is possible to substitute Eqs. (10.50) into
Eqs. (10.48) to get

$$\Delta s = \Delta s_1 = \Delta s_2 = C_1 h(\sigma_1 + \sigma_2) \qquad\qquad (10.51)$$

These results show that a single caustic is formed since $\Delta s_1 = \Delta s_2$. An example
of the single and double caustic patterns formed by the surface deformation of a
specimen containing a crack is shown in Fig. 10.28.

10.8.2 Caustics due to Stress Singularities

The formation of an optical pattern in a caustic arrangement is dependent on the
gradient of the stresses [see Eq. (10.44)]. Higher gradients produce larger deflections of the light rays and images with distinguishing characteristics. This sensitivity of the method of caustics to stress gradients permits an effective approach
to stress analysis of plane bodies with stress singularities. The application described
here pertains to a plane body containing a crack with opening-mode loading;
however, the method can be applied to many other problems where singular
stresses occur.

Consider a plane specimen with a crack of length a loaded in the opening
mode as shown in Fig. 4.12. The point P_s is taken in the very near field where the
stresses are represented by Eqs. (4.36). Substituting Eqs. (4.36) into the stress

equations of transformation (1.11) and using Eq. (4.35) to relate K_1 with A_0 gives

$$\sigma_{rr} = \frac{K_1}{4\sqrt{2\pi r}} \left[5 \cos \frac{\theta}{2} - \cos \frac{3\theta}{2} \right]$$

$$\sigma_{\theta\theta} = \frac{K_1}{4\sqrt{2\pi r}} \left[3 \cos \frac{\theta}{2} + \cos \frac{3\theta}{2} \right] \qquad (10.52)$$

$$\tau_{r\theta} = \frac{K_1}{4\sqrt{2\pi r}} \left[\sin \frac{\theta}{2} + \sin \frac{3\theta}{2} \right]$$

The mapping of point P_s on the image plane is dependent on the specimen material. Consider an optically isotropic material which produces a single caustic. Substituting Eqs. (10.52) into Eqs. (10.43) and (10.44) leads to

$$x_i = r \cos \theta + \frac{K_1}{\sqrt{2\pi}} z_0 C_1 h r^{-3/2} \cos \frac{3\theta}{2}$$

$$y_i = r \sin \theta + \frac{K_1}{\sqrt{2\pi}} z_0 C_1 h r^{-3/2} \sin \frac{3\theta}{2} \qquad (10.53)$$

If the point P_s is moved over the very near field with many values of (r, θ), then a complete family of light rays is generated. These rays form a shadow "cone" downstream from the specimen plane. The surface of this "cone" is called the *caustic surface*, and the intersection of the caustic surface with the image plane produces the caustic curve. The caustic pattern for this case, presented in Fig. 10.29, shows the caustic curve as the line between the dark shadow zone and the bright band. This transition occurs when

$$\frac{\partial x_i}{\partial r} \frac{\partial y_i}{\partial \theta} - \frac{\partial x_i}{\partial \theta} \frac{\partial y_i}{\partial r} = 0 \qquad (10.54)$$

FIGURE 10.29
A caustic pattern produced by a crack in an optically isotropic material loaded in mode I. (*Courtesy of J. F. Kalthoff.*)

From Eqs. (10.53) and Eq. (10.54) we obtain a relation for r which locates the loci of points P_s on the specimen that corresponds to the caustic curve on the image plane. This relation

$$r = \left[\frac{3K_1}{2\sqrt{2\pi}} z_0 C_1 h \right]^{2/5} \equiv r_0 \tag{10.55}$$

shows that the initial curve is a circle with a radius r_0. Note that the location of the initial circle on the model can be varied by changing the position of the screen (z_0) or by changing the load (K_1).

For a fixed K_1 and z_0, we locate the initial circle according to Eq. (10.55). Substituting this value for r into Eq. (10.54) gives the description of the caustic curve as an image of the initial curve. Thus

$$x' = r_0 \left[\cos \theta + \tfrac{2}{3} \operatorname{sgn} (z_0 C_1) \cos \frac{3\theta}{2} \right]$$

$$y' = r_0 \left[\sin \theta + \tfrac{2}{3} \operatorname{sgn} (z_0 C_1) \sin \frac{3\theta}{2} \right] \tag{10.56}$$

where

$$\operatorname{sgn} (z_0 C_1) = \begin{cases} 1 & \text{if} (z_0 C_1) > 0 \\ 0 & \text{if} (z_0 C_1) = 0 \\ -1 & \text{if} (z_0 C_1) < 0 \end{cases}$$

Equations (10.56) generate a caustic curve which is classified as a generalized epicycloid. The maximum diameter D of the caustic and the radius r_0 of the initial circle are shown in Fig. 10.30. The diameter D is related to r_0 by

$$D = 3.17 r_0 \tag{10.57}$$

Finally, substituting Eq. (10.57) into Eq. (10.55) and solving for K_1 gives

$$K_1 = 0.0934 \frac{D^{5/2}}{z_0 C_1 h} \tag{10.58}$$

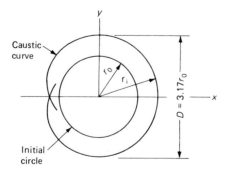

FIGURE 10.30
The initial circle and the caustic curve (mode I).

It is clear from Eq. (10.58) that the opening-mode stress intensity factor K_I can be determined simply by measuring the major diameter D of the caustic since z_0, C_1, and h are known parameters in a typical experiment. Care should be taken to ensure that $r_0 \geq h/2$ so that the plane-stress assumption made in developing all of the relations shown here is valid. If $r_0 < h/2$, then the radius of the initial curve is so close to the crack tip that measurement is made in a region where the stress state is three-dimensional. If $r_0 < h/2$, the use of Eq. (10.58) will give results for K_I which are much lower than the true values.

When the model material is optically anisotropic, two caustics are formed, as shown in Fig. 10.28. Both caustics, the inner and outer, provide diameters D_i and D_o. In this case two estimates of K_I are possible by using a relation similar to Eq. (10.58) which is written as

$$K_I = \frac{2\sqrt{2\pi}D_{o/i}^{5/2}}{3f_{o/i}^{5/2} z_0 C_1 h} \tag{10.59}$$

where $f_{o/i}$ is a linear function of the constant C_2 shown in Fig. 10.31. Note that

$$D_{o/i} = f_{o/i} r_0 \tag{10.60}$$

The restriction $r_o < h/2$ is also required for Eq. (10.59).

The advantage of the caustic method in experimental studies of fracture parameters such as the stress intensity factors K_I or K_{II} is its simplicity. The optical bench has relatively few and inexpensive components. The analysis to determine K_I involves knowledge of the experimental variables z_0, C_1, and h and the measurement of the major diameter D of the caustic.

A complete coverage of the method of caustics is beyond the scope of this text. We have attempted a logical extension by including several exercises which require the development of the mapping equations for the use of caustics in reflection. We have also included exercises dealing with the shearing mode in fracture mechanics. For a more complete treatment of this relatively new and interesting experimental method, the reader is referred to Ref. 14.

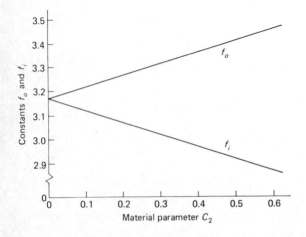

FIGURE 10.31
Numerical constants f_o and f_i used in Eq. (10.31).

EXERCISES

10.1. Verify Eqs. (10.11) and (10.12).

10.2. The wavelength of light from a helium-neon laser is 632.8 nm. Determine:
(a) The frequency of this light
(b) The wavelength of this light in a glass plate ($n = 1.522$)
(c) The velocity of propagation in the glass plate
(d) The linear phase shift (in terms of wavelength in free space) after the light has passed through a 25-mm-thick glass plate

10.3. Verify Eq. (10.15).

10.4. Show that the waves described by Eqs. (10.9) are the real part of the expression for $\bar{E}$ as given in Eq. (10.16).

10.5. Verify Eq. (10.17).

10.6. A plate of glass having an index of refraction of 1.57 with respect to air is to be used as a polarizer. Determine the polarizing angle and the angle of refraction of the transmitted ray.

10.7. Unpolarized light is directed onto a plane glass surface ($n = 1.57$) at an angle of incidence of 60°. Determine reflection coefficients associated with the parallel and perpendicular components of the reflected beam.

10.8. A light source is located 15 m below the surface of a body of water ($n = 1.33$). Determine the maximum distance (measured from a point directly above the source) at which the source will be visible from the air side of the air-water interface.

10.9. A ray of monochromatic light is directed at oblique incidence onto the surface of a glass plate. The ray emerges from the opposite side of the plate in a direction parallel to its initial direction but with a transverse displacement. Develop an expression for this transverse displacement in terms of the plate thickness h, the index of refraction of the glass n, and the angle of the incidence α of the light beam.

10.10. The radius of a concave spherical mirror is 500 mm. An object is located 1000 mm from the mirror. Determine the image location and the magnification. Show the results in a sketch similar to Fig. 10.13.

10.11. Solve Exercise 10.10 if the object is located 750 mm from the mirror.

10.12. Solve Exercise 10.10 if the object is located 100 mm from the mirror.

10.13. A concave mirror will be used to focus the image of an object onto a screen 1.50 m from the object. If a magnification of -3 is required, what radius of curvature must the mirror have?

10.14. Determine the image location and the magnification for an object located 800 mm from a mirror if the mirror is (a) a plane mirror and (b) a convex mirror with a radius of curvature of 1000 mm.

10.15. A convex mirror has a radius of curvature of 2500 mm. Determine the magnification and the image location for an object located 1000 mm from the mirror. Show the results in a sketch similar to Fig. 10.13.

10.16. A thin convex lens has a focal length of 600 mm. An object is located 900 mm to the left of the lens. Determine the image location and the magnification. Show the results in a sketch similar to Fig. 10.14.

10.17. Solve Exercise 10.16 if the object is located 300 mm from the lens.

10.18. An object is placed 500 mm to the left of a thin convex lens having a focal length of 250 mm. A second lens having a focal length of 300 mm is placed 600 mm to the right of the first lens. Determine the location and magnification of the resulting image. Show the results in a sketch similar to Fig. 10.16.

10.19. An object is located 225 mm to the left of a thin convex lens having a focal length of 300 mm. A second lens having a focal length of 250 mm is located 150 mm to the right of the first lens. Determine the location and magnification of the resulting image. Show the results in a sketch similar to Fig. 10.16.

10.20. An object is located 125 mm to the left of a thin convex lens having a focal length of 250 mm. A spherical concave mirror having a radius of curvature of 500 mm is located 1000 mm to the right of the lens. Determine the location and magnification of the image resulting from this lens-mirror combination. Show the results in a sketch similar to Fig. 10.16.

10.21. Two thin lenses having focal lengths f_1 and f_2 are placed in contact with one another. Develop an expression for the equivalent focal length of this series combination.

10.22. Verify Eqs. (10.28) through (10.31).

10.23. Parallel light of wavelength 546.1 nm is directed at normal incidence onto a slit having a width of 0.050 mm. A lens having a focal length of 1000 mm is positioned behind the slit and is used to focus the light passing through the slit onto a screen. Determine the distance (a) from the center of the diffraction pattern to the first diffraction fringe (minimum intensity) and (b) between the third and fourth diffraction fringes.

10.24. Solve Exercise 10.23 if light from a helium-neon laser is used which has a wavelength of 632.8 nm.

10.25. In a single-slit diffraction pattern the distance from the first minimum on the left to the first minimum on the right is 25 mm. The screen on which the pattern is displayed is 5 m from the slit. The wavelength of the light is 589.3 nm. Determine the slit width.

10.26. Verify Eq. (10.32).

10.27. If two linear polarizers are arranged in series with an arbitrary angle θ between their planes of polarization, plot the amplitude of the transmitted light as a function of θ as it varies from 0 to 90°.

10.28. What will be the relative angular retardation Δ in a quarter-wave plate designed for operation at $\lambda = 546.1$ nm if it is employed with (a) sodium light where $\lambda = 589.3$ nm, and (b) with a helium-neon laser where $\lambda = 632.8$ nm.

10.29. Show that two uniformly imperfect quarter-wave plates can be arranged in series with an appropriate angle between the two plates to produce the effect of a perfect quarter-wave plate. Solve for this angle by assuming that both imperfect quarter-wave plates have an angular retardation of $\Delta = \pi/2 + \epsilon$.

10.30. Determine the magnitude and direction of the light vector emerging from a series combination of a linear polarizer and a half-wave plate oriented at an arbitrary angle θ with respect to the plane of vibration of the linear polarizer.

10.31. Show that the four possible arrangements for a circular polariscope listed in Table 10.3 produce either a light field or a dark field.

10.32. Design the lens system for a 200-mm-diameter lens-type polariscope. Specify a range of magnification M available within the limits of the design parameters f_1, f_2, u_1, and u_2.

10.33. Light having a wavelength of 546.1 nm is directed at normal incidence onto a thin film of transparent material having an index of refraction of 1.63. Ten dark and nine bright fringes are observed over a 25-mm length of the film. Determine the thickness variation over this length.

10.34. A model is placed in a Mach-Zehnder interferometer, as shown in Fig. 10.25. The model is loaded after adjusting the reference and active beams to produce a null field on the screen. Upon loading, an interference pattern is observed with several fringe orders. Develop the relation for the change in thickness Δh of the model in terms of the fringe order m.

10.35. Continue Exercise 10.34 to show the relation for the sum of the principal stresses $(\sigma_1 + \sigma_2)$ in terms of the fringe order m.

10.36. Verify Eqs. (10.48) and then let the model material be optically isotropic and verify Eq. (10.51).

10.37. Beginning with Eqs. (4.36), verify Eqs. (10.52) and (10.53).

10.38. A caustic due to a stress singularity at a crack tip is formed on a screen located 1 m from a 4-mm-thick model fabricated from PMMA (plexiglas). The diameter D_0 of the shadow spot is 10 mm. Determine K_1 and then the radius r_0 of the initial curve.

10.39. If the screen in Exercise 10.38 is moved to a new position so that $z_0 = 800$ mm, determine the diameter D_0 of the shadow spot. Also, find the new radius r_0 of the initial curve.

10.40. For mode II loading of a crack tip, show that the mapping relation [similar to Eqs. (10.53)] is given by

$$x' = r \cos\theta - \frac{K_{\mathrm{II}}}{\sqrt{2\pi}} z_0 C_1 h r^{-3/2} \sin\frac{3\theta}{2}$$

$$y' = r \sin\theta + \frac{K_{\mathrm{II}}}{\sqrt{2\pi}} z_0 C_1 h r^{-3/2} \cos\frac{3\theta}{2}$$

10.41. Show that the crack-tip caustic curve for pure mode II loading is of the form shown in Fig. E10.41.

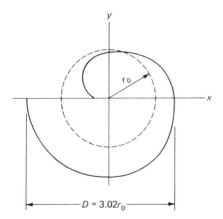

$D = 3.02 r_0$ **FIGURE E10.41**

10.42. Show for pure mode II loading that

$$r_0 = \left[\frac{3}{2}\frac{K_{\mathrm{II}}}{\sqrt{2\pi}} z_0 C_1 h\right]^{2/5}$$

10.43. Show for pure mode II loading that

$$K_{II} = \frac{2\sqrt{2\pi}}{3(3.02)^{5/2} z_0 C_1 h} D^{5/2}$$

Hint: Note that $D = 3.02 r_0$ as defined in Fig. E10.41.

10.44. Sketch the shape of a mixed-mode caustic.

REFERENCES

1. Born, M., and E. Wolf: *Principles of Optics*, Pergamon Press, New York, 1970.
2. Jenkins, F. A., and H. E. White: *Fundamentals of Optics*, 4th ed., McGraw-Hill, New York, 1976.
3. Shurcliff, W. A.: *Polarized Light*, Harvard University Press, Cambridge, Mass., 1962.
4. Grabau, M.: Optical Properties of Polaroid for Visible Light, *J. Opt. Soc. Am.*, vol. 27, pp. 420–424, 1937.
5. Land, E. H.: Some Aspects of the Development of Sheet Polarizers, *J. Opt. Soc. Am.*, vol. 41, pp. 957–963, 1951.
6. Jerrard, H. G.: The Calibration of Quarter-Wave Plates, *J. Opt. Soc. Am.*, vol. 42, pp. 159–165, 1952.
7. Tuzi, Z., and H. Oosima: On the Artificial Quarter Wave Plate for Photoelasticity Apparatus, *Sci. Pap. Inst. Phys. Chem. Res. (Tokyo)*, vol. 36, pp. 72–81, 1939.
8. Jessop, H. T.: The Optical System in Photo-Elastic Observations, *J. Sci. Instrum.*, vol. 25, p. 124, 1948.
9. Mindlin, R. D.: A Reflection Polariscope for Photoelastic Analysis, *Rev. Sci. Instrum.*, vol. 5, pp. 224–228, 1934.
10. Durelli, A. J.: discussion of paper entitled The Photoelastic Laboratory at the Newport News Shipbuilding and Dry Dock Company, *Proc. SESA*, vol. VI, no. 1, pp. 106–110, 1948.
11. Lee, R. R., R. Meadows, Jr., and W. F. Taylor: The Photoelastic Laboratory at the Newport News Shipbuilding and Dry Dock Company, *Proc. SESA*, vol. VI, no. 1, pp. 83–106, 1948.
12. Manogg, P.: Schattenoptische Messung der spezifischen Bruchenergie während des Bruchvorgangs bei Plexiglas, *Proc. Int. Conf. Phys. Non-Crystalline Solids, Delft, The Netherlands*, pp. 481–490, 1964.
13. Theocaris, P. S., and N. Joakimides, Some Properties of Generalized Epicycloids Applied to Fracture Mechanics, *J. Appl. Mech.*, vol. 22, pp. 876–890, 1971.
14. Kalthoff, J. F.: "Shadow Optical Method of Caustics," chap. 9 in A. S. Kobayashi (ed.), *Handbook on Experimental Mechanics*, Prentice-Hall, Englewood Cliffs, N.J., 1987, pp. 430–500.

CHAPTER
11

MOIRÉ
METHODS

11.1 INTRODUCTION [1–8]

The word "moiré" is the French name for a fabric known as watered silk, which exhibits patterns of light and dark bands. This moiré effect occurs whenever two similar but not quite identical arrays of equally spaced lines or dots are arranged so that one array can be viewed through the other. Almost everyone has seen the effect in two parallel snow fences or when two layers of window screen are placed in contact.

The first practical application of the moiré effect may have been its use in judging the quality of line rulings used for diffraction gratings or halftone screens. In this application, the moiré fringes provide information on errors in spacing, parallelism, and straightness of the lines in the ruling. All of these factors contribute to the quality of the ruling.

Elimination of the moiré effect has always been a major problem associated with screen photography in the printing industry. In multicolor printing, for example, where several screened images must be superimposed, the direction of screening must be carefully controlled to minimize moiré effects.

Considerable insight into the moiré effect can be gained by studying the relationships which exist between the spacings and inclinations of the moiré fringes in a pattern and the geometry of the two interfering line arrays which produced the pattern. This geometrical interpretation of the moiré effect was first published by Tollenaar [1] in 1945. Later Morse, Durelli, and Sciammarella [2] presented a complete analysis of the geometry of moiré fringes in strain analysis.

A second method of analysis, in which moiré fringes are used to measure displacements, was presented by Weller and Shepard [3] in 1948. Dantu [4] followed the same approach in 1954 and introduced the interpretation of moiré fringes as components of displacements for plane-elasticity problems. Sciammarella and Durelli [5] extended this approach into the region of large strains in a paper in 1961.

Moiré fringes have also been used by Theocaris [6, 7] to measure out-of-plane displacements and by Ligtenberg [8] to measure slopes and moment distributions in flat slabs. More recently, Post [9] has developed advanced methods using moiré interferometry, which greatly improves the sensitivity of the method. The previous discussion serves to illustrate the broad field of application of the moiré method in the determination of displacements and strains.

In the following sections of this chapter, both the geometrical and the displacement-field approaches to moiré fringe analysis will be outlined. The advantages and limitations of the approaches will be discussed. The more modern methods of moiré interferometry will be introduced at the end of the chapter.

11.2 MOIRÉ FRINGES PRODUCED BY MECHANICAL INTERFERENCE [2–5, 13]

The arrays used to produce moiré fringes may be a series of straight parallel lines, a series of radial lines emanating from a point, a series of concentric circles, or a pattern of dots. In stress-analysis work, arrays consisting of straight parallel lines (ideally, opaque bars with transparent interspaces of equal width) are the most commonly used. Such arrays are frequently referred to as *grids*, *gratings*, or *grills*. In this book, the term "grid" has been used to denote a coarse array (10 lines per inch or less) of perpendicular lines that does not produce a moiré effect (see Sec. 5.5). Image analysis techniques are normally employed to measure changes in spacing between the intersecting points of a grid network before and after loading, and from these data, displacements and strains can be determined. In this book, the term "grating" will be used to denote a parallel-line array suitable for moiré work (50 to 1000 lines per inch, or 2 to 40 lines per millimeter). When two perpendicular line arrays are used on a specimen, the term "cross-grating" will be employed.

When two gratings are overlaid, moiré fringes are produced. The overlaying can be accomplished by mechanical or optical means. In the following discussions, the two gratings will be referred to as the *model*, or *specimen*, *grating* and the *master*, or *reference*, *grating*. Quite frequently, the model grating is applied by coating the specimen with a photographic emulsion and contact-printing through the master grating. In this way, the model and master gratings are essentially identical (matched) when the specimen is in the undeformed state. Model arrays can also be applied by bonding, etching, ruling, etc.

A typical moiré fringe pattern, obtained using transmitted light through the model and master gratings, is shown in Fig. 11.1. In this instance, both the master grating and the model grating before deformation had 40 lines per millimeter (1000 lines per inch). The number of lines per unit length is frequently referred to as the *density of the grating*.

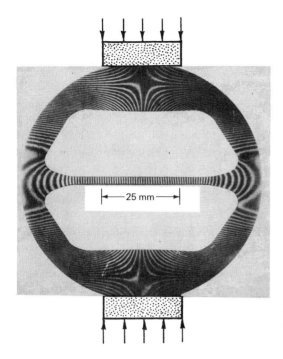

FIGURE 11.1
Moiré fringe pattern in a special tensile-strength specimen. Primary direction of grating was vertical. (*Courtesy of A. J. Durelli.*)

In the discussion which follows, the center-to-center distance between the master grating lines will be referred to as the *pitch of the grating* (reciprocal of the density) and will be designated by the symbol p. The center-to-center distance on the model grating in the deformed state will be designated by p'. The direction perpendicular to the lines of the master grating will be referred to as a *primary direction*. The direction parallel to the lines of the master grating will be referred to as a *secondary direction*.

The mechanism of formation of moiré fringes can be illustrated by considering the transmission of a beam of light through model and reference arrays, as shown in Fig. 11.2*a*. If the model and master gratings are identical, and if they are aligned such that the opaque bars of one grating coincide exactly with the opaque bars of the other grating, the light will be transmitted as a series of bands having a width equal to one-half the pitch of the gratings. However, due to the effects of diffraction and the resolution capabilities of the eye, this series of bands appears as a uniform gray field with an intensity equal to approximately one-half the intensity of the incident beam.

If the model is then subjected to a uniform deformation, like the one shown in the central tensile bar of Fig. 11.1, the model grating will exhibit a deformed pitch p', as shown in Fig. 11.2*b*. The transmission of light through the two gratings will now occur as a series of bands of different width, with the width of the band depending on the overlap of an opaque bar with a transparent interspace. If the intensity of the emerging light is averaged over the pitch length of the master

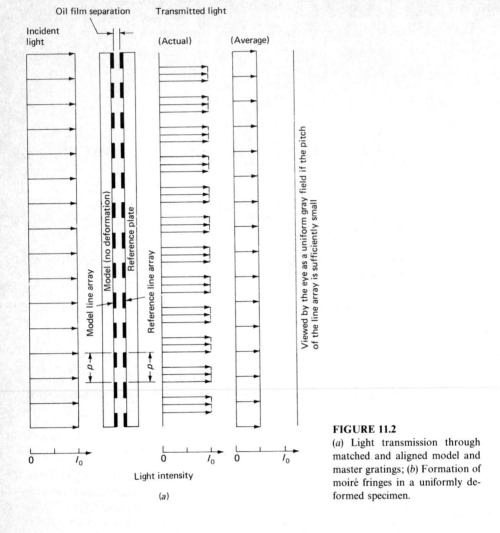

FIGURE 11.2
(a) Light transmission through matched and aligned model and master gratings; (b) Formation of moiré fringes in a uniformly deformed specimen.

grating, to account for diffraction effects and the resolution capabilities of the eye, the intensity is observed to vary as a staircase function of position. The peaks of the function occur at positions where the transparent interspaces of the two gratings are aligned. A light band is perceived by the eye in these regions. When an opaque bar of one grating is aligned with the transparent interspace of the other grating, the light transmitted is minimum and a dark band known as a *moiré fringe* is formed.

Inspection of the opaque bars in Fig. 11.2b indicates that a moiré fringe is formed each time the model grating undergoes a deformation in the primary direction equal to the pitch p of the master grating. Specimen deformations in the secondary direction do not produce moiré fringes. In the case illustrated in Fig.

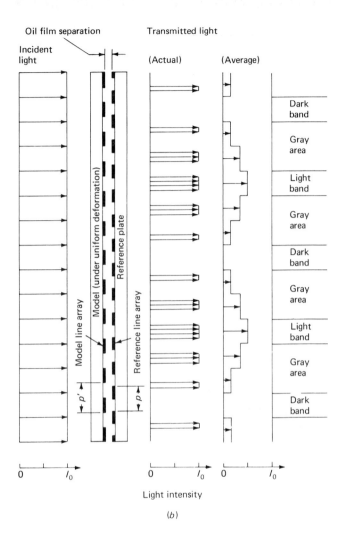

Oil film separation

Transmitted light

Incident light

(Actual)

(Average)

Dark band

Gray area

Light band

Gray area

Dark band

Gray area

Light band

Gray area

Dark band

Model line array

Model (under uniform deformation)

Reference plate

Reference line array

p'

p

0   l_0

0   l_0

0   l_0

Light intensity

(b)

11.1, 32 fringes have formed in the 25-mm gage length indicated on the specimen. Thus, the change in length of the specimen in this 25-mm interval is

$$\Delta l = np = 32(0.025) = 0.8 \text{ mm}$$

It should be noted in this illustration that the 25-mm gage length represents the final or deformed length of the specimen rather than the original length. Thus, the engineering strain in this interval could be expressed as

$$\epsilon = \frac{\Delta l}{l_0} = \frac{np}{l_g - np} = \frac{0.8}{25 - 0.8} = 0.033$$

The results of the previous observations can be generalized to write expressions for either tensile or compressive strains over an arbitrary gage length for cases involving uniform elongation or contraction but no rotation as

$$\epsilon = \begin{cases} +\dfrac{\Delta l}{l_0} = +\dfrac{np}{l_g - np} & \text{for tensile strains} \qquad (11.1) \\[3ex] -\dfrac{\Delta l}{l_0} = -\dfrac{np}{l_g + np} & \text{for compressive strains} \quad (11.2) \end{cases}$$

where p = pitch of master grating and undeformed model grating
n = number of moiré fringes in gage length
l_g = gage length

On the horizontal bar of Fig. 11.1, the moiré fringes were formed by elongations or contractions of the specimen in a direction perpendicular to the lines of the master grating. Simple experiments with a pair of identical gratings indicate that moiré fringes can also be formed by pure rotations (no elongations or contractions), as illustrated in Fig. 11.3. In this illustration, two gratings having a line density of 2.3 lines per millimeter have been rotated through an angle θ with respect to one another. Note that the moiré fringes have formed in a direction which bisects the obtuse angle between the lines of the two gratings. The relationship between angle of rotation θ and angle of inclination ϕ of the moiré fringes, both measured in the same direction and with respect to the lines of the master grating, can be expressed as

$$\phi = \frac{\pi}{2} + \frac{\theta}{2} \qquad (11.3)$$

or

$$\theta = 2\phi - \pi \qquad (11.4)$$

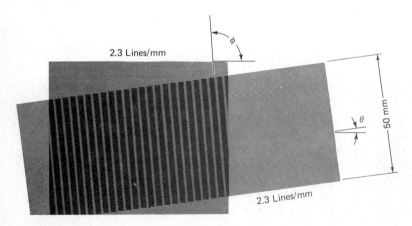

FIGURE 11.3
Moiré fringes formed by rotation of one grating with respect to the other.

11.3   THE GEOMETRICAL APPROACH TO MOIRÉ FRINGE ANALYSIS [2]

In Sec. 11.2, it was shown that moiré fringes are produced either by rotation of the specimen grating with respect to the master grating or by changes in pitch of the specimen grating as a result of load-induced deformations. At a general point in a stressed specimen, these two effects occur simultaneously and produce a fringe pattern similar to the one shown in Fig. 11.4. The information readily available from such a pattern is the angle of inclination ϕ of the fringes with respect to the lines of the master grating and the distance between fringes δ. The quantities to be determined at the point of interest are the angle of rotation θ of the specimen grating with respect to the lines of the master grating and the pitch p' of the specimen grating in the deformed state. This method of analysis, which is very convenient when information is desired at only a few selected points, has become known as the *geometrical approach*. Since this approach gives rotations and strains that are average values between two fringes, such analyses should be limited to uniform strain fields or to very small regions of nonuniform fields.

Relationships between the master grating pitch p, the deformed specimen grating pitch p', the angle of rotation θ of the specimen grating with respect to the lines of the master grating, and the data available from a moiré fringe pattern can be obtained from a geometric analysis of the intersections of the lines of the two gratings, as shown in Fig. 11.5a and b. In both parts of this figure the opaque bars of the gratings are defined by their centerlines, and light rather than dark bands are referred to as "fringes" since they are easier to locate accurately.

Consider first the distance δ between fringes as shown in Fig. 11.5a where it is clear that

$$\frac{p}{\sin \theta} = \frac{\delta}{\sin (\phi - \theta)} \tag{11.5}$$

Solving for the angle of rotation θ in terms of the pitch p of the master grating

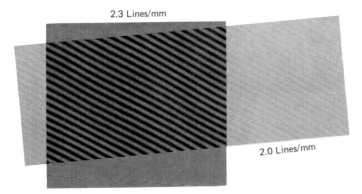

2.3 Lines/mm

2.0 Lines/mm

FIGURE 11.4
Moiré fringes formed by a combination of rotation and difference in pitch.

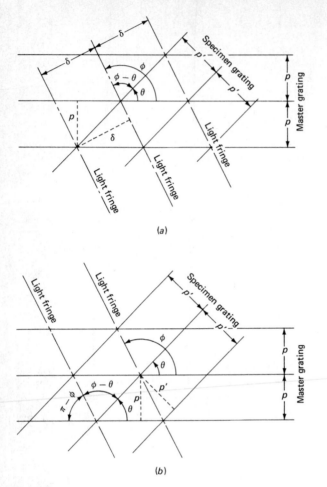

(a)

(b)

FIGURE 11.5
Geometry of moiré fringes: (a) in terms of fringe spacing δ. (b) in terms of fringe inclination φ.

and the quantities δ and φ which can be obtained from the moiré fringe pattern gives

$$\tan \theta = \frac{\sin \phi}{\delta/p + \cos \phi} \tag{11.6}$$

If gratings with a very fine pitch are used to measure small angles of rotation, then $\phi \approx \pi/2$ and Eq. (11.6) reduces to

$$\theta \approx \tan \theta = \frac{p}{\delta} \tag{11.7}$$

In a similar manner, it is observed in Fig. 11.5*b* that

$$\frac{p}{\sin(\pi - \phi)} = \frac{p'}{\sin(\phi - \theta)} \qquad (11.8)$$

which reduces to

$$p' = \frac{\delta \sin \theta}{\sin \phi} \qquad (11.9)$$

The angle θ can be eliminated from Eq. (11.9) by using the trigonometric identity

$$\sin \theta = \frac{\tan \theta}{\sqrt{1 + \tan^2 \theta}}$$

Then by using Eqs. (11.9) and (11.6), the deformed specimen pitch p' can be expressed in terms of the pitch p of the master grating and the quantities δ and ϕ as

$$p' = \frac{\delta}{\sqrt{1 + (\delta/p)^2 + 2(\delta/p)\cos\phi}} \qquad (11.10)$$

In many instances, moiré fringe patterns will be evaluated in regions where rotations are small. In these cases, $\phi \approx \pi/2$, and the binomial expansion with Eq. (11.10) reduces to

$$p' = \frac{p\delta}{p \pm \delta} \qquad (11.11)$$

Once the deformed specimen pitch p' has been determined, the component of normal strain in a direction perpendicular to the lines of the master grating can be written as

$$\epsilon = \frac{p' - p}{p} \qquad (11.12)$$

11.4   THE DISPLACEMENT-FIELD APPROACH TO MOIRÉ FRINGE ANALYSIS [3–5, 10–12]

Moiré fringe patterns can also be interpreted by relating them to a displacement field. In Sec. 11.2 it was shown that a moiré fringe is formed within a given gage length l_g in a uniformly deformed specimen (the central tensile bar of Fig. 11.1, for example) each time the specimen grating within the gage length is extended (or shortened) in a direction perpendicular to the lines of the master grating by an amount equal to the pitch p of the master grating. Deformations in a direction parallel to the lines of the master grating have no effect on the moiré fringe pattern.

The displacement-field concept can be extended to the general case of

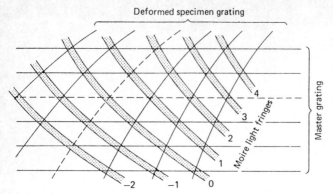

FIGURE 11.6
Moiré fringes at an arbitrary point in a stressed specimen.

extension or contraction combined with rotation. In Fig. 11.6 the lines of a deformed specimen grating are shown superimposed on a master grating. One of the lines of each of the gratings is shown dotted for easy reference, and it will be assumed that these lines coincided in the undeformed state. Both gratings are assumed to have had an initial pitch p. Interference between the master grating and the deformed specimen grating produces the light moiré fringes. In this pattern it should be noted that the intersection of the two dotted lines has been used to locate the zero-order fringe. This point on the specimen did not displace in the vertical direction as the specimen deformed. Similar intersections of specimen and master grating lines which initially coincided are also located on the zero-order fringe. The intersection of the dotted line on the specimen with the line above the dotted line on the master grating lies on the fringe of order 1. This point on the specimen has moved a distance p in the vertical direction from its original position. Similarly, a point lying on the fringe of order 4 has moved a distance $4p$ in the vertical direction from its original position. Thus, a moiré fringe is a locus of points exhibiting the same component of displacement in a direction perpendicular to the lines of the master grating.

For purposes of analysis, the moiré fringe pattern can be visualized as a displacement surface where the height of a point on the surface above a plane of reference represents the displacement of the point in a direction perpendicular to the lines of the master grating. A simple illustration of this concept is shown in Fig. 11.7. Similar patterns and displacement surfaces can be obtained for any other orientation of the specimen and master gratings.

Once u and v displacement surfaces have been established by using line arrays perpendicular to the x and y axes of a specimen, respectively, the cartesian components of strain can be computed from the derivatives of the displacements (slopes of the displacement surfaces). Two models are normally required for these determinations unless an axis of symmetry exists in the specimen so that mutually perpendicular arrays can be placed on the two halves of the specimen. For the case of large strains, the relationships between displacements and strains are given

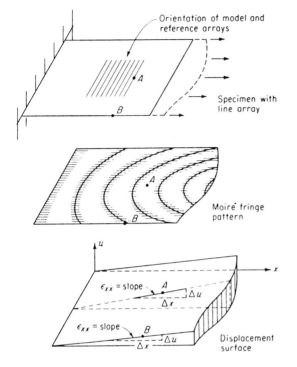

FIGURE 11.7
Moiré fringe pattern and associated displacement surface for a nonuniform strain field.

by Eqs. (2.2) and (2.3) as

$$\epsilon_{xx} = \sqrt{1 + 2\frac{\partial u}{\partial x} + \left(\frac{\partial u}{\partial x}\right)^2 + \left(\frac{\partial v}{\partial x}\right)^2 + \left(\frac{\partial w}{\partial x}\right)^2} - 1$$

$$\epsilon_{yy} = \sqrt{1 + 2\frac{\partial v}{\partial y} + \left(\frac{\partial v}{\partial y}\right)^2 + \left(\frac{\partial w}{\partial y}\right)^2 + \left(\frac{\partial u}{\partial y}\right)^2} - 1$$

$$\gamma_{xy} = \arcsin \frac{\dfrac{\partial u}{\partial y} + \dfrac{\partial v}{\partial x} + \dfrac{\partial u}{\partial x}\dfrac{\partial u}{\partial y} + \dfrac{\partial v}{\partial x}\dfrac{\partial v}{\partial y} + \dfrac{\partial w}{\partial x}\dfrac{\partial w}{\partial y}}{(1 + \epsilon_{xx})(1 + \epsilon_{yy})} \tag{11.13}$$

where u, v, and w are the displacement components in the x, y, and z directions, respectively. When products and powers of derivatives are small enough to be neglected, Eqs. (11.13) reduce to

$$\epsilon_{xx} = \frac{\partial u}{\partial x} \qquad \epsilon_{yy} = \frac{\partial v}{\partial y} \qquad \gamma_{xy} = \frac{\partial v}{\partial x} + \frac{\partial u}{\partial y} \tag{2.4}$$

The displacement gradients $\partial u/\partial x$ and $\partial v/\partial y$ are obtained from the slopes of the two displacement surfaces in a direction perpendicular to the lines of the master gratings. The displacement gradients $\partial u/\partial y$ and $\partial v/\partial x$ are obtained from the slopes of the displacement surfaces in a direction parallel to the lines of the master gratings. The displacement gradients $\partial w/\partial x$ and $\partial w/\partial y$ are not considered in moiré analysis of in-plane deformation fields.

In application of the moiré method, two moiré fringe patterns are usually obtained with specimen and master gratings oriented perpendicular to the x and y axes. A schematic illustration of one of these fringe patterns is shown in Fig. 11.8. Lines along the x and y axes, say AB and CD, are drawn, and displacements u and v along each of these lines are plotted by noting that

$$u, v = np \qquad (11.15)$$

where n is the order of the moiré fringe at the point and p is the pitch of the master grating. Tangents drawn to these curves give $\partial u/\partial x$ and $\partial u/\partial y$, as shown in Fig. 11.8. Similar moiré fringe patterns with an x-oriented model array are used to determine $\partial v/\partial x$ and $\partial v/\partial y$. Equations (11.13) or (2.4) can then be used to determine the strains ϵ_{xx}, ϵ_{yy}, and γ_{xy}. If lines similar to AB and CD are drawn at all critical areas of the specimen, a complete description of the strain over the entire field of the model can be achieved.

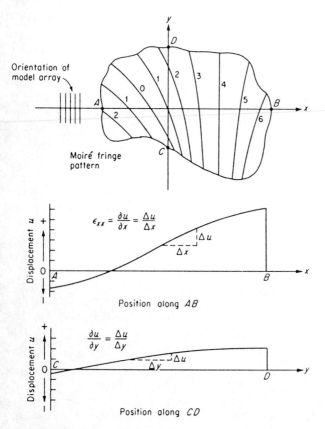

FIGURE 11.8
Displacement-position graphs used to determine $\partial u/\partial x$ and $\partial u/\partial y$.

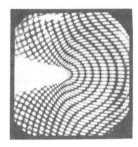

FIGURE 11.9
Moiré fringe pattern with crossed gratings of different pitch on the master and the specimen. (*Courtesy of D. Post.*)

The displacement-field approach to moiré fringe analysis, as previously outlined, is based on an accurate determination of the displacement derivatives $\partial u/\partial x$, $\partial u/\partial y$, $\partial v/\partial x$, and $\partial v/\partial y$. In practice, when the u and v moiré fringe patterns are obtained from separate models, from the two halves of a symmetric model, or with separate u and v master gratings on a crossed model grating (an array of orthogonal lines), the direct derivatives and $\partial u/\partial x$ and $\partial v/\partial y$ can usually be obtained with acceptable accuracy but the cross derivatives $\partial u/\partial y$ and $\partial v/\partial x$ cannot. The error in the cross derivatives is due to slight errors in alignment of either the specimen or master gratings with the x or y axes. Misalignment produces a fringe pattern due to rotation (see Fig. 11.3) in addition to the load-induced pattern.

One method proposed to eliminate shear-strain error makes use of crossed gratings on both the specimen and master to obtain simultaneous displays of the u and v displacement fields. Since any rotational misalignment is then equal for the two fields, its contribution to the cross derivatives is equal in magnitude but opposite in sign and thus cancels in the shear-strain determination.

In 1948, Weller and Shepard [3] recognized the possibility of displaying two moiré fringe patterns simultaneously with crossed gratings but recommended against its use because of the interweaving between the two families of fringes. Post [11] finally resolved this difficulty by using crossed gratings with slightly different pitches on the specimen and master to produce moiré fringe patterns with two distinct families, as illustrated in Fig. 11.9.

The problem of shear-strain errors can also be solved by using the strain-rosette concept commonly employed with electrical-resistance strain gages. With this method, model gratings are employed perpendicular to the x, n (usually $45°$ with respect to the x axis), and y axes. Once the three normal strains ϵ_{xx}, ϵ_n, and ϵ_{yy} are determined at a point, the complete state of strain at the point can be calculated by means of the rosette equations of Chap. 9.

11.5 OUT-OF-PLANE DISPLACEMENT MEASUREMENTS [6–7]

The moiré fringe methods discussed in the previous sections of this chapter have been concerned with the determination of in-plane displacements u and v, rigid-body rotations θ_z, and the strains ϵ_{xx}, ϵ_{yy}, and γ_{xy}. In select plane-stress problems

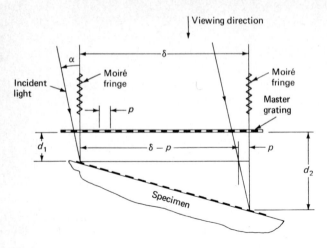

FIGURE 11.10
Moiré method for measuring out-of-plane displacements.

and in a wide variety of problems involving laterally loaded plates, out-of-plane displacements w become important considerations. A moiré method for determining out-of-plane displacements w has been developed by Theocaris [7, 8] and applied to a number of these problems. The essential features of the method are illustrated in Fig. 11.10.

For out-of-plane displacement measurements, a master grating is placed in front of the specimen, and a collimated beam of light is directed at oblique incidence through the master grating and onto the surface of the specimen. The shadow of the master grating on the surface of the specimen serves as the specimen grating. When the specimen is viewed at normal incidence, moiré fringes form as a result of interference between the lines of the master and the shadows. Use of a matte surface on the specimen ensures distinct shadows and improves the quality of the moiré fringe patterns.

From the geometry illustrated in Fig. 11.10 it can be seen that the difference in distance between the master grating and the specimen surface at two adjacent fringe locations can be expressed as

$$\Delta w = d_2 - d_1 = \frac{p}{\tan \alpha}$$

where p is the pitch of the master grating and α is the angle of incidence of the collimated light beam.

In practice, the master grating is located a small distance away from the specimen to accommodate any surface displacements toward the master grating and to serve as a datum plane for the measurement of load-induced, out-of-plane displacements. Any distribution of moiré fringes appearing with the master grating in this initial position will represent irregularities in the surface of the specimen.

The presence of any irregularity must be accounted for in the final determination of the out-of-plane displacements.

If a point of zero out-of-plane displacement is known to exist at some location on the specimen, e.g., from theoretical considerations, the master grating can be positioned to center a moiré fringe, identified as the *zero-order fringe*, over this point. At all other fringe locations, the out-of-plane displacement w can then be expressed as

$$w = \frac{np}{\tan \alpha} \qquad (11.16)$$

where n is the order of the moiré fringe at the point of measurement.

If no point exists in the specimen where $w = 0$, it will be necessary to measure the displacement at some convenient point by another experimental method. Then, the displacements at all other points on the surface can be referred to this point of reference.

11.6 OUT-OF-PLANE SLOPE MEASUREMENTS [9]

The determination of stress distributions and deflections in laterally loaded plates is a difficult but important engineering problem. From the theory of elasticity it is known that the stresses at a point in the plate due to bending moments can be expressed in terms of the local curvatures of the plate as

$$\sigma_x = \frac{Ez}{1 - v^2} \left(\frac{1}{\rho_x} + v \frac{1}{\rho_y} \right)$$
$$\sigma_y = \frac{Ez}{1 - v^2} \left(\frac{1}{\rho_y} + v \frac{1}{\rho_x} \right) \qquad (11.17)$$

The deflections are related to the curvatures by the approximate expressions

$$\frac{1}{\rho_x} = -\frac{\partial^2 w}{\partial x^2}$$
$$\frac{1}{\rho_y} = -\frac{\partial^2 w}{\partial y^2} \qquad (11.18)$$

In theory, the out-of-plane displacement-measuring technique discussed in Sec. 11.5 can provide the required curvatures for a solution to the stress problem. However, in practice, double differentiations cannot be performed with sufficient accuracy to provide suitable values for the curvatures. To overcome this experimental difficulty, Ligtenberg [9] has developed a moiré method for measuring the partial slopes $\partial w/\partial x$ and $\partial w/\partial y$. A single differentiation of these slopes then provides reasonably accurate values of the second derivatives needed to determine the curvatures, $1/\rho$.

The essential features of the Ligtenberg method are illustrated in Fig. 11.11.

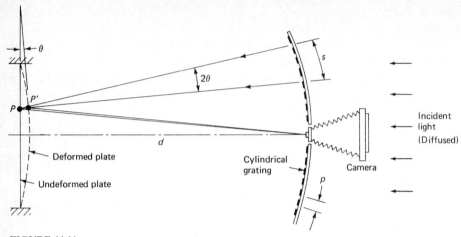

FIGURE 11.11
Moiré method for measuring out-of-plane slopes.

The equipment consists of a fixture for holding and loading the plate, a large cylindrical surface with a coarse line grating, and a camera for recording the moiré fringe patterns. The cylinder segment with the coarse grating is fabricated from a transparent sheet of plastic and the incident light passes through this shell. The surface of the plate is polished or coated to make it reflecting since the camera views the image of the grating on the surface of the plate. Because the image does not depend on the angle of incidence of the light, collimated light is not required for this method. The moiré fringe pattern is formed by superimposing grating images before and after loading. Double-exposure photography is useful for performing the superposition of images.

From the geometry of Fig. 11.11, it can be seen that the location on the grating being viewed by the camera, as a result of reflections from a typical point P on the surface of the specimen, shifts as the plate deflects under load. The moiré fringe pattern formed by the superposition of the two images provides a measure of this shift. Observe in Fig. 11.11 that the shift can be expressed in terms of the local slope of the plate as

$$s = 2\theta d \tag{11.19}$$

where s = magnitude of shift
θ = local slope of plate at point P'
d = distance between plate and grating

A moiré fringe will form upon superposition of the two images if the shift s is equal to the pitch p of the grating. Thus, the order of the moiré fringe can be expressed as

$$n = \frac{2\theta d}{p}$$

or

$$\theta = \frac{np}{2d} \qquad (11.20)$$

The separation distance d should be large to minimize the effects of out-of-plane displacements w on the shift distance s. Ligtenberg has also shown that the cylindrical grating should have a radius of the order of $3.5d$.

The angle θ given by Eq. (11.20) is the partial slope $\partial w/\partial x$ or $\partial w/\partial y$, depending on the orientation of the grating relative to the specimen. Two moiré patterns give the two slopes that are needed to solve the plate problem completely. The second pattern is obtained by rotating the grating $90°$ after the first pattern is recorded.

11.7 SHARPENING AND MULTIPLICATION OF MOIRÉ FRINGES [13–19]

Application of moiré methods to the study of deformations and strains in the elastic range of material response is usually limited by the lack of sensitivity of the method with respect to other methods, e.g., electrical-resistance strain gages. Both the geometrical and displacement-field approaches to moiré fringe analysis, which were discussed in Secs. 11.3 and 11.4, require accurate determinations of either the fringe spacings or the fringe gradients at the point of interest in the specimen. In a typical moiré fringe pattern obtained with line gratings having a density of 40 lines/mm or less, only a few fringes are normally present; therefore, the spacings or gradients cannot be established with the required accuracy. Since it is not practical to increase the line densities of the gratings much beyond 40 lines/mm when mechanical methods are used to form the moiré fringe pattern, several methods have been introduced to improve the sensitivity of the moiré method. Moiré-fringe-sharpening and moiré-fringe-multiplication methods, which enhance the sensitivity of mechanically formed images, will be discussed in this section. Moiré interferometry, a method which uses very-high-density gratings, will be covered in Sec. 11.8.

Moiré fringe sharpening and moiré fringe multiplication were first discussed by Post [14] in 1967. Moiré fringe sharpening permits more precise location of the position of the fringe and enhances the accuracies which can be achieved with a limited number of fringes. In previous descriptions of the formation of moiré fringe patterns, it was assumed that the master and model gratings were identical (before deformation) and that the opaque bars and transparent space in the gratings were the same width $p/2$. Post has shown that moiré fringe patterns can also be formed when the opaque bars and the transparent spaces in the gratings do not have equal widths and that under certain conditions such gratings can be used to produce very desirable effects. For example, take the case where one grating has wide opaque bars and narrow transparent spaces while the other one has narrow opaque bars and wide transparent spaces. Such gratings are said to be *complementary* if the ratio R (opaque width to transparent width) of one of the

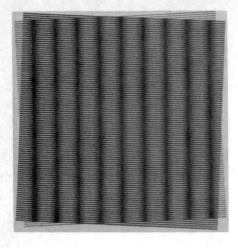

FIGURE 11.12
Moiré fringe sharpening with complementary
gratings. (*Courtesy of D. Post.*)

gratings is the reciprocal of the other. A typical example of a sharpened moiré
fringe pattern, obtained with complementary gratings, is shown in Fig. 11.12. The
mechanism of formation of the sharpened fringes is illustrated in Fig. 11.13, where
the intensity of light transmitted through the pair of gratings is plotted as a
function of position along a line which is perpendicular to the moiré fringes. This
type of intensity distribution is characteristic of complementary gratings.

The methods of moiré fringe formation and sharpening discussed previously
are often called *geometrical* or *mechanical moiré* since the formation of the moiré
pattern is based on a simple geometric-optics (or ray-optics) treatment of the
passage of light through superimposed gratings. Other methods of fringe sharpening and fringe multiplication introduced by Holister [14], and developed by Post
[15, 16], Sciammarella [17], and Chiang [18], utilize the diffraction effects
associated with light passage through the gratings.

The diffraction of light from a narrow slit was considered in Sec. 10.5. When
light consisting of a series of plane wavefronts is used to illuminate a grating, each
of the transparent spaces in the grating acts as a slit. In accordance with Huygens'

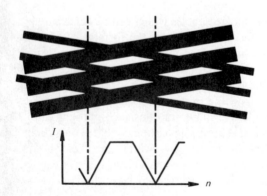

FIGURE 11.13
Mechanism of formation of sharpened
fringes with complementary gratings, $R = 2$
and $\frac{1}{2}$.

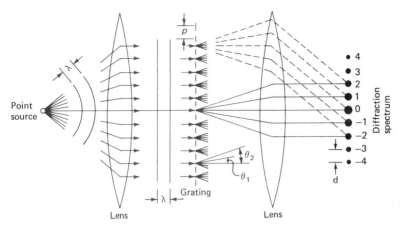

FIGURE 11.14
Mechanism of formation of the diffraction spectrum of a grating.

principle, the transparent space in the grating emits light in the form of a series of secondary cylindrical wavelets. A number of wavefronts are possible and are referred to as the *different diffraction orders*. For example, the zero-order beam is formed by wavelets which have identical phase. This wavefront is parallel to the incident plane wavefront and propagates along the axis of the optical system as shown in Fig. 11.14. The first-order beams, which are deflected to both sides of the zero-order beam, are formed by adjoining wavelets which are out of phase by one wavelength. Similarly, the second-order beams are formed by adjoining wavelets which are out of phase by two wavelengths, and so on. The diffraction of a plane wavefront by a grating which contains many slits is illustrated in Fig. 11.14. The rays emerging from each of the slits of the grating are associated with the different diffraction orders. The image formed by collecting all the diffraction orders from the grating with a lens is known as the *diffraction spectrum* of the grating. A photograph of an actual diffraction spectrum of a 12 line/mm grating is shown in Fig. 11.15. The dots of the diffraction spectrum are images of the point

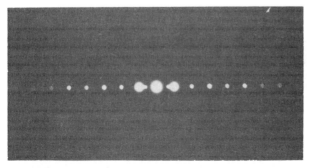

FIGURE 11.15
The diffraction spectrum of a 12 line/mm grating. (*Courtesy of Fu-pen Chiang.*)

source, and the intensity of the light associated with a dot decreases with the diffraction order.

Since the wavelets forming the first-order beam are out of phase by one wavelength, the angle θ_1 defined in Fig. 11.14 can be determined from the expression

$$\sin \theta_1 = \theta_1 = \frac{\lambda}{p}$$

since the angles are small. Similarly for the other diffraction orders

$$\sin \theta_n = \theta_n = \frac{n\lambda}{p} \tag{11.21}$$

The distance d between any two dots is then given simply as

$$d = f\theta_1 = \frac{f\lambda}{p} \tag{11.22}$$

where f is the focal length of the second lens.

One of the methods used to produce moiré fringe sharpening and moiré fringe multiplication is based on the diffraction phenomenon illustrated in Fig. 11.14. When both a model grating and a master grating are inserted in series in the collimated light beam between the two lenses, the plane wavefronts generated by the first grating are diffracted again by the second grating to produce a diffraction spectrum associated with the superimposed pair. If the complete diffraction spectrum is collected with a camera lens (an ideal situation not realized in practice), the image recorded by the camera will be an exact reproduction of the grating pair and their moiré pattern. However, if all the diffraction orders are not collected by the camera lens, the image recorded on the film plane will be modified. The theory associated with these modifications is beyond the scope of this book, but images recorded with certain individual diffraction orders can be shown to provide both moiré fringe sharpening and moiré fringe multiplication.

An optical system used for moiré fringe sharpening and moiré fringe multiplication is illustrated in Fig. 11.16. Note the presence of the light stop, or aperture, which is used to isolate the particular diffraction order being permitted to enter the camera lens. A selection of photographs recorded using different diffraction orders is shown in Fig. 11.17. The first photograph shows the image recorded if a large number of diffraction orders is permitted to enter the camera. The moiré fringes and the lines of the gratings are visible. This pattern is similar to the ones previously shown, which were obtained with two coarse gratings in contact using diffused light (mechanical moiré). The second photograph shows the image recorded with the zero diffraction order. The moiré fringes in this image have been broadened and the grating lines have been eliminated. The third photograph shows the image produced by the $+1$ diffraction order. In this instance, the moiré fringes are sharpened and the lines of the gratings remain extinguished. It can be shown that at least two diffraction orders must enter the camera before a crude gratinglike image is produced. In the fourth, fifth, and sixth photographs, the images produced by the $+2$, $+3$, and $+5$ diffraction orders,

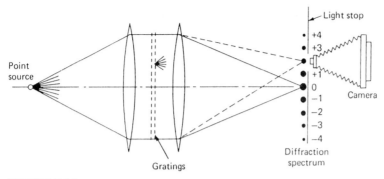

FIGURE 11.16
Basic optical system for moiré fringe multiplication.

respectively, are shown. In each of these photographs, moiré fringe multiplication (2×, 3×, and 5×, respectively) is indicated. Higher levels of fringe multiplication can be achieved with the higher diffraction orders, but the intensity of light associated with these orders is very small.

In utilizing the diffraction method for forming the moiré images and increasing the sensitivity, light rays from several different diffraction orders are combined, as indicated in Fig. 11.18. In this figure three different order groups (of many possible groups) are shown focused at the first spot. In a similar manner, multiple beams from several different diffraction orders are combined at each of

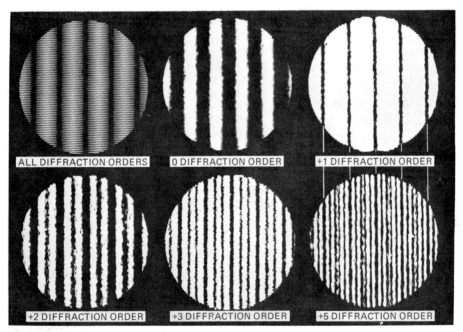

FIGURE 11.17
Moiré patterns produced by different diffraction orders. (*Courtesy of A. J. Durelli.*)

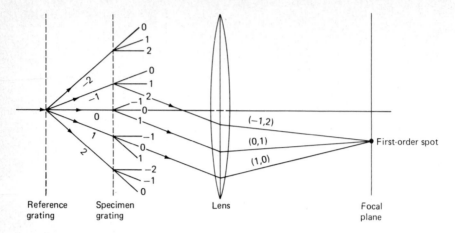

FIGURE 11.18
The ray optics representation of the formation of order groups by diffraction from two gratings in series.

the focus spots. The apparent fringe multiplication depends on the diffraction order of focus spots, as indicated in Fig. 11.17. Recently, Graham [20] has studied the effect of imperfections of the diffraction gratings and showed that if both the specimen and reference gratings have a 50 percent transmission ratio (i.e., the line width and space width are equal) then the added fringes are correctly positioned between the original (nonmultiplied) fringes. An intensity-position graph for two series gratings with 50 percent transmission and a multiplication factor of 5, presented in Fig. 11.19, shows that the four additional fringes due to multiplication are uniformly spaced between the first-order fringes. However, if the transmission ratio of the gratings is different from 50 percent, the fidelity of the multiplication is impaired and error is introduced even in the location of the first-crder fringes.

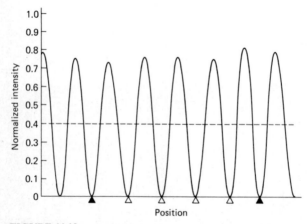

FIGURE 11.19
Interference pattern for two gratings with a transmission ratio of 50 percent and a multiplication factor of 5. △, Locations of fringes due to multiplication; ▲, locations of first-order fringes.

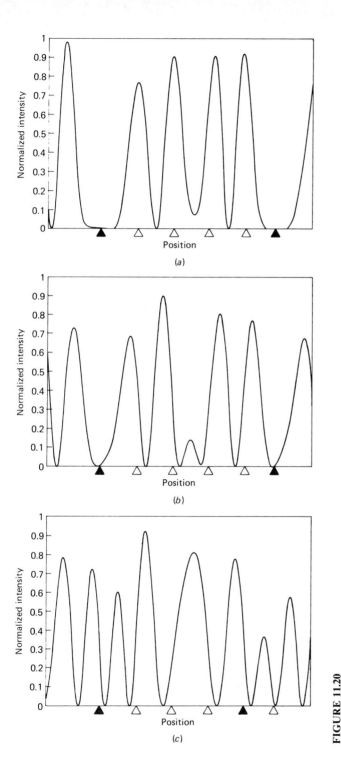

FIGURE 11.20
The influence of transmission ratio on the moiré fringe locations for two gratings with a fringe multiplication factor of 5 and (*a*) a transmission ratio of 40 percent; (*b*) a transmission ratio of 55 percent; (*c*) a transmission ratio of 45 percent. △, theoretical locations of fringes due to multiplication; ▲, locations of first-order fringes.

Intensity-position curves for grating pairs with 40, 45, and 55 percent transmission are shown in Fig. 11.20. Note first that the periodic sinusoidal form of these curves has been seriously distorted. More importantly, the positions of minimum intensity, which locate the fringes, are not at the theoretical locations. The effect of imperfections (even uniform deviations in the transmission ratio) is to produce serious errors in location of the zero-order and the higher-order moiré fringes. One observes the correct number of fringes but they are not at the correct positions. Indeed, a close examination of the third and fifth diffraction order fringes in Fig. 11.17 shows lack of uniformity in fringe width due to the distortion in the sinusoidal intensity-position relation. Also, the fringe spacing of the third and fifth diffraction orders does not represent a uniform displacement field as it should.

11.8 EXPERIMENTAL PROCEDURE AND TECHNIQUE [21–26]

Master gratings for moiré work with 12, 20, 40, and 80 lines/mm are readily available from several commercial suppliers. Indeed, with the availability of high-quality photoplotters used extensively in the electronics industry, custom master gratings can be procured at relatively low cost. Because of the fragility of the master grating (glass) and the risk involved in its use, high-quality duplicates are usually made on film for routine stress-analysis work. If proper photographic procedures are followed, little difficulty is encountered in producing good-quality duplicates on high-resolution film.

Printing or etching the grating on the model, on the other hand, is often difficult. Work on the development of techniques to apply gratings to the various model materials is constantly in progress, and some interesting procedures have been developed for specific applications. Two factors are extremely important in the production of a satisfactory model array: (1) The photographic emulsion must adhere well to the model material and have a high resolution; (2) the exposure must be carefully controlled to produce an array with proper line width and spacing from the master array.

Since materials, techniques, and procedures associated with moiré work are constantly being improved, no attempt will be made to outline specific procedures. For the latest materials, procedures, and techniques, the interested reader should consult the current technical literature. In particular, the papers by Holister and Luxmoore [21], Chiang [22], and Zandman [23] will in general be helpful. The books on moiré by Durelli and Parks [24] and Theocaris [25] and the handbook edited by A. S. Kobayashi [26] should be consulted for more detail on many of the topics discussed in this chapter.

11.9 MOIRÉ INTERFEROMETRY [27–32]

11.9.1 Two-Beam Interference

In Sec. 10.2.2, the superposition of two waves with the same frequency ω propagating along the z axis was described. It was shown that the two waves combine and that the intensity of light, when the amplitudes a_1 and a_2 of the two waves are the same, is

$$I = 4a_1^2 \cos^2 \frac{\pi\delta}{\lambda} \qquad (10.13)$$

The ratio of δ/λ, the relative phase, indicates the position of one wave with respect to the other. When the ratio equals an integer value n,

$$\frac{\delta}{\lambda} = n \qquad n = 0, 1, 2, \ldots \qquad (11.23a)$$

then $I = 4a_1^2$, which is the maximum intensity. In this case, the waves reinforce one another and the interference effect is constructive. However, if the ratio δ/λ is

$$\frac{\delta}{\lambda} = \frac{n+1}{2} \qquad n = 0, 1, 2, \ldots \qquad (11.23b)$$

then $I = 0$, which is the minimum intensity. In this instance, the waves acted to cancel one another and the interference effect is destructive. Since both of the waves are parallel to the z axis, the interference is the same at every point in space (x, y, z) where both beams of light exist.

A second example of two-beam interference involves two nonparallel beams (either diverging or intersecting) as illustrated in Fig. 11.21. The two beams are produced by a single coherent light source and emerge from the mirror arrangement, used to orient the beams, with the same phase ($\delta = 0$). At some point z_0 downstream from the mirror arrangement, the two waves interfere with a phase difference δ that is given by the equation

$$\delta = 2y \sin \theta \qquad (11.24)$$

When δ/λ is an integer value, constructive interference occurs and a light fringe is formed. On the other hand, when δ/λ is a half-order value, destructive interference occurs and a dark fringe is formed. The resulting interference fringe pattern is an array of alternating light and dark fringes, as shown in Fig. 11.22.

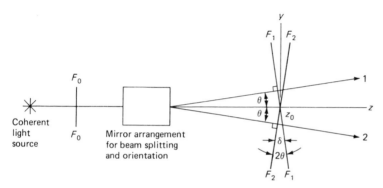

FIGURE 11.21
Interference of two wave fronts F_1 and F_2 associated with diverging beams.

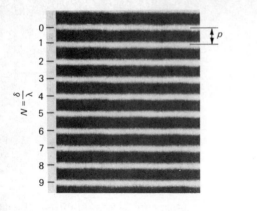

FIGURE 11.22
Interference fringe pattern produced by intersecting wavefronts F_1 and F_2. (*Courtesy of D. Post.*)

The distance p between fringes is determined from Eq. (11.24) by noting that $y = p$ when $\delta = \lambda$ so that

$$p = \frac{\lambda}{2 \sin \theta} \tag{11.25}$$

The fringe gradient is representative of a spatial frequency f given by

$$f = \frac{1}{p} = \frac{2 \sin \theta}{\lambda} \tag{11.26}$$

where f is expressed in terms of lines/mm.

The two intersecting beams are used to produce high-frequency gratings which are employed as the specimen gratings in moiré interferometry. The optical arrangement recommended by Post [27] is shown in Fig. 11.23. Since the frequency of the interference pattern is very high (1000 to 3000 lines/mm), production of the gratings is far from routine. An excellent vibration isolation table is required to

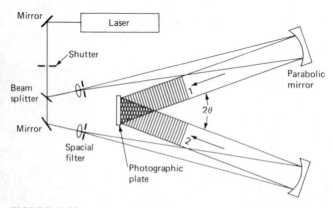

FIGURE 11.23
Optical arrangement used to produce high-frequency moiré gratings.

maintain stability of the optical elements during the exposure interval. Also, forming the image of the grating on the film plane requires a very-high-quality process lens. Finally, glass plates with a very-high-resolution emulsion must be used to record the images.

The theoretical limit of the frequency f_t of the interference pattern produced with the two intersecting beams is given by Eq. (11.26) as

$$f_t = \frac{2}{\lambda} \tag{11.27}$$

Thus, the theoretical frequency f_t depends on the wavelength of the light source employed. For a helium-neon laser with $\lambda = 632.8$ nm, $f_t = 3160$ lines/mm, and for an argon laser with $\lambda = 488.0$ nm, $f_t = 4098$ lines/mm. It is not possible to achieve the theoretical frequency because of difficulties encountered in forming the interference pattern when $\theta \rightarrow \pi/2$ and the two beams interfere at grazing incidence. Post has achieved 97.6 percent of the theoretical limit (4000 lines/mm) by using $\theta = 77.4°$ with $\lambda = 488.0$ nm.

11.9.2 Specimen Gratings

Moiré interferometry requires a specimen grating which is similar to the grating used in mechanical moiré. The primary difference between these two gratings is in the frequency or pitch of the line arrays employed. With moiré interferometry, the frequency of the grating is usually in the range from 1000 to 2000 lines/mm, whereas with mechanical moiré, the frequency rarely exceeds 80 lines/mm.

The placement of these very-high-frequency gratings on the surface of a specimen is accomplished by using a replication technique. The grating is first produced on a glass photographic plate by using the optical arrangement described in Fig. 11.23. In developing the photographic emulsion, the alternating bands of exposed and unexposed silver halide produce a wave pattern which is due to nonuniform shrinkage of the gelatin-based emulsion (see Fig. 11.24). The waved surface of the photographic plate is coated with a very thin coating of either gold or aluminum using a high-vacuum deposition process.

The metal-coated photographic plate serves as a mold in the replication process used to transfer the grating to the specimen surface. This process, illustrated in Fig. 11.25, involves pressing a small quantity of an adhesive (epoxy) between the mold and the specimen. After the adhesive has cured, the mold is stripped from the specimen surface. The gelatin-metal interface is the weakest link

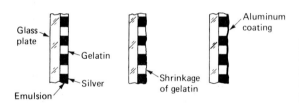

Glass plate
Gelatin
Silver
Emulsion
Shrinkage of gelatin
Aluminum coating

FIGURE 11.24
Behavior of gelatin-based photographic emulsion which produces a surface wave pattern representing the interference fringe pattern.

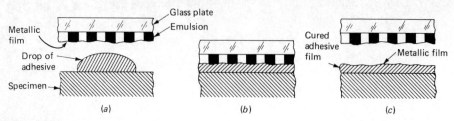

FIGURE 11.25
(a), (b), and (c) Replication process used in transferring the grating with a reflective metallic film to the specimen.

in the mold-specimen bonding chain, and separation occurs at this interface. The result of the replication process is a thin film (about 0.025 mm thick) with a reflective high-frequency phase-type diffraction grating as the exposed surface bonded to the surface of the specimen.

11.9.3 Moiré Interference

Moiré interferometry is identical to mechanical moiré in the sense that a moiré image is formed by light passing through two gratings—one on the specimen and the other, a reference grating, usually located adjacent to the specimen. The difference in moiré interferometry, in addition to the frequency of the gratings discussed previously, is with the reference grating. In mechanical moiré, the reference grating is real, consisting of an array of lines on a sheet of film placed on contact with the specimen grating as illustrated in Fig. 11.2a. With moiré interferometry, the reference grating is imaginary, consisting of a virtual image of an interference pattern produced with mirrors.

An optical arrangement that is used to develop the virtual reference grating is shown in Fig. 11.26. The frequency of the reference grating f_r follows directly from the derivation of Eq. (11.26) which gives

$$f_r = 2 \sin \frac{\alpha}{\lambda} \tag{11.28}$$

where α is the angle of the incident light.

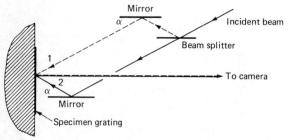

FIGURE 11.26
Optical arrangement for developing a virtual reference array.

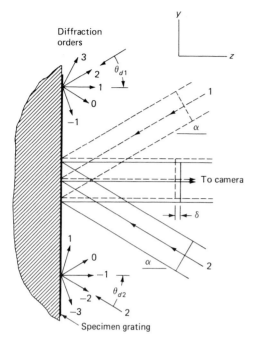

Diffraction
orders

Specimen grating

FIGURE 11.27
Diffraction of the light beams as they interact
with the specimen diffraction grating.

A more detailed description of a moiré interferometer is illustrated in Fig.
11.27. The two beams of the interferometer are incident to the surface of the
specimen at symmetrical angles $\pm\alpha$. The beams are reflected from the diffraction
grating of frequency f_s which has been replicated on the surface of the specimen.
Upon reflection, the light is diffracted and several diffraction orders are produced,
as indicated in Fig. 11.27. It is important to consider select combinations of these
diffraction orders because emerging light rays parallel to the z axis are combined
to produce interference which gives the moiré fringe pattern.

Reference to Eq. (10.34a) shows that the angle θ_d between the incident ray
and the reflected ray associated with the first diffraction order is

$$\sin \theta_d = \frac{\lambda}{b} \tag{11.29a}$$

In this case, the width b of the slit in the diffraction grating is equivalent to the
pitch p in the specimen grating. Recall that $f_s = 1/p$ and rewrite Eq. (11.29a) to give

$$\sin \theta_d = f_s \lambda \tag{11.29b}$$

Furthermore, if

$$f_s = \frac{f_r}{2} \tag{11.30a}$$

then it is clear from Eqs. (11.28) and (11.29b) that

$$\theta_{d1} = \theta_{d2} = \alpha \tag{11.30b}$$

Equation (11.30b) shows that the first diffraction order from incident beam (1) propagates along the z axis. Similarly, the negative first diffraction order from incident beam (2) also propagates along the z axis. These two parallel beams produce a field of uniform intensity—the null field.

Next, consider that the specimen is subjected to a strain ϵ_{yy}. This strain deforms the diffraction grating on the specimen and the new frequency f'_s becomes

$$f'_s = \frac{f_s}{1 + \epsilon_{yy}} \tag{11.31}$$

With the new frequency of the specimen grating, the angles θ_{d1} and θ_{d2} are changed. From Eq. (11.29b) it is evident that

$$\sin \theta_{d1} = \sin \theta_{d2} = \frac{f_s \lambda}{1 + \epsilon_{yy}} \tag{11.32}$$

These two beams make an angle $\pm \theta$ relative to the z axis where

$$\theta = \alpha - \theta_d \tag{11.33}$$

If it is assumed that α and θ_d are small so that $\sin \alpha \approx \alpha$ and $\sin \theta_d \approx \theta_d$, then substituting Eqs. (11.32) and (11.28) into Eq. (11.33) gives

$$\theta = \frac{\lambda f_r \epsilon_{yy}}{2} \tag{11.34}$$

This angle θ of the emerging beam is the same as that shown in Fig. 11.21. Indeed, the two beams emerge as divergent or intersecting beams and produce an interference pattern like the one illustrated in Fig. 11.22. The fringe gradient $\partial N / \partial y$ is the same as the frequency f_m of this interference (moiré) pattern that is given by Eq. (11.26) as

$$\frac{\partial N}{\partial y} = f_m = \frac{2 \sin \theta}{\lambda} \tag{11.35}$$

Since θ is small, Eq. (11.35) can be combined with Eq. (11.34) to give

$$\epsilon_{yy} = \frac{1}{f_r} \frac{\partial N}{\partial y} = \frac{f_m}{f_r} \tag{11.36}$$

This result shows that the strain can be determined from the frequency of the moiré pattern relative to the frequency of the reference pattern. Equation (11.36) can also be written in terms of pitch as

$$\epsilon_{yy} = \frac{p_r}{p_s} \tag{11.37}$$

where p_s is the distance between two adjacent fringes with orders N and $N \pm 1$ on the moiré fringe pattern.

In practice, the angle θ is very small. For example, with $\lambda = 488$ nm, $f_r = 2000$ lines/mm, and $\epsilon_{yy} = 0.002$, Eq. (11.34) gives $\theta = 0.0559°$. The corresponding moiré fringe pattern frequency or fringe gradient is 4 lines/mm. This example shows the validity of the small-angle assumption and clearly indicates the enhancement of the sensitivity of the moiré method by using very-high-density diffraction gratings. Indeed, moiré fringe patterns developed using interferometry methods often have fringe densities so high that they are difficult to reproduce using conventional lithographic techniques.

When the specimen is subjected to a general state of deformation, where $\bar{u} = \bar{u}(x, y)$, the frequency of the specimen grating changes over the entire field. The previous analysis is valid for any given point $P(x, y)$ in the field if the gradient of the fringes $\partial N/\partial y$ can be determined at the point. Generally, both the u and v displacement fields are required to provide sufficient data for determining the complete strain field $\epsilon_{xx}(x, y)$, $\epsilon_{yy}(x, y)$, and $\gamma_{xy}(x, y)$ as described in Sec. 11.4.

Post [27] has used a four-beam interferometer with a crossed diffraction grating (see Fig. 11.28) to produce moiré fringe patterns for both the u and v displacement fields. The light in the incident beam is partitioned into four regions, as indicated in the illustration. Light from regions A_1 and A_2 is reflected from mirrors A_1 and A_2 onto the specimen with angles of incidence of $\pm\alpha$. These two beams combine to form a virtual reference grating with its lines parallel to the x axis. This reference grating interacts with the lines parallel to the x axis in the specimen grating to form the moiré pattern representative of the v displacement field. The light from regions B_1 and B_2 of the incident beam reflects from mirrors B_1 and B_2 to form a reference grating with lines parallel to the y axis. This reference grating interacts with lines parallel to the y axis in the specimen grating to form the moiré pattern which gives the u displacement field.

The two moiré fringe patterns are obtained separately with two different exposures. On the first exposure the light over regions B_1 and B_2 is blocked and on the second exposure the light over regions A_1 and A_2 is blocked. On each

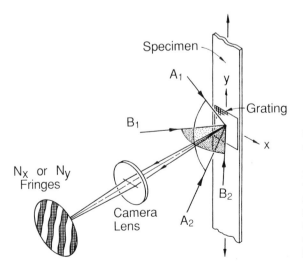

FIGURE 11.28
Schematic diagram of a four-beam moiré interferometer used to produce u and v moiré fringe patterns. Beams A_1 and A_2 create a virtual reference grating which interacts with the horizontal lines of the specimen grating to create the N_y fringe pattern. Beams B_1 and B_2 interact with the vertical lines to create the N_x fringe pattern. (*Courtesy of D. Post.*)

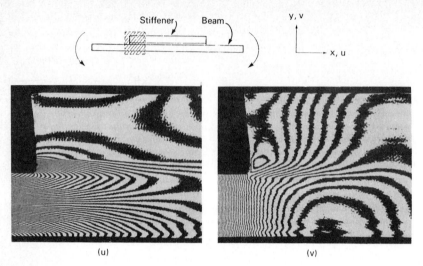

FIGURE 11.29

Moiré fringe patterns N_x and N_y showing the in-plane u and v displacement fields in the critical corner region of a bonded joint between two quasi-isotropic, graphite-epoxy members subjected to bending. $f = 2400$ l/mm. (*Courtesy of D. Post.*)

exposure the camera is adjusted to focus the image of the moiré pattern on the film plane. An example of fringe patterns for u and v displacement fields is shown in Fig. 11.29. The specimen, a beam with a bonded stiffener, was subjected to pure bending. Both the beam and stiffener were fabricated from a graphite-epoxy composite. The two elements were bonded together with a thick layer of an epoxy adhesive. The fringe patterns were obtained for the critical corner region where the load is transferred from the beam to the stiffener. The high density of the fringe patterns permits an accurate determination of both the displacement and the strain fields.

EXERCISES

11.1. Determine the change in length ΔD of the horizontal diameter of the theta specimen (θ-shaped ring) shown in Fig. 11.1. The pitch of the vertical grid is 0.025 mm.

11.2. Equations (11.1) and (11.2) give the relations for engineering strain where l_0 is the original length. Write the corresponding relations for true strain.

11.3. Beginning with Eq. (11.5) verify Eq. (11.7).

11.4. Determine the angle of rotation θ from the moiré fringe pattern shown in Fig. 11.3.

11.5. Verify Eqs. (11.10) and (11.11).

11.6. In many moiré fringe patterns, the perpendicular distance δ between fringes is difficult to measure accurately while the distance δ_p between fringes in a direction perpendicular to the master grating lines is easy to measure. For this case, develop an expression for the deformed-specimen grating pitch p' in terms of the distance δ_p, the angle of inclination ϕ of the fringes, and the master grating pitch p.

11.7. In many moiré fringe patterns, the perpendicular distance δ between fringes is difficult to measure accurately while the distance δ_s between fringes in a direction parallel to the master grating lines is easy to measure. For this case, develop an expression for the deformed-specimen grating pitch p' in terms of the distance δ_s, the angle of inclination ϕ of the fringes, and the master grating pitch p.

11.8. Use the moiré fringe pattern shown in Fig. E11.8 to prepare a plot of the displacement v as a function of position along the vertical axis of symmetry of the disk. Note that $v = 0$ at the center of the disk. The grating lines are horizontal and have a density of 12 lines/mm. The original diameter of the disk was 100 mm. From the displacement-position curve, determine the distribution of strain ϵ_{yy} along the vertical axis of symmetry of the disk.

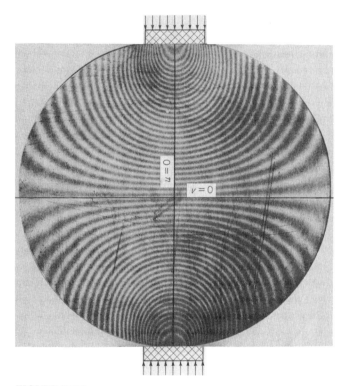

FIGURE E11.8.

11.9. Determine the distribution of strain ϵ_{yy} along the horizontal axis of symmetry of the disk shown in Fig. E11.8.

11.10. Using the geometry defined in Fig. 11.10 derive the expression for the out-of-plane displacement w given by Eq. (11.16).

11.11. A circular plate of radius R_0 and thickness h with a transverse load P applied at the center undergoes a displacement

$$w = \frac{P}{16\pi D}\left[\frac{3+v}{1+v}(R_0^2 - r^2) + 2r^2 \log \frac{r}{R_0}\right]$$

where $D = Eh^3/12(1 - v^2)$ is the flexural rigidity of the plate. Sketch the moiré fringe pattern observed if the optical arrangement of Fig. 11.10 is used in the displacement analysis.

11.12. Repeat Exercise 11.11 by assuming that the Ligtenberg method, illustrated in Fig. 11.11, is used to produce the moiré fringe pattern.

11.13. When the fringe multiplication method illustrated in Fig. 11.16 is used, what requirement is placed on both moiré gratings? Discuss the errors involved if this requirement is not satisfied.

11.14. Design a moiré grating for use in analyzing one-dimensional axisymmetric problems where r is the only variable of significance. Write the specifications for this grating.

11.15. Derive the equation for the spatial frequency f of the interference pattern produced by two diverging beams of coherent light as shown in Fig. 11.21.

11.16. Prepare a graph showing the frequency f of the interference pattern of Exercise 11.14 as a function of angle θ with the wavelength λ of light as a display parameter. Use helium-neon and argon laser light.

11.17. Describe the replication process used to produce a diffraction grating on a model that is to be analyzed using moiré interferometry.

11.18. What is a virtual reference grating?

11.19. A diffraction grating with a frequency f of 2000 lines/mm is placed on an aluminum ($E = 71$ GPa and $v = 0.33$) specimen with the lines oriented in the x direction. If the specimen is subjected to a uniaxial stress $\sigma_{xx} = 50$ MPa, determine the new frequency f' of the grating.

11.20. Beginning with Eq. (11.28) derive Eq. (11.37).

11.21. Write an engineering brief describing the operating features of Post's four-beam interferometer shown in Fig. 11.28.

11.22. Use the moiré interference patterns shown in Fig. 11.29 to determine the strain distribution at select locations in the adhesive joints and adherents. Note $f = 2400$ lines/mm.

11.23. Discuss the concept of gage length when displacement data from moiré interferometry is used for the measurement of strain. In the discussion consider strain fields with magnitudes of 1, 10, 100, and 1000 $\mu\epsilon$. Compare the moiré "gage length" with the gage length used for electrical-resistance strain gages.

REFERENCES

1. Tollenaar, D.: Moiré: *Interferentieverschijnselen bij Rasterdruk*, Amsterdam Instituut voor Grafische Techniek, Amsterdam, 1945.
2. Morse, S., A. J. Durelli, and C. A. Sciammarella: Geometry of Moiré Fringes in Strain Analysis, *J. Eng. Mech. Div.*, *ASCE*, vol. 86, no. EM4, pp. 105–126, 1960.
3. Weller, R., and B. M. Shepard: Displacement Measurement by Mechanical Interferometry, *Proc. SESA*, vol. VI, no. 1, pp. 35–38, 1948.
4. Dantu, P.: Recherches diverses d'extensometrie et de determination des contraintes, *Anal. Constraintes*, *Mem. GAMAC*, tome II, no. 2, pp. 3–14, 1954.
5. Sciammarella, C. A., and A. J. Durelli: Moiré Fringes as a Means of Analyzing Strains, *J. Eng. Mech. Div.*, *ASCE*, vol. 87, no. EM1, pp. 55–74, 1961.
6. Theocaris, P. S.: Isopachic Patterns by the Moiré Method, *Exp. Mech.*, vol. 4, no. 6, pp. 153–159, 1964.
7. Theocaris, P. S.: Moiré Patterns of Isopachics, *J. Sci. Instrum.*, vol. 41, pp. 133–138, 1964.

8. Ligtenberg, F. K.: The Moiré Method: A New Experimental Method for the Determination of Moments in Small Slab Models, *Proc. SESA*, vol. XII, no. 2, pp. 83–98, 1954.
9. Post, D.: Moiré Interferometry in White Light, *Appl. Opt.*, vol. 18, no. 24, pp. 4163–4167, 1979.
10. Parks, V. J., and A. J. Durelli: Various Forms of the Strain Displacement Relations Applied to Experimental Strain Analysis, *Exp. Mech.*, vol. 4, no. 2, pp. 37–47, 1964.
11. Post, D.: The Moiré Grid-Analyzer Method for Strain Analysis, *Exp. Mech.*, vol. 5, no. 11, pp. 368–377, 1965.
12. Dantu, P.: Extension of the Moiré Method to Thermal Problems, *Exp. Mech.*, vol. 4, no. 3, pp. 64–69, 1964.
13. Post, D.: Sharpening and Multiplication of Moiré Fringes, *Exp. Mech.*, vol. 7, no. 4, pp.154–159, 1967.
14. Holister, G. S.: Moiré Method of Surface Strain Measurement, *Engineer*, vol. 223, no. 5792, pp. 149–152, 1967.
15. Post, D.: Analysis of Moiré Fringe Multiplication Phenomena, *Appl. Opt.*, vol. 6, no. 11, pp. 1039–1942, 1967.
16. Post, D.: New Optical Methods of Moiré Fringe Multiplication, *Exp. Mech.*, vol. 8, no. 2, pp. 63–68, 1968.
17. Sciammarella, C. A.: Moiré-Fringe Multiplication by Means of Filtering and a Wave-Front Reconstruction Process, *Exp. Mech.*, vol. 9, no. 4, pp. 179–185, 1969.
18. Chiang, F.: Techniques of Optical Spatial Filtering Applied to the Processing of Moiré-Fringe Patterns, *Exp. Mech.*, vol. 9, no. 11, pp. 523–526, 1969.
19. Durelli, A. J., and V. J. Parks: *Moiré Analysis of Strain*, chap. 16, Prentice-Hall, Englewood Cliffs, N.J., 1970.
20. Graham, S. M.: "Stress Intensity Factors for Bodies Containing Initial Stress," Ph.D. dissertation, Mechanical Engineering Department, University of Maryland, College Park, 1988.
21. Holister, G. S., and A. R. Luxmoore: The Production of High-Density Moiré Grids, *Exp. Mech.*, vol. 8, no. 5, pp. 210–216, 1968.
22. Chiang, F.: Discussion of the Production of High-Density Moiré Grids, *Exp. Mech.*, vol. 9, no. 6, pp. 286–288, 1969.
23. Zandman, F.: The Transfer-Grid Method: A Practical Moiré Stress-Analysis Tool, *Exp. Mech.*, vol. 7, no. 7, pp. 19A–22A, 1967.
24. Durelli, A. J., and V. J. Parks: in op. cit., chaps. 14 and 15.
25. Theocaris, P. S.: *Moiré Fringes in Strain Analysis*, chap. 11, Pergamon Press, New York, 1969.
26. Kobayashi, A. S. (ed.): *Handbook on Experimental Mechanics*, Prentice-Hall, Englewood Cliffs, N.J., 1987.
27. Post, D.: "Moiré Interferometry," chap. 7 in ibid.
28. Walker, C. A., J. McKelvie, and A. McDonach: Experimental Study of Inelastic Strain Patterns in a Model of a Tube-Plate Ligament Using an Interferometric Moiré Technique, *Exp. Mech.*, vol. 23, no. 1, pp. 21–29, 1983.
29. Weissman, E. M., and D. Post: Moiré Interferometry near the Theoretical Limit, *Appl. Opt.*, vol. 21, no. 9, pp. 1621–1623, 1982.
30. Post, D.: High Sensitivity Displacement Measurements by Moiré Interferometry, *Proc. 7th Int. Conf. Exp. Stress Anal., Haifa, Israel, 1982*, pp. 397–408.
31. Smith, C. W., D. Post, and G. Nicoletto: Experimental Stress-Intensity Factors in Three-Dimensional Cracked-Body Problems, *Exp. Mech.*, vol. 23, no. 4, pp. 378–381, 1983.
32. Baschore, M. L., and D. Post: High-Frequency, High-Reflectance Transferable Moiré Gratings, *Exp. Tech.*, vol. 8, no. 5, pp. 29–31, 1984.

CHAPTER
12

THEORY OF PHOTOELASTICITY

12.1 INTRODUCTION [1, 2]

In Chap. 10, the working optical instrument of photoelasticity, a polariscope, was described in detail. The purpose of this chapter is to discuss photoelasticity, i.e., what happens in the polariscope when a photoelastic model is placed in the field and loaded. This discussion will be kept as simple as possible, yet it will be sufficiently complete to describe most of the photoelastic effects that can be observed in a transmission polariscope.

Many transparent noncrystalline materials that are optically isotropic when free of stress become optically anisotropic and display characteristics similar to crystals when they are stressed. These characteristics persist while loads on the material are maintained but disappear when the loads are removed. This behavior, known as *temporary double refraction*, was first observed by Sir David Brewster in 1816. The method of photoelasticity is based on this physical behavior of transparent noncrystalline materials.

The optical anisotropy (temporary double refraction) which develops in a material as a result of stress can be represented by an ellipsoid, known in this case as the *index ellipsoid*. The semiaxes of the index ellipsoid represent the principal indices of refraction of the material at the point, as shown in Fig. 12.1. Any radius of the ellipsoid represents a direction of light propagation through the point. A plane through the origin, which is perpendicular to the radius, intersects the ellipsoid as an ellipse. The semiaxes of the ellipse represent the indices of refraction associated with light waves having planes of vibration which contain the radius vector and an axis of the ellipse. For a material which is optically isotropic, the

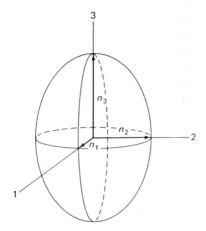

FIGURE 12.1
The index ellipsoid.

three principal indices of refraction are equal, and the index ellipsoid becomes a sphere. The index of refraction is then the same for all directions of light propagation through the material.

The similarities which exist between the stress ellipsoid for the state of stress at a point and the index ellipsoid for the optical properties of a material exhibiting temporary double refraction suggest the presence of a relationship between the two quantities. This relationship, which forms the basis for an experimental determination of stresses (or strains), is known as the *stress-optic law*.

12.2   THE STRESS-OPTIC LAW [3–7]

The theory which relates changes in the indices of refraction of a material exhibiting temporary double refraction to the state of stress in the material is due to Maxwell, who reported the phenomenon in 1853. Maxwell noted that the changes in the indices of refraction were linearly proportional to the loads and thus to stresses or strains for a linearly elastic material. The relationships can be expressed in equation form as

$$n_1 - n_0 = c_1\sigma_1 + c_2(\sigma_2 + \sigma_3)$$

$$n_2 - n_0 = c_1\sigma_2 + c_2(\sigma_3 + \sigma_1) \qquad (12.1)$$

$$n_3 - n_0 = c_1\sigma_3 + c_2(\sigma_1 + \sigma_2)$$

where $\sigma_1, \sigma_2, \sigma_3$ = principal stresses at point

$\quad n_0$ = index of refraction of material in unstressed state

$\quad n_1, n_2, n_3$ = principal indices of refraction which coincide with the principal stress directions

$\quad c_1, c_2$ = constants known as *stress-optic coefficients*

Equations (12.1) are the fundamental relationships between stress and optical effects and are known as the *stress-optic law*. These equations indicate that the complete state of stress at a point can be determined by measuring the three principal indices of refraction and establishing the directions of the three principal optical axes. Since the measurements are extremely difficult to make in the three-dimensional case, practical application has been limited to cases of plane stress ($\sigma_3 = 0$). For plane-stress situations, Eqs. (12.1) reduce to

$$n_1 - n_0 = c_1\sigma_1 + c_2\sigma_2$$
$$n_2 - n_0 = c_1\sigma_2 + c_2\sigma_1 \tag{12.2}$$

Favre [6] has used a Mach-Zehnder interferometer to make measurements of absolute retardations. Such measurements permit direct determination of the individual principal stresses at interior points of a loaded two-dimensional model. However, measurements with an interferometer are difficult and time-consuming and for this reason are seldom used. A better approach is to use photoelasticity, which measures relative retardations ($n_2 - n_1$) by using simple equipment which is easy to operate.

12.3 THE STRESS-OPTIC LAW IN TERMS OF RELATIVE RETARDATION [2, 6–11]

Equations (12.1) describe the changes in index of refraction due to applied stress experienced by a material exhibiting temporary double refraction. The method of photoelasticity makes use of relative changes in index of refraction which can be written by eliminating n_0 from Eqs. (12.1) as

$$n_2 - n_1 = (c_2 - c_1)(\sigma_1 - \sigma_2) = c(\sigma_1 - \sigma_2)$$
$$n_3 - n_2 = (c_2 - c_1)(\sigma_2 - \sigma_3) = c(\sigma_2 - \sigma_3) \tag{12.3}$$
$$n_1 - n_3 = (c_2 - c_1)(\sigma_3 - \sigma_1) = c(\sigma_3 - \sigma_1)$$

where $c = c_2 - c_1$ is the relative stress-optic coefficient first expressed in terms of brewsters (1 brewster = 10^{-13} cm^2/dyn = 10^{-12} m^2/N = 6.895×10^{-9} in^2/lb). Photoelastic materials are considered to exhibit positive birefringence when the velocity of propagation of the light wave associated with the principal stress σ_1 is greater than the velocity of the wave associated with the principal stress σ_2. Since the principal stresses are ordered such that $\sigma_1 \geq \sigma_2 \geq \sigma_3$, the principal indices of refraction of a positive doubly refracting material can be ordered such that $n_3 \geq n_2 \geq n_1$. The form of Eqs. (12.3) has been selected to make the relative stress-optic coefficient c a positive constant.

Since a stressed photoelastic model behaves like a temporary wave plate, Eq. (10.37) can be used to relate the relative angular phase shift Δ (or relative retardation) to changes in the indices of refraction in the material resulting from the stresses. For example, consider a slice of material (thickness h) oriented perpendicular to one of the principal-stress directions at the point of interest in the model. If a beam of plane-polarized light is passed through the slice at normal incidence, the relative retardation Δ accumulated along each of the principal-stress directions can be obtained by substituting Eq. (10.37) into each of Eqs. (12.3) to yield

$$\Delta_{12} = \frac{2\pi hc}{\lambda}(\sigma_1 - \sigma_2)$$

$$\Delta_{23} = \frac{2\pi hc}{\lambda}(\sigma_2 - \sigma_3) \qquad (12.4)$$

$$\Delta_{31} = \frac{2\pi hc}{\lambda}(\sigma_3 - \sigma_1)$$

where Δ_{12} is the magnitude of the relative angular phase shift (relative retardation) developed between components of a light beam propagating in the σ_3 direction. Similar interpretations can be applied to the retardations Δ_{23} and Δ_{31}.

The relative retardation Δ is linearly proportional to the difference between the two principal stresses having directions perpendicular to the path of propagation of the light beam. The third principal stress, having a direction parallel to the path of propagation of the light beam, has no effect on the relative retardation. Also, the relative retardation Δ is linearly proportional to the model or slice thickness h and inversely proportional to the wavelength λ of the light being used.

The relative stress-optic coefficient c is usually assumed to be a material constant that is independent of the wavelength of the light being used. A study by Vandaele-Dossche and van Geen [10] has shown, however, that this coefficient may depend on wavelength in some cases as the model material passes from the elastic to the plastic state. The dependence of the relative stress-optic coefficient c on the wavelength of the light being used is referred to as *photoelastic dispersion* or *dispersion of birefringence*.

From an analysis of the general three-dimensional state of stress at a point, and from an analysis of the change in index of refraction with direction of light propagation in the stressed material, it can be shown that Eqs. (12.4) apply not only for principal stresses but also for secondary principal stresses σ_1' and σ_2'. Thus,

$$\Delta' = \frac{2\pi hc}{\lambda}(\sigma_1' - \sigma_2') \qquad (12.5)$$

The secondary principal stresses $(\sigma_1' \geq \sigma_2')$ at the point of interest lie in the plane whose normal vector is coincident with the path of the light beam. The stress-optic law in terms of secondary principal stresses is widely used in applications of three-dimensional photoelasticity.

For two-dimensional plane-stress bodies where $\sigma_3 = 0$, the stress-optic law for light at normal incidence to the plane of the model can be written without the subscripts on the retardation simply as

$$\Delta = \frac{2\pi hc}{\lambda}(\sigma_1 - \sigma_2) \qquad (12.6)$$

Here it is understood that σ_1 and σ_2 are in-plane principal stresses and that σ_1 is greater than σ_2 but not greater than $\sigma_3 = 0$ if both in-plane stresses are compressive. The form of Eq. (12.6) is usually simplified to

$$\sigma_1 - \sigma_2 = \frac{N f_\sigma}{h} \qquad (12.7)\dagger$$

where

$$N = \frac{\Delta}{2\pi} \qquad (12.8)$$

is the relative retardation in terms of cycles of retardation, and counted as the fringe order. The term f_σ is

$$f_\sigma = \frac{\lambda}{c} \qquad (12.9)$$

The material fringe value f_σ is a property of the model material for a given wavelength λ, and h is the model thickness.

It is immediately apparent from Eq. (12.7) that the stress difference $\sigma_1 - \sigma_2$ in a two-dimensional model can be determined if the relative retardation N can be measured and if the material fringe value f_σ can be established by means of calibration. Actually, the function of the polariscope is to determine the value of N at each point in the model.

If a photoelastic model exhibits a perfectly linear elastic behavior, the difference in the principal strains $\epsilon_1 - \epsilon_2$ can also be measured by establishing the fringe order N. The stress-strain relations for a two-dimensional or plane state of stress are given by Eqs. (2.19) as

$$\epsilon_1 = \frac{1}{E}(\sigma_1 - \nu\sigma_2)$$

$$\epsilon_1 = \frac{1}{E}(\sigma_2 - \nu\sigma_1)$$

or

$$\epsilon_1 - \epsilon_2 = \frac{1+\nu}{E}(\sigma_1 - \sigma_2)$$

Substituting these results into Eq. (12.7) yields

$$\frac{N f_\sigma}{h} = \frac{E}{1+\nu}(\epsilon_1 - \epsilon_2) \qquad (12.10)$$

---

† For some references Eq. (12.7) is written as $\tau = (\sigma_1 - \sigma_2)/2 = N f_\sigma^*/h$, where f_σ^* is the material fringe value in terms of shear and is equal to one-half the f_σ value defined here.

which is rewritten

$$\frac{Nf_\epsilon}{h} = \epsilon_1 - \epsilon_2 \qquad (12.11)$$

It is clear that

$$f_\epsilon = \frac{1 + \nu}{E} f_\sigma \qquad (12.12)$$

where f_ϵ is the material fringe value in terms of strain.

For a perfectly linear elastic photoelastic model, the determination of N is sufficient to establish both $\sigma_1 - \sigma_2$ and $\epsilon_1 - \epsilon_2$ if any three of the material properties E, ν, f_σ, or f_ϵ are known. However, many photoelastic materials exhibit viscoelastic properties, and Eqs. (2.19) and (12.12) are not valid. The viscoelastic behavior of photoelastic materials is discussed in Sec. 13.3.

12.4   EFFECTS OF A STRESSED MODEL IN A PLANE POLARISCOPE [2, 6–9]

It is clear that the principal-stress difference $\sigma_1 - \sigma_2$ can be determined in a two-dimensional model if the fringe order N is measured at each point in the model. Moreover, the optical axes of the model (a temporary wave plate) coincide with the principal-stress directions. These two facts can be effectively utilized to determine $\sigma_1 - \sigma_2$ if a method to measure the optical properties of a stressed model has been established.

Consider first the case of a plane-stressed model inserted into the field of a plane polariscope with its normal coincident with the axis of the polariscope, as illustrated in Fig. 12.2. Note that the principal-stress direction at the point under

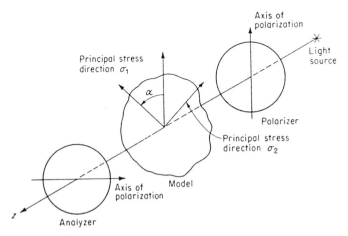

FIGURE 12.2
A stressed photoelastic model in a plane polariscope.

consideration in the model makes an angle α with the axis of polarization of the polarizer.

We know that a plane polarizer resolves an incident light wave into components which vibrate parallel and perpendicular to the axis of the polarizer. The component parallel to the axis is transmitted, and the component perpendicular to the axis is internally absorbed. Since the initial phase of the wave is not important in the development which follows, the plane-polarized light beam emerging from the polarizer can be represented by the simple expression

$$E_{py} = k \cos \omega t \tag{a}$$

After leaving the polarizer, this plane-polarized light wave enters the model, as shown in Fig. 12.3. Since the stressed model exhibits the optical properties of a wave plate, the incident light vector is resolved into two components E_1 and E_2 with vibrations parallel to the principal stress directions at the point. Thus

$$E_1 = k \cos \alpha \cos \omega t$$
$$E_2 = k \sin \alpha \cos \omega t \tag{b}$$

Since the two components propagate through the model with different velocities ($c > v_1 > v_2$), they develop phase shifts Δ_1 and Δ_2 with respect to a wave in air. The waves upon emerging from the model can be expressed as

$$E_1' = k \cos \alpha \cos (\omega t - \Delta_1)$$
$$E_2' = k \sin \alpha \cos (\omega t - \Delta_2) \tag{c}$$

where

$$\Delta_1 = \frac{2\pi h}{\lambda} (n_1 - 1)$$

$$\Delta_2 = \frac{2\pi h}{\lambda} (n_2 - 1)$$

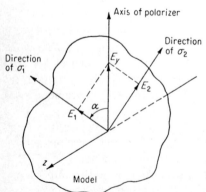

FIGURE 12.3
Resolution of the light vector as it enters a stressed model in a plane polariscope.

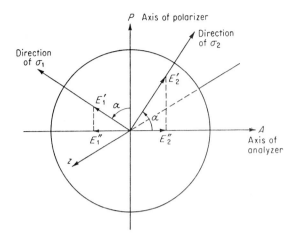

FIGURE 12.4
Components of the light vectors which are transmitted through the analyzer of a plane polariscope.

After leaving the model, the two components continue to propagate without further change and enter the analyzer in the manner shown in Fig. 12.4. The light components E_1' and E_2' are resolved when they enter the analyzer into horizontal components E_1'' and E_2'' and into vertical components. Since the vertical components are internally absorbed in the analyzer, they have not been shown in Fig. 12.4.

The horizontal components transmitted by the analyzer combine to produce an emerging light vector $\mathbf{E}_{ax}$, which is given by

$$\mathbf{E}_{ax} = E_2'' - E_1'' = E_2' \cos \alpha - E_1' \sin \alpha \qquad (d)$$

Substituting Eqs. (c) into Eq. (d) yields

$$\mathbf{E}_{ax} = k \sin \alpha \cos \alpha \left[\cos (\omega t - \Delta_2) - \cos (\omega t - \Delta_1) \right]$$

$$= k \sin 2\alpha \sin \frac{\Delta_2 - \Delta_1}{2} \sin \left(\omega t - \frac{\Delta_2 + \Delta_1}{2} \right) \qquad (12.13)$$

It is interesting to note in Eq. (12.13) that the average angular phase shift $(\Delta_2 + \Delta_1)/2$ affects the phase of the light wave emerging from the analyzer but not the amplitude (coefficient of the time-dependent term). It has no influence on the intensity (intensity is proportional to the square of the amplitude) of the light emerging from the analyzer. The relative retardation $\Delta = \Delta_2 - \Delta_1$ appears in the amplitude of the wave; therefore, it is one of the two important parameters that control the intensity of light emerging from the analyzer. Since the average angular phase shift $(\Delta_2 + \Delta_1)/2$ had no effect on the intensity, it does not contribute to the optical patterns observed in a photoelastic model. In future developments of photoelasticity theory only relative retardations will be considered.

Since the intensity of light is proportional to the square of the amplitude of the light wave, the intensity of the light emerging from the analyzer of a plane polariscope is given by

$$I = K \sin^2 2\alpha \sin^2 \frac{\Delta}{2} \tag{12.14}$$

where

$$\Delta = \Delta_2 - \Delta_1 = \frac{2\pi h}{\lambda}(n_2 - n_1) = \frac{2\pi h c}{\lambda}(\sigma_1 - \sigma_2)$$

Examination of Eq. (12.14) indicates that extinction ($I = 0$) occurs either when $\sin^2 2\alpha = 0$ or when $\sin^2(\Delta/2) = 0$. Clearly, one condition for extinction is related to the principal-stress directions and the other is related to the principal-stress difference.

12.4.1   Effect of Principal-Stress Directions

When $2\alpha = n\pi$, where $n = 0, 1, 2, \ldots$, $\sin^2 2\alpha = 0$ and extinction occurs. This relation indicates that, when one of the principal-stress directions coincides with the axis of the polarizer ($\alpha = 0$, $\pi/2$, or any exact multiple of $\pi/2$), the intensity of the light is zero. Since the analysis of the optical effects produced by a stressed model in a plane polariscope was conducted for an arbitrary point in the model, the analysis is valid for all points of the model. When the entire model is viewed in the polariscope, a fringe pattern is observed; the fringes are loci of points where the principal-stress directions (either σ_1 or σ_2) coincide with the axis of the polarizer. The fringe pattern produced by the $\sin^2 2\alpha$ term in Eq. (12.14) is the isoclinic fringe pattern. Isoclinic fringe patterns are used to determine the principal-stress directions at all points of a photoelastic model. Since isoclinics represent a very important segment of the data obtained from a photoelastic model, the topic of isoclinic-fringe-pattern interpretation will be treated in more detail in Sec. 13.2.2.

12.4.2   Effect of Principal-Stress Difference

When $\Delta/2 = n\pi$, where $n = 0, 1, 2, 3, \ldots$, $\sin^2(\Delta/2) = 0$ and extinction occurs. When the principal-stress difference is either zero ($n = 0$) or sufficient to produce an integral number of wavelengths of retardation ($n = 1, 2, 3, \ldots$), the intensity of light emerging from the analyzer is zero. When a model is viewed in the polariscope, this condition for extinction yields a second fringe pattern where the fringes are loci of points exhibiting the same order of extinction ($n = 0, 1, 2, 3, \ldots$). The fringe pattern produced by the $\sin^2(\Delta/2)$ term in Eq. (12.14) is the isochromatic fringe pattern. The nature of the optical effect producing the isochromatic fringe pattern requires some additional discussion.

Recall from Eq. (12.6) that the relative retardation Δ may be expressed as

$$\Delta = \frac{2\pi h c}{\lambda}(\sigma_1 - \sigma_2) \tag{12.6}$$

and

$$n = \frac{\Delta}{2\pi} = \frac{hc}{\lambda}(\sigma_1 - \sigma_2) \qquad (e)$$

Examination of Eq. (e) indicates that the order of extinction n depends on both the principal-stress difference $\sigma_1 - \sigma_2$ and the wavelength λ of the light. When a model is viewed in monochromatic light, the isochromatic fringe pattern appears as a series of dark bands since the intensity of light is zero when $n = 0, 1, 2, 3, \ldots$. However, when a model is viewed with white light (all wavelengths of the visible spectrum present), the isochromatic fringe pattern appears as a series of colored bands. The intensity of light is zero, and a black fringe appears only when the principal-stress difference is zero and a zero order of extinction occurs for all wavelengths of light. No other region of zero intensity is possible since the value of $\sigma_1 - \sigma_2$ required to produce a given order of extinction is different for each of the wavelengths. For nonzero values of $\sigma_1 - \sigma_2$ only one wavelength can be extinguished from the white light. The colored bands form in regions where $\sigma_1 - \sigma_2$ produces extinction of a particular wavelength of the white light. For example, when $\sigma_1 - \sigma_2$ produces extinction of the green wavelengths, the complementary color, red, appears as the isochromatic fringe. At the higher levels of principal-stress difference, where several wavelengths of light can be extinguished simultaneously, e.g., second-order red and third-order violet, the isochromatic fringes become pale and very difficult to identify, and should not be used for analysis.

With monochromatic light, the individual fringes in an isochromatic fringe pattern remain sharp and clear to very high orders of extinction. Since the wavelength of the light is fixed, Eq. (e) can be written in terms of the material fringe value f_σ and the isochromatic fringe order N as

$$n = N = \frac{h}{f_\sigma}(\sigma_1 - \sigma_2) \qquad (f)$$

The number of fringes appearing in an isochromatic fringe pattern is controlled by the magnitude of the principal-stress difference $\sigma_1 - \sigma_2$, by the thickness h of the model, and by the sensitivity of the photoelastic material, as denoted by the material fringe value f_σ.

In general, the principal-stress difference $\sigma_1 - \sigma_2$ and the principal-stress directions vary from point to point in a photoelastic model. As a result, the isoclinic fringe pattern and the isochromatic fringe pattern are superimposed, as shown in Fig. 12.5. Separation of the patterns requires special techniques which will be discussed in Sec. 13.2.5.

Theoretically, the isoclinic and isochromatic fringes should be lines of zero width; however, the photograph in Fig. 12.5 shows the fringes as bands with considerable width. Also, direct visual examination of the fringe pattern in a polariscope will show again that fringes are bands and not lines. In both instances, the width of the fringes is due to the recording characteristics of the eye and the photographic film and not to inaccuracies in the previous development. If the

FIGURE 12.5
Superimposed isochromatic and isoclinic fringe patterns for a ring loaded in diametral compression.

intensity of light emerging from the analyzer is measured with a photodiode, a minimum intensity is recorded at some point near the center of the fringe that coincides with the exact extinction line.

12.4.3   Frequency Response of a Polariscope

The frequency f for light in the visible spectrum is approximately 10^{15} Hz (see Fig. 10.2). As a result of this very high frequency, neither the eye nor any type of existing high-speed photographic equipment can detect the periodic extinction associated with the ωt term in the expression for the light wave. The light wave represents the carrier of the information (signal) in a polariscope. Since the carrier frequency should be about 10 times larger than the signal frequency for fidelity in dynamic recording, the frequency response of a polariscope is of the order of 10^{14} Hz.

12.5   EFFECTS OF A STRESSED MODEL IN A CIRCULAR POLARISCOPE (DARK FIELD, ARRANGEMENT A) [2, 6–9, 12, 13]

When a stressed photoelastic model is placed in the field of a circular polariscope with its normal coincident with the z axis, the optical effects differ significantly from those obtained in a plane polariscope. The use of a circular polariscope eliminates the isoclinic fringe pattern while it maintains the isochromatic fringe

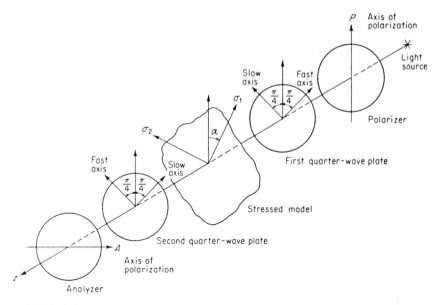

FIGURE 12.6
A stressed photoelastic model in a circular polariscope (arrangement A, crossed polarizer and analyzer, crossed quarter-wave plates).

pattern. To illustrate this effect, consider the stressed model in the circular polariscope (arrangement A) shown in Fig 12.6.

The plane-polarized light beam emerging from the polarizer can be repre-sented by the same simple expression used for the plane polariscope, namely,

$$E_{py} = k \cos \omega t \tag{a}$$

As the light enters the first quarter-wave plate, it is resolved into components E_f and E_s with vibrations parallel to the fast and slow axes, respectively. Since the axes of the quarter-wave plate are oriented at 45° with respect to the axis of the polarizer, we can write

$$E_f = \frac{\sqrt{2}}{2} k \cos \omega t$$

$$E_s = \frac{\sqrt{2}}{2} k \cos \omega t$$

As the components propagate through the plate, they develop a relative angular phase shift $\Delta = \pi/2$, and the components emerging from the plate are out of phase. Thus,

$$E'_f = \frac{\sqrt{2}}{2} k \cos \omega t$$

$$E'_s = \frac{\sqrt{2}}{2} k \cos \left(\omega t - \frac{\pi}{2} \right) = \frac{\sqrt{2}}{2} k \sin \omega t \tag{b}$$

Recall Sec. 10.2.2 where it was shown that these two plane-polarized beams represent circularly polarized light with the light vector rotating counterclockwise as it propagates between the quarter-wave plate and the model.

After leaving the quarter-wave plate, the components of the light vector enter the model in the manner illustrated in Fig. 12.7. Since the stressed model exhibits the characteristics of a temporary wave plate, the components E'_f and E'_s are resolved into components E_1 and E_2, which have directions coincident with principal-stress directions in the model. Thus

$$E_1 = E'_f \cos\left(\frac{\pi}{4} - \alpha\right) + E'_s \sin\left(\frac{\pi}{4} - \alpha\right)$$

$$E_2 = E'_s \cos\left(\frac{\pi}{4} - \alpha\right) - E'_f \sin\left(\frac{\pi}{4} - \alpha\right)$$

(c)

Substituting Eqs. (b) into Eqs. (c) yields

$$E_1 = \frac{\sqrt{2}}{2} k \cos\left(\omega t + \alpha - \frac{\pi}{4}\right)$$

$$E_2 = \frac{\sqrt{2}}{2} k \sin\left(\omega t + \alpha - \frac{\pi}{4}\right)$$

The two components E_1 and E_2 propagate through the model with different velocities. The additional relative retardation Δ accumulated during passage

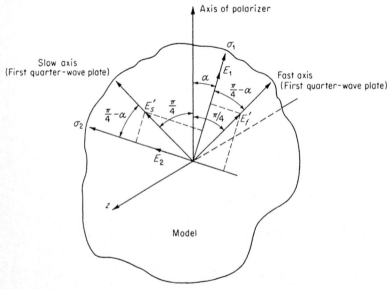

FIGURE 12.7
Resolution of the light components as they enter the stressed model.

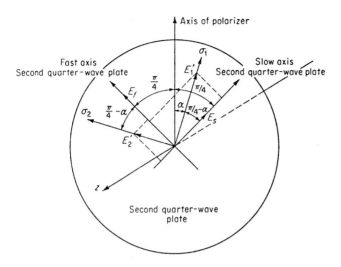

FIGURE 12.8
Resolution of the light components as they enter the second quarter-wave plate.

through the model is given by Eq. (12.6) and the waves upon emerging from the model can be expressed as

$$E_1' = \frac{\sqrt{2}}{2} k \cos\left(\omega t + \alpha - \frac{\pi}{4}\right)$$

$$E_2' = \frac{\sqrt{2}}{2} k \sin\left(\omega t + \alpha - \frac{\pi}{4} - \Delta\right)$$

(d)

The light emerging from the model propagates to the second quarter-wave plate and enters it according to the diagram shown in Fig. 12.8. The components associated with the fast and slow axes of the second quarter-wave plate are

$$E_f = E_1' \sin\left(\frac{\pi}{4} - \alpha\right) + E_2' \cos\left(\frac{\pi}{4} - \alpha\right)$$

$$E_s = E_1' \cos\left(\frac{\pi}{4} - \alpha\right) - E_2' \sin\left(\frac{\pi}{4} - \alpha\right)$$

(e)

Substituting Eqs. (d) into Eqs. (e) yields

$$E_f = \frac{\sqrt{2}}{2} k\left[\cos\left(\omega t + \alpha - \frac{\pi}{4}\right)\sin\left(\frac{\pi}{4} - \alpha\right) + \sin\left(\omega t + \alpha - \frac{\pi}{4} - \Delta\right)\cos\left(\frac{\pi}{4} - \alpha\right)\right]$$

$$E_s = \frac{\sqrt{2}}{2} k\left[\cos\left(\omega t + \alpha - \frac{\pi}{4}\right)\cos\left(\frac{\pi}{4} - \alpha\right) - \sin\left(\omega t + \alpha - \frac{\pi}{4} - \Delta\right)\sin\left(\frac{\pi}{4} - \alpha\right)\right]$$

As the light passes through the second quarter-wave plate, a relative phase

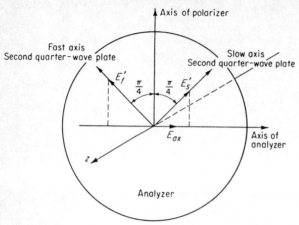

FIGURE 12.9
Components of the light vectors which are transmitted through the analyzer (dark field).

shift of $\Delta = \pi/2$ develops between the fast and slow components. The waves emerging from the plate can be expressed as

$$E'_f = \frac{\sqrt{2}}{2} k \left[\cos\left(\omega t + \alpha - \frac{\pi}{4}\right) \sin\left(\frac{\pi}{4} - \alpha\right) + \sin\left(\omega t + \alpha - \frac{\pi}{4} - \Delta\right) \cos\left(\frac{\pi}{4} - \alpha\right) \right]$$

$$E'_s = \frac{\sqrt{2}}{2} k \left[\sin\left(\omega t + \alpha - \frac{\pi}{4}\right) \cos\left(\frac{\pi}{4} - \alpha\right) + \cos\left(\omega t + \alpha - \frac{\pi}{4} - \Delta\right) \sin\left(\frac{\pi}{4} - \alpha\right) \right]$$

$$(12.15)$$

Finally, the light enters the analyzer, as shown in Fig. 12.9. The vertical components of E'_f and E'_s are absorbed in the analyzer while the horizontal components are transmitted to give

$$\mathbf{E}_{ax} = \frac{\sqrt{2}}{2} (E'_s - E'_f) \qquad (f)$$

Substituting Eqs. (12.15) into Eq. (f) gives an expression for the light emerging from the analyzer of a circular polariscope (arrangement A). Thus

$$\mathbf{E}_{ax} = k \sin \frac{\Delta}{2} \sin\left(\omega t + 2\alpha - \frac{\Delta}{2}\right) \qquad (12.16)$$

Since the intensity of light is proportional to the square of the amplitude of the light wave, the light emerging from the analyzer of a circular polariscope (arrangement A) is given by

$$I = K \sin^2 \frac{\Delta}{2} \qquad (12.17)$$

(a)

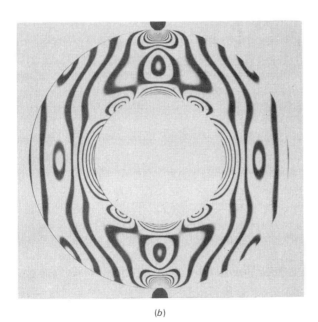

(b)

FIGURE 12.10
Isochromatic fringe patterns of a ring loaded in diametral compression: (a) dark field; (b) light field.

This result indicates that the intensity of the light beam emerging from the circular polariscope is a function only of the principal-stress difference since the angle α does not appear in the expression for the amplitude of the wave. This fact indicates that the isoclinics have been eliminated from the fringe pattern observed with the circular polariscope. From the $\sin^2(\Delta/2)$ term in Eq. (12.17) it is clear that extinction will occur when $\Delta/2 = n\pi$, where $n = 0, 1, 2, 3, \ldots$. This type of extinction is identical with that previously described for the plane polariscope for the isochromatic fringe pattern. An example of this fringe pattern is shown in Fig. 12.10a.

12.5.1 Effects of a Stressed Model in a Circular Polariscope (Light Field, Arrangement B)

A circular polariscope is usually employed with both the dark- and light-field arrangements (A and B). The circular polariscope can be converted from dark field (arrangement A) to light field (arrangement B) simply by rotating the analyzer through $90°$. The advantage of employing both light- and dark-field arrangements is that twice as many data are obtained for the whole-field determination of $\sigma_1 - \sigma_2$. Recall that the order of the fringes N coincides with n for the plane polariscope and for the dark-field circular polariscope; therefore, the fringes are counted in the sequence $0, 1, 2, 3, \ldots$. However, with the light-field arrangement of the circular polariscope, N and n do not coincide. Instead, it can be shown that the intensity I is given by

$$I = K \cos^2 \frac{\Delta}{2} \tag{12.18}$$

Equation (12.18) shows that extinction ($I = 0$) will occur when

$$\frac{\Delta}{2} = \frac{1 + 2n}{2}\pi \qquad \text{for } n = 0, 1, 2, 3, \ldots$$

and

$$N = \frac{\Delta}{2\pi} = \frac{1}{2} + n$$

which implies that the order of the first fringe observed in a light-field polariscope is $\frac{1}{2}$, which corresponds to $n = 0$. An example of a light-field isochromatic fringe pattern is shown in Fig. 12.10b.

By using the circular polariscope with both light- and dark-field arrangements, it is possible to obtain two photographs of the resulting isochromatic fringe patterns. The data from the light and dark fields give a whole-field representation of the fringes to the nearest $\frac{1}{2}$ order. Interpolation between fringes often permits an estimate of the order of fringes to ± 0.1, giving accuracies for $\sigma_1 - \sigma_2$ of $\pm 0.1 f_\sigma/h$. If more accurate determinations are necessary, compensation techniques described in Sec. 12.6 can be used.

12.6 EFFECTS OF A STRESSED MODEL IN A CIRCULAR POLARISCOPE (ARBITRARY ANALYZER POSITION, TARDY COMPENSATION) [14–18]

The analysis for the dark- and light-field arrangements of the circular polariscope can be carried one step further to include rotation of the analyzer through some arbitrary angle. The purpose of such a rotation is to provide a means for determining fractional fringe orders.

In the previous derivations, the optical effects produced by a stressed model in both plane and circular polariscopes were studied by using a trigonometric representation of the light wave. For more complicated situations, the derivations with this representation quickly become an exercise in the manipulation of trigonometric identities and very little of the physical significance of the problem is retained. Under such circumstances, an exponential representation of the light wave can be used to simplify the derivations; therefore, it will be used in all future developments.

Consider first the passage of light through the optical elements of a circular polariscope (arrangement A). With the exponential representation, the light wave emerging from the polarizer can be expressed as

$$E_{py} = k e^{i\omega t} \tag{a}$$

As the light enters the first quarter-wave plate, it is resolved into components parallel to the fast and slow axes of the plate. The components develop a phase shift $\Delta = \pi/2$ as they propagate through the plate and emerge as

$$E'_f = \frac{\sqrt{2}}{2} k e^{i\omega t}$$

$$E'_s = -i \frac{\sqrt{2}}{2} k e^{i\omega t} \tag{b}$$

As the light enters the model, it is resolved into components parallel to the principal-stress directions. The components develop an additional phase shift Δ which depends on the principal-stress difference and emerge from the model as

$$E'_1 = \frac{\sqrt{2}}{2} k e^{i(\omega t + \alpha - \pi/4)}$$

$$E'_2 = -i \frac{\sqrt{2}}{2} k e^{i(\omega t + \alpha - \pi/4 - \Delta)} \tag{c}$$

As the light enters the second quarter-wave plate, it is again resolved into components parallel to the fast and slow axes of the plate. The components

experience a phase shift of $\Delta = \pi/2$ as they propagate through the plate and emerge
as

$$E'_f = \frac{\sqrt{2}}{2}\, k\left[\sin\left(\frac{\pi}{4} - \alpha\right) - ie^{-i\Delta}\cos\left(\frac{\pi}{4} - \alpha\right)\right]e^{i(\omega t + \alpha - \pi/4)}$$

$$E'_s = \frac{\sqrt{2}}{2}\, k\left[e^{-i\Delta}\sin\left(\frac{\pi}{4} - \alpha\right) - i\cos\left(\frac{\pi}{4} - \alpha\right)\right]e^{i(\omega t + \alpha - \pi/4)} \tag{12.19}$$

Finally, as the light passes through the analyzer, the vertical components of E'_f
and E'_s are absorbed while the horizontal components are transmitted. Thus

$$\mathbf{E}_{ax} = \frac{k}{2}\left[(e^{-i\Delta} - 1)\sin\left(\frac{\pi}{4} - \alpha\right) + i(e^{-i\Delta} - 1)\cos\left(\frac{\pi}{4} - \alpha\right)\right]e^{i(\omega t + \alpha - \pi/4)}$$

$$= \frac{k}{2}(e^{-i\Delta} - 1)e^{i(\omega t + 2\alpha)} \tag{12.30}$$

Recall from Eq. (10.17) that the square of the amplitude of a wave in exponential
notation is the product of the amplitude and its complex conjugate. Thus

$$I \approx \mathbf{E}_{ax}\mathbf{E}_{ax}^* = K\,\sin^2\frac{\Delta}{2} \tag{12.21}$$

This expression for the intensity of the light emerging from the analyzer of a
circular polariscope (arrangement A) is identical with that previously determined
using a trigonometric representation of the light wave and presented in Eq. (12.17).

   To establish the effect of a stressed model in a circular polariscope with the
analyzer oriented at an arbitrary angle γ with respect to its dark-field position, it
is only necessary to consider the light components emerging from the second
quarter-wave plate, as represented by Eqs. (12.19), and their transmission through
the analyzer when it is positioned as shown in Fig. 12.11. Thus,

$$\mathbf{E}_{ay} = E'_s \cos\left(\frac{\pi}{4} + \gamma\right) - E'_f \sin\left(\frac{\pi}{4} + \gamma\right) \tag{d}$$

Substituting Eqs. (12.19) into Eq. (d) and combining terms yields

$$\mathbf{E}_{ay} = \frac{\sqrt{2}}{2}\, k\left\{\sin\left(\frac{\pi}{4} - \alpha\right)\left[e^{-i\Delta}\cos\left(\frac{\pi}{4} + \gamma\right) - \sin\left(\frac{\pi}{4} + \gamma\right)\right]\right.$$

$$\left. + i\cos\left(\frac{\pi}{4} - \alpha\right)\left[e^{-i\Delta}\sin\left(\frac{\pi}{4} + \gamma\right) - \cos\left(\frac{\pi}{4} + \gamma\right)\right]\right\}e^{i(\omega t + \alpha - \pi/4)} \tag{12.22}$$

The intensity of the light emerging from the analyzer is given by Eq. (10.17) as

$$I \approx \mathbf{E}_{ay}\mathbf{E}_{ay}^* \tag{e}$$

Substituting Eq. (12.22) and its complex conjugate into Eq. (e) and combining

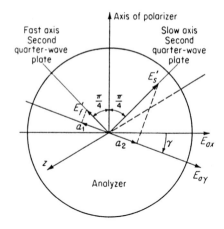

FIGURE 12.11

Rotation of the analyzer to obtain extinction in the Tardy method of compensation.

terms through the use of suitable trigonometric identities yields

$$I = K(1 - \cos 2\gamma \cos \Delta - \cos 2\alpha \sin 2\gamma \sin \Delta) \qquad (12.23)$$

For a given angle of analyzer rotation γ, values of α and Δ required for maximum intensity or minimum intensity are obtained from

$$\frac{\partial I}{\partial \alpha} = K(2 \sin 2\alpha \sin 2\gamma \sin \Delta) = 0 \qquad (f)$$

$$\frac{\partial I}{\partial \Delta} = K(\cos 2\gamma \sin \Delta - \cos 2\alpha \sin 2\gamma \cos \Delta) = 0 \qquad (g)$$

Values of α and Δ, satisfying Eqs. (f) and (g) simultaneously, which give minimum intensity are

$$\alpha = \frac{n\pi}{2} \qquad \text{and} \qquad \Delta = 2\gamma \pm 2n\pi \qquad n = 0, 1, 2, 3, \ldots$$

This result for $I = 0$ indicates that one of the principal-stress directions must be parallel to the axis of the polarizer ($\alpha = 0, \pi/2, \ldots$). The fringe order is then

$$N = \frac{\Delta}{2\pi} = n \pm \frac{\gamma}{\pi} \qquad (12.24)$$

Rotation of the analyzer through an angle γ (Tardy method of compensation) is widely used to determine fractional fringe orders at select points on a photo-elastic model. A plane polariscope is employed first so that isoclinics can be used to establish the directions of the principal stresses at the point of interest, as illustrated in Fig. 12.12. The axis of the polarizer is then aligned with a principal-stress direction ($\alpha = 0$ or $\alpha = \pi/2$), and the other elements of the polariscope are oriented to produce a standard dark-field circular polariscope. The analyzer is

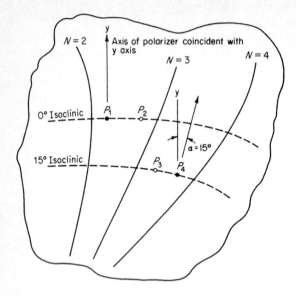

FIGURE 12.12
Locations of the points of interest relative to the isoclinic and isochromatic fringe patterns.

then rotated until extinction occurs at the point of interest, as indicated by Eq. (12.24).

To illustrate the procedure, consider the hypothetical dark-field fringe pattern and points of interest shown in Fig. 12.12. At point P_1, which lies between fringes of orders 2 and 3, the value assigned to n is 2. As the analyzer is rotated through an angle γ, the second-order fringe will move toward point P_1 until extinction is obtained. The fringe order at P_1 is then given by $N = 2 + \gamma/\pi$. For point P_2 the value of n is also taken as 2, and the analyzer is rotated through an angle γ_1 until the second-order fringe produces extinction, giving a value for the fringe order of $n = 2 + \gamma_1/\pi$. In this instance n could also be taken as 3, and the analyzer rotated in the opposite direction through an angle $-\gamma_2$ until the third-order fringe produced extinction at point P_2. In this instance the fringe order would be given by $N = 3 - \gamma_2/\pi$, which should check the value of $N = 2 + \gamma_1/\pi$ obtained previously.

The Tardy method of compensation can be quickly and effectively employed to determine fractional fringe orders at arbitrary points in a model, provided isoclinic parameters are used to obtain the directions of the principal stresses. The accuracy of the method depends upon the quality of the quarter-wave plates employed in the polariscope; however, accuracies of ± 0.02 fringes can usually be achieved.

12.7   PHOTOELASTIC PHOTOGRAPHY
[19–22]

In most photoelastic analyses, photographs are taken of the isochromatic and isoclinic fringe patterns to establish a permanent record of the test. For this reason, it is important that the basic principles of photography be established and that the differences between landscape and photoelastic photography be understood.

A sheet of photographic film is prepared with a coating containing certain silver halides. When this coating is exposed to light, the silver halides undergo a latent change that is permanently distinguishable on the film after a photographic development process. The change is a darkening produced by the formation of metallic silver. The amount of darkening is called the *density*. The density of a portion of a piece of exposed and developed film is simply a measure of the ability of the deposited silver to prevent the transmission of light. The density of a given type of film is a function of the exposure (light intensity × time) presented to the film. The characteristics of a density-exposure function were first established by Hurter and Driffield and represented in the manner shown in Fig. 12.13.

The curve presented in Fig. 12.13 defines four important characteristics of a photographic film: namely, the fog density D_0, the exposure inertia E_0, the slope of the curve which establishes the gamma number of the film, and the maximum density D_1 which occurs with an exposure E_1. Each of these characteristics is important and should be considered in selecting a film for a particular photoelastic analysis. In a photoelastic photograph, zero exposures occur whenever the intensity goes to a minimum, i.e., when $N = 0, 1, 2, \ldots$, in a dark-field polariscope; however, the film coating records values of the ranges of exposure above the inertia value E_0. It is this exposure inertia $E < E_0$ which produces the fringe width on a negative when in theory the fringe is a line. The slope of the density-versus-log-exposure curve given by γ determines the latitude of the film. The usual landscape film incorporates an emulsion with $\gamma < 1$. This relatively low value of γ gives a wide range of exposure values over which the film emulsion will be effective in producing a satisfactory negative. This feature is of course important in landscape

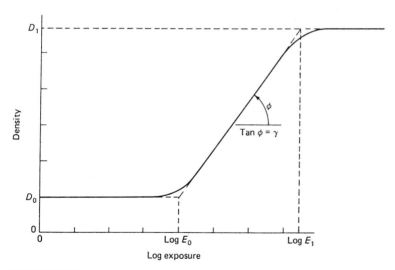

FIGURE 12.13
Hurter-Driffield graph relating exposure and density.

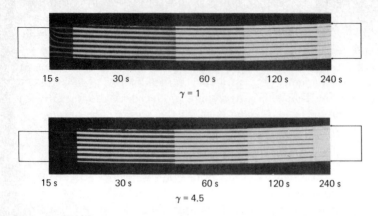

15 s   30 s   60 s   120 s   240 s

$\gamma = 1$

15 s   30 s   60 s   120 s   240 s

$\gamma = 4.5$

FIGURE 12.14
Influence of exposure time and gamma number on the appearance of the isochromatic fringe pattern
for a beam in pure bending.

photography, where the proper exposure time cannot be precisely established. For
photoelastic photographs, film emulsions with $3 < \gamma < 6$ are often employed since
this type of film gives a high-contrast negative; i.e., blacks are very black, whites
are very white, with very little gray. This is a desirable situation since the fringes
tend to sharpen and become better defined. The exposure time is of course more
critical, but this time can be accurately established in preliminary exposures for
any given polariscope. The fog density D_0 is less important since it implies that
there is a thin coating uniformly distributed over the film which absorbs light.
This of course detracts from the sharpness of a negative, but since it is a relatively
small influence it is not objectionable. As the exposure time is increased, the
exposure $E = It$ increases and the density of exposed regions on the negative is
enhanced. The effect of increased exposure reduces the fringe width and improves
the contrast, as illustrated in Fig. 12.14 for both high-contrast and landscape
film.

12.8 FRINGE MULTIPLICATION WITH PARTIAL MIRRORS [23–25]

Post has shown that partial mirrors can be usefully employed to multiply the
number of fringes which can be observed in a photoelastic model. As pointed out
previously, fringe multiplication is quite important since the standard methods of
compensation used to evaluate fractional fringe orders are time-consuming. Fringe
multiplication is, in a sense, a whole-field compensation technique where the
fractional orders of the fringes can be determined simultaneously over the entire
field of the model.

When partial mirrors are used in fringe multiplication, they are inserted into
a lens polariscope on both sides of the model with one of the mirrors inclined
slightly, as illustrated in Fig. 12.15. The effect of the inclined partial mirror is to

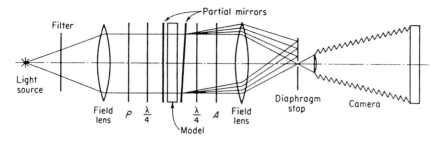

FIGURE 12.15
Partial mirrors as employed in a circular lens polariscope for fringe multiplication.

reflect the light back and forth through the model, as illustrated in Fig. 12.16. From this figure it is clear that each ray of light emerges from the mirror system at an angle which depends on the number of times the light ray has traversed the model. For instance, rays 1, 3, 5, and 7, which have traversed the model the same number of times as their ray number, emerge at angles 0, 2ϕ, 4ϕ, and 6ϕ. Although the rays do not pass through the same point, the inclination angle ϕ used in the illustration was greatly exaggerated. The length of the path over which the photoelastic effect is averaged depends upon the angle of inclination ϕ, the ray number, and the separation distance between the mirrors. In practice, multiplication by factors of 5 to 7 can be achieved without introducing objectionable errors due to the averaging process which is inherent in this method. Typical patterns obtained from a slice from a three-dimensional model are illustrated in Fig. 12.17.

The fact that different rays of light are inclined at different angles with respect to the axis of the polariscope permits each ray to be isolated. The rays are all

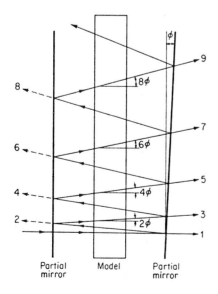

FIGURE 12.16
Light reflection and transmission between two slightly inclined partial mirrors.

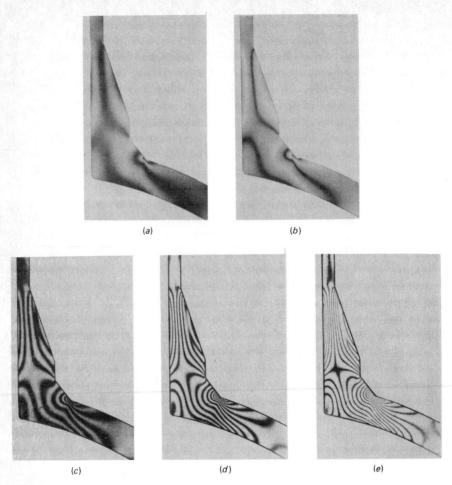

FIGURE 12.17
Isochromatic fringe patterns of a three-dimensional slice obtained by using a lens polariscope equipped with partial mirrors: (a) normal fringe pattern; (b) sharpened fringe pattern; (c) after fringe multiplication 3 × ; (d) after fringe multiplication 5 × ; (e) after fringe multiplication 7 × . (*Courtesy of C. E. Taylor.*)

collected by the field lens but focused at different points in the focal plane of the field lens (see Fig. 12.15). Any one of these rays can be observed by placing the eye or a camera lens at the proper image point. A diaphragm stop is used to eliminate all images except the one under observation.

In practice, the isochromatic fringe patterns associated with rays 1, 3, 5, 7, etc., can be observed and photographed for both light- and dark-field settings. Suppose, for instance, that the light- and dark-field photographs are obtained for rays 1, 3, and 5. The fringe patterns recorded from ray 1 may be interpreted in the conventional sense where the orders of the fringes are sequenced as $0, \frac{1}{2}, 1, \frac{3}{2}, 2, \frac{5}{2}, \ldots$. However, for ray 3, the light has passed through the model three times and the orders of the fringes are sequenced as $0, \frac{1}{6}, \frac{1}{3}, \frac{1}{2}, \frac{2}{3}, \frac{5}{6}, 1, \ldots$. Finally, for ray 5, the orders of the fringes are sequenced as $0, \frac{1}{10}, \frac{1}{5}, \frac{3}{10}, \frac{2}{5}, \frac{1}{2}, \ldots$. The superposition of the results obtained from these three rays is sufficient to determine the fringe order to the nearest one-tenth of an order over the entire model. The fringe-multiplication technique can be interpreted as a whole-field compensation method where fractional orders of the fringes can be determined with a high degree of accuracy.

The intensity relationship for the mth ray, where $m = 1, 2, 3$, etc., as shown in Fig. 12.16, can be established by modifying Eq. (12.17) to account for the loss in intensity and the added thickness effects as the light traverses the model m times. The intensity of each ray can be written as

$$I_1 = KT^2 \sin^2 \frac{\Delta}{2}$$

$$I_3 = KT^2R^2 \sin^2 \frac{3\Delta}{2} \qquad (12.25)$$

$$\ldots\ldots\ldots\ldots\ldots\ldots\ldots\ldots\ldots\ldots$$

$$I_m = KT^2R^{m-1} \sin^2 \frac{m\Delta}{2}$$

Fringe multiplication, by Post's partial-mirror method, is accompanied by a considerable loss of light intensity. The intensity of the multiplied fringe pattern as compared with the ordinary fringe pattern is decreased by the term T^2R^{m-1}, which is always much less than 1. The loss in intensity for a particular ray can be minimized by properly selecting the mirror coefficients R and T. Assuming that the mirrors are perfect,

$$T + R = 1 \qquad (12.26)$$

Substituting Eq. (12.26) into Eqs. (12.25) yields

$$I_m = K(R - 1)^2 R^{m-1} \sin^2 \frac{m\Delta}{2} \qquad m = 1, 2, 3, \ldots \qquad (a)$$

Differentiating this expression with respect to R yields

$$\frac{dI_m}{dR} = K \sin^2 \frac{m\Delta}{2} \{(R - 1)[(R - 1)(m - 1)R^{m-2} + 2R^{m-1}]\} \qquad (b)$$

Setting Eq. (b) equal to zero and solving for R gives the reflection coefficient for the mirrors which minimize the intensity loss. The result is

$$R = \frac{m - 1}{m + 1} \qquad (12.27)$$

Since R is a function of m (the number of light traverses through the model), it is not possible to optimize the mirrors for all rays simultaneously. A multiplication factor of 5 is usually sufficient for most applications and, for this case, the mirrors should have $R = 0.667$ and $T = 0.333$ to optimize the minimum intensity condition.

EXERCISES

12.1. Derive Eqs. (12.4) beginning with Eqs. (12.1).

12.2. A secondary plane is defined by an $Ox'y'z'$ coordinate system where the z' axis is normal to the plane. If the $Oxyz$ coordinate system is a principal system and the angles between the two coordinate systems are known, write the equations for the secondary principal stresses which lie in the $Ox'y'$ plane.

12.3. If a particular point in a photoelastic model is examined in a polariscope with a mercury light source ($\lambda = 546.1$ nm) and a fringe order of 4.00 is established, what fringe order would be observed if a sodium light source ($\lambda = 589.3$ nm) were used in place of the mercury source?

12.4. The stress fringe value f_σ for a material was determined to be 35 kN/m when sodium light ($\lambda = 589.3$ nm) was used in its determination. What would the stress fringe value for the same material be if mercury light ($\lambda = 546.1$ nm) were used in place of the sodium light?

12.5. Derive the equations for light passing through a stressed model in a plane polariscope with the polarizer and analyzer in parallel positions. Under what conditions does extinction ($I = 0$) occur?

12.6. Derive the equations for light passing through a stressed model in a plane polariscope (polarizer and analyzer crossed). Use an exponential representation for the light wave.

12.7. Compare the frequency response of a polariscope with that of an oscilloscope.

12.8. Prepare sketches similar to Fig. 12.6 showing the four different arrangements (see Table 10.3) for the circular polariscope.

12.9. Derive the equations for light passing through a stressed model in a circular polariscope (arrangement C of Table 10.3). Use a trigonometric representation for the light wave.

12.10. Derive the equations for light passing through a stressed model in a circular polariscope (arrangement D of Table 10.3). Use an exponential representation for the light wave.

12.11. Determine the optical effects produced by light passing through a stressed model in a circular polariscope with imperfect quarter-wave plates. Use arrangements A and B of Table 10.3 for the polariscope and assume $\Delta = \pi/2 + \epsilon$ for the imperfect quarter-wave plates.

12.12. Determine the optical effects produced by light passing through a stressed model in a circular polariscope with imperfect quarter-wave plates. Use arrangements C and D of Table 10.3 for the polariscope and assume $\Delta = \pi/2 + \epsilon$ for the imperfect quarter-wave plates.

12.13. With Tardy compensation, the fractional fringe order can be determined at an arbitrary point in a stressed model by aligning a principal-stress direction at the point with the axis of the polarizer and rotating the analyzer for extinction. What effect would be produced by rotating the polarizer instead of the analyzer?

12.14. Investigate the error introduced in fractional fringe-order determinations by improper setting of the isoclinic angle ($\alpha = 0 + \epsilon$ or $\pi/2 + \epsilon$) in Tardy compensation. Numerically evaluate for $\epsilon = \pm 10°$.

12.15. Investigate the error introduced in fractional fringe-order determinations by imperfect quarter-wave plates ($\Delta = \pi/2 + \epsilon$) in Tardy compensation. Make a numerical evaluation for the case of quarter-wave plates for mercury light ($\lambda = 546.1$ nm) being used with

(a) Sodium light ($\lambda = 589.3$ nm)
(b) Helium-neon laser light ($\lambda = 632.8$ nm)
(c) Argon laser light ($\lambda = 488.0$ nm)

12.16. The following procedure has been proposed for establishing the fractional fringe order at a point in a stressed model. Establish the validity of the procedure.
1. Begin with a circular polariscope (arrangement A).
2. Remove the first quarter-wave plate.
3. Position the model so that the principal stress directions at the point of interest are oriented $\pm 45°$ with respect to the axis of the polarizer.
4. Align the fast axis of the second quarter-wave plate with the axis of the polarizer.
5. Rotate the analyzer for extinction.

12.17. Establish a procedure for aligning a plane polariscope to obtain a dark field.

12.18. Establish a procedure for aligning a circular polariscope to give arrangement A. Then add to the procedure to yield arrangement B.

12.19. During the course of a photoelastic investigation you instruct your technician to photograph the isoclinic patterns in the model under study at $5°$ intervals. To save film, the technician decides to record two sets of patterns on the same film by using the following procedure: (1) first exposing the film for half the normal time with the plane polariscope set at $0°$ position; (2) then rotating both the polarizer and the analyzer $45°$ counterclockwise but not moving the model and exposing the film for the remaining half of the time; (3) on other sheets of film, recording the $5°$ and $50°$ isoclinics, the $10°$ and $55°$ isoclinics, etc., using the same procedure. When developing the film, what does the technician see? *Hint*: Write equations for the Hunter-Driffield relation between exposure and density.

12.20. Discuss the influence of a uniform preexposure of a given photographic film which would be sufficient to overcome the inertia exposure E_0. Consider in this discussion the width of the fringes and the exposure time required in the polariscope.

12.21. If perfect partial mirrors are used in the design of a photoelastic fringe multiplier, what reflection and transmission coefficients should be specified to minimize the intensity loss for multiplication factors of 1, 3, 5, 7, and 9?

REFERENCES

1. Jenkins, F. A., and H. E. White: Fundamentals of Optics, 4th ed., McGraw-Hill, New York, 1976.
2. Coker, E. G., and L. N. G. Filon: *A Treatise on Photoelasticity*, Cambridge University Press, New York, 1931.
3. Maxwell, J. C.: On the Equilibrium of Elastic Solids, *Trans. R. Soc. Edinburgh*, vol. XX, part 1, pp. 87–120, 1853.
4. Neumann, F. E.: Die Gesetze der Doppelbrechung des Lichts in comprimierten oder ungleichformig erwarmten unkristallinischen Korpern, *Abh. K. Acad. Wiss. Berlin*, pt. II, pp. 1–254, 1841.

5. Favre, H.: Sur une nouvelle method optique de determination des tensions interieures, *Rev. Opt.*, vol. 8, pp. 193–213, 241–261, 289–307, 1929.
6. Frocht, M. M.: *Photoelasticity*, vol. 1, John Wiley & Sons, New York, 1941; vol. 2, 1948.
7. Durelli, A. J., and W. F. Riley: *Introduction to Photomechanics*, Prentice-Hall, Englewood Cliffs, N.J., 1965.
8. Kuske, A., and G. Robertson: *Photoelastic Stress Analysis*, John Wiley & Sons, New York, 1974.
9. Mindlin, R. D.: A Review of the Photoelastic Method of Stress Analysis, *J. Appl. Phys.*, vol. 10, pp. 222–241, 273–294, 1939.
10. Vandaele-Dossche, M., and R. van Geen: La birefringence mecanique en lumiere ultra-violette et ses applications, *Bull. Class Sci. Acad. R. Belg.*, vol. 50, no. 2, 1964.
11. Monch, E., and R. Lorek: A Study of the Accuracy and Limits of Application of Plane Photoplastic Experiments, *Photoelasticity*, M. M. Frocht (ed.), Pergamon Press, New York, 1963, pp. 169–184.
12. Mindlin, R. D.: Distortion of the Photoelastic Fringe Pattern in an Optically Unbalanced Polariscope, *J. Appl. Mech.*, vol. 4, pp. A170–A172, 1937.
13. Mindlin, R. D.: Analysis of Doubly Refracting Materials with Circularly and Elliptically Polarized Light, *J. Opt. Soc. Am.*, vol. 27, pp. 288–291, 1937.
14. Tardy, M. H. L.: Methode pratique d'examen de mesure de la birefringence des verres d'optique, *Rev. Opt.*, vol. 8, pp. 59–69, 1929.
15. Jessop, H. T.: On the Tardy and Senarmont Methods of Measuring Fractional Relative Retardations, *Brit. J. Appl. Phys.*, vol. 4, pp. 138–141, 1953.
16. Flynn, P. D.: Theorems for Senarmont Compensation, *Exp. Mech.*, vol. 10, no. 8, pp. 343–345, 1970.
17. Chakrabati, S. K., and K. E. Machin: Accuracy of Compensation Methods in Photoelastic Fringe-Order Measurements, *Exp. Mech.*, vol. 9, no. 9, pp. 429–431, 1969.
18. Sathikh, S. M., and G. W. Bigg: On the Accuracy of Goniometric Compensation Methods in Photoelastic Fringe-Order Measurements, *Exp. Mech.*, vol. 12, no. 1, pp. 47–49, 1972.
19. Flanagan, J. H.: Photoelastic Photography, *Proc. SESA*, vol. XV, no. 2, pp. 1–10, 1958.
20. Koch, W. M., and P. A. Szego: Use of Double-Exposure Photography in Photoelasticity, *J. Appl. Mech.*, vol. 21, p. 198, 1954.
21. Dally, J. W., and F. J. Ahimaz: Photographic Method to Sharpen and Double Isochromatic Fringes, *Exp. Mech.*, vol. 2, no. 6, pp. 170–175, 1962.
22. Das Talukder, N. K., and P. Ghosh: On Fringe Multiplication by Superposition of Negatives, *Exp. Mech.*, vol. 15, no. 6, pp. 137–239, 1975.
23. Post, D.: Isochromatic Fringe Sharpening and Fringe Multiplication in Photoelasticity, *Proc. SESA*, vol. XII, no. 2, pp. 143–156, 1955.
24. Post, D.: Fringe Multiplication in Three-Dimensional Photoelasticity, *J. Strain Anal.*, vol. 1, no. 5, pp. 380–388, 1966.
25. Post, D.: Photoelastic-Fringe Multiplication: For Tenfold Increase in Sensitivity, *Exp. Mech.*, vol. 10, no. 8, pp. 305–312, 1970.

CHAPTER
13

APPLIED
PHOTOELASTICITY:
TWO- AND THREE-
DIMENSIONAL
STRESS
ANALYSIS

13.1 INTRODUCTION

Applications of photoelasticity are usually divided into two broad classes depending on the geometry of the body under analysis. If the body is plane, in the sense that it can be represented by plane stress or plane strain, then two-dimensional methods of photoelasticity are employed. If the body is not plane, then the more complex and time-consuming methods of three-dimensional photoelasticity are required. Both two- and three-dimensional applications of photoelasticity will be described in this chapter.

13.2 TWO-DIMENSIONAL PHOTOELASTIC STRESS ANALYSIS [1–5]

In conventional two-dimensional photoelastic analysis, a suitable model is fabricated, loaded, and placed in a polariscope, and the fringe pattern is examined and photographed. The next step in the analysis is the interpretation of the fringe patterns which, in reality, represent the raw test data. The purpose of this section is to discuss the interpretation of the isoclinic and isochromatic fringe patterns, compensation techniques, separation methods, and scaling of the stresses between the model and prototype in a typical stress analysis.

13.2.1 Isochromatic Fringe Patterns [1–6]

The isochromatic fringe pattern obtained from a two-dimensional model gives lines along which the principal-stress difference $\sigma_1 - \sigma_2$ is equal to a constant. A typical example of a light-field isochromatic fringe pattern, which will be utilized to describe the analysis, is shown in Fig. 13.1. This photoelastic model represents a chain link subjected to tensile loads applied axially through roller pins. First, it is necessary to determine the fringe order at each point of interest in the model. In this example the assignment of the fringe order is relatively simple since 15 rather obvious $\frac{1}{2}$-order fringes can quickly be identified. The two oval-like fringes located on the flanks of the teeth (labeled A) are $\frac{1}{2}$-order fringes since the flank, because of its geometry, cannot support high stresses. The four fringes located at points B on the pinholes can be identified if the model is viewed with white light since zero-order fringes appear black while higher-order fringes are colored. The irregularly shaped fringe labeled C near the center of the link is also a $\frac{1}{2}$-order

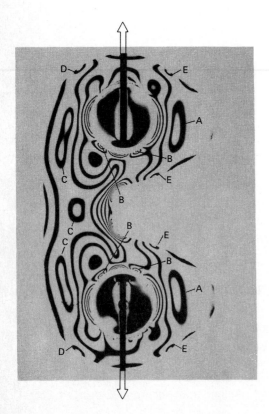

FIGURE 13.1
Light-field isochromatic fringe pattern of a chain link subjected to axial tensile loads through the roller pins.

fringe. Explanation of the $\frac{1}{2}$-order fringes at points D and E is covered in the exercises.

With these $\frac{1}{2}$-order fringes established, it is a simple matter to determine the fringe order at any point in the model by progressively counting the fringes outward from the $\frac{1}{2}$-order location. For instance, the order of the fringe at the flank fillets is $7\frac{1}{2}$. When the fringe order at any point on the model has been established, it is possible to compute $\sigma_1 - \sigma_2$ from Eq. (12.7):

$$\sigma_1 - \sigma_2 = \frac{Nf_\sigma}{h} \tag{12.7}$$

where σ_1 and σ_2 are the principal stresses in the plane of the model. The maximum shear stress is given by

$$\tau_{max} = \tfrac{1}{2}(\sigma_1 - \sigma_2) = \frac{Nf_\sigma}{2h} \tag{13.1}$$

provided σ_1 and σ_2 are of opposite sign and $\sigma_3 = 0$; otherwise,

$$\tau_{max} = \begin{cases} \tfrac{1}{2}(\sigma_1 - \sigma_3) = \tfrac{1}{2}\sigma_1 & \text{if } \sigma_1 \text{ and } \sigma_2 \text{ are positive} \\ \tfrac{1}{2}(\sigma_3 - \sigma_2) = \tfrac{1}{2}\sigma_2 & \text{if } \sigma_1 \text{ and } \sigma_2 \text{ are negative} \end{cases} \tag{13.2}$$

The difference between Eq. (13.1) and Eqs. (13.2) is presented graphically in Fig. 13.2, where Mohr stress circles for two cases are given. When $\sigma_1 > 0$ and

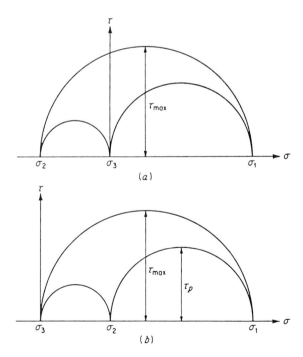

(a)

(b)

FIGURE 13.2
Mohr's circle for the state of stress at a point (a) $\sigma_1 > 0$, $\sigma_2 < \sigma_3 = 0$; (b) $\sigma_1 > \sigma_2 > \sigma_3 = 0$.

$\sigma_2 < \sigma_3 = 0$, the maximum shear stress is one-half the value of $\sigma_1 - \sigma_2$ and can be determined directly from the isochromatic fringe pattern according to Eq. (13.1). However, when $\sigma_1 > \sigma_2 > \sigma_3 = 0$, the maximum shear stress does not lie in the plane of the model, and Eq. (13.1) gives τ_p (see Fig. 13.2) and not τ_{max}. To establish τ_{max} in this case it is necessary to determine σ_1 individually. This is an important point since the maximum-shear theory of failure is often used in the design of machine components.

On the free boundary of the model either σ_1 or σ_2 is equal to zero and the stress tangential to the boundary is

$$\sigma_1 \text{ or } \sigma_2 = \frac{Nf_\sigma}{h} \tag{13.3}$$

The sign can usually be determined by inspection, particularly in the critical areas where the boundary stresses are a maximum. By referring to Fig. 13.1, it is apparent that the stresses in the flank fillets are tensile while the stresses along the back of the plate are compressive. On the free surface of the pinholes, the stresses on the horizontal diameter are tensile and the stresses on the vertical diameter are compressive (see Sec. 3.14). The stresses on that portion of the boundary of the hole where the pin is in contact cannot be determined by applying Eq. (13.3) since the boundary is not free and, in general, σ_1 or σ_2 will not approach zero. In this case neither σ_1 nor σ_2 is known, a priori, and it is necessary to separate the stresses, i.e., individually determine the values of σ_1 and σ_2, at this region of contact on the pinhole. Methods to employ in separating the stresses are presented later in this section.

As another example of the interpretation of isochromatic fringe patterns, consider the photograph shown in Fig. 13.3. In this instance, a photoelastic model

FIGURE 13.3
Dark-field isochromatic fringe pattern for a pressurized section of square tubing with a circular bore.

of a square conduit with a pressurized circular borehole was analyzed. A uniformly distributed load (the pressure) was applied to the circular hole of the model. The stresses along the outside edges of the model can be determined directly from Eq. (13.3) since these edges represent free boundaries of the model. Along the boundary of the circular hole, the surface is not free; however, in this case the stress acting normal to the boundary is known ($\sigma_2 = -p$, the distributed load). From Eq. (12.7) we write

$$\sigma_1 - \sigma_2 = \sigma_1 + p = \frac{Nf_\sigma}{h}$$

or   (13.4)

$$\sigma_1 = \frac{Nf_\sigma}{h} - p$$

Note that the applied pressure p is considered as a positive quantity.

Some rules for interpreting isochromatic fringe patterns are:

1. The principal stress difference $\sigma_1 - \sigma_2$ can be determined at any point in the model by using Eq. (12.7).
2. If $\sigma_1 > 0$ and $\sigma_2 < 0$, $\sigma_1 - \sigma_2$ can be related to the maximum-shear stress through Eq. (13.1).
3. If $\sigma_1 > \sigma_2 > 0$ or if $0 > \sigma_1 > \sigma_2$, $\sigma_1 - \sigma_2$ cannot be related to the maximum-shear stress and it is necessary to determine σ_1 and σ_2 and relate τ_{max} to σ_1 or σ_2 by Eqs. (13.2).
4. If the boundary can be considered free (that is, σ_1 or $\sigma_2 = 0$), the other principal stress can be determined directly from Eq. (13.3).
5. If the boundary is not free, but the applied normal load is known, the unknown tangential boundary stress can be determined by applying Eqs. (13.4).
6. If the boundary is not free and the applied load is not known, separation techniques discussed later must be applied to determine the boundary stresses.

13.2.2 Isoclinic Fringe Patterns [1–6]

The isoclinic fringe pattern obtained in the plane polariscope is used to give the directions of the principal stresses at any point in the model. In practice, this may be accomplished in one of two ways. First, a number of isoclinic patterns at different polarizer orientations can be obtained and combined to give a composite picture of the isoclinic parameters over the entire field of the model. Second, the isoclinic parameter can be determined at individual points of interest.

An example of a series of isoclinic fringe patterns is shown in Fig. 13.4, where a thick-walled ring has been loaded in diametral compression. The data presented in this series of photographs are combined to give the composite

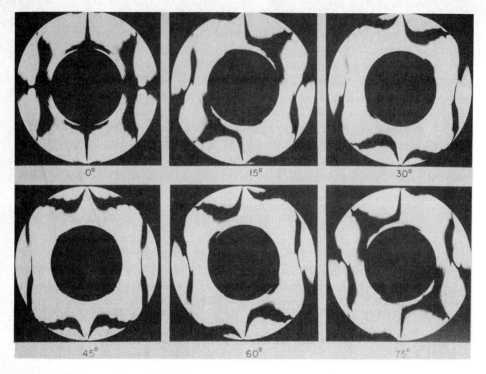

FIGURE 13.4
Isoclinic fringe patterns for a circular ring subjected to a diametral compressive load.

isoclinic pattern illustrated in Fig. 13.5. Several rules can be followed in sketching the composite isoclinic patterns from the individual patterns. These rules are:

1. Isoclinics of all parameters must pass through isotropic or singular points.
2. An isoclinic of one parameter must coincide with any axis of symmetry which exists.
3. The parameter of an isoclinic intersecting a free boundary is determined by the slope of the boundary at the point of intersection.
4. Isoclinics of all parameters pass through points of concentrated load since they are singular points.

Inspection of Figs. 13.4 and 13.5 shows that isoclinics of all parameters pass through the isotropic points labeled A through J. At isotropic points $(\sigma_1 = \sigma_2)$, all directions are principal; therefore, isoclinics of all parameters pass through these points. Again, from Fig. 13.5, it is clear that the parameter of an isoclinic can be established from the slope of the boundary at its point of intersection. The reason for this behavior of isoclinics at free boundaries is based on the fact that the boundaries are isostatics or stress trajectories. That is, the

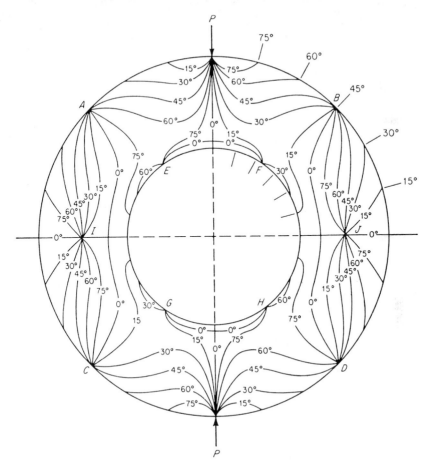

FIGURE 13.5
Composite isoclinic pattern for a circular ring subjected to a diametral compressive load.

tangential stress at the boundary is principal, and as such, the isoclinic parameter must identify the slope of the boundary at the point of intersection. Also, axes of symmetry are principal directions; therefore, the isoclinics must identify them. The horizontal and vertical axes of the ring shown in Fig. 13.4 are axes of symmetry and are included in the 0° isoclinic family. Finally, at points where a concentrated load is applied, the stress system in the local neighborhood of the loading point is principal in r and θ. Thus, the principal-stress directions will vary from 0 to 180° in this local region; therefore, isoclinics of all parameters will converge at the point of load application, as illustrated in Fig. 13.5.

The isoclinics, lines along which the principal stresses have a constant inclination, give the principal-stress directions in a form that is not well understood. To improve interpretation, the principal-stress directions can be presented in the form of an isostatic or stress trajectory diagram where the principal stresses are tangent or normal to the isostatic lines at each point. The isostatic diagram can be constructed directly from the composite isoclinic pattern by utilizing the procedure illustrated in Fig. 13.6. In this construction technique, the stress trajectories are initiated on the 0° isoclinic at arbitrarily spaced points. Lines labeled (1) and oriented 0° from the normal are drawn through each of these arbitrary points until they intersect the 10° isoclinic line. The lines (1) are bisected, and a new set of lines (2) is drawn, inclined at 10° to the vertical to the next isoclinic parameter. Again these lines are bisected, and another set of construction lines (3) is drawn oriented at an angle of 20° to the vertical. The process is repeated until the entire field is covered. The stress trajectories are then drawn by using lines 1, 2, 3, etc., as guides. The stress trajectories are tangent to the construction lines at each isoclinic intersection, as illustrated in Fig. 13.6.

The isoclinic parameters are also employed to determine the shear stresses on an arbitrary plane. By recalling that the isoclinic parameter gives the direction between the x axis of the coordinate system and the direction σ_1 or σ_2, and referring to Eqs. (1.16), it is clear that

$$\tau_{xy} = -\frac{\sigma_1 - \sigma_2}{2} \sin 2\theta_2 = \frac{Nf_\sigma}{2h} \sin 2\theta_2 \qquad (13.5a)$$

or

$$\tau_{xy} = -\frac{\sigma_1 - \sigma_2}{2} \sin 2\theta_1 = \frac{Nf_\sigma}{2h} \sin 2\theta_1 \qquad (13.5b)$$

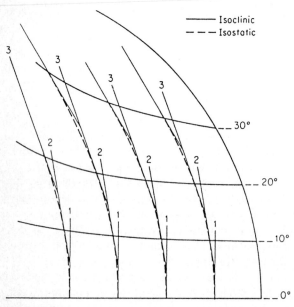

FIGURE 13.6
Construction technique for converting isoclinics to isostatics.

where θ_1 or θ_2 is the angle between the x axis and the direction of σ_1 or σ_2 as given by the isoclinic parameter.

The combined isochromatic and isoclinic data represented by Eqs. (13.5) yield τ_{xy}, which is used in the application of the shear-difference method (see Sec. 13.2.5) for individually determining the values of σ_1 and σ_2.

13.2.3 Compensation Techniques [7–10]

The isochromatic fringe order can be determined to the nearest $\frac{1}{2}$ order by employing both the light- and dark-field fringe patterns. Further improvements on the accuracy of the fringe-order determination can be made by employing mixed-field patterns or by using Post's method of fringe multiplication. However, in certain instances even greater accuracies are required, and point-per-point compensation techniques are employed to establish the fringe order N.

The Tardy method of compensation is very commonly employed to determine the order of the fringe at any arbitrary point on the model. Actually, the Tardy method is often preferred over other compensation techniques since no auxiliary equipment is necessary and the analyzer of the polariscope serves as the compensating device. To employ the Tardy method, the axis of the polarizer is aligned with the principal direction of σ_1 at the point in question, and all other elements of the polariscope are rotated relative to the polarizer so that a standard dark-field polariscope exists. The analyzer is then rotated to produce extinction at the select point. The interpretation of the angle of rotation of the analyzer in terms of fractional fringe order was discussed in Sec. 12.6.

13.2.4 Calibration Methods [1]

In most photoelastic analyses the stress distribution in a complex model is sought as a function of the load. To determine this stress distribution accurately requires the careful calibration of the material fringe value f_σ. The values of f_σ given in the technical literature are typical. The material fringe value of photoelastic materials varies with the supplier, the batch of resin, temperature, and age, and it is necessary to calibrate each sheet of photoelastic material at the time of the test. The calibration method presented here is a simple and accurate way to determine f_σ.

In any calibration technique one must select a body for which the theoretical stress distribution is accurately known. The model should also be easy to machine and simple to load. The calibration model is loaded in increments, and the fringe order and the loads are noted. From these data the material fringe value f_σ can be determined.

The circular disk subjected to a diametral compressive load P is often employed as a calibration model since it is easy to machine and to load. The stress distribution along the horizontal diameter (that is, $y = 0$) is given by the following expressions:

$$\sigma_{xx} = \sigma_1 = \frac{2P}{\pi h D} \left(\frac{D^2 - 4x^2}{D^2 + 4x^2}\right)^2$$

$$\sigma_{yy} = \sigma_2 = -\frac{2P}{\pi h D} \left[\frac{4D^4}{(D^2 + 4x^2)^2} - 1\right] \qquad (13.6)$$

$$\tau_{xy} = 0$$

where D = diameter of disk
h = thickness of disk
x = distance along the horizontal diameter measured from center of disk

The difference in the principal stresses σ_1 and σ_2 is

$$\sigma_1 - \sigma_2 = \frac{8P}{\pi h D}\left[\frac{D^4 - 4D^2 x^2}{(D^2 + 4x^2)^2}\right] = \frac{Nf_\sigma}{h} \qquad (13.7)$$

or

$$f_\sigma = \frac{8P}{\pi D N}\left[\frac{D^4 - 4D^2 x^2}{(D^2 + 4x^2)^2}\right] \qquad (13.8)$$

Equation (13.8) can be employed to calibrate photoelastic materials if only one load P is applied to the disk. In this case, the fringe order N is determined as a function of position x along the horizontal diameter. These values of N and x are then substituted into Eq. (13.8) to give several values of f_σ, which in turn are averaged to reduce errors in the reading of the fringe order N.

More often, however, the center point of the disk, that is, $x = y = 0$, is used for the calibration point, and several values of load are applied to the model. In this instance, Eq. (13.8) reduces to

$$f_\sigma = \frac{8}{\pi D}\left(\frac{P}{N}\right) \qquad (13.9)$$

Note that the value of f_σ is independent of the model thickness h. The value of P/N used is determined by plotting a line through several points of P versus N and establishing its slope.

13.2.5 Separation Methods [1–5, 11–14]

At interior regions of the model, individual values for the principal stresses σ_1 and σ_2 cannot be obtained directly from the isochromatic and isoclinic patterns without using supplementary data or employing numerical methods. The separation methods discussed here have been limited to those which are commonly used and include methods based on the equilibrium equations and the comparability equations.

THE SHEAR-DIFFERENCE METHOD [1, 2, 5]. The equations of equilibrium [Eqs. (1.3)] when applied to the plane-stress problem in the absence of body forces reduce to

$$\frac{\partial \sigma_x}{\partial x} + \frac{\partial \tau_{yx}}{\partial y} = 0$$

$$\frac{\partial \sigma_y}{\partial y} + \frac{\partial \tau_{xy}}{\partial x} = 0$$

(13.10)

Integration of the equilibrium equations leads to

$$\sigma_x = (\sigma_x)_0 - \int \frac{\partial \tau_{yx}}{\partial y} \, dx$$

$$\sigma_y = (\sigma_y)_0 - \int \frac{\partial \tau_{xy}}{\partial x} \, dy$$

which can be closely approximated by the finite-difference expressions

$$\sigma_x = (\sigma_x)_0 - \sum \frac{\Delta \tau_{yx}}{\Delta y} \Delta x$$

$$\sigma_y = (\sigma_y)_0 - \sum \frac{\Delta \tau_{xy}}{\Delta x} \Delta y$$

(13.11)

In the above expressions, the terms $(\sigma_x)_0$ and $(\sigma_y)_0$ represent known stresses at boundary locations that have been selected as starting points for the integration process. Usually, these points are selected on free boundaries where the nonzero stress can be computed directly from the isochromatic data. The term τ_{xy} can be computed at any interior point of the model by using Eq. (13.5b). When using this expression, care must be taken to maintain the proper algebraic sign for τ_{xy}. The expression gives the sign of the shear stress in accordance with the theory-of-elasticity sign convention outlined in Chap. 1. The sign should be verified by inspection whenever possible. The term $\Delta \tau_{xy}$ is determined from a graph of the shear-stress distributions along two auxiliary lines (parallel and symmetric to the line of interest), as illustrated in Fig. 13.7. From the figure it can be noted that the area (shown shaded) between the τ_{xy} curves represents the quantity $\sum \Delta \tau_{xy} \Delta x$. The accumulated area between the origin and a point at a distance x from the origin along the line of interest can be used to compute the difference between $(\sigma_x)_0$ and $(\sigma_x)_x$ simply by dividing by the distance Δy between the two auxiliary lines. Once σ_x is known at a given point, the value for σ_y can be computed from

$$\sigma_y = \sigma_x - (\sigma_1 - \sigma_2) \cos 2\theta_1$$

(13.12)

Equations (1.8), (1.9), and (12.7) are combined to give the two principal stresses as follows:

$$\sigma_1, \sigma_2 = \tfrac{1}{2}(\sigma_x + \sigma_y) \pm \tfrac{1}{2}(\sigma_1 - \sigma_2) = \tfrac{1}{2}(\sigma_x + \sigma_y) \pm \frac{Nf_\sigma}{2h}$$

(13.13)

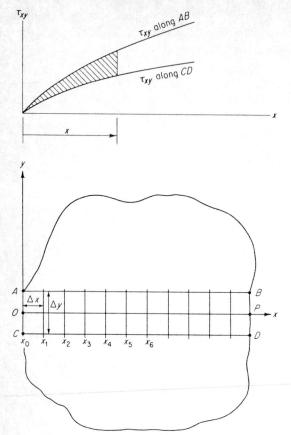

FIGURE 13.7
Grid system used with the shear-difference method.

With the principal stresses and their orientation known at every point along the line, the state of stress is completely specified. The procedure can be repeated for any line of interest in the specimen.

THE ANALYTIC SEPARATION METHOD [11–12]. The compatibility equation for plane stress or plane strain in terms of cartesian stress components is given by

$$\frac{\partial^2}{\partial x^2}(\sigma_{xx} + \sigma_{yy}) + \frac{\partial^2}{\partial y^2}(\sigma_{xx} + \sigma_{yy}) = 0 \tag{13.14}$$

With constant or zero body forces, equations of this form are known as *Laplace's equation*, and any function that satisfies this equation is said to be a *harmonic function*. In photoelasticity, interest in Laplace's equation arises from the fact that the value of the function is uniquely determined at all interior points of a region if the boundary values are known. It was shown previously that the photoelastic isochromatics provide an accurate means for determining both the principal stress difference $\sigma_1 - \sigma_2$ at all interior points of a two-dimensional model and in many instances complete boundary-stress information. Knowledge of the principal-stress sum $\sigma_1 + \sigma_2$ throughout the interior, together with the principal-stress difference $\sigma_1 - \sigma_2$, provides an effective means for evaluating the individual principal stresses.

Solution of Laplace's equation by the method of separation of variables yields a sequence of harmonic functions which can be added together in a linear combination to give a series representation H that is an approximation of the first stress invariant I. Terms which are solutions to $\nabla^2 H = 0$ referred to cartesian and polar coordinate systems are presented in Table 13.1.

If the region of the model conforms to a regular coordinate system, determining the coefficient for each term in the series representation is considerably simplified. The sequence of harmonic functions reduces to a Fourier series on the boundaries of the region. The unknown coefficients in these cases are the Fourier coefficients obtained by integrating the prescribed boundary values.

If the region of the model does not conform to a particular coordinate system, Fourier analysis cannot be employed to determine the coefficients which satisfy the prescribed boundary conditions. Instead, the method of least squares is used to determine the coefficients. A finite number of harmonic functions is selected to form a truncated series solution for H. Coefficients for the terms in the series are then chosen such that the mean-square difference between the prescribed boundary values and the evaluation of the series along the boundary is minimized. If N harmonic functions $F_1, F_2, \ldots, F_N$ are selected and the associated unknown coefficients are denoted as $C_1, C_2, \ldots, C_N$, the truncated series for H is

$$H = \sum_{n=1}^{N} C_n F_n \tag{13.15}$$

If $I(s)$ is used to represent the distribution of the first stress invariant along a boundary of total length L, the approximation H must be selected such that

$$\int_0^L \left[I(s) - \sum_{n=1}^{N} C_n F_n \right]^2 ds = \text{minimum} \tag{13.16}$$

TABLE 13.1
Solution of $\nabla^2 H = 0$ in cartesian and polar coordinate systems

Coordinate system	Sequence of harmonic functions		
Cartesian: $\dfrac{\partial^2 H}{\partial x^2} + \dfrac{\partial^2 H}{\partial y^2} = 0$	1 x y xy	$\sinh kx \sin ky$ $\sinh kx \cos ky$ $\cosh kx \sin ky$ $\cosh kx \cos ky$	$\sinh ky \sin kx$ $\sinh ky \cos kx$ $\cosh ky \sin kx$ $\cosh ky \cos kx$
Polar: $\dfrac{\partial^2 H}{\partial r^2} + \dfrac{1}{r}\dfrac{\partial H}{\partial r} + \dfrac{1}{r^2}\dfrac{\partial^2 H}{\partial \theta^2} = 0$	1 $\ln r$	$x = r \cos \theta$ $r^n \cos n\theta$ $r^{-n} \cos n\theta$	$y = r \sin \theta$ $r^n \sin n\theta$ $r^{-n} \sin n\theta$

The N unknown coefficients of this series can be evaluated by using the method of least squares where the partial derivative of the integral with respect to each of the coefficients is set equal to zero to obtain

$$\frac{\partial}{\partial C_k} \int_0^L \left[I(s) - \sum_{n=1}^N C_n F_n \right]^2 ds = 0 \qquad k = 1, 2, \ldots, N$$

which can be reduced to

$$\sum_{n=1}^N C_n \int_0^L F_n F_k \, ds = \int_0^L I(s) F_k \, ds \qquad k = 1, 2, \ldots, N \tag{13.17}$$

Equation (13.17) yields N simultaneous equations in terms of the N unknown coefficients. Solution of this set of equations gives the coefficients which provide the best match of boundary values possible with the initial selection of N harmonic functions in the truncated series. By increasing the number of functions in the series for H, the fit can be made as accurate as the original photoelastic determination of $I(s)$.

13.2.6 Scaling Model-to-Prototype Stresses [15–18]

In the analysis of a photoelastic model fabricated from a polymeric material, the question of applicability of the results is often raised since the prototype is usually fabricated from metal. Obviously, the elastic constants of the photoelastic model are greatly different from those of the metallic prototype. However, the stress distribution obtained for a plane-stress or plane-strain problem by a photoelastic analysis is usually independent of the elastic constants, and the results for an elastic analysis are applicable to a prototype constructed from any material. This statement is established most readily by reference to the stress equation of compatibility for the plane-stress case. From Eqs. (3.15) and (3.17c) it is clear that

$$\nabla^2(\sigma_{xx} + \sigma_{yy}) = -(\nu + 1)\left(\frac{\partial F_x}{\partial x} + \frac{\partial F_y}{\partial y} \right) \tag{13.18}$$

This equation of stress compatibility is independent of the modulus of elasticity E and shows that the modulus of the model does not influence the stress distribution. The influence of Poisson's ratio depends on the nature of the body-force distribution. If $\partial F_x/\partial x + \partial F_y/\partial y = 0$, the stress distribution is independent of Poisson's ratio, which implies that there will be no influence due to Poisson's ratio when

1. $F_x = F_y = 0$ (the absence of body forces)
2. $F_x = C_1$, $F_y = C_2$ (the uniform body-force field, i.e., gravitational)
3. $F_x = C_1 x$, $F_y = -C_1 y$ (a linear body-force field in x and y)

There are two exceptions to this general law of similarity of stress distributions in two-dimensional bodies. First, if the two-dimensional photoelastic model is multiply connected, Eq. (13.18) does not apply. The multiply-connected body has a hole or series of holes, and the influence of Poisson's ratio will depend upon the loading on the boundary of the hole. If the resultant force acting on the boundary of the hole is zero, the stress distribution is again independent of Poisson's ratio. However, if the resultant force applied to the boundary of the hole is not zero, Poisson's ratio influences the distribution of the stresses. Fortunately, in those examples where the effect of Poisson's ratio has been evaluated, its influence on the maximum principal stress is usually small (less than about 7 percent).

The second exception to the laws of similitude is when the photoelastic model undergoes appreciable distortion under the action of the applied load. Local distortions are a source of error in notches, for example, since curvatures are modified and the stress-concentration factors are decreased. These model distortions can be minimized by selecting a model material with a high figure of merit and reducing the applied load to the lowest value consistent with adequate model response.

Since the photoelastic model may differ from the prototype in respect to scale, thickness, and applied load, as well as the elastic constants, it is important to extend this treatment to include the scaling relationships. The literature abounds with scaling relationships employing dimensionless ratios and the Buckingham π theory. However, in most photoelastic applications, scaling the stresses from the model to the prototype is a relatively simple matter where the pertinent dimensionless ratios can be written directly. For instance, for a two-dimensional model with an applied load P, the dimensionless ratio for stress is $\sigma h L/P$ and for displacements $\delta E h/P$. Thus the prototype stresses σ_p are written as

$$\sigma_p = \sigma_m \frac{P_p}{P_m} \frac{h_m}{h_p} \frac{L_m}{L_p} \tag{13.19}$$

and the prototype displacements δ_p as

$$\delta_p = \delta_m \frac{P_p}{P_m} \frac{E_m}{E_p} \frac{h_m}{h_p} \tag{13.20}$$

where h = thickness

L = typical lateral dimension

and subscripts p and m refer to the prototype and the model, respectively.

It is clear that scaling between model and prototype is accomplished easily in most two-dimensional problems encountered by the photoelastician. The modulus of elasticity is never a consideration in determining the stress distribution unless the loading deforms the model and changes the load distribution, e.g., contact stresses. Also, Poisson's ratio need not be considered when the body is simply connected and the body-force field is either absent or uniform, i.e., gravity loading.

13.3 MATERIALS FOR TWO-DIMENSIONAL PHOTOELASTICITY

One of the most important factors in a photoelastic analysis is the selection of the proper material for the photoelastic model. Unfortunately, an ideal photoelastic material does not exist, and the investigator must select from the list of available polymers the one which most closely fits the investigation's needs. The quantity of photoelastic plastic used each year is not sufficient to entice a chemical company into the development and subsequent production of a polymer especially designed for photoelastic applications. As a consequence, the photoelastician must select a model material which is usually employed for some purpose other than photoelasticity.

The following list gives properties which an ideal photoelastic material should exhibit. The material should

1. Be transparent to the light employed in the polariscope
2. Be sensitive to either stress or strain, as indicated by a low material fringe value in terms of either stress f_σ or strain f_ϵ
3. Exhibit linear characteristics with respect to (a) stress-strain properties, (b) stress-fringe-order properties, and (c) strain-fringe-order properties
4. Exhibit mechanical and optical isotropy and homogeneity
5. Not exhibit viscoelastic behavior
6. Have a high modulus of elasticity and a high proportional limit
7. Have sensitivities f_σ or f_ϵ that are essentially constant with small variations in temperature
8. Be free of time-edge effects
9. Be capable of being machined by conventional means
10. Be free of residual stresses
11. Be available at reasonable cost

These criteria are discussed individually in the following subsections.

13.3.1 Transparency

In most applications, the materials selected for photoelastic models are transparent plastics. These plastics must be transparent to visible light, but they need not be crystal clear. This transparency requirement is not difficult to meet since most polymeric materials are colored or made opaque by the addition of fillers. The basic polymer, although not crystal clear, is usually transparent.

In certain special applications which involved a study of the residual stresses in normally opaque materials, e.g., germanium or silicon, an infrared polariscope has been used. A few materials are transparent in either the ultraviolet region or the infrared region of the radiant-energy spectrum. Polariscopes can be constructed to operate in either of these regions if advantages can be gained by employing light with very short or very long wavelengths. However, for most conventional stress-analysis purposes, visible-light polariscopes are preferred.

13.3.2 Sensitivity

A highly sensitive photoelastic material is preferred since it increases the number of fringes which can be observed in the model. If the value of f_σ for a model material is low, a satisfactory fringe pattern can be achieved in the model with relatively low loads. This feature reduces the complexity and size of the loading fixture and limits the distortion of the model. In the case of birefringent coatings, which will be discussed later, a material with a low value of f_ε is essential to reduce the effects due to the coating thickness.

Photoelastic materials are available with values of f_σ which range from 0.035 to 350 kN/m. The situation regarding values of f_ε is not so satisfactory since materials with the desired low value of f_ε are not yet available. Values of f_ε are usually in the range from 0.005 to 0.50 mm. A material with a strain sensitivity of 0.0005 mm would greatly enhance the applicability of the birefringent coating method of photoelasticity.

13.3.3 Linearity

Photoelastic models are normally employed to predict the stresses that occur in a metallic prototype. Since model-to-prototype scaling must be used to establish prototype stresses, the model material must exhibit linear stress-strain, optical-stress, and optical-strain properties. Very few data are available in the technical literature on optical-strain relationships, but fortunately the photoelastic method is usually employed to determine stress differences and the lack of data on strain behavior is not serious. The typical stress-strain and stress-fringe-order curves presented in Fig. 13.8 show the characteristic behavior of polymeric photoelastic

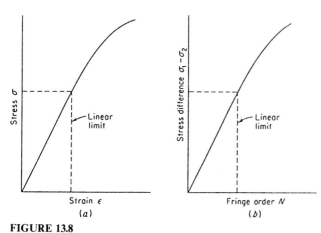

FIGURE 13.8
Typical (a) stress-strain and (b) stress-optical response curves for polymers used in photoelastic models.

materials. Most polymers exhibit linear stress-strain and stress-fringe-order curves for the initial portion of the curve. However, at higher levels of stress, the material may exhibit nonlinear effects. For this reason the higher stress levels are to be avoided in photoelastic tests associated with these nonlinear materials.

13.3.4   Isotropy and Homogeneity

Many photoelastic materials are prepared from liquid polymers by casting between two glass plates which form the mold. When the photoelastic materials are prepared by a casting process, the molecular chains of the polymer are randomly oriented and the materials are essentially isotropic and homogeneous. However, certain plastics are rolled, stretched, or extruded during the production process. In these production processes the molecular chains are oriented in the direction of rolling, stretching, or extrusion. These materials will exhibit anisotropic properties (both mechanical and optical) and must be annealed to randomly orient the molecular chains to give a material with isotropic properties.

13.3.5   Creep

Unfortunately, most polymers creep both mechanically and optically over the time associated with a photoelastic analysis. Because of the effects of mechanical and optical creep, polymers cannot be truly characterized as elastic materials but must be considered viscoelastic.

One of the first attempts to formulate a mathematical theory of photoviscoelasticity was made by Mindlin [19], who considered a generalized model with both elastic elements and viscous elements. By assuming that the photoelastic effect resulted from only the deformation of the elastic elements of the model, Mindlin showed that the relative retardation could be related to the stress and strain as

$$\sigma_1 - \sigma_2 = \frac{N}{h} f_\sigma(t)$$

$$\epsilon_1 - \epsilon_2 = \frac{N}{h} f_\epsilon(t)$$

(13.21)

where f_σ and f_ϵ are written as functions of time rather than as constants.

The results of Eqs. (13.21) are significant since they show that viscoelastic model materials can be employed to perform elastic-stress analyses. Because of the viscoelastic nature of the model materials, the stress and strain material fringe values are functions of time; however, over the short time interval needed to photograph a fringe pattern, either f_σ or f_ϵ can be treated as a constant. A typical graph of the variation in material fringe value with time is shown in Fig. 13.9. For most polymers, the stress fringe value f_σ decreases rapidly with time immediately after loading but then tends to stabilize after about 30 to 60 min. In practice, the load is maintained on the model until the fringe pattern stabilizes. The pattern is

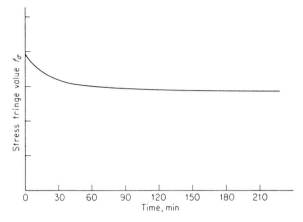

FIGURE 13.9
Typical curve showing the time-after-loading dependence of the stress fringe value in a viscoelastic polymeric material.

then photographed, and the material fringe value associated with the time of the photograph is used for the analysis. It should be noted that photoelastic materials vary from batch to batch; hence, each sheet of material must be calibrated at the time of the photoelastic analysis to determine $f_\sigma(t)$. Also, in most photoelastic materials, the polymerization process continues, and f_σ changes with time on a scale of months. Hence, material stored for a year will, in general, exhibit a higher value of f_σ than newly processed material.

13.3.6 Modulus of Elasticity and Proportional Limit

The modulus of elasticity is important in the selection of a material for a photoelastic analysis because the modulus controls the distortion of the model resulting from the applied loads. If the geometry of a boundary changes due to distortion, the photoelastic solution will be in error. For example, a strip with a very sharp notch will have a very high stress concentration at the root of the notch. If the model distorts under load, the sharpness of the notch is decreased and the experimentally determined stress concentration is reduced. Another case where model distortion influences the photoelastic results is the contact problem. In this instance, the stress distribution is a function of the bearing area between the two components and, obviously, the modulus of elasticity is influential in controlling this quantity.

The factor which can be used to judge different photoelastic materials in regard to their ability to resist distortion is $1/f_\epsilon$ or $E/f_\sigma(1 + v)$. The best photoelastic materials to resist distortion will exhibit low values for the material fringe value in terms of strain. Since Poisson's ratio for most polymers varies over the limited range between 0.36 and 0.42, the ratio $Q = E/f_\sigma$ is sometimes used to evaluate the merits of the materials. This factor E/f_σ is known as the *figure of merit*.

The proportional limit σ_{pl} of a photoelastic material is important in two respects. First, a material with a high proportional limit can be loaded to a higher level without endangering the integrity of the model (it should be noted here that the polymers normally employed in photoelasticity exhibit a brittle fracture and as such fail catastrophically when the ultimate load is reached). Second, a material with a high proportional limit can produce a higher-order fringe pattern which tends to improve the accuracy of the stress determinations. A sensitivity index for a model material can be defined as

$$S = \frac{\sigma_{pl}}{f_\sigma} \tag{13.22}$$

Superior model materials exhibit high values for both the sensitivity index S and the figure of merit Q.

13.3.7   Temperature Sensitivity [20]

If the material fringe value f_σ changes markedly with temperature, errors can be introduced in a photoelastic analysis by minor temperature variations during the experiment. Typical changes in the material fringe value f_σ with temperature are shown in Fig. 13.10. For most polymers there is a linear region of the curve where f_σ decreases slightly with temperature. For one commonly used epoxy, the change in f_σ is only 0.022 kN/(m)(°C) in this linear region. However, at temperatures in excess of 150°F (65°C), the value of f_σ begins to drop sharply with any increase in temperature, as shown in Fig. 13.10. For conventional two-dimensional photo-elastic studies conducted at room temperature (75°F or 24°C), the slope of the curve in the linear region is the important characteristic. For most of the commonly used materials, the slope is modest at room temperature and variations in f_σ can be neglected if temperature fluctuations are limited to ± 5°F or ± 3°C.

13.3.8   Time-Edge Effect [21]

When a photoelastic model is machined from a sheet of material and examined under no-load for a period of time, it is observed that a time-edge stress is induced on the boundary of the model which produces a fringe or a series of fringes parallel

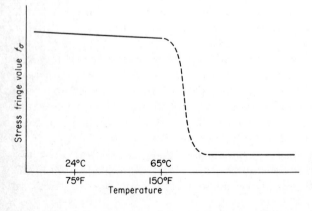

FIGURE 13.10
Typical curve showing the change in stress fringe value with temperature.

to the boundary. An example of a model with a severe case of time-edge stress is illustrated in Fig. 13.11.

The influence of these time-edge stresses on a photoelastic analysis is quite important. The fringe pattern observed is due to the superposition of two states of stress, the first associated with the load and the second a result of the time-edge stresses. Since the time-edge stresses are large on the boundary, the errors introduced may be quite large in determining the extremely important boundary stresses.

It has been established that the time-edge stresses are caused by diffusion of water vapor from the air into the plastic or from the plastic into the air. For many plastics, the diffusion process is so slow at room temperature that it requires years to reach an equilibrium state. For this reason, a freshly machined edge of a model usually will be in a condition to accept water from the air (its central region has not been saturated), and time-edge stresses will begin to develop. The rate at which the time-edge stresses develop for a particular photoelastic polymer will depend upon the relative humidity of the air and the temperature.

The epoxy resins are somewhat different from most other polymers in that their diffusion rate is sufficiently high that a saturated condition can be established after 2 to 3 months. If a two-dimensional model is machined from a sheet of material that has been maintained at a constant humidity for several months so that it is in a state of equilibrium (concentration of water is uniform through the thickness of the sheet), and if the model is tested under these same humidity conditions, then time-edge stresses will not develop.

13.3.9 Machinability

Photoelastic materials must be machinable in order to form the complex models employed in photoelastic analyses. Ideally, it should be possible to turn, mill, rout, drill, and grind these plastics. Although machinability properties may appear to be a trivial requirement, it is often extremely difficult to machine a high-quality photoelastic model properly. The action of the cutting tool on the plastic often produces heat coupled with relatively high cutting forces. As a consequence, boundary stresses due to machining can be introduced and locked into the model, making it unsuitable for quantitative photoelastic analysis.

In machining photoelastic models, care must be taken to avoid high cutting forces and the generation of excessive amounts of heat. These requirements can best be accomplished by using sharp carbide-tipped tools, air or oil cooling, and light cuts coupled with a relatively high cutting speed. For two-dimensional applications, complex models may be routed from almost any thermosetting plastic. In this machining method a router motor (20,000 to 40,000 r/min) is used to drive a carbide rotary file. The photoelastic model is mounted to a metal template which describes the exact shape of the final model. The plastic is rough-cut with a jigsaw to within about 3 mm of the template boundary. The final machining operation is accomplished with the router, as illustrated in Fig. 13.12. The metal template is guided by an oversize-diameter pin which is coaxial with

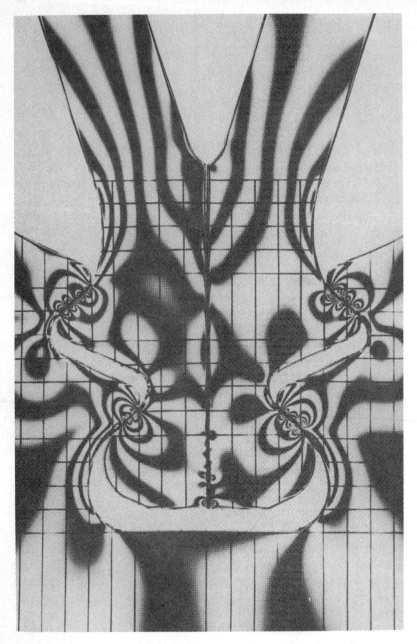

FIGURE 13.11
An example of fringe distortion near the boundary due to time-edge stresses in a photoelastic model of a turbine-blade dovetail joint.

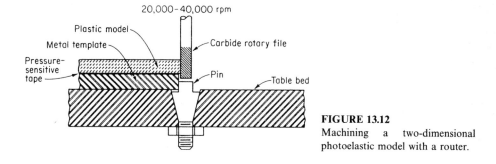

FIGURE 13.12
Machining a two-dimensional photoelastic model with a router.

the rotary file. The rate of feed along the boundary of the model is carefully controlled by moving the model along the pin by hand. Successive cuts are taken by reducing the diameter of the stationary pin until it finally coincides with the cutter diameter. By using this technique, satisfactory two-dimensional models can be produced in less than 1 h by a skilled technician.

13.3.10 Residual Stresses

Residual stresses are sometimes introduced into photoelastic plastics during casting and curing operations and almost always by rolling or extrusion processes. They can be observed simply by inserting the sheet of material into a polariscope and noting the order of the fringes in the sheet. The presence of residual stresses in photoelastic models is detrimental since they are superimposed on the true stress distribution produced by loading the model. Since it is difficult, if not impossible, to subtract out the contribution due to the residual-stress distribution, the presence of residual stresses in the model material often introduces serious errors into a photoelastic analysis. In certain cases it is possible to reduce the level of the residual stresses by thermally cycling the sheet above its softening point on a flat plate or in an oil bath.

13.3.11 Cost of Material

Normally the cost of the model material in a photoelastic analysis represents a very small percentage of the total cost of the investigation. For this reason, the cost of the materials should not be overemphasized, and the most suitable material should be selected on the basis of other parameters, regardless of the apparent difference in cost of the material on a pound or kilogram basis.

13.4 PROPERTIES OF COMMONLY EMPLOYED PHOTOELASTIC MATERIALS
[22–29]

A brief examination of the photoelastic literature will show that almost all polymers exhibit temporary double refraction and that many different materials have been employed in photoelastic analyses. The list includes several types of glass, celluloid, gelatin, the glyptal resins, natural and synthetic rubber, fused silica, the phenolformaldehydes, polycarbonate, allyl diglycol (CR-39), and several compositions of the epoxies and the polyesters. Today, most elastic-stress analyses are conducted by employing one of the following materials:

1. Homalite 100
2. Polycarbonate
3. Epoxy resin
4. Urethane rubber

13.4.1 Homalite 100

Homalite 100 is a polyester resin which is cast between two plates of glass to form very large sheets. The surfaces of the commercially available sheets are of optical quality, and the material is free of residual stresses. Models can be machined by routing; however, since the material is extremely brittle, edge chipping can be a problem.

Homalite 100 does not exhibit appreciable creep; therefore, the material fringe value can be treated as a constant for loading times in excess of 10 to 15 min. Since moisture absorption is very slow in this material, time-edge effects do not become apparent for several days even under very humid test conditions. The material exhibits both a low figure of merit Q and a low sensitivity index S. High fringe orders cannot be achieved without fracturing the model.

13.4.2 Polycarbonate [29]

Polycarbonate is an unusually tough and ductile polymer which yields and flows prior to fracture. It is known by the trade name Lexan in the United States and as Makrolan in Europe. Polycarbonate exhibits both a high figure of merit Q and a high sensitivity index S. It is relatively free of time-edge effects and exhibits very little creep at room temperature.

Polycarbonate is a thermoplastic and is produced in sheet form by an extrusion process. It is available in large sheets with reasonably good surface characteristics. Unfortunately, the extrusion process usually produces some residual birefringence in the sheets. Annealing for an extended period of time at or near the softening temperature is required to eliminate this residual birefringence. Polycarbonate is also difficult to machine. Any significant heat produced by the cutting tool will cause the material to soften and deform under the tool. Routing can be performed only under water, and side milling is possible only with a continuous flow of coolant at the tool-workpiece interface. Band sawing and hand filing are often required to produce satisfactory model boundaries. Since the material exhibits both yield and flow characteristics, it can also be employed for photoplastic studies. The birefringence introduced in the plastic state is permanent and is locked into the material on a molecular scale. This behavior makes the material suitable for three-dimensional photoplasticity studies.

13.4.3   Epoxy Resin [22, 24, 26, 27]

Epoxy resins were first introduced in photoelastic applications in the mid-1950s, when they were employed predominantly as materials for three-dimensional photoelasticity. However, a brief review of their properties indicates that they are also quite suitable for use in a wide variety of two-dimensional applications. The commercial epoxy resins are condensation products of epichlorohydrin and a polyhydric phenol. The basic monomer can be polymerized by using acid anhydrides, polyamides, or polyamines. In general, curing with the acid anhydrides requires higher temperatures than curing with the polyamides or polyamines.

A wide variety of epoxy materials can be cast into sheet form. The type of basic monomer, the curing agent, and the percentage of the curing agent relative to the basic monomer can be varied to give an almost infinite number of epoxy materials. The epoxies are usually characterized as brittle materials, but they are easier to machine than the polyesters and polycarbonate. Most of the epoxies exhibit better optical sensitivity than Homalite 100, but they are less sensitive than polycarbonate.

Although the material is susceptible to time-edge effects, the rate of diffusion of water into epoxy is sufficiently high to permit a saturation condition to be achieved in about 2 months. If the sheets are stored until saturated, the model can be cut from the conditioned sheet and little or no time-edge effect will be noted as long as the humidity is held constant. Finally, the material creeps approximately the same amount as polycarbonate or Homalite 100.

13.4.4   Urethane Rubber [23]

Urethane rubber is an unusual photoelastic material in that it exhibits a very low modulus of elasticity (three orders of magnitude lower than that of the other materials listed) and a very high sensitivity, $f_\sigma \approx 0.175$ kN/m. The material can be cast between glass plates to produce an amber-colored sheet with optical-quality surfaces. Except for its very low figure of merit, the material ranks relatively well in comparison with the other materials listed. Its strain sensitivity is so low that time-edge effects are negligible; moreover, the material exhibits little mechanical or optical creep. The material can readily be machined on a high-speed router, but it must be frozen at liquid-nitrogen temperatures before its surfaces can be turned or milled.

The material is particularly suited for demonstration models. Loads applied by hand are sufficient to produce well-defined fringe patterns, and the absence of time-edge effects permits the models to be stored for years. Also, the material is so sensitive to stress that it can be used to study body-force problems if fringe-multiplication techniques are employed with thick models. Finally, urethane rubber can be used for models in dynamic photoelasticity, where its low modulus of elasticity has the effect of lowering the velocity of the stress wave to less than 90 m/s as compared with 2000 m/s in rigid polymers. The low-velocity stress waves in urethane-rubber models are easy to photograph with moderate-speed framing cameras (10,000 frames per second), which are common, while the high-speed stress waves in rigid polymers require high-speed cameras (200,000 frames per second or more) to produce satisfactory fringe patterns for analysis.

13.4.5 Model Material Summary

A summary of the mechanical and optical properties of the four photoelastic materials is presented in Table 13.2.

It is clear by comparing the figure of merit Q and the sensitivity index S that polycarbonate and the epoxies exhibit superior properties. Unfortunately, the polycarbonate material is difficult to machine, and the epoxy resin materials require special precautions to minimize time-edge effects.

Homalite 100 with its low sensitivity can be used in applications where high precision and low model distortion are not required. Homalite 100 has the

TABLE 13.2
Summary of the optical and mechanical properties of several photoelastic materials

Property	Homolite 100	Polycarbonate	Epoxy†	Urethane rubber‡
Time-edge effect	Excellent	Excellent	Good	Excellent
Creep	Excellent	Excellent	Good	Excellent
Machinability	Good	Poor	Good	Poor
Modulus of elasticity E:				
psi	560,000	360,000	475,000	450
MPa	3860	2480	3275	3
Poisson's ratio v	0.35	0.38	0.36	0.46
Proportional limit σ_{pl}:				
psi	7000	5000	8000	20
MPa	48.3	34.5	55.2	0.14
Stress fringe value§ f_σ:				
lb/in	135	40	64	1
kN/m	23.6	7.0	11.2	0.18
Strain fringe value§ f_ε:				
in	0.00033	0.00015	0.00018	0.00324
mm	0.0084	0.0038	0.0046	0.082
Figure of merit Q:				
1/in	4150	9000	7400	450
1/mm	163	354	292	17
Sensitivity index S:				
1/in	52	125	125	20
1/mm	2.05	4.92	4.92	0.78

† ERL-2774 with 42 parts per hundred phthalic anhydride and 20 parts per hundred hexahydrophthalic anhydride.
‡ 100 parts by weight Hysol 2085 with 24 parts by weight Hysol 3562.
§ For green light ($\lambda = 546.1$ nm).

advantage of being available in large sheets with optical-quality surfaces and exhibits low creep and little time-edge effect.

Urethane rubber is extremely useful in special-purpose applications such as demonstration models for instructional purposes. Because of its low material stress fringe value it is also useful for modeling where body forces due to gravity produce the loads. Finally, urethane rubber can be used to great advantage in dynamic photoelastic studies, where its low modulus of elasticity results in low-velocity stress waves which are easy to photograph.

13.5 THREE-DIMENSIONAL PHOTOELASTICITY—STRESS FREEZING
[30–32]

The applications of photoelasticity described in previous sections have been limited to two-dimensional techniques for determining stresses in plane models. However, many bodies exist which are three-dimensional in character and cannot be effectively approached by using two-dimensional techniques. For example, it is not practical to relate the integrated optical effects in a complicated three-dimensional model to the stresses in the model. It is possible, however, to construct and load a three-dimensional model and to analyze interior planes of the model photoelastically by using frozen-stress methods. With the frozen-stress method, model deformations and the associated optical response are locked into a loaded three-dimensional model. Once the stress-freezing process is completed, the model can be sliced and photoelastically analyzed to obtain interior-stress information.

The frozen-stress method for three-dimensional photoelasticity was initiated by Oppel [30] in Germany in 1936. The foundation for the method is the process by which deformations are permanently locked in the model. The four different techniques for locking deformations in the loaded model include stress-freezing, creep, curing, and gamma-ray irradiation methods. In all these methods, the deformations are locked into the model on a molecular scale, permitting the models to be sliced without relieving the locked-in deformations. Since the stress-freezing process is by far the most effective and most popular technique for locking the deformations in the model, it will be the only technique described.

The stress-freezing method of locking in the model deformations is based on the diphase behavior of many polymeric materials when they are heated. The polymeric materials are composed of long-chain hydrocarbon molecules, as illustrated in Fig. 13.13. Some of the molecular chains are well bonded into a three-dimensional network of primary bonds. However, a large number of molecules are loosely bonded together into shorter secondary chains. When the polymer is at room temperature, both sets of molecular bonds, the primary and the secondary, act to resist deformation due to applied load. However, as the temperature of the polymer is increased, the secondary bonds break down and the primary bonds in effect carry the entire applied load. Since the secondary bonds constitute a very large portion of the polymer, the deflections which the primary bonds undergo are quite large yet elastic in character. If the temperature

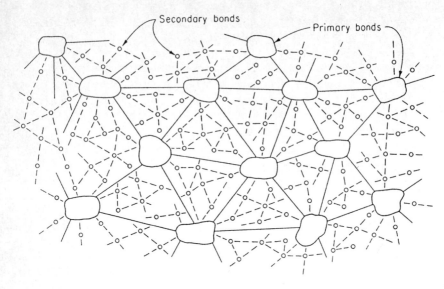

FIGURE 13.13
Primary and secondary molecular chains in a diphase polymer.

of the polymer is lowered to room temperature while the load is maintained, the secondary bonds will re-form between the highly elongated primary bonds and serve to lock them into their extended positions. When the load is removed, the primary bonds relax slightly, but a significant portion of their deformation is not recovered. The elastic deformation of the primary bonds is permanently locked into the body by the re-formed secondary bonds. Moreover, these deformations are locked in on a molecular scale; thus the deformation and accompanying birefringence are maintained in any small section cut from the original body.

This diphase behavior of polymeric materials constitutes the basis of the stress-freezing process so often employed in three-dimensional photoelasticity. This process can be illustrated by considering the simple tensile specimen shown in Fig. 13.14. The specimen shown in Fig. 13.14a is first loaded at room temperature with an axially applied force P, and a displacement Δl_1 is produced, as shown in Fig. 13.14b. Next, the temperature is increased until the secondary bonds break down and the tensile specimen elongates by an additional amount Δl_2 (see Fig. 13.14c). The temperature is then reduced while the load is maintained until the secondary bonds re-form. If thermal expansions and contractions are neglected, the tensile specimen does not change length during the cooling cycle. Finally, the load is removed and the specimen contracts by an amount Δl_1 while retaining a permanently locked-in deformation Δl_2, as illustrated in Fig. 13.14e.

After stress-freezing, a photoelastic model with its locked-in deformations and attendant fringe pattern can be carefully cut or sliced without disturbing the character of either the deformation or the fringe pattern. This fact is due to the

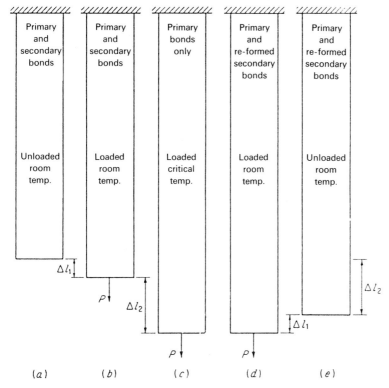

FIGURE 13.14
Stress freezing of a simple uniaxial tensile specimen.

molecular nature of the locking process, where the extended primary bonds are locked into place by the re-formed secondary bonds. The cutting or slicing process may relieve a molecular layer on each face of a slice cut from a model, but this relieved layer is so thin relative to the thickness of the slice that the effect cannot be observed. An example of a cut across a locked-in fringe pattern is shown in Fig. 13.15.

   The temperature to which the polymer is heated to break down the secondary bonds is called the *critical temperature*. Actually, this terminology is somewhat unfortunate because the temperature required to break down the secondary bonds is not critical. Instead, degradation of the secondary bonds depends upon both the temperature and the time under load. If a load is applied to a tensile specimen and maintained at temperatures somewhat below the critical temperature, the deflection or fringe-order response will vary with time under load, as illustrated in Fig. 13.16. At temperatures greater than 95 percent of the critical temperature, the maximum fringe order or deflection is obtained in about 1 min. At the so-called critical temperature the response of the model is almost immediate, with the

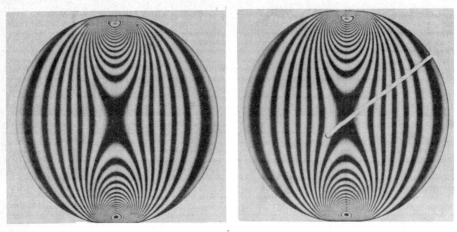

FIGURE 13.15
Illustration showing that careful cutting does not disturb the locked-in fringe pattern.

maximum fringe order or deflection attained in less than 0.1 min. For temperatures between 85 and 90 percent of the critical temperature, maximum response can be achieved, but the load must be maintained for several hours.

The stress-freezing process is used today to lock in the fringe pattern in most three-dimensional photoelastic analyses. The method is extremely simple to apply, as temperature control to ± 5 percent of the critical temperature is usually sufficient. The model response is entirely elastic in character, and the model can

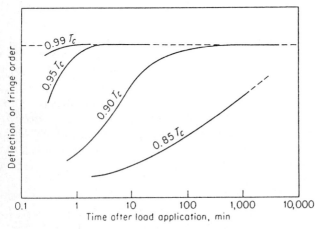

FIGURE 13.16
Deflection or fringe order as a function of time of load application for temperatures below the critical value.

be sliced into plane sections for analysis without disturbing the fringe pattern. The procedure for the stress-freezing process is described below:

1. Place the model in an oven with programmed temperature control.
2. Heat the model relatively rapidly until the critical temperature is attained.
3. Apply the required loads.
4. Soak the model 2 to 4 h until a uniform temperature throughout the model is obtained.
5. Cool the model slowly to minimize temperature gradients.
6. Remove the load and slice the model.

13.6 MATERIALS FOR THREE-DIMENSIONAL PHOTOELASTICITY [33–37]

Since three-dimensional photoelasticity became a reality late in the 1930s with the discovery of the stress-freezing process, a number of different model materials have been introduced and used. These materials include Catalin 61-893, Fosterite, Kriston, Castolite, and the epoxy resins. Of these five materials, only the epoxy-based resins will be discussed here since Fosterite and Kriston are no longer commercially available, and the superiority of the epoxies precludes both Catalin 61-893 and Castolite from practical consideration.

In the discussion of two-dimensional photoelastic materials, 11 requirements of an ideal material were listed. Two additional requirements should be added to this list when considering three-dimensional photoelastic materials: (1) that the material be castable in large sizes and to a final configuration and (2) that the material be cementable. In three-dimensional analyses, the model often is rather large and intricate, and the need for large castings of the epoxy resin becomes very real. Also, in complex models it is very often desirable to cement parts together much as steel components are welded together to form a complex structure or to cast them to the final size in a precision mold. A complex multicomponent model is shown in Fig. 13.17.

The epoxies should be considered as a large family of resins since they cannot be characterized by a single molecular structure. Basically, they consist of condensation products of epichlorohydrin and a polyhydric phenol. The base monomer is commercially available from several sources in both the United States and Europe. The monomer is usually polymerized by employing either an amine or an acid anhydride curing agent. A very large number of these curing agents are commercially available. It is clear, then, that a very large number of polymerized epoxy resins can be produced by varying the type of monomer and the type of hardening agent as well as the percentage of each of these constituents.

Leven [33, 34] investigated the problem of compounding epoxy resins for the specific purpose of optimizing their three-dimensional photoelastic properties.

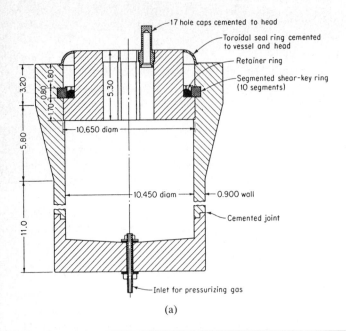

(a)

(b)

FIGURE 13.17

(a) The essential features of a complex three-dimensional photoelastic model; (b) epoxy-resin components used in fabricating the model. (*Courtesy of M. M. Leven, Westinghouse Electric Corporation.*)

From the results of his investigations, it appears that the choice of the basic monomer does not greatly influence the photoelastic properties of the polymerized resin. The basic monomer should be a liquid in which the hardening agent can easily be dissolved and then mixed. Also, the viscosity of the monomer should be low to permit both easy pouring and the release of bubbles produced by stirring. Finally, the monomer should be slow in reacting to the curing agents so that heat generated during curing can be minimized.

The selection of the hardening agent, on the other hand, is extremely important since it influences the photoelastic properties of the resin to an appreciable extent. Of the two general types of hardening agent, the anhydrides are superior to the amines. The amines are not suitable for large castings because of their extremely exothermic reactions. The heat generated in even modest-size castings of amine-cured epoxies is often sufficient to destroy the casting. Moreover, the locked-in fringe pattern in amine-cured epoxies tends to relax with time, and a decrease in the maximum fringe order of 10 to 20 percent in 1 year due to creep of the primary bonds is not uncommon.

The anhydrides (phthalic or hexahydrophthalic) are recommended for the curing of large castings of epoxy resins because of their low exothermic reaction and the low susceptibility of the castings to time-edge stresses. Of the numerous anhydrides available for curing epoxy resins, Leven [34] has recommended the following composition:

1. Liquid epoxy, 100 parts by weight
2. Phthalic anhydride, 42 parts by weight
3. Hexahydrophthalic anhydride, 20 parts by weight

The pertinent three-dimensional photoelastic properties of this particular epoxy resin are:

Critical temperature: 324 to 347°F (162 to 175°C)
Effective modulus: 5300 to 6500 psi (37 to 45 MPa)
Effective material fringe value: 2.48 to 2.84 lb/in (435 to 500 N/m)
Figure of merit: 2100 to 2450 in^{-1} (83 to 96 mm^{-1})

These properties are compared with the photoelastic properties of other resin materials in Table 13.3, which clearly shows the superiority of epoxy resins over all other materials except Kriston, which is no longer commercially available.

The epoxies cured with acid anhydride agents require an extended time in an oven to polymerize. Also, the surfaces of the casting exhibits a rind effect due to a reaction between the epoxy and the atmosphere in the oven. The birefringence associated with this rind effect eliminates the possibility of producing "cast to size and shape" models. These two problems (long curing periods and the rind effect) were circumvented with the development of a new material by Cernosek [35].

TABLE 13.3
Properties of three-dimensional photoelastic materials at their critical temperature

Material	T_{cr} °F	°C	$f_{effective}$† lb/in	N/m	$E_{effective}$ psi	MPa	$Q = E/f$ in^{-1}	mm^{-1}
Catalin 61-893	230	110	3.20	560	1100	7.6	344	13.6
Fosterite	189	87	3.85	674	2320	16.0	603	23.7
Kriston	273	134	6.25	1094	13800	95.2	2208	87.0
Castolite	244	118	8.30	1453	4060	28.0	489	19.3
Epoxy (Cernosek)	285	141	2.00	350	2900	20.0	1450	57.1
Epoxy (Leven)	338	170	2.68	469	6450	44.5	2407	94.9
PLM-4B‡	248	120	2.20	385	2500	17.2	1136	44.7

† For green light ($\lambda = 546.1$ nm).
‡ Available from Photoelastic, Inc., Malvern, Pa.

This epoxy, comprised of

1. Liquid epoxy (Epon 828)
2. Phthalic anhydride (50 percent of epoxy weight)
3. Hardener CA-1 (0.67 percent of epoxy weight)

is a slight modification of the materials developed by Leven. However, the addition of hardener CA-1 (a proprietary blend of aromatic amines) accelerates the cure so that the casting cures in about 12 h. In addition to markedly shortening the curing period, the addition of amines eliminates the rind effect and permits casting of photoelastic models to exact "size and shape." The amine-modified-epoxy is less brittle and can be machined easily using standard tooling.

The procedure for mixing and then casting the amine-modified-epoxy involves the following steps. First, the constituents are heated, mixed, and partially cured to initiate polymerization. After initial polymerization, the batch is mixed again to disperse the polymerization centers (called *mottles*) and then the entrapped air is removed by vacuum. Next, the mixture is poured into a precision mold and cured for about 8 h. The epoxy solidifies and can be easily stripped from the mold. Any machining that is required is performed at this time because the material is not brittle and relatively easy to mill or turn. Finally, the epoxy, now a completed model, is postcured for an additional 6 h to enhance the modulus of elasticity and figure of merit.

Properties of the Cernosek material listed in Table 13.3 show that it has a higher figure of merit than the commercially available PLM-4B but a significantly lower figure of merit than the epoxy developed by Leven.

The anhydride-cured epoxy resins exhibit a time-edge-stress behavior which is unique in comparison with the behavior of other plastics. As discussed previously, the time-edge effect is due to the diffusion of water vapor from the air into the plastic. The rate at which the water vapor diffuses into the plastic depends upon the diffusion constant and the humidity of the air, i.e., the concentration of water vapor in the air. For most plastics the diffusion process is so slow that a state of equilibrium is not reached in a period of years; however, for the anhydride-cured epoxy resins, the rate of diffusion is sufficiently rapid to produce saturation in about 2 months. This ability of the anhydride-cured epoxy resins to saturate can be used to control the time-edge stresses in a slice taken from a three-dimensional model. The stress-freezing process drives off the water vapor stored in the model and, when the model is sliced in preparation for the photoelastic examination, the water vapor begins to diffuse into the dry plastic. The concentration gradient produces time-edge stresses, which increase to a maximum after about 5 days and then decreases to zero in about 2 months as the slice becomes saturated, i.e., gradients of concentration of water go to zero. Any changes in the humidity conditions upset the state of equilibrium and create a new set of time-edge stresses. The procedure employed in controlling the time-edge stresses in slices is to heat the slices to 130°F (55°C) for 1 to 2 days to drive off the absorbed water vapor. The slices must be examined immediately upon their removal from the oven and stored in a desiccator to avoid reabsorption of the water vapor.

13.7 SLICING THE MODEL AND INTERPRETATION OF THE FRINGE PATTERNS [38–41]

If a three-dimensional photoelastic model is observed in a polariscope, the resulting fringe pattern cannot, in general, be interpreted. The conditioned light passing through the thickness of the model integrates the stress difference $\sigma'_1 - \sigma'_2$ over the length of the path of the light so that little can be concluded regarding the state of stress at any point.

To circumvent this difficulty, the three-dimensional model is usually sliced to remove planes of interest which can then be examined individually to determine the stresses existing in that particular plane or slice. In studies of this type it is assumed that the slice is sufficiently thin in relation to the size of the model to ensure that the stresses do not change in either magnitude or direction through the thickness of the slice.

The particular slicing plan employed in sectioning a three-dimensional photoelastic model will depend upon the geometry of the model and the information being sought in the analysis. There are, however, some general principles which can be followed in slicing the model, as described below.

13.7.1 Surface Slices

The free surfaces of a three-dimensional model are principal surfaces since both the stress normal to the surface and the shearing stresses acting on the surface are zero. As an example of a surface slice, consider the flat head on a thick-walled pressure vessel, illustrated schematically in Fig. 13.18. In this example, a surface slice of thickness h is removed from the head and examined at normal incidence, in the polariscope. The fringe pattern observed can be interpreted to give

$$\sigma_1 - \sigma_2 = \frac{N_z f_\sigma}{h}$$

or   (13.23)

$$\sigma_{\theta\theta} - \sigma_{rr} = \frac{N_z f_\sigma}{h}$$

Because of symmetry, $(\sigma_{\theta\theta} - \sigma_{rr}) = (\sigma_1 - \sigma_2)$. The application of Eqs. (13.23) tacitly assumes that the value of $\sigma_{\theta\theta} - \sigma_{rr}$ is constant through the thickness of the slice and that the directions of the principal stresses do not rotate along the z axis. To determine the accuracy of these assumptions or to correct for the errors introduced by changes in the stress distribution with slice thickness, the shaving method is often employed. The shaving method consists essentially in removing a thick slice,

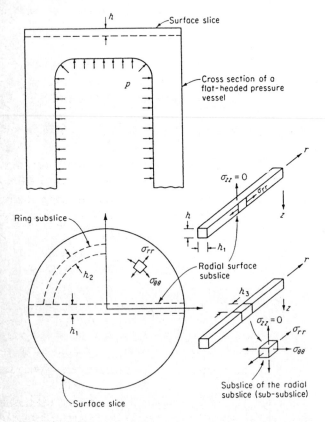

FIGURE 13.18
Surface slicing on a flat-headed, thick-walled pressure vessel.

determining the fringe pattern associated with this slice, and then progressively decreasing the thickness of the slice and establishing the fringe pattern corresponding to each thickness. The results of the analysis are then plotted as shown in Fig. 13.19 and extrapolated to $h = 0$ to establish the surface stresses.

It is often advantageous to subslice the surface slice to obtain the individual values of σ_{rr} or $\sigma_{\theta\theta}$. Two such subslices, a radial subslice and a ring subslice, are illustrated in Fig. 13.18. The radial subslice is examined in the polariscope with the light passing through the subslice in the y or θ direction. The resulting fringe pattern gives

$$\sigma_{rr} - \sigma_{zz} = \frac{N_\theta f_\sigma}{h_1}$$

Because $\sigma_{zz} = 0$ at $z = 0$,

$$\sigma_{rr} = \frac{N_\theta f_\sigma}{h_1} \tag{13.24}$$

Combining Eqs. (13.23) and (13.24) yields

$$\sigma_{\theta\theta} = f_\sigma\left(\frac{N_z}{h} + \frac{N_\theta}{h_1}\right) \tag{13.25}$$

It should be noted in the analysis of the radial subslice that the influence of the $\sigma_{\theta\theta}$ stress, which is coincident with the direction of light, is not effective. Only the stresses which lie in the plane of the slice normal to the direction of light (that is, σ_{zz} or σ_{rr}) influence the fringe pattern. A value of $\sigma_{\theta\theta}$ will occur in this slice; however, this stress will not influence the nature of the fringe pattern.

An alternative procedure which can be employed in the sectioning of the model is the sub-subslice technique. This technique is illustrated in Fig. 13.18, which shows a small cube removed by sectioning the radial subslice. The cube has dimensions h in the z direction, h_1 in the θ direction, and h_3 in the r direction. By viewing the cube in the polariscope in all three possible directions, the following

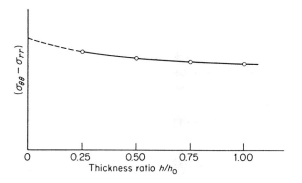

FIGURE 13.19
The shaving method to determine the surface-stress distribution.

three relationships can be obtained:

$$\sigma_{\theta\theta} - \sigma_{rr} = \frac{N_z f_\sigma}{h}$$

$$\sigma_{\theta\theta} = \sigma_{\theta\theta} - \sigma_{zz} = \frac{N_r f_\sigma}{h_3} \qquad (13.26)$$

$$\sigma_{rr} = \sigma_{rr} - \sigma_{zz} = \frac{N_\theta f_\sigma}{h_1}$$

where the subscript on N indicates the direction of the incident light.

This example shows that surface stresses can be determined by employing various slicing techniques with a three-dimensional model. Since more than one technique is often available, it is advisable to employ at least two different methods and cross-check the results obtained. In this manner the results can be averaged and the errors minimized.

13.7.2 Principal Slices Other Than Surface Slices

Often the three-dimensional model will contain planes of symmetry or other planes which are known to be principal. The flat-headed, thick-walled pressure vessel can also be used as an example to illustrate this topic. The meridional plane presented in Fig. 13.20 represents a plane of symmetry which is also known to be a principal plane. Moreover, in the cylindrical portion of the pressure vessel, the transverse or hoop planes are also known to be principal planes where the theoretical solution presented in Sec. 3.13 will apply. Thus, it is reasonable to section the three-dimensional model of the pressure vessel to obtain a meridional slice and several hoop or transverse slices along the axis of the cylinder. If these slices are examined in the polariscope at normal incidence, the resulting fringe patterns will provide the following data. For the meridional slice

$$\sigma_1 - \sigma_2 = \frac{N f_\sigma}{h} \qquad (12.7)$$

where $(\sigma_1 - \sigma_2) = (\sigma_a - \sigma_{rr})$ in the cylindrical portion of the model and $(\sigma_1 - \sigma_2) = (\sigma_{rr} - \sigma_{zz})$ in the head portion of the model. In the transition region between the cylinder and the head, the principal directions are not known since they differ from the axial or radial directions. Isoclinic data show the extent of the transition region and the directions of the principal stresses in this region.

On the external boundaries of the meridional slice shown in Fig. 13.20:

$$\sigma_a = \frac{N f_\sigma}{h} \qquad \text{on the vertical boundary}$$

$$\sigma_{rr} = \frac{N f_\sigma}{h} \qquad \text{on the horizontal boundary}$$

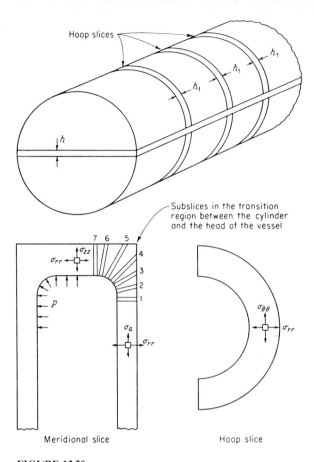

FIGURE 13.20
Slicing a thick-walled pressure vessel along the planes of principal stress.

On the internal boundaries of the meridional slice:

$$\sigma_a = \frac{Nf_\sigma}{h} - p \qquad \text{on the vertical boundary}$$

$$\sigma_{rr} = \frac{Nf_\sigma}{h} - p \qquad \text{on the horizontal boundary}$$

At all interior points on the meridional plane, the value of the maximum shear stress $\tau_{\max}$ is given by

$$\tau_{\max} = \frac{Nf_\sigma}{2h}$$

since one of the two principal stresses is always negative.

The hoop slices in the cylindrical portion of the pressure vessel provide fringe patterns which can be related to the circumferential and radial stresses. Thus,

$$\sigma_{\theta\theta} - \sigma_{rr} = \frac{N f_\sigma}{h_1} \tag{12.7}$$

On the external boundary $\sigma_{rr} = 0$; therefore,

$$\sigma_{\theta\theta} = \frac{N f_\sigma}{h_1}$$

On the internal boundary $\sigma_{rr} = -p$; therefore,

$$\sigma_{\theta\theta} = \frac{N f_\sigma}{h_1} - p$$

and at interior points in the hoop slice, the maximum shear is

$$\tau_{\max} = \frac{\sigma_{\theta\theta} - \sigma_{rr}}{2} = \frac{N f_\sigma}{2 h_1}$$

It is clear that the meridional and hoop slices provide sufficient data to give the individual values of the principal stresses on the surfaces of the pressure vessel and the maximum shear stresses on these two planes. In the transitional region between the head and cylinder, it is frequently advisable to subslice the meridional slice (see Fig. 13.20) to determine the circumferential stress $\sigma_{\theta\theta}$ on the boundary. Although the subslices are not principal over their entire length, they are principal at the region near the interior boundary and give data for $\sigma_{\theta\theta}$ on the interior boundary.

Separation of the principal stresses at interior points in this model is quite involved and requires use of the shear-difference method in three dimensions, which will be described in Sec. 13.9. If the prototype material follows the maximum-shear-stress theory of failure, the pressure required to fail the vessel in shear can be predicted without separating the stresses since the maximum shears can be obtained directly from fringe patterns of the meridional and hoop slices.

13.7.3   The General Slice

In some instances it is necessary to remove and analyze a slice which does not coincide with a principal plane. For nonprincipal slices the stresses in the plane of the slice are secondary principal stresses σ_1' and σ_2', which do not coincide with the principal stresses σ_1, σ_2, or σ_3, as illustrated in Fig. 13.21. The secondary principal stresses in the plane of the slice are given by Eqs. (1.12) as

$$\sigma_1', \sigma_2' = \frac{\sigma_{xx} + \sigma_{yy}}{2} \pm \sqrt{\left(\frac{\sigma_{xx} - \sigma_{yy}}{2}\right)^2 + \tau_{xy}^2} \tag{1.12}$$

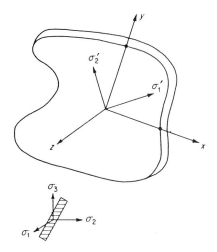

FIGURE 13.21
The general case where the secondary principal stresses σ'_1 and σ'_2 in the plane of the slice do not coincide with the principal stresses σ_1, σ_2, or σ_3.

Also, the angle which σ'_1 makes with the x axis is given by

$$\tan 2\theta = \frac{2\tau_{xy}}{\sigma_{xx} - \sigma_{yy}} \qquad (1.14)$$

If this slice is observed at normal incidence, with the light along the z axis, both the isoclinic and isochromatic fringe patterns will be due to stresses in the xy plane and will not be influenced by the z components of stress. The isochromatic fringe order is interpreted by employing

$$\left(\frac{f_\sigma N_z}{h}\right)^2 = (\sigma'_1 - \sigma'_2)^2_{xy} = (\sigma_{xx} - \sigma_{yy})^2 + 4\tau^2_{xy} \qquad (13.27)$$

The isoclinic parameters are related to the stresses in the xy plane by Eqs. (1.14). Thus,

$$\tan^2 2\theta_z = \frac{4\tau^2_{xy}}{(\sigma_{xx} - \sigma_{yy})^2}$$

$$\cos^2 2\theta_z = \frac{(\sigma_{xx} - \sigma_{yy})^2}{(\sigma_{xx} - \sigma_{yy})^2 + 4\tau^2_{xy}} \qquad (13.28)$$

$$\sin^2 2\theta_z = \frac{4\tau^2_{xy}}{(\sigma_{xx} - \sigma_{yy})^2 + 4\tau^2_{xy}}$$

where θ_z is the angle between σ'_1 and the x axis when the light passes through the model along the z axis.

Determining the complete state of stress at an arbitrary interior point on a general slice is very involved and requires use of the shear-difference method in three dimensions described in Sec. 13.9. The material presented here represents only the interpretation of the fringe patterns obtained in a normal-incidence examination of a nonprincipal slice.

13.8 EFFECTIVE STRESSES [42–43]

The data sought in a three-dimensional analysis depend to a great degree upon the specific problem being investigated. In certain instances boundary stresses will be sufficient, and the methods presented in Sec. 13.7 can be applied. In other cases a more complete solution is required, and an extensive analysis of a nonprincipal slice will be necessary to obtain the six cartesian components of stress. However, the decision to determine the individual interior stresses should be given a great deal of consideration since application of the three-dimensional shear-difference method is expensive, time-consuming, and often inaccurate.

One alternative to separating the individual stresses in three-dimensional problems is to determine the effective stress σ_e, which is defined as

$$\sigma_e^2 = \tfrac{1}{2}[(\sigma_1 - \sigma_2)^2 + (\sigma_2 - \sigma_3)^2 + (\sigma_3 - \sigma_1)^2] \tag{13.29}$$

The effective stress σ_e is widely accepted as a criterion (Von Mises) for plastic yielding or fatigue failures and is often much more significant than the separate shear and normal stresses.

The effective stress at a point can be determined from a cube of material, cut from the model at any random orientation, at the point, as illustrated in Fig. 13.22. The cube is observed at normal incidence on its three mutually orthogonal faces, and the fringe orders N_x, N_y, N_z and the isoclinic parameters θ_x, θ_y, and θ_z are recorded. Next, consider the following expansion of σ_e:

$$\sigma_e^2 = (\sigma_1^2 + \sigma_2^2 + \sigma_3^2) - (\sigma_1\sigma_2 + \sigma_2\sigma_3 + \sigma_1\sigma_3) \tag{a}$$

and note from Eqs. (1.9) that

$$\sigma_1^2 + \sigma_2^2 + \sigma_3^2 = I_1^2 - 2I_2 \quad \text{and} \quad \sigma_1\sigma_2 + \sigma_2\sigma_3 + \sigma_3\sigma_1 = I_2 \tag{b}$$

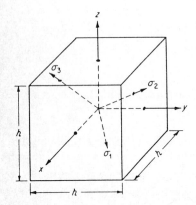

FIGURE 13.22
Cube removed from any interior point in a three-dimensional photoelastic model. The x, y, and z axes are at arbitrary angles relative to the principal coordinate system.

By combining Eqs. (*a*) and (*b*) and using Eqs. (1.8), it is clear that

$$\sigma_e^2 = I_1^2 - 3I_2$$

$$= \sigma_{xx}^2 + \sigma_{yy}^2 + \sigma_{zz}^2 - \sigma_{xx}\sigma_{yy} - \sigma_{yy}\sigma_{zz} - \sigma_{zz}\sigma_{xx} + 3(\tau_{xy}^2 + \tau_{yz}^2 + \tau_{zx}^2) \quad (13.30)$$

It is evident from an inspection of Eqs. (13.27) and (13.30) that the effective stress may be expressed as

$$\sigma_e^2 = \frac{1}{4}\left[\left(\frac{f_\sigma N_x}{h}\right)^2 (2 + \sin^2 2\theta_x) + \left(\frac{f_\sigma N_y}{h}\right)^2 (2 + \sin^2 2\theta_y)\right.$$

$$\left. + \left(\frac{f_\sigma N_z}{h}\right)^2 (2 + \sin^2 2\theta_z)\right] \quad (13.31)$$

The value of σ_e obtained through the use of the cube technique with Eq. (13.31) permits failure by fatigue or plastic yielding to be predicted for a wide variety of engineering materials employed in machine components.

13.9 THE SHEAR-DIFFERENCE METHOD IN THREE DIMENSIONS [44–45]

In certain instances it is necessary to determine the complete state of stress (that is, σ_{xx}, σ_{yy}, σ_{zz}, τ_{xy}, τ_{yz}, τ_{zx}) at an arbitrary point in a three-dimensional model. The shear-difference method is the most practical technique available to completely determine the state of stress at interior points. However, it should be noted that the method involves a stepwise numerical-integration procedure that tends to accumulate error. Great care must be exercised in collecting the data to ensure that its error is minimized.

The shear-difference method is based on the numerical integration of one of the stress equations of equilibrium. For example,

$$\frac{\partial \sigma_{xx}}{\partial x} + \frac{\partial \tau_{xy}}{\partial y} + \frac{\partial \tau_{xz}}{\partial z} = 0 \quad (1.3a)$$

If a perfectly general line (*OP*) is selected in the model, as illustrated in Fig. 13.23, Eq. (1.3a) can be integrated to obtain the stress σ_{xx} at the interior point x_1. The integration procedure, which is identical to that presented in Sec. 13.2.5, is

$$\int_{x_0}^{x_1} \frac{\partial \sigma_{xx}}{\partial x} dx + \int_{x_0}^{x_1} \frac{\partial \tau_{xy}}{\partial y} dx + \int_{x_0}^{x_1} \frac{\partial \tau_{xz}}{\partial z} dx = 0 \quad (a)$$

If finite but small values of Δx, Δy, and Δz are substituted for the partial differentials, it is possible to write

$$\sigma_{xx}|_{x_1} = \sigma_{xx}|_{x_0} - \frac{\Delta \tau_{xy}}{\Delta y} \Delta x|_{x_0}^{x_1} - \frac{\Delta \tau_{xz}}{\Delta z} \Delta x|_{x_0}^{x_1} \quad (13.32)$$

The value of $\Delta\tau_{xy}/\Delta y$ is obtained by measuring τ_{xy} along lines *AB* and *CD* (shown in Fig. 13.23), finding the difference, and dividing by Δy. In a similar manner

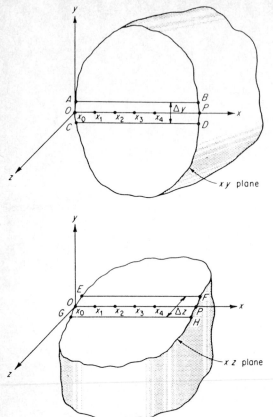

FIGURE 13.23
An arbitrary line OP and associated auxiliary lines in a three-dimensional body.

$\Delta\tau_{xz}/\Delta z$ is obtained by determining τ_{xz} along lines EF and GH, finding the difference, and dividing by Δz. If, in particular, $\Delta x = \Delta y = \Delta z$, Eq. (13.32) reduces to

$$\sigma_{xx}|_{x_1} = \sigma_{xx}|_{x_0} - \Delta\tau_{xy}|_{(x_0 + x_1)/2} - \Delta\tau_{xz}|_{(x_0 + x_1)/2} \qquad (13.33a)$$

By continuing this integration in a stepwise procedure:

$$\sigma_{xx}|_{x_2} = \sigma_{xx}|_{x_1} - \Delta\tau_{xy}|_{(x_1 + x_2)/2} - \Delta\tau_{xz}|_{(x_1 + x_2)/2} \qquad (13.33b)$$

$$\cdots = \cdots$$

In concept, the shear-difference method expressed in terms of Eqs. (13.33) is extremely simple; however, in application, the method requires data obtained at many points along lines OP, AB, CD, EF, and GH. To show the procedure for collecting these data, consider a slice taken from the model which contains the xy plane indicated in Fig. 13.23. This slice is then observed in normal incidence, and the isoclinic parameters and isochromatic fringe orders are established along lines

OP, *AB*, and *CD*. These data can be employed to obtain the shear stresses τ_{xy} along these three lines from

$$\tau_{xy} = \tfrac{1}{2}(\sigma_1' - \sigma_2')_{xy} \sin 2\theta_z = \frac{1}{2}\frac{N_z f_\sigma}{h} \sin 2\theta_z \qquad (13.34)$$

where N_z is the isochromatic fringe order observed by passing light through the *xy* slice in the *z* direction and θ_z is the angle which σ_1' makes with the *x* axis as provided by the isoclinic parameter.

Next a subslice is removed from the *xy* slice, as illustrated in Fig. 13.24, and observed in the polariscope with the light in the *y* direction. Isoclinic parameters and isochromatic fringe orders along lines *OP*, *EF*, and *GH* give the shear stress τ_{xz} as

$$\tau_{xz} = \tfrac{1}{2}(\sigma_1' - \sigma_2')_{xz} \sin 2\theta_y = \frac{1}{2}\frac{N_y f_\sigma}{h} \sin 2\theta_y \qquad (13.35)$$

At this stage τ_{xy} and τ_{xz} have been established along *OP*, and sufficient data have been obtained to employ Eqs. (13.33) to arrive at σ_{xx}. The other two normal

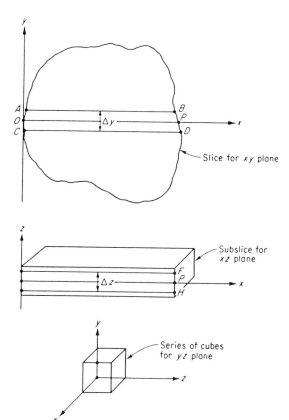

FIGURE 13.24
The slicing plan normally used with the shear-difference method in three dimensions.

stresses σ_{yy} and σ_{zz} can be established once σ_{xx} is known by utilizing the following equations, which are apparent from an examination of Mohr's circle (see Fig. 1.12):

$$\sigma_{yy} = \sigma_{xx} - (\sigma'_1 - \sigma'_2)_{xy} \cos 2\theta_z = \sigma_{xx} - \frac{N_x f_\sigma}{h} \cos 2\theta_z$$

$$\sigma_{zz} = \sigma_{xx} - (\sigma'_1 - \sigma'_2)_{xz} \cos 2\theta_y = \sigma_{xx} - \frac{N_y f_\sigma}{h} \cos 2\theta_y$$

(13.36)

To evaluate the stress τ_{yz}, the subslice may be reduced to a series of cubes, each containing an evaluation point x_1, x_2, x_3, etc., as its center, as shown in Fig. 13.24. The yz plane of these cubes is examined in the polariscope with the light in the x direction. The isochromatic and isoclinic data are then employed to give τ_{yz}:

$$\tau_{yz} = \tfrac{1}{2}(\sigma'_1 - \sigma'_2)_{yz} \sin 2\theta_x = \frac{1}{2} \frac{N_x f_\sigma}{h} \sin 2\theta_x$$

(13.37)

To summarize, the shear-difference method requires isoclinic parameters θ_y and θ_z and isochromatic fringe orders N_y and N_z along selected lines. As indicated in Fig. 13.24, θ_y and N_y are determined along OP, AB, and CD, whereas θ_z and N_z are determined along OP, EF, and GH. Utilizing these data with Eqs. (13.33) to (13.35) gives σ_{xx}, τ_{xy}, and τ_{xz}. The normal stresses σ_{yy} and σ_{zz} are determined from existing data with Eqs. (13.36). Finally, a cube technique is presented for determining the final stress components τ_{yz} given in Eq. (13.37). The six cartesian

FIGURE 13.25
The assembled closure head ready for testing.

components of stress completely define the state of stress acting at any point along OP. The principal stresses and their directions are obtained from Eq. (1.7).

13.10 APPLICATION OF THE FROZEN-STRESS METHOD†

Perhaps one of the best examples of the use of three-dimensional photoelasticity in application to problems related to pressure-vessel design is the analysis of a reactor head. From the point of view of the photoelastician there are two factors which make this type of problem difficult. First, the model must be made unusually large in order to scale all the important structural components. Second, the model is extremely complex. The size and complexity of the photoelastic model of the reactor head described in this analysis are shown in Fig. 13.25.

The model material used for the study was ERL-2774 with 50 parts by weight of phthalic anhydride. Standard procedures for stress-freezing with this material involve heating the model to a critical temperature of about 330°F (165°C). The bolts on the closure head were tightened against the compression springs to give a clamping force equal to 1.57 times the pressure thrust of the head when 34.5-kPa (5.0 psi) pressure was applied to the model. The model was slowly cooled at a rate of 0.7°C/h (1.25°F/h) to room temperature with the bolt and pressure loads acting on the model. The stresses and strains representing the elastic stress distribution due to these forces were permanently locked into the model upon completion of this stress-freezing cycle.

After stress-freezing, eight slices were removed from the closure head, as illustrated in Fig. 13.26. Four radial slices, two outer-surface slices, and two inner-surface slices were adequate to provide the necessary photoelastic data.

The isochromatic fringe patterns obtained from the radial slices, as shown in Fig. 13.27a and b, were interpreted by employing Eqs. (12.7) and (13.4) to determine the meridional stresses σ_m on the boundary. The circumferential stresses $\sigma_{\theta\theta}$ were obtained by using surface subslices cut from the radial slices, as shown in Fig. 13.26. These subslices give data for N_z and

$$\frac{\sigma_m - \sigma_{\theta\theta}}{p} = \frac{N_z f_\sigma}{hp} \tag{13.38}$$

Since σ_m is known from the examination of the radial slice, the data obtained from the radial subslice, when used with Eq. (13.38), will give $\sigma_{\theta\theta}$ directly.

The distribution of the meridional and circumferential stresses in the transitional region of the closure head is shown in Fig. 13.27c. The maximum tensile stress occurs on the outer surface at the knuckle between the head and the flange (see point A in Fig. 13.27c). The maximum compressive stress occurs at point D on the interior surface of the model. The maximum tensile stress at the knuckle

---

† The data and figures included in this section have been provided through the courtesy of M. M. Leven, Westinghouse Electric Corp.

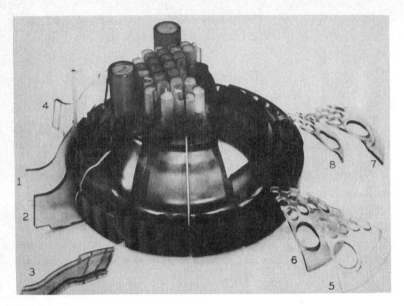

FIGURE 13.26
The closure head after stress-freezing, showing the eight slices removed.

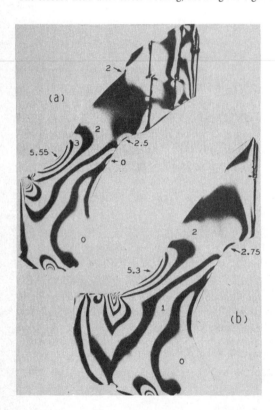

FIGURE 13.27
Isochromatic fringe patterns for (a) radial slice no. 3 (0.254 in thick) and (b) radial slice no. 4 (0.248 in thick); (c) meridional σ_m and circumferential σ_θ stresses along inner and outer surfaces of the closure head, as obtained from radial slice no. 3 and subslices.

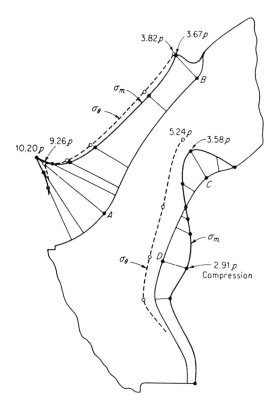

FIGURE 13.27   (c)

fillet *A* is equal to 10.2 times the pressure, or 175.8 MPa (25.5 ksi) in the prototype based on a design pressure of 17.2 MPa (2500 psi).

A typical example of the fringe pattern obtained from one of the surface slices is shown in Fig. 13.28. A maximum fringe order of 9.8 occurred in this slice at refueling penetration *V.* This proved to be the point of maximum stress (19 times the pressure) in the entire reactor head.

This example problem shows the applicability of the photoelastic method in solving extremely complex design problems. The whole-field potential of the photoelastic method is quite advantageous in examining the region of the penetrations. It was possible to examine large symmetric portions of the head, select critical locations from among the many possible points of stress concentration, and precisely determine the magnitude of both the model stresses and the prototype stresses. In this example, photoelasticity was employed to verify the validity of a particular design. The method could also be employed to improve the design. The procedure would involve testing of a number of models after contour changes had been introduced to increase radii of curvature in regions of high stress.

FIGURE 13.28
Isochromatic fringe pattern for inner surface slice no. 6 (0.188 in thick).

EXERCISES

13.1. Plot the fringe orders as a function of position across the horizontal centerline of the chain-link model shown in Fig. 13.1.

13.2. Determine the fringe orders associated with the tensile and compressive stress concentrations at the pinholes of the chain-link model shown in Fig. 13.1. Determine the stress concentration at these locations based on the maximum stress across the horizontal centerline of the chain link.

13.3. For a state of plane stress with $\sigma_1 > 0 > \sigma_2$, plot a Mohr's circle and show the plane upon which the maximum shear stress acts.

13.4. For a state of plane stress with $\sigma_1 > \sigma_2 > 0$, plot a Mohr's circle and show the plane upon which the maximum shear stress acts.

13.5. Determine the distribution of stress $\sigma_{\theta\theta}$ on the boundary of the pinhole of the chain link shown in Fig. 13.1 if $f_\sigma = 20$ kN/m and $h = 6$ mm. Prepare a graph of this distribution which shows $\sigma_{\theta\theta}$ as a function of θ.

13.6. Plot the fringe orders as a function of position across the horizontal centerline of the pressurized square conduit shown in Fig. 13.3. Determine an approximate distribution for the stress σ_{yy} along this line if $f_\sigma = 20\,\text{kN/m}$, $h = 6\,\text{mm}$, and $p = 2.00\,\text{MPa}$.

13.7. Plot the fringe orders as a function of position across the diagonal of the pressurized square conduit shown in Fig. 13.3.

13.8. Plot the fringe orders as a function of position along the outer edge of the pressurized square conduit shown in Fig. 13.3.

13.9. Determine the stress distribution on the boundary of the circular hole of the pressurized square conduit shown in Fig. 13.3 if $f_\sigma = 20\,\text{kN/m}$, $h = 6\,\text{mm}$, and $p = 2.0\,\text{MPa}$.

13.10. Construct the isostatics for one quadrant of a circular ring subjected to a diametral compressive load P by using the isoclinic data shown in Fig. 13.5.

13.11. Determine the maximum tensile and compressive stresses on the inner boundary of the thick-ring specimen shown in Fig. 12.10 if $f_\sigma = 20\,\text{kN/m}$ and $h = 6\,\text{mm}$. The specimen is subjected to concentrated compressive loads at the ends of the vertical diameter.

13.12. Given a fringe order of 6, a model thickness of 8 mm, a material fringe value of 20 kN/m, and an isoclinic parameter of $30°$ defining the angle between the x axis and σ_1, determine the shear stress τ_{xy} and show the direction of the shear stress on the face of a small element.

13.13. Explain how a simple tension or compression specimen could be used as a compensator.

13.14. Describe how the Tardy method of compensation could be employed to determine the fractional fringe order at points P_3 and P_4 of Fig. 12.12.

13.15. If the circular disk shown in Fig. 13.15 has an outside diameter $D = 100\,\text{mm}$ and a thickness $h = 10\,\text{mm}$, determine the material fringe value f_σ if a load $P = 4.0\,\text{kN}$ produced the fringe pattern.

13.16. A plane-stress model with an axis of symmetry is placed in a circular polariscope. A fringe pattern recorded at normal incidence gives the fringe order distribution $N_0(y)$ along the axis of symmetry. The model is then rotated about the axis of symmetry by an angle θ, as shown in Fig. E13.16, and the fringe-order distribution

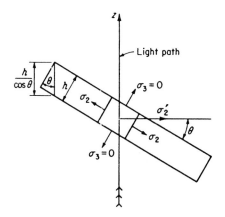

FIGURE E13.16.

$N_\theta(y)$ is recorded. If diffraction effects are eliminated by using an immersion tank, show that this oblique incidence technique gives

$$\sigma_1 = \frac{f_\sigma \cos \theta}{h \sin^2 \theta} (N_\theta - N_0 \cos \theta)$$

and

$$\sigma_2 = \frac{f_\sigma}{h \sin^2 \theta} (N_\theta \cos \theta - N_0)$$

13.17. If the rotation angle θ of Exercise 13.16 is 15, 30, and 45°, determine the corresponding relations for σ_1 and σ_2 for these three cases.

13.18. A load $P = 1200$ N is applied to a circular disk with a diameter $D = 40$ mm and a thickness $h = 6$ mm. The Tardy method of compensation is used to determine that the fringe order N at the center of the disk is 5.4. Determine the material fringe value f_σ.

13.19. Write a computer program for plane-stress models based on the shear difference method which will give as output σ_{xx}, σ_{yy}, τ_{xy}, σ_1, σ_2, and ϕ along an arbitrary line in the x direction. Data to be read into the program include the fringe order N and the isoclinic parameter ϕ along the three lines shown in Fig. 13.7.

13.20. Write a computer program for plane-stress models based on the analytic separation method. The output of the program should give $\sigma_{xx} + \sigma_{yy}$, σ_1, and σ_2 over the entire field. The data to be read into the program must permit determination of $\sigma_{xx} + \sigma_{yy}$ on the boundary of the model.

13.21. Write an engineering brief to your manager (who is a business school major) explaining how a plastic photoelastic model $100 \times 200 \times 4$ mm in size can be used to predict the stress distribution in a very large structural component fabricated from steel.

13.22. Write a test procedure describing techniques to be used to minimize error due to viscoelastic effects in a photoelastic model study of an elastic-stress problem.

13.23. You are planning to conduct a photoelastic analysis of a series of deep beams subjected to bending produced by concentrated loads. From the four materials listed in Sec. 13.4, select the most suitable model material and write an engineering brief justifying your selection.

13.24. You are planning to conduct a photoelastic analysis of a very large concrete structure where the loads are due to body forces. From the four materials listed in Sec. 13.4 select the most suitable model material and write an engineering brief justifying your selection.

13.25. Determine the changes in length Δl_1 and Δl_2 for the tension strut shown in Fig. 13.14 if $l_0 = 150$ mm and the modulus of elasticity is 3500 MPa at room temperature and 35 MPa at elevated temperature. The stress applied to the strut was 350 kPa.

13.26. Prepare a slicing plan for a three-dimensional photoelastic model of a thick-walled pressure vessel with a hemispherical head. The cylindrical region of the model has a length-to-diameter ratio of 3. The thickness-to-diameter ratio for both the head and the cylinder is 1:8. List the equations which hold for each slice and/or subslice.

13.27. Derive Eq. (13.31).

13.28. Write a computer program for determining the effective (Von Mises) stress from photoelastic data taken from a cube.

13.29. Write a test procedure which can be followed to determine the data necessary for running the program of Exercise 13.28.

13.30. Write a computer program for determining the principal stresses σ_1, σ_2, and σ_3 along an arbitrary line in the x direction from data taken from a three-dimensional photoelastic model.

REFERENCES

1. Durelli, A. J., and W. F. Riley: *Introduction to Photomechanics*, Prentice-Hall, Englewood Cliffs, N.J., 1965.
2. Frocht, M. M.; *Photoelasticity*, vol. 1, John Wiley & Sons, New York, 1941; vol. 2, 1948.
3. Coker, E. G., and L. N. G. Filon: *A Treatise on Photoelasticity*, Cambridge University Press, London, 1931.
4. Jessop, H. T., and F. C. Harris: *Photoelasticity: Principles and Methods*, Dover Publications, New York, 1950.
5. Kuske, A., and G. Robertson: *Photoelastic Stress Analysis*, John Wiley & Sons, New York, 1974.
6. Durelli, A. J., E. A. Phillips, and C. H. Tsao: *Introduction to the Theoretical and Experimental Analysis of Stress and Strain*, chap. 8, McGraw-Hill, New York, 1958.
7. Tardy, M. H. L.: Methode pratique d'examen de mesure de la birefringence des verres d'optique, *Rev. Opt.*, vol. 8, pp. 59–69, 1929.
8. Chakrabarti, S. K., and K. E. Machin: Accuracy of Compensation Methods in Photoelastic Fringe-Order Measurements, *Exp. Mech.*, vol. 9, no. 9, pp. 429–431, 1969.
9. Flynn, P. D.: Theorems for Senarmont Compensation, *Exp. Mech.*, vol. 10, no. 8, pp. 343–345, 1970.
10. Sathikh, S. M., and G. W. Bigg: On the Accuracy of Goniometric Compensation Methods in Photoelastic Fringe-Order Measurements, *Exp. Mech.*, vol. 12, no. 1, pp. 47–49, 1972.
11. Dally, J. W., and E. R. Erisman: An Analytic Separation Method for Photoelasticity, *Exp. Mech.*, vol. 6, no. 10, pp. 493–499, 1966.
12. Shortly, G. H., and R. Weller: The Numerical Solution of Laplace's Equation, *J. Appl. Phys.*, vol. 9, pp. 334–348, 1938.
13. Post, D.: A New Photoelastic Interferometer Suitable for Static and Dynamic Measurements, *Proc. SESA*, vol. XII, no. 1, pp. 191–202, 1954.
14. Drucker, D. C.: The Method of Oblique Incidence in Photoelasticity, *Proc. SESA*, vol. VIII, no. 1, pp. 51–66, 1950.
15. Clutterbuck, M.: The Dependence of Stress Distribution on Elastic Constants, *Brit. J. Appl. Phys.*, vol. 9, pp. 323–329, 1959.
16. Young, D. F.: Basic Principles and Concepts of Model Analysis, *Exp. Mech.*, vol. 11, no. 7, pp. 325–336, 1971.
17. Sanford, R. J.: The Validity of Three-Dimensional Photoelastic Analysis of Non-Homogeneous Elastic Field Problems, *Brit. J. Appl. Phys.*, vol. 17, pp. 99–108, 1966.
18. Dundurs, J.: Dependence of Stress on Poisson's Ratio in Plane Elasticity, *Int. J. Solids Struct.*, vol. 3, pp. 1013–1021, 1967.
19. Mindlin, R. D.: A Mathematical Theory of Photoviscoelasticity, *J. Appl. Phys.*, vol. 20, pp. 206–216, 1949.
20. Lee, G. H., and C. W. Armstrong: Effect of Temperature on Physical and Optical Properties of Photoelastic Material, *J. Appl. Mech.*, vol. 5, pp. A11–A12, 1938.
21. Frocht, M. M.: On the Removal of Time Stresses in Three-Dimensional Photoelasticity, *Proc. SESA*, vol. V, no. 2, pp. 9–13, 1948.
22. Leven, M. M.: "Epoxy Resins for Photoelastic Use," in *Photoelasticity*, M. M. Frocht (ed.), Pergamon Press, New York, 1963, pp. 145–165.
23. Dally, J. W., W. F. Riley, and A. J. Durelli: A Photoelastic Approach to Transient Stress Problems Employing Low Modulus Materials, *J. Appl. Mech.*, vol. 26, no. 4, pp. 613–620, 1959.

24. Leven, M. M.: A New Material for Three-Dimensional Photoelasticity, *Proc. SESA*, vol. VI, no. 1, pp. 19–28, 1948.
25. Kolsky, H.: Stress-Birefringence in Polystyrene, *Nature*, vol. 166, pp. 235–236, 1950.
26. Spooner, H., and L. D. McConnell: An Ethoxylene Resin for Photoelastic Work, *Brit. J. Appl. Phys.*, vol. 4, pp. 181–184, 1953.
27. D'Agostino, J., D. C. Drucker, C. K. Liu, and C. Mylonas: Epoxy Adhesives and Casting Resins as Photoelastic Plastics, *Proc. SESA*, vol. XII, no. 2, pp. 123–128, 1955.
28. Bayley, H. G.: Gelatin as a Photoelastic Material, *Nature*, vol. 183, pp. 1757–1758, 1959.
29. Ito, K.: New Model Material for Photoelasticity and Photoplasticity, *Exp. Mech.*, vol. 2, no. 12, pp. 373–376, 1962.
30. Oppel, G.: Polarisationsoptische Untersuchung raumlicher Spannungs- und Dehnungszustande, *Forsch. Geb. Ingenieurw.*, vol. 7, pp. 240–248, 1936.
31. Hetenyi, M.: The Application of the Hardening Resins in Three-Dimensional Photoelastic Studies, *J. Appl. Phys.*, vol. 10, pp. 295–300, 1939.
32. Hetenyi, M.: The Fundamentals of Three-Dimensional Photoelasticity, *J. Appl. Mech.*, vol. 5, no. 4, pp. 149–155, 1938.
33. Leven, M. M.: A New Material for Three-Dimensional Photoelasticity, *Proc. SESA*, vol. VI, no. 1, pp. 19–28, 1948.
34. Leven, M. M.: Epoxy Resins for Photoelastic Use, in *Photoelasticity*, M. M. Frocht (ed.), Pergamon Press, New York, 1963, pp. 145–165.
35. Cernosek, J. "Three-Dimensional Photoelasticity by Stress Freezing," *Exp. Mech.*, vol. 20, pp. 417–426, 1980.
36. Johnson, R. L.: Model Making and Slicing for Three-Dimensional Photoelasticity, *Exp. Mech.*, vol. 9, no. 3, pp. 23N–32N, 1969.
37. Nikola, W. E., and M. J. Greaves: Construction of Complex Photoelastic Models Using Thin Molded-Epoxy Sheets, *Exp. Mech.*, vol. 10, no. 4, pp. 23N–30N, 1970.
38. Drucker, D., and R. D. Mindlin: Stress Analysis by Three-Dimensional Photoelasticity Methods, *J. Appl. Phys.*, vol. 11, pp. 724–732, 1940.
39. Mindlin, R. D., and L. E. Goodman: The Optical Equations of Three-Dimensional Photoelasticity, *J. Appl. Phys.*, vol. 20, pp. 89–95, 1949.
40. Drucker, D. C., and W. B. Woodward: Interpretation of Photoelastic Transmission Patterns for Three-Dimensional Models. *J. Appl. Phys.*, vol. 25, no. 4, pp. 510–512, 1954.
41. Leven, M. M.: Quantitative Three-Dimensional Photoelasticity, *Proc. SESA*, vol. XII, no. 2, pp. 157–171, 1955.
42. Leven, M. M., and A. M. Wahl: Three-Dimensional Photoelasticity and Its Application in Machine Design, *Trans. ASME*, vol. 80, pp. 1683–1694, 1958.
43. Brock, J. S.: The Determination of Effective Stress and Maximum Shear Stress by Means of Small Cubes Taken from Photoelastic Models, *Proc. SESA* vol. XVI, no. 1, pp. 1–8, 1958.
44. Frocht, M. M., and R. Guernsey, Jr.: Studies in Three-Dimensional Photoelasticity, *Proc. 1st U.S. Natl. Congr. Appl. Mech.*, 1951, pp. 301–307.
45. Frocht, M. M., and R. Guernsey, Jr.: Further Work on the General Three-Dimensional Photoelastic Problem, *J. Appl. Mech.*, vol. 22, pp. 183–189, 1955.

CHAPTER
14

OPTICAL METHODS FOR DETERMINING FRACTURE PARAMETERS

14.1 INTRODUCTION

When a structure or machine component contains a flaw, such as a crack, stresses in the local neighborhood of the crack tip are singular. Because these local stresses exceed both the yield strength and the ultimate tensile strength of the material for very small loads, the usual approach for predicting failure loads based on Von Mises or Tresca failure theories is not feasible. Instead, one determines a fracture parameter, such as a stress intensity factor K_I, and compares this parameter to the toughness of the material K_{Ic}, as described previously in Sec. 4.1. This method of structural analysis uses fracture mechanics to predict if the crack will be stable under an applied load or if it will become unstable by initiating and then propagating to cause an abrupt failure.

In applying fracture mechanics in structural analysis, it is necessary to determine the pertinent fracture parameter, such as K_I, as a function of the applied load. In some simple bodies, K_I can be determined from data previously ascertained and given in handbooks such as that listed in Ref. 1. For more complex structures, it is necessary to use either finite-element methods or experimental procedures to determine the fracture parameter-load relationship. The use of strain

```
0        50        100        150
              mm
```

FIGURE 14.1
Isochromatic fringes for an SEN
specimen.

gages in determining K_I and K_{II} was covered in Secs. 9.6 through 9.8. In this chapter, optical methods for determining the fracture parameters will be described. In this treatment, emphasis is placed on the photoelastic method because the isochromatic fringe pattern in the local region near the crack tip (see Fig. 14.1) provides a rich field of data that enables an accurate determination of the fracture parameter to be made.

14.2 REVIEW OF IRWIN'S METHOD TO DETERMINE K_I FROM ISOCHROMATIC FRINGE PATTERNS

Post [2] and Wells and Post [3] were the first researchers to show the application of photoelasticity to fracture mechanics. However Irwin [4], in a discussion of Ref. 3, developed a relation for the opening-mode stress intensity factor K_I in terms of the geometric characteristics of the fringe loops near the tip of a crack, as

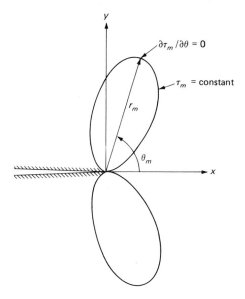

$\partial \tau_m / \partial \theta = 0$

τ_m = constant

r_m

θ_m

FIGURE 14.2
Characteristic geometry of an isochromatic fringe loop near the crack tip. Opening-mode loading.

illustrated in Fig. 14.2. Irwin began by modifying the very-near-field relations, Eqs. (4.36),

$$\sigma_{xx} = \frac{K_I}{\sqrt{2\pi r}} \cos \frac{\theta}{2} \left(1 - \sin \frac{\theta}{2} \sin \frac{3\theta}{2} \right) - \sigma_{0x}$$

$$\sigma_{yy} = \frac{K_I}{\sqrt{2\pi r}} \cos \frac{\theta}{2} \left(1 + \sin \frac{\theta}{2} \sin \frac{3\theta}{2} \right) \tag{14.1}$$

$$\tau_{xy} = \frac{K_I}{\sqrt{2\pi r}} \sin \frac{\theta}{2} \cos \frac{\theta}{2} \cos \frac{3\theta}{2}$$

where the stress $\sigma_{0x} = \sigma_{yy}^{\infty} - \sigma_{xx}^{\infty}$ was subtracted from the expression for σ_{xx} to provide another degree of freedom in bringing the theoretical fringe loops in correspondence with the experimentally observed fringe loops.

The maximum shear stress τ_m is expressed in terms of the cartesian stress components as

$$(2\tau_m)^2 = (\sigma_{yy} - \sigma_{xx})^2 + (2\tau_{xy})^2 \tag{14.2}$$

From Eqs. (14.1) and (14.2), it is apparent that

$$(2\tau_m)^2 = \frac{K_I^2}{2\pi r} \sin^2 \theta + \frac{2\sigma_{0x} K_I}{\sqrt{2\pi r}} \sin \theta \sin \frac{3\theta}{2} + \sigma_{0x}^2 \tag{14.3}$$

Next, Irwin observed the geometry of the fringe loops and noted that

$$\frac{\partial \tau_m}{\partial \theta} = 0 \tag{14.4}$$

at the extreme position on the fringe loop where $r = r_m$ and $\theta = \theta_m$. Differentiating Eq. (14.3) with respect to θ and using Eq. (14.4) gives

$$\sigma_{0x} = \frac{-K_1}{\sqrt{2\pi r_m}} \left[\frac{\sin \theta_m \cos \theta_m}{\cos \theta_m \sin (3\theta_m/2) + \frac{3}{2} \sin \theta_m \cos (3\theta_m/2)} \right] \tag{14.5}$$

The two unknown parameters K_1 and σ_{0x} are determined from the complete solution of Eqs. (14.3) and (14.5) as

$$\sigma_{0x} = \frac{-2\tau_m \cos \theta_m}{\cos (3\theta_m/2)[\cos^2 \theta_m + \frac{9}{4} \sin^2 \theta_m]^{1/2}} \tag{14.6}$$

and

$$K_1 = \frac{2\tau_m \sqrt{2\pi r_m}}{\sin \theta_m} \left[1 + \left(\frac{2}{3 \tan \theta_m} \right)^2 \right]^{-1/2} \left[1 + \frac{2 \tan (3\theta_m/2)}{3 \tan \theta_m} \right] \tag{14.7}$$

The term τ_m in Eq. (14.7) is determined from the isochromatic data since the maximum shear stress is given by

$$\tau_m = \frac{N f_\sigma}{2h} \tag{14.8}$$

Etheridge and Dally [5] showed that Irwin's two-parameter (K_1 and σ_{0x}) method predicts K_1 to within 5 percent of the solution for the central crack problem provided $73° < \theta_m < 139°$ and the data used in Eqs. (14.6) and (14.7) are exact. Outside this range for θ_m, the errors increase rapidly and the Irwin approach is not applicable because Eqs. (14.1) do not represent the stress field.

Irwin's approach is sometimes called the *apogee method* and requires only a single data point where N, r_m, and θ_m are prescribed for a given load. Unfortunately, it is difficult to measure r_m and θ_m with precision and small errors in these dimensions can lead to large errors in determining K_1.

14.3   MODIFICATIONS OF IRWIN'S TWO-PARAMETER METHOD

There have been several modifications of Irwin's original approach to improve accuracy by using data from more than one fringe loop. Bradley and Kobayashi [6] let $\sigma_{0x} = \delta K_1/\sqrt{\pi a}$, which permitted Eq. (14.3) to be simplified to

$$2\tau_m = K_1 g(\theta, r, a) \tag{14.9}$$

where

$$g(\theta, r, a) = \left[\sin^2 \theta + 2\delta \left(\frac{2r}{a} \right)^{1/2} \sin \theta \sin \left(\frac{3\theta}{2} \right) + \frac{2r\delta^2}{a} \right]^{1/2}$$

and $2a$ is the length of the crack, as shown in Fig. 4.1.

Data are then taken from two fringe loops (r_1, θ_1) and (r_2, θ_2) with the fringe order $N_2 > N_1$. Substituting these data into Eq. (14.9) and solving for K_1 gives

$$K_1 = \frac{f_\sigma(2\pi r_1 r_2)^{1/2}(N_2 - N_1)}{h(g_2\sqrt{r_1} + g_1\sqrt{r_2})} \tag{14.10}$$

The relation for $g(\theta, r, a)$ is often simplified by letting $\delta = 1$; however, this simplification has never been justified.

A different modification of the Irwin method was developed by Schroedl and Smith [7] by restricting the data to a line defined by $\theta = 90°$. With this restriction Eq. (14.3) reduces to

$$(2\tau_m)^2 = \frac{K_1^2}{2\pi r} + \frac{\sqrt{2}K_1\sigma_{0x}}{\sqrt{2\pi r}} + \sigma_{0x}^2 \tag{14.11}$$

Solving Eq. (14.11) for K_1 and retaining only the positive root from the quadratic formula gives

$$K_1 = \sqrt{\pi r}\left[(8\tau_m^2 - \sigma_{0x}^2)^{1/2} - \sigma_{0x}\right] \tag{14.12}$$

Smith simplified Eq. (14.12) by neglecting σ_{0x}^2 relative to $8\tau_m^2$ to obtain

$$K_1 = \sqrt{\pi r}\left[\sqrt{2}(2\tau_m) - \sigma_{0x}\right] \tag{14.13}$$

By adopting the Bradley-Kobayashi differencing technique, Smith uncouples the K_1 and σ_{0x} relation. Using τ_m from the ith and jth fringe loops gives

$$K_1 = \sqrt{2\pi r_i}\frac{(2\tau_m)_i - (2\tau_m)_j}{1 - (r_i/r_j)^{1/2}} \tag{14.14}$$

In application, K_1 is determined from Eq. (14.14) for all possible permutations of pairs of fringe loops to give several different values of K_1. Then the mean $\bar{K}_1$ and standard deviation S_K are determined. The results are conditioned by eliminating all the values of K_1 outside $\pm S_K$ and recomputing $\bar{K}_1$ from the remaining K_1 values.

All three of the two-parameter methods are applicable for determining K_1 in the range $73° < \theta_m < 139°$ provided $r_m/a < 0.03$. If no measurement errors are made in r_m or θ_m, the two-parameter methods will predict K_1 with an accuracy of ± 5 percent.

When two or more fringe loops occur at the crack tip and the radii can be measured with better than 2 percent accuracy, then the Bradley-Kobayashi shear-stress differencing method provides the most accurate results. When the measurement errors exceed 2 percent, the differencing technique magnifies these errors and Irwin's method produces more accurate estimates of K_1. However, the Irwin method is slightly more sensitive to errors in the measurement of θ_m than the Bradley-Kobayashi method.

The errors in θ_m determinations can be eliminated in both of the differencing methods since any fixed value of θ can be employed. However, at least part of this advantage is offset by the error introduced in making the second r measurement.

If only one fringe loop is available for analysis, Irwin's method must be used because it is the only method which uses data from a single fringe loop.

14.4   HIGHER-PARAMETER METHODS FOR DETERMINING K_1 FROM ISOCHROMATIC FRINGE PATTERNS

The use of the two-parameter methods requires data to be taken close to the crack tip with $r_m/a < 0.03$ so that Eqs. (14.1) are a valid representation of the stress field. There are two difficulties associated with restricting $r_m/a < 0.03$. First, for the region $r/h < 0.5$ the stress state is three-dimensional and the plane-stress assumptions used in deriving Eqs. (14.1) and (14.3) are not valid. To avoid errors due to the three-dimensional state of stress at the crack tip, $0.03a > r_m > h/2$. Clearly, the requirement on model thickness ($h < 0.06a$) for a valid region restricts where data can be taken. This concept of a valid region for the two-parameter methods of analysis is illustrated in Fig. 14.3.

The second difficulty is associated with the measurement of the position coordinates r_i and θ_i locating a specific data point. Both the width of the fringe and the poor definition of the origin lead to errors in measuring r_i and θ_i. For fringes very near the crack tip with r_i small, the relative errors $\Delta r_i/r_i$ and $\Delta \theta_i/\theta_i$ can be large. Examination of Fig. 14.1 illustrates the width of the fringes on the loops and the uncertainties in locating the origin at the tip of the crack.

To circumvent these difficulties, a higher-order representation of the stress field is utilized and Eqs. (14.1) are replaced with

$$
\begin{aligned}
\sigma_{xx} = {} & A_0 r^{-1/2} \cos \frac{\theta}{2} \left(1 - \sin \frac{\theta}{2} \sin \frac{3\theta}{2}\right) + 2B_0 \\
& + A_1 r^{1/2} \cos \frac{\theta}{2} \left(1 + \sin^2 \frac{\theta}{2}\right) + 2B_1 r \cos \theta \\
& + A_2 r^{3/2} \left(\cos \frac{3\theta}{2} - \frac{3}{2} \sin \theta \sin \frac{\theta}{2}\right) + 2B_2 r^2 (\sin^2 \theta + \cos 2\theta) \\
\sigma_{yy} = {} & A_0 r^{-1/2} \cos \frac{\theta}{2} \left(1 + \sin \frac{\theta}{2} \sin \frac{3\theta}{2}\right) \\
& + A_1 r^{1/2} \cos \frac{\theta}{2} \left(1 - \sin^2 \frac{\theta}{2}\right) \\
& + A_2 r^{3/2} \left(\cos \frac{3\theta}{2} + \frac{3}{2} \sin \theta \sin \frac{\theta}{2}\right) + 2B_2 r^2 \sin^2 \theta \\
\tau_{xy} = {} & A_0 r^{-1/2} \cos \frac{\theta}{2} \sin \frac{\theta}{2} \cos \frac{3\theta}{2} \\
& - A_1 r^{1/2} \sin \frac{\theta}{2} \cos^2 \frac{\theta}{2} - 2B_1 r \sin \theta \\
& - 3A_2 r^{3/2} \sin \frac{\theta}{2} \cos^2 \frac{\theta}{2} - 2B_2 r^2 \sin 2\theta
\end{aligned}
$$

(4.48)

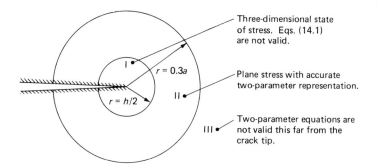

FIGURE 14.3
Concept of three regions near a crack tip. Only data taken from region II can be analyzed.

These equations provide six terms to represent the stress field, and with the additional terms, it is possible to extend significantly the boundary between regions II and III of Fig. 14.3. This representation permits larger values of r to be employed in selecting data from the valid field (region II) near the crack tip.

The coefficients A_0, A_1, A_2, B_0, B_1, and B_3 are unknown and must be determined. Recall from Eq. (4.35) that

$$K_I = \sqrt{2\pi A_0} \tag{14.15a}$$

and

$$\sigma_{0x} = -2B_0 \tag{14.15b}$$

The other coefficients are not used in any fracture-mechanics-based analysis. They are included only to improve the accuracy of the determination of K_I. Substituting Eqs. (4.48) into Eqs. (14.2) and (14.8) leads to

$$\left(\frac{Nf_\sigma}{2h}\right)^2 = \left(\frac{\sigma_{yy} - \sigma_{xx}}{2}\right)^2 + \tau_{xy}^2 = D^2 + T^2 \tag{14.16}$$

where

$$D = \frac{\sigma_{xx} - \sigma_{yy}}{2} = \sum_{n=0}^{2} (n - \tfrac{1}{2})A_n r^{n-1/2} \sin \theta \sin (n - \tfrac{3}{2})\theta$$

$$+ \sum_{m=0}^{2} B_m r^m [m \sin \theta \sin (m\theta) + \cos (m\theta)] \tag{14.17}$$

$$T = \tau_{xy} = -\sum_{n=0}^{2} (n - \tfrac{1}{2})A_n r^{n-1/2} \sin \theta \cos (n - \tfrac{3}{2})\theta$$

$$- \sum_{m=0}^{2} B_m r^m [m \sin \theta \cos (m\theta) + \sin (m\theta)] \tag{14.18}$$

When Eqs. (14.17) and (14.18) are substituted into Eq. (14.16), one obtains a higher-order equivalent to Eq. (14.3) which is to be solved for the unknown coefficients. Unfortunately, Eq. (14.16) is nonlinear in the unknown coefficients, and the matrix methods associated with linear algebra cannot be applied. In Sec. 14.5 an overdeterministic method capable of solving nonlinear equations will be developed.

The use of overdeterministic methods is important in this application because these methods provide a means of statistically averaging the results from many data points. This averaging process improves the accuracy in K_1 by accommodating for random error in the measurement of r_i and θ_i.

14.5 AN OVERDETERMINISTIC METHOD TO SOLVE NONLINEAR EQUATIONS [8, 9]

The unknown coefficients A_n and B_m in Eq. (14.16) appear as nonlinear terms (i.e., A_0^2, A_0B_0, etc.) and this nonlinearity complicates the approach used to solve the equation. Moreover, we want to use many data points from region II so that the results for K_1 are statistically averaged. The use of many data points (more than the number of unknowns) leads to a set of overdetermined relations, each of the form of Eq. (14.16). One then seeks a least-squares type of solution to these relations.

The solution is obtained by defining a function g based on Eq. (14.16) so that

$$g_k = D_k^2 + T_k^2 - \left(\frac{N_k f_\sigma}{2h}\right)^2 = 0 \tag{14.19}$$

where the subscript k indicates the value of g evaluated at a point (r_k, θ_k) with a fringe order N_k in region II. Since D_k and T_k are both dependent on A_n and B_m, the correct values for these constants will give $g_k = 0$ for all values of k. One initially makes an estimate of the coefficients and computes g_k only to find $g_k \neq 0$. The initial estimates of A_n and B_m are in error and need to be corrected. The correction process involves an iterative equation, based on a Taylor-series expansion of g_k, which is

$$g_k^{i+1} = g_k^i + \frac{\partial g_k^i}{\partial A_0}\Delta A_0 + \frac{\partial g_k^i}{\partial A_1}\Delta A_1 + \cdots + \frac{\partial g_k^i}{\partial B_0}\Delta B_0 + \frac{\partial g_k^i}{\partial B_1}\Delta B_1 + \cdots \tag{14.20}$$

where the superscript i shows the iteration step and $\Delta A_0, \Delta A_1, \ldots, \Delta B_0, \Delta B_1, \ldots$ are corrections to the previous estimates of $A_0, A_1, \ldots, B_0, B_1, \ldots$, respectively.

Since Eq. (14.19) indicates $g_k = 0$, it is clear that the correction terms in Eq. (14.20) can be written as a system of linear equations of the form

$$-g_k^i = \frac{\partial g_k^i}{\partial A_0}\Delta A_0 + \frac{\partial g_k^i}{\partial A_1}\Delta A_1 + \cdots + \frac{\partial g_k^i}{\partial B_0}\Delta B_0 + \frac{\partial g_k^i}{\partial B_1}\Delta B_1 + \cdots \tag{14.21}$$

In matrix notation, Eq. (14.21) becomes

$$[\mathbf{g}] = [\mathbf{c}][\Delta] \tag{14.22}$$

where

$$[\mathbf{g}] = \begin{bmatrix} -g_1 \\ -g_2 \\ \vdots \\ -g_L \end{bmatrix} \qquad [\mathbf{\Delta}] = \begin{bmatrix} \Delta A_0 \\ \vdots \\ \Delta A_N \\ \Delta B_0 \\ \vdots \\ \Delta B_M \end{bmatrix} \qquad \text{and} \qquad [\mathbf{c}] = \begin{bmatrix} \dfrac{\partial g_1}{\partial A_0} \cdots \dfrac{\partial g_1}{\partial A_N} & \dfrac{\partial g_1}{\partial B_0} \cdots \dfrac{\partial g_1}{\partial B_M} \\ \vdots & \vdots \\ \dfrac{\partial g_L}{\partial A_0} \cdots \dfrac{\partial g_L}{\partial A_N} & \dfrac{\partial g_L}{\partial B_0} \cdots \dfrac{\partial g_L}{\partial B_M} \end{bmatrix}$$

$$(14.23)$$

L is the total number of data points used, and N and M are the upper limits of the truncated series. Usually it is sufficient to take $N = M = 2$ to give a six-term-series representation with $L = 5(N + M + 2)$ data points.

Matrix $[\mathbf{c}]$ involves the evaluation of $L(M + N)$ partial derivatives; however, the task is made easy by noting that the functional form of each column is identical. Only the coordinates (r_k, θ_k) used to evaluate the row elements are changed. Moreover, the partial derivatives are obtained from Eq. (14.19) as

$$\frac{\partial g_k}{\partial A_n} = 2D_K \frac{\partial D_k}{\partial A_n} + 2T_k \frac{\partial T_k}{\partial A_n} \qquad (14.24)$$

Since the functions D_k and T_k are linear in A_n and B_m, the partial derivatives in Eq. (14.24) are obtained from Eqs. (14.17) and (14.18) by inspection. The implementation of this approach is a straightforward algebraic exercise which yields relations for each element in the $[\mathbf{c}]$ and $[\mathbf{g}]$ matrices.

The overdetermined system, given in Eq. (14.22), is solved in a least-squares sense. In matrix notation, the least-squares minimization process is accomplished by multiplying, from the left, both sides of Eq. (14.22) by the transpose of matrix $[\mathbf{c}]$ to give

$$[\mathbf{c}]^T[\mathbf{g}] = [\mathbf{c}]^T[\mathbf{c}][\mathbf{\Delta}] \qquad (14.25)$$

Next define

$$[\mathbf{a}] = [\mathbf{c}]^T[\mathbf{c}] \qquad (14.26)$$

and then solve Eq. (14.25) for $[\mathbf{\Delta}]$. Thus,

$$[\mathbf{\Delta}] = [\mathbf{a}]^{-1}[\mathbf{c}]^T[\mathbf{g}] \qquad (14.27)$$

The solution of Eq. (14.27) yields corrections to be applied to the previous estimates of A_n and B_m. As indicated in Eq. (14.20), iteration is used with the corrections repeated until a sufficiently accurate set of coefficients is obtained. The steps in the iterative process include:

1. From the fringe pattern, in region II, select L data points which characterize the distinguishing features of the pattern. Record (r_k, θ_k, N_k) for each of the L points.
2. Assume initial values for the unknown constants $A_0, A_1, \ldots, A_N, B_0, B_1, \ldots, B_M$. The algorithm is not sensitive to the accuracy of the initial guesses.

3. Calculate the elements of $[\mathbf{c}]$ and $[\mathbf{g}]$.

4. Solve Eq. (14.27) and correct the estimates of the unknown constants by using the expressions

$$A_0^{L+1} = A_0^L + \Delta A_0$$
$$\vdots$$
$$A_N^{i+1} = A_N^i + \Delta A_N$$
$$B_0^{L+1} = B_0^L + \Delta B_0 \tag{14.28}$$
$$\vdots$$
$$B_M^{i+1} = B_M^i + \Delta B_M$$

5. Repeat steps 3 and 4 until the corrections $\Delta A_0, \ldots, \Delta B_M$ all become sufficiently small. Convergence is rapid and good results are usually obtained in fewer than 10 iterations.

6. Determine K_{I} from A_0 by using Eq. (14.15a).

An example of a fringe pattern which has been analyzed by using this method is shown in Fig. 14.4. The isochromatic pattern was obtained from a beam in three-point bending. The crack length to beam depth ratio a/W was 0.6. The fringe pattern in the region near the crack tip was enlarged as shown in Fig. 14.5a. The data points used in the analysis (about 80 points) are shown in Fig. 14.5b. Note the priority given in selecting points close (but not too close) to the crack tip. The small circle about the crack tip defines region I, which is excluded from the data field.

The analysis employed a six-term-series representation with $N = M = 2$, and the six unknown coefficients were determined using the overdeterministic method.

FIGURE 14.4
Fringe pattern for a beam subjected to three-point bending. Crack length to beam depth ratio $a/w = 0.6$. (*Courtesy of R. J. Sanford.*)

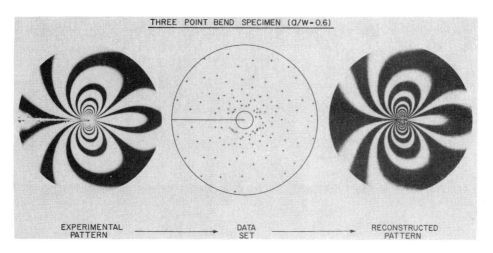

FIGURE 14.5
Development of the analysis in the application of the overdetermined method of data analysis. (*Courtesy of R. Chona.*)

These coefficients were then substituted into Eqs. (14.16), (14.17), and (14.18), and the fringe orders were determined over region II. A plotting program was written and these computer-generated fringes were reconstructed as shown in Fig. 14.5c. A comparison of the experimental pattern with the computer-reconstructed pattern shows the adequacy of the data analysis method in representing the data. A good match between the two fringe patterns gives one confidence in the accuracy of the K_I determination.

14.6   DETERMINING K_I FROM ISOPACHIC FRINGE PATTERNS [9–11]

A typical isopachic fringe pattern in the region of a crack tip is shown in Fig. 14.6. This pattern was recorded using double-exposure interferometry with an optical arrangement that responds to out-of-plane displacements w. Since the model was a sheet of Plexiglas loaded in plane stress, the out-of-plane displacements correspond to the change in thickness Δh of the model. Recall Eq. (10.42) and note that the sum of the in-plane stresses $\sigma_{xx} + \sigma_{yy}$ is

$$\sigma_{xx} + \sigma_{yy} = -\frac{E}{v}\,\epsilon_{zz} = -\frac{E}{v}\,\frac{\Delta h}{h} \tag{14.29}$$

In the holographic interferometer, the fringe order N is proportional to Δh and the corresponding stress-optic law is

$$\sigma_{xx} + \sigma_{yy} = \frac{Nf_p}{h} \tag{14.30}$$

FIGURE 14.6
An isopachic fringe pattern representing $(\sigma_{xx} + \sigma_{yy})$ in the region near a crack tip.

where f_p is an optical constant. The fringes in the pattern shown in Fig. 14.6 are called *isopachics* and they are contour lines representing a constant value of $\sigma_{xx} + \sigma_{yy}$.

The isopachic fringe patterns give full-field data (in the region near the crack tip) which can be analyzed to determine the opening-mode stress intensity factor. The approach is similar to that previously described for isochromatic fringe patterns. Begin with Eqs. (4.48) and develop the relation for $\sigma_{xx} + \sigma_{yy}$ and then use Eq. (14.30) to obtain

$$\frac{\sigma_{xx} + \sigma_{yy}}{2} = \sum_{n=0}^{N} A_n r^{n-1/2} \cos\left(n - \tfrac{1}{2}\right)\theta$$

$$+ \sum_{m=0}^{M} B_m r^m \cos m\theta = \frac{N f_p}{2h} \tag{14.31}$$

Inspection of Eq. (14.31) shows that the unknown coefficients A_n, B_m are linear. This is an important feature of the analysis of isopachic patterns since it simplifies the mathematical methods involved in an overdetermined analysis.

Dudderar and Gorman [10] described a simple approach for extracting K_1 from Eq. (14.31). In this simplification, the series was truncated at $N = M = 0$ and data were selected only along the line $\theta = 0$. With these two restrictions, Eq. (14.31) reduces to

$$A_0 r^{-1/2} + B_0 = \frac{N f_p}{2h} \tag{14.32}$$

Following Irwin's approach, both A_0 and B_0 are treated as unknowns in this two-parameter representation. The unknown constant B_0 is eliminated by using fringe orders N_1 and N_2 located at positions r_1 and r_2 along the x axis. Writing

Eq. (14.32) twice (once for each data point), subtracting one from the other, and using Eq. (4.35) gives

$$K_1 = \left(\frac{\pi}{2}\right)^{1/2} \frac{(r_1 r_2)^{1/2}}{r_2^{1/2} - r_1^{1/2}} \left[\frac{(N_1 - N_2)f_p}{h}\right] \tag{14.33}$$

Again, this relatively simple approach is effective if the data are taken from a region where the two-parameter representation of $\sigma_{xx} + \sigma_{yy}$ is valid and if very accurate measurements of r_1 and r_2 are made.

Accuracy in the determination of K_1 from isopachic patterns can be improved by using the overdeterministic method which uses whole-field data and permits error to be minimized in a least-squares sense. In applying the overdeterministic method, data N_k, r_k, θ_k are collected from region II (see Fig. 14.3) and then each data set is substituted into Eq. (14.31). The result is an overdetermined system of equations that are solved by using a least-squares approach. In matrix notation, the system of Eqs. (14.31) is written as

$$[N] = [c][U] \tag{14.34}$$

where [N] = a column matrix of the fringe orders
    [c] = a matrix of the coefficients of the unknown constants
    [U] = a column matrix of the unknown constants A_n, B_m

To square the matrix, multiply through Eq. (14.34) from the left with the transpose of [c] to obtain

$$[c]^T[N] = [c]^T[c][U] = [a][U]$$

where

$$[a] = [c]^T[c]$$

Solving Eq. (14.35) for [U] by using ordinary methods of matrix inversion gives

$$[U] = [a]^{-1}[c]^T[N] \tag{14.36}$$

Solution of Eq. (14.36) is easily accomplished with a short computer program which can be executed on a modern desktop computer. Of the six to eight unknown coefficients usually employed in an analysis, only A_0 is retained. It is converted to K_1 by using the relation $K_1 = \sqrt{2\pi A_0}$.

14.7 DETERMINING K_1 FROM MOIRÉ FRINGE PATTERNS [12]

A typical pair of moiré fringe patterns representing the in-plane displacements u and v in the region near a crack tip are shown in Fig. 14.7. Recall that the moiré fringe orders n are related to the displacement field by

$$u, v = np \tag{11.15}$$

where p is the pitch of the master grating. The moiré fringe pattern provides the

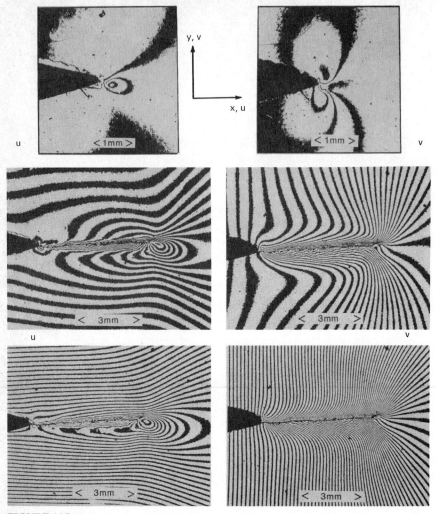

FIGURE 14.7
Moiré fringe patterns N_x and N_y representing the in-plane u and v displacement fields around a fatigue crack in 7075-T6 aluminum. Top and center patterns show residual deformations; lower patterns show live load plus residual deformations. (*Courtesy of D. Post.*)

data necessary to determine K_I. The relation between the displacements and the unknown constants A_n, B_m is obtained by integrating Eqs. (4.49) to obtain:

For the u field:

$$n_u E p = \sum_{n=0}^{N} A_n \frac{r^{n+1/2}}{n + \frac{1}{2}} \left[(1 - v) \cos \left(n + \tfrac{1}{2} \right)\theta - (1 + v)\left(n + \tfrac{1}{2} \right) \sin \theta \sin \left(n - \tfrac{1}{2} \right)\theta \right]$$

$$+ \sum_{m=0}^{M} B_m \frac{r^{m+1}}{m + 1} \left[2 \cos \left(m + 1 \right)\theta - (1 + v)(m + 1) \sin \theta \sin m\theta \right]$$

$$(14.37)$$

For the v field:

$$n_v E p = \sum_{n=1}^{N} A_n \frac{r^{n+1/2}}{n + \frac{1}{2}} [2 \sin (n + \tfrac{1}{2})\theta - (1 + v)(n + \tfrac{1}{2}) \sin \theta \cos (n - \tfrac{1}{2})\theta]$$

$$+ \sum_{m=0}^{M} B_m \frac{r^{m+1}}{m + 1} [(1 - v) \sin (m + 1)\theta - (1 + v)(m + 1) \sin \theta \cos m\theta]$$

$$(14.38)$$

These relations assume a symmetrical loading of the cracked body and a fixed origin at the crack tip. In producing moiré fringe patterns in the laboratory, it is very difficult to produce a symmetrical fringe pattern because of rigid-body motions. These motions, which usually involve both a rotation and translation of the origin, produce parasitic fringes that are not due to strain. This contribution of the rigid-body motion to the fringe field may be written as

$$T = PR \cos \theta + Qr \sin \theta + R \qquad (14.39)$$

where P, Q, and R are unknown constants that describe the parasitic fringe field.

In applying the overdeterministic method, either Eq. (14.37) or Eq. (14.38) can be employed. Usually the fringe pattern corresponding to the v displacement field is selected because this pattern exhibits a higher-order fringe pattern. Regardless of the choice of patterns, the analysis requires that the term T of Eq. (14.39) be added to the right-hand side of either Eq. (14.37) or (14.38) to account for the effect of rigid-body motions. The superposition of the term T adds three additional unknowns P, Q, and R to the unknown constants A_n, B_m; however, the overdeterministic method can accommodate these additional unknowns.

In solving the modified form of say Eq. (14.38), we note that the unknowns A_n, B_m, P, Q, and R are all linear. This fact implies that the matrix manipulations described in Eqs. (14.34) to (14.36) can be applied to determine the unknowns in the column matrix $[\mathbf{U}]$.

14.8 METHODS FOR DETERMINING K_1 IN THREE-DIMENSIONAL BODIES [13–15]

Sections 14.3 through 14.7 have covered methods for determining the opening-mode stress intensity factor for two-dimensional bodies under plane-stress conditions with through thickness cracks. However, in many engineering problems, the components are three-dimensional and the cracks only extend part of the way into the body. This part-through crack in a three-dimensional body is a very difficult problem. The most effective experimental method used to determine K_1 involves the stress-freezing process of photoelasticity (see Sec. 13.5). A three-dimensional model of the body containing a crack is loaded and the temperature is cycled to lock in the deformations on a molecular scale. The model is then sliced so that the crack front is essentially perpendicular to the plane of the slice. If the crack front is curved, several slices are removed at various locations along the front of the crack, as shown in Fig. 14.8.

The slices containing the crack-front segments are examined in a standard

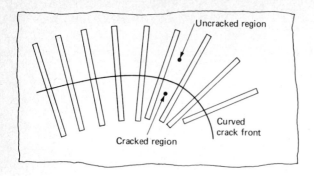

FIGURE 14.8
Slices removed from a three-dimensional photoelastic model that are oriented so that the plane of the slice is normal to the crack front.

circular polariscope and fringe patterns are recorded as illustrated in Fig. 14.9. Smith and his associates use an extrapolation technique that is based on Eq. (14.13). In the extrapolation technique, an apparent stress intensity factor K_{AP}^* is defined as

$$K_{AP}^* = \sqrt{2\pi r}(2\tau_n) = \sqrt{2\pi r}\,\frac{Nf_\sigma}{h} \tag{14.40}$$

where the effect of σ_{0x} in Eq. (14.13) has been dropped in the definition of K_{AP}^*. Data along the $\theta = 90°$ line are taken to give (N_i, r_i) at many different locations. Using Eq. (14.40) and experimental information regarding the loading σ on the model and the crack length a, Smith determines $K_{AP}^*/\sigma(\pi a)^{1/2}$ for each (N_i, r_i) and then plots these values as a function of $(r/a)^{1/2}$, as illustrated in Fig. 14.10.

FIGURE 14.9
Isochromatic fringe pattern for a slice removed from a three-dimensional model with a crack. (*Courtesy of C. W. Smith.*)

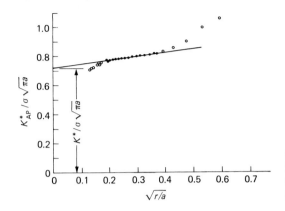

FIGURE 14.10
Extrapolation of $K^*_{AP}/\sigma(\pi a)^{1/2}$ to determine $K_1/\sigma(\pi a)^{1/2}$ from the intercept. (*From data by C. W. Smith.*)

Experience has shown that the data in Fig. 14.10 fall into three regions. For small values of r, the points fall under the extrapolation line. This is probably due to the effects of crack-tip blunting which occurs during the stress-freezing process. For intermediate values of r (the solid points in Fig. 14.10), $K^*_{AP}/\sigma(\pi a)^{1/2}$ increases linearly with $(r/a)^{1/2}$. These points define the linear region of the response and provide the basis for the extrapolation procedure. Finally, for large values of r, the data points fall above the extrapolation line. This deviation is probably due to the inadequacies of the two-parameter method used in the formulation of Eq. (14.13).

A second approach is to treat the slice as if it were a principal plane and to use the overdeterministic method described in Secs. 14.4 and 14.5 to determine K_1. Hyde and Warrior [15] have evaluated the accuracies of several different methods of determining K_1 from slices taken from frozen-stress photoelastic models. They concluded that the multiple-point overdeterministic method was simple to apply and provided very accurate results.

14.9 MIXED-MODE STRESS INTENSITY FACTORS K_I AND K_{II} [16–18]

In many instances, the loading on a body with a crack is not symmetric and the crack is subjected to either shear loading or a combination of shear- and opening-mode loading. In situations where only the shearing mode exists, the behavior of the crack is assessed using the stress intensity factor K_{II}. However, if the body is subjected to mixed-mode loading with both opening and shearing components of the load, then both K_I and K_{II} must be determined to assess the stability of the crack.

Several experimental methods can be employed to determine K_I, K_{II}, and other stress parameters such as σ_{0x}. The photoelastic method, which yields isochromatic fringe patterns, is an excellent approach since the character of the fringe pattern is markedly affected by these three quantities. Computer-reconstructed fringe patterns, presented in Fig. 14.11, show the effects of varying K_I,

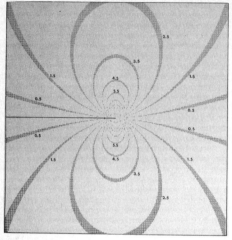

(a) Pure mode I, $K_I = C$, $K_{II} = \sigma_{ox} = 0$

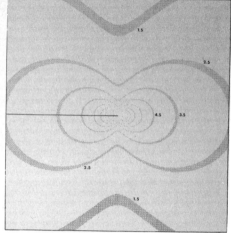

(b) Pure mode II, $K_{II} = C$, $K_I = \sigma_{ox} = 0$

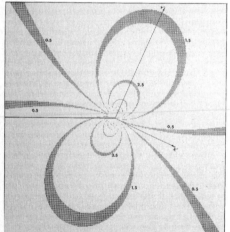

(c) Mixed mode, $K_I/K_{II} = 1/4$, $\sigma_{ox} = 0$

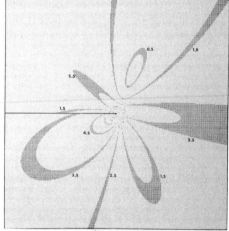

(d) Mixed mode, $K_I/K_{II} = 1/4$,
$\sigma_{ox}/K_I = -3/4$ in.$^{-1/2}$

FIGURE 14.11
Computer reconstructed fringe patterns showing the effects of K_I, K_{II}, and σ_{0x} on the characteristic shape of the isochromatic fringes.

K_{II}, and σ_{0x} on the shape of the classical fringe loops. In Fig. 14.11a, the loading is pure mode I and the fringe loops are symmetric about both the x and y axes. The fringe loops close at the origin, which is located at the tip of the crack. The fringe pattern for pure mode II loading, shown in Fig. 14.11b, is symmetric again relative to both the x and y axes. Note that the fringe loops appear to be continuous across the crack line, and that they do not close on the crack tip.

As indicated in Fig. 14.11c, the application of K_{II} destroys symmetry relative to the x axis. The application of the shear loading has rotated the two sets of fringe loops by the same angle, and the pattern is symmetric about a rotated coordinate system $Ox'y'$. The final example, presented in Fig. 14.11d, shows the effect of the uniform field stress σ_{0x} on a mixed-mode fringe pattern. All aspects of symmetry are destroyed by σ_{0x}. Moreover, the order of the fringes above and below the crack line is markedly different.

When the isochromatic fringe pattern is not symmetric relative to the crack line (i.e., the x axis), the stress state is due to a combination of both opening- and shear-mode loading. Two approaches are described to treat the mixed-mode case. The first approach relates the stress field to three unknowns, namely, K_I, K_{II}, and σ_{0x}. The second approach describes the stress field in terms of many unknowns A_n, B_m, C_n, D_m, where K_I, K_{II} are defined in terms of A_0 and C_0.

Consider the first approach, which is valid only in the very-near field as described in Fig. 4.12, and superimpose Eqs. (4.36) and (4.43) to write

$$
\sigma_{xx} = (2\pi r)^{-1/2}\left[K_I \cos\frac{\theta}{2}\left(1 - \sin\frac{\theta}{2}\sin\frac{3\theta}{2}\right)\right.
$$

$$
\left. - K_{II}\sin\frac{\theta}{2}\left(2 + \cos\frac{\theta}{2}\cos\frac{3\theta}{2}\right)\right] - \sigma_{0x}
$$

$$
\sigma_{yy} = (2\pi r)^{-1/2}\left[K_I \cos\frac{\theta}{2}\left(1 + \sin\frac{\theta}{2}\sin\frac{3\theta}{2}\right)\right.
$$

$$
\left. + K_{II}\sin\frac{\theta}{2}\cos\frac{\theta}{2}\cos\frac{3\theta}{2}\right] \qquad (14.41)
$$

$$
\tau_{xy} = (2\pi r)^{-1/2}\left[K_I \sin\frac{\theta}{2}\cos\frac{\theta}{2}\cos\frac{3\theta}{2}\right.
$$

$$
\left. + K_{II}\cos\frac{\theta}{2}\left(1 - \sin\frac{\theta}{2}\sin\frac{3\theta}{2}\right)\right]
$$

Substituting Eqs. (14.41) and (14.8) into Eq. (14.2) gives the stress-optic relation for the limited-parameter, mixed-mode condition as

$$
\left(\frac{Nf_\sigma}{h}\right)^2 = (2\pi r)^{-1}[(K_I \sin\theta + 2K_{II}\cos\theta)^2 + (K_{II}\sin\theta)^2]
$$

$$
+ 2\sigma_{0x}(2\pi r)^{-1/2}\sin\frac{\theta}{2}[K_I \sin\theta(1 + 2\cos\theta)
$$

$$
+ K_{II}(1 + 2\cos^2\theta + \cos\theta)] + \sigma_{0x}^2 \qquad (14.42)
$$

This relation is nonlinear in terms of the three unknowns K_I, K_{II}, and σ_{0x}, and the solution is more difficult because of the nonlinear terms. A method of solution of Eq. (14.42) involves selecting data along one or two lines which constrains θ

and simplifies Eq. (14.42). For example, consider the case where $K_{II} < 0$ and $\sigma_{0x} \neq 0$. The isochromatic fringes do not form closed loops, but instead they intersect the upper crack boundary as illustrated in Fig. 14.12. Along this selected line ($\theta = \pi$), Eq. (14.42) reduces to:

$$\left(\frac{Nf_\sigma}{h}\right)^2 = (2\pi r)^{-1}(4K_{II}^2) + (2\pi r)^{-1/2}(4K_{II}\sigma_{0x}) + \sigma_{0x}^2 \qquad (a)$$

Note that Eq. (a) is independent of K_I. Rewriting Eq. (a) gives

$$\frac{Nf_\sigma}{h} = \pm\left(\frac{2K_{II}}{\sqrt{2\pi r}} + \sigma_{0x}\right) \qquad (14.43)$$

The choice of signs in Eq. (14.43) is based on the fact that the isochromatic fringes intersect the $\theta = \pi$ line when $K_{II} < 0$ and intersect the $\theta = -\pi$ line when $K_{II} > 0$. Consider two different fringes $N_1 > N_2$, both of which intercept the upper edge of the crack ($\theta = \pi$) at positions r_1, r_2, respectively, as shown in Fig. 14.12. Taking the negative sign option in Eq. (14.43) and eliminating σ_{0x} by substitution leads to

$$K_{II} = \frac{f_\sigma}{h}\sqrt{\frac{\pi}{2}}\left(\frac{\sqrt{r_1 r_2}}{\sqrt{r_1} - \sqrt{r_2}}\right)(N_1 - N_2) \qquad (14.44)$$

Note that $K_{II} < 0$ as required since $N_1 < N_2$ and $r_2 < r_1$. Next, σ_{0x} is obtained from Eqs. (14.43) and (14.44) as

$$\sigma_{0x} = -\left(\frac{N_1 f_\sigma}{h}\right) - \frac{2K_{II}}{\sqrt{2\pi r_1}} \qquad (14.45)$$

To determine K_I, reevaluate Eq. (14.42) with $\theta = \pi/2$ to obtain

$$\left(\frac{N_3 f_\sigma}{h}\right)^2 = \frac{1}{2\pi r_3}(K_I^2 + K_{II}^2) + \frac{\sigma_{0x}}{\sqrt{\pi r_3}}(K_I + K_{II}) + \sigma_{0x}^2 \qquad (14.46)$$

Equation (14.46) is quadratic in terms of K_I and thus

$$K_I = \frac{-b + \sqrt{b^2 - 4ac}}{2a} \qquad (14.47)$$

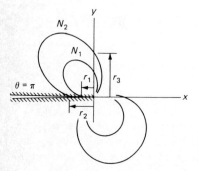

FIGURE 14.12
Isochromatic fringes intersect the upper crack line ($\theta = \pi$) when $K_{II} < 0$.

where

$$a = \frac{1}{(2\pi r_3)}$$

$$b = \frac{\sigma_{0x}}{\sqrt{\pi r_3}} \qquad (14.48)$$

$$c = \frac{K_{II}^2}{2\pi r_3} + \frac{K_{II}\sigma_{0x}}{\sqrt{\pi r_3}} + \sigma_{0x}^2 - \left(\frac{N_3 f_\sigma}{h}\right)^2$$

The terms N_3 and r_3 are defined in Fig. 14.12. The plus sign is selected in front of the radical in Eq. (14.47) because $K_1 > 0$.

This example illustrates the selected-line method where the data were constrained to two lines $\theta = \pi$ and $\theta = \pi/2$ to simplify Eq. (14.42). Other selections of lines are possible as other choices of θ exist which simplify Eq. (14.42). However, care should be exercised in selecting θ so as to define lines which intersect one or more fringes as nearly normal as possible. Near-normal intersections improve the accuracy in determining the radial position r of the intersection point.

The second approach to solve for K_1 and K_{II} in a mixed-mode loading situation is to use the near-field equations which describe the stresses in region 2 of Fig. 4.12. Superposition of Eqs. (4.48) for the opening mode and Eqs. (4.52) for the shearing mode gives series equations for σ_{xx}, σ_{yy}, and τ_{xy}. The series equations contain unknown coefficients A_n, B_m, C_n, and D_m. The number of unknowns depends upon the complexity of the mixed-mode pattern. In most analyses, three terms in each series for a total of 11 terms (recall D_0 does not occur) is more than sufficient to accurately represent the stresses in the near field.

The relations for the superimposed stresses

$$\sigma_{xx} = (\sigma_{xx})_I + (\sigma_{xx})_{II} \qquad (14.49)$$

and similarly for σ_{yy} and τ_{xy} are substituted into Eq. (14.16) to obtain expressions for D and T. Thus,

$$D = \frac{\sigma_{yy} - \sigma_{xx}}{2}$$

$$= \sum_{n=0}^{2} (n - \tfrac{1}{2}) A_n r^{n-1/2} \sin \theta \sin (n - \tfrac{3}{2})\theta$$

$$+ \sum_{m=0}^{2} B_m r^m [m \sin \theta \sin m\theta + \cos m\theta]$$

$$- \sum_{n=0}^{2} C_n r^{n-1/2} [(n - \tfrac{1}{2}) \sin \theta \cos (n - \tfrac{3}{2})\theta + \sin (n - \tfrac{1}{2})\theta]$$

$$- \sum_{m=0}^{2} D_m r^m \sin \theta \cos (m - 1)\theta \qquad (14.50)$$

and

$$T = \tau_{xy} = - \sum_{n=0}^{2} (n - \tfrac{1}{2})A_n r^{n-1/2} \sin \theta \cos (n - \tfrac{3}{2})\theta$$

$$- \sum_{m=0}^{2} B_m r^m [m \sin \theta \cos m\theta + \sin m\theta]$$

$$+ \sum_{n=0}^{2} C_n r^{n-1/2} [\cos (n - \tfrac{1}{2})\theta - (n - \tfrac{1}{2}) \sin \theta \sin (n - \tfrac{3}{2})\theta]$$

$$- \sum_{m=0}^{2} D_m r^m [m \sin \theta \sin (m - 1)\theta] \tag{14.51}$$

Substituting Eqs. (14.50) and (14.51) into Eq. (14.16) yields a very long and tedious stress-optic relation. This relation is solved for the 11 unknown coefficients by using the overdeterministic methods described in Sec. 14.5. The values of A_0 and C_0 are related to stress intensity factors K_I and K_{II} by

$$K_I = \sqrt{2\pi} A_0 \tag{14.15a}$$

$$K_{II} = \sqrt{2\pi} C_0 \tag{14.52}$$

14.10   THE USE OF BIREFRINGENT COATINGS IN FRACTURE MECHANICS [19–25]

The use of birefringent coatings in fracture mechanics has been limited for a variety of different reasons. In the classical two-dimensional elasto-static problem there is little motivation to use coatings because the problem can be modeled accurately and either transmission photoelasticity or finite-element methods provide more direct solutions. In the two-dimensional elasto-dynamic problem, numerical methods are difficult to implement and modeling is not possible. Experiments to measure initiation toughness K_{Ic} and propagation toughness K_{ID} are possible using either strain gages [21] or birefringent coatings [20]. Of the two experimental approaches, the strain-gage methods are preferred over the birefringent coating method because of equipment requirements.

It appears that there are two technical areas where birefringent coatings have significant advantage over other experimental approaches. The first is in the development of optical gages (which sense, say, K_I) that can be used to monitor flawed structures over extended periods of time. The second is with static problems where elastic modeling is not possible but where surface measurements are still sufficient to give the data necessary to characterize the fracture process. This is an important category of problems (ductile fracture) where the yield zone near the crack tip has enlarged and the normal procedure of characterizing fracture with a stress intensity factor K_1 is not valid. Instead, other fracture parameters such as the J integral are required to describe the plastic field at the crack tip.

14.10.1 Birefringent Coatings for K_1 Determinations [22]

When a birefringent coating is applied to cover the near-field region, the coating should be cut as shown in Fig. 14.13. This cut prevents the coating from bridging the crack and avoids the response due to the crack-opening displacements. The isochromatic fringe loops which form in the coating are related to K_1. Note from Eqs. (15.7b) and (15.10b) that

$$\sigma_1^s - \sigma_2^s = \frac{E^s}{E^c} \frac{1 + v^c}{1 + v^s} \frac{Nf_\sigma}{2h^c} = \frac{E_s}{1 + v^s} \frac{Nf_\epsilon}{2h^c} \tag{14.53}$$

Since $2\tau_m^s = \sigma_1^s - \sigma_2^s$, it is clear that

$$\tau_m = \frac{E_s}{1 + v^s} \frac{Nf_\epsilon}{4h^c} \tag{14.54}$$

and that Eq. (14.16) holds if it is modified to read

$$\left(\frac{E_s}{1 + v^s} \frac{Nf_\epsilon}{4h^c} \right)^2 = D^2 + T^2 \tag{14.55}$$

where D and T are given by Eqs. (14.17) and (14.18), respectively.

The stress intensity factor K_1 is contained in D and T, as described previously in Secs. 14.4 and 14.5. Since the unknowns in Eq. (14.55) appear as nonlinear terms, this approach with birefringent coatings is not recommended. A much easier approach is possible if the continuous coating is slit or if a coating strip is employed, as shown in Fig. 14.14. We will describe the advantage of using either slit or strip coatings in Sec. 14.10.2.

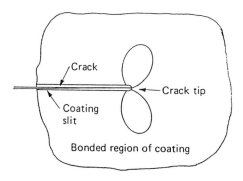

FIGURE 14.13
Placement of a slit birefringent coating to avoid bridging the crack.

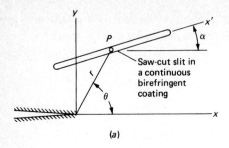

(a)

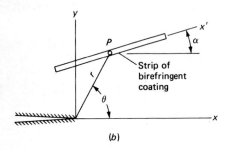

(b)

FIGURE 14.14
Discontinuous coatings in the near-field region. (a) Coating with a straight slit oriented at an angle α. (b) A thin birefringent strip oriented at an angle α.

14.10.2 Birefringent Strip or Slit Coatings [22]

At the edge of a slit, the stress $\sigma^c_{\parallel}$ is uniaxial and in the direction of the slit since $\sigma^c_{\perp} = 0$. The same condition holds for the birefringent strip providing the strip is thin compared to its thickness h^c. This fact implies that

$$\epsilon^c_{x'x'} = \epsilon^s_{x'x'} \tag{14.56}$$

and $\sigma^c_{y'y'} = 0$ even when $\sigma^s_{y'y'} \neq 0$. Clearly, Eq. (14.56) shows that the slit and/or strip acts like a row of strain gages spaced on infinitely close centers.

From Eq. (4.49) and Eq. (2.6a) it is clear that

$$2\mu\epsilon_{x'x'} = A_0 r^{-1/2}\left[k\cos\frac{\theta}{2} - \tfrac{1}{2}\sin\theta\sin\frac{3\theta}{2}\cos 2\alpha + \tfrac{1}{2}\sin\theta\cos\frac{3\theta}{2}\sin 2\alpha\right]$$
$$+ B_0(k + \cos 2\alpha) + A_1 r^{1/2}\cos\frac{\theta}{2}\left(k + \sin^2\frac{\theta}{2}\cos 2\alpha - \tfrac{1}{2}\sin\theta\sin 2\alpha\right) \tag{14.57}$$

where α is defined in Fig. 14.14, $\mu = E/2(1 + \nu)$ is the shear modulus, and

$$k = \frac{1 - \nu}{1 + \nu} \tag{14.58}$$

Equation (14.57) is a three-term series representing $\epsilon_{x'x'}$ that is linear in the unknowns. By using slits and/or strips and restricting the field of data to these lines, we have avoided measuring $\sigma_1 - \sigma_2$ or $\epsilon_1 - \epsilon_2$ which gives rise to the nonlinear terms A_0^2, B_0^2, $A_0 B_0$, etc.

Next, examine Eq. (14.57) and note that the coefficient of the B_0 term $k + \cos 2\alpha$ will vanish if

$$\cos 2\alpha = -k = -\frac{1 - v^s}{1 + v^s} \tag{14.59}$$

Also, the coefficient of the A_1 term in Eq. (14.57) vanishes if

$$\tan \frac{\theta}{2} = -\cot 2\alpha \tag{14.60}$$

These restrictions on α and θ depend entirely on Poisson's ratio of the specimen material, as indicated in Table 14.1.

If the slit and/or strip is positioned with α according to Eq. (14.59) and the fringe order read at point P which gives an angle θ defined by (14.60), then at this point (and only this point)

$$2\mu\epsilon_{x'x'} = A_0 r^{-1/2}\left[k \cos \frac{\theta}{2} + \frac{k}{2} \sin \theta \sin \frac{3\theta}{2} + \tfrac{1}{2} \sin \theta \cos \frac{3\theta}{2} \sin 2\alpha \right] \tag{14.61}$$

Equation (15.9) for a birefringent coating reduces to

$$\epsilon_{x'x'} = \frac{N f_\epsilon}{2h^c} \tag{14.62}$$

Substituting Eqs. (14.62) and (14.15a) into Eq. (14.61) yields

$$K_1 = \frac{\sqrt{2\pi}\,\mu(N f_\epsilon/h^c) r^{1/2}}{k \cos \dfrac{\theta}{2} + \dfrac{k}{2} \sin \theta \sin \dfrac{3\theta}{2} + \tfrac{1}{2} \sin \theta \cos \dfrac{3\theta}{2} \sin 2\alpha} \tag{14.63}$$

where N is measured at point $P(r, \theta)$ with restrictions on θ as defined in Table 14.1. Inspection of Eq. (14.63) shows that the opening-mode stress intensity factor K_1 is a linear function of the fringe order N. Application of Eq. (14.63) to a direct-reading K_1 gage is discussed in Sec. 14.10.3.

TABLE 14.1
Angles α and θ as a function of Poisson's ratio v^s

v^s	θ, deg	α, deg
0.250	73.74	63.43
0.300	65.16	61.43
0.333	60.00	60.00
0.400	50.76	57.69
0.500	38.97	54.74

14.10.3 A Direct-Reading K_I Gage [22]

Consider a slit and/or strip positioned on an aluminum plate ($v = \frac{1}{3}$) with the orientation $\alpha = 60°$. Note also from Table 14.1 that the angle θ, which defines the reading point P, is also $60°$. This fact that $\alpha = \theta$ implies that any point on a strip or on the edge of a slit can be used to indicate K_I. With an aluminum ($\mu = 26.2$ GPa, $v = \frac{1}{3}$, $k = \frac{1}{2}$, $\theta = \alpha = 60°$) specimen and a polycarbonate coating ($f_\epsilon = 40 \times 10^{-4}$ mm/fringe), Eq. (14.63) reduces to

$$K_I = 2.355 \frac{Nr^{1/2}}{h} \tag{14.64}$$

where K_I is given in units of MPa $\cdot \sqrt{m}$.

The results shown in Eq. (14.64) suggest the use of a scale to measure K_I directly on the specimen without instrumentation (other than a bonded circular polarizer and a light source). Consider a coating with $h = 1$ mm and observe two fringe orders $N = 1$ and $N = 2$. Equation (14.64) reduces to

$$\begin{aligned} K_I &= 2.355r^{1/2} \quad \text{for } N = 1 \\ K_I &= 4.710r^{1/2} \quad \text{for } N = 2 \end{aligned} \tag{14.65}$$

where r is measured in mm. These two equations permit the scales (two) for the strip/slit to be determined as shown in Fig. 14.15.

The range of the K_I gage shown in Fig. 14.15 varies from about 30 to 140 MPa $\cdot \sqrt{m}$. Of course, strips or slits covering different ranges can be fabricated by changing the coating thickness. More sensitive gages measuring lower values of K_I employ thicker strips and less sensitive gages employ thinner strips. It is important to select the thickness so that the fringe orders used to scale K_I are located at

$$\frac{h^s}{2} < r < \frac{a}{5} \tag{14.66}$$

where a is the crack length. If r is larger than half the specimen thickness, then the plane-stress equations used in developing Eq. (14.57) are valid. If r is less than one-fifth the crack length and far from other boundaries and points of load

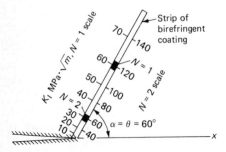

FIGURE 14.15

Scaling of the birefringent strip for a direct-reading K_I gage. ($h^c = 1$ mm, $\alpha = \theta = 60°$, $v = \frac{1}{3}$, $f_\epsilon = 40 \times 10^{-4}$ mm/fringe.)

application, then Eq. (14.57), with its three-term series representation, gives a close approximation to the infinite series solution.

14.10.4   Determining J in Power-Law Hardening Materials

Hutchinson [23] and Rice and Rosengren [24] considered materials with power-law hardening where the uniaxial stress-strain relation is defined by

$$\frac{\epsilon}{\epsilon_0} = \alpha \left(\frac{\sigma}{\sigma_0} \right)^n \tag{14.67}$$

where $\sigma_0 =$ the yield stress
  $\epsilon_0 =$ the strain at yield given by (σ_0/E)
  $\alpha =$ a material constant
  $n =$ the strain hardening coefficient

These authors developed the HRR theory, applicable for ductile materials, which gives the stress and strain distributions in the J-dominated region near the crack tip, as illustrated in Fig. 14.16. These relations for stress and strain are

$$\sigma_{ij} = \sigma_0 \left(\frac{J}{\alpha \sigma_0 \epsilon_0 I_n r} \right)^{1/(n+1)} S_{ij}(n, \theta)$$

$$\epsilon_{ij} = \alpha \epsilon_0 \left(\frac{J}{\alpha \sigma_0 \epsilon_0 I_n r} \right)^{n/(n+1)} E_{ij}(n, \theta) \tag{14.68}$$

where I_n is a dimensionless constant which varies with n, ranging from 5 to 2.57 as n varies from 1 to ∞. S_{ij} and E_{ij} are functions of n and θ. These functions I_n, S_{ij}, and E_{ij} are tabulated by Shih in Ref. 25.

Next, examine the strain as expressed by Eqs. (14.68) in the radial and tangential directions along the x axis. First, consider the distribution of

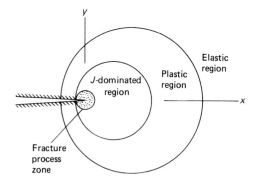

Fracture
process
zone

FIGURE 14.16
Schematic illustration of the J-dominated region embedded in the plastic region.

strain along the positive x axis, where it is evident that $\epsilon_{xx} = \epsilon_{rr}$, $\theta = 0$, and $E_{rr}(n, \theta) = E_{rr}(n, 0)$. Substituting these equalities into Eqs. (14.68) gives

$$\epsilon_{xx} = \alpha\epsilon_0\left[\frac{1}{(\alpha\sigma_0\epsilon_0 I_n)}\right]^{n/(1+n)}\left(\frac{J}{x}\right)^{n/(1+n)} E_{rr}(n, 0) \qquad (14.69)$$

For a given material α, ϵ_0, σ_0, and n are fixed and we may simplify the expressions for the strains by defining a material constant Q as

$$Q = \alpha\epsilon_0\left[\frac{1}{(\alpha\sigma_0\epsilon_0 I_n)}\right]^{n/(n+1)} \qquad (14.70)$$

Substituting Eq. (14.70) into Eq. (14.69) gives

$$\epsilon_{xx} = QE_{rr}(n, 0)\left(\frac{J}{x}\right)^{n/(n+1)} \qquad (14.71)$$

which is valid along the positive x axis.

The strain ϵ_{yy} along the same x axis is

$$\epsilon_{yy} = \epsilon_{\theta\theta} = QE_{\theta\theta}(n, 0)\left(\frac{J}{x}\right)^{n/(n+1)} \qquad (14.72)$$

Note that the x axis is a principal axis and that $\epsilon_1 = \epsilon_{yy}$ and $\epsilon_2 = \epsilon_{xx}$ for the opening-mode loading used in the HRR theory.

Consider next a patch of birefringent coating positioned in front of the crack, as shown in Fig. 14.17. The coating will respond to the strains developed in the plastic and J-dominated regions and give the fringe order N as a function of position x. To interpret this data in terms of J (see Sec. 4.9.3 for a description of the J integral), substitute Eqs. (14.71) and (14.72) into Eq. (15.19) to obtain

$$\frac{J}{x} = \left\{\frac{Nf_\epsilon}{2h^cQ[E_{\theta\theta}(n, 0) - E_{rr}(n, 0)]}\right\}^{(n+1)/n} \qquad (14.73)$$

If N is recorded at a number of locations along the x axis, J can be determined at a number of points in the plastic zone in front of the crack. However, only those values of N measured in the J-dominated region are valid. It is possible to extend this analysis to determine the bounds of the J-dominated region by taking

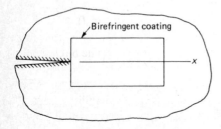

Birefringent coating

FIGURE 14.17
Patch of birefringent coating located in front of the crack covering both the J-dominated region and the plastic region.

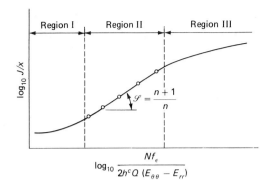

FIGURE 14.18
A log-log graph of J/x as a function of N/C shows a linear region II which is J-dominated and where the HRR theory is valid.

the logarithm of both sides of Eq. (14.73) to obtain

$$\log_{10}\left(\frac{J}{x}\right) = \frac{(n+1)}{n}\log_{10}\left\{\frac{Nf_{\epsilon}}{2hQ[E_{\theta\theta}(n,0) - E_{rr}(n,0)]}\right\} \qquad (14.74)$$

This relation is the basis for the graph shown in Fig. 14.18, where three regions are defined. Region I corresponds to low values of N which occur at large x and relatively low values of J/x. These measurements, close to the origin in Fig. 14.18, are taken at locations beyond the J-dominated region. Region III corresponds to high-order fringes, with small x and large J/x. Locations of these measurements are too close to the crack tip for the HRR theory to be valid. Region II is between these extremes, and the results shown on the log-log graph are linear with a slope $\mathscr{S} = (n+1)/n$. The range of the linear portion of the curve of Fig. 14.18 defines the bounds of the J-dominated region along the x axis.

14.10.5 Discussion

In the analysis of specimens containing cracks, the form of the solution for the stresses and strains in the local neighborhood of the crack tip is known in terms of some theory. In this coverage, the generalized Westergaard theory was used to represent the stress and strain distributions for the elastic or small-scale yielding examples. The HRR theory was used to give the plastic strain relations near the crack tip where the loading produced larger plastic zones. Since the theory describes the form of the solution, the experiment is only necessary to establish the validity of the theory and to determine the scaling constants used to fit the theory to the observed behavior, either elastic or plastic (i.e., K_I or J).

There are many experimental methods which may be employed in checking the validity of the theory and in determining the scaling constants. The birefringent coatings, strips/slits, and patches described here have the advantage of simplicity since no instrumentation is necessary and the analysis is straightforward in all but the overdeterministic approaches. A second advantage is the availability of nearly continuous data over the region of interest. This is particularly important in ductile fractures where the plastic zone size is large and the available theories for the strain distribution are valid only over relatively small regions within the plastic zone.

EXERCISES

14.1. Verify Eq. (14.3).

14.2. Verify Eqs. (14.6) and (14.7) by using Eq. (14.3) together with the fact that $(\delta\tau_m/\partial\theta) = 0$.

14.3. From the fringe pattern shown in Fig. 14.1, determine K_1, if $f_\sigma = 7.0\,\text{kN/m}$ and $h = 6.35\,\text{mm}$. Use the method developed by Irwin which uses a single data point.

14.4. From the fringe pattern shown in Fig. 14.1, determine K_1 if $f_\sigma = 7.0\,\text{kN/m}$ and $h = 6.35\,\text{mm}$. Use a two-point deterministic approach and determine K_1 and σ_{0x} without using the fact that $(\delta\tau_m/\delta\theta) = 0$.

14.5. From the fringe pattern shown in Fig. 14.1, determine K_1 if $f_\sigma = 7.0\,\text{kN/m}$ and $h = 6.35\,\text{mm}$. Use the two-fringe differencing method of Bradley and Kobayashi.

14.6. From the fringe pattern shown in Fig. 14.1, determine K_1 if $f_\sigma = 7.0\,\text{kN/m}$ and $h = 6.35\,\text{mm}$. Use C. W. Smith's approach, which uses data along the line defined by $\theta = 90°$.

14.7. Write an engineering brief describing the difficulties encountered in measuring N, r, and θ from Fig. 14.1 as required in Exercises 14.4, 14.5, and 14.6.

14.8. Develop a technique for measuring K_1 based on data taken along the line $\theta = 60°$.

14.9. Expand the relations for D and T in Eqs. (14.17) and (14.18) for $\theta = 0, 30, 45, 60$, and $90°$.

14.10. Write a computer program for determining K_1 by using the overdeterministic method described in Sec. 14.5.

14.11. Use the computer program written for Exercise 14.10 together with 40 data points taken from Fig. 14.1 to determine K_1. Use $M = N = 2$.

14.12. Write a computer program which takes, as input, the results from an overdeterministic analysis for A_0, A_1, A_2, B_0, B_1, and B_2 and gives, as output, a reconstructed pattern similar to that shown in Fig. 14.5.

14.13. Use the program from Exercise 4.12 and the results from Exercise 14.11 to reconstruct a fringe pattern. Compare this reconstructed pattern with the fringe pattern of Fig. 4.1, which is the experimentally determined input data.

14.14. Verify Eqs. (14.32) and (14.33).

14.15. From the fringe pattern shown in Fig. 14.6, determine the quantity $(K_1 h/f_p)$ by using the method of Dudderar and Gorman.

14.16. Consider a six-parameter expansion of Eq. (14.31) and write a definition of each matrix $[N]$, $[c]$, and $[U]$ in Eq. (14.34).

14.17. From the fringe pattern shown in Fig. 14.6, determine the quantity $(K_1 h/f_p)$ by using the overdeterministic method outlined in Sec. 14.6.

14.18. Develop a simple method for determining K_1 from moiré data which uses only two data points, one from the u field and the other from the v field.

14.19. Repeat Exercise 14.18 but select both data points from the u field.

14.20. Repeat Exercise 14.18 but select both data points from the v field.

14.21. You have a slice taken from a three-dimensional frozen-stress model which yields a fringe pattern similar to that shown in Fig. 14.9 when the slice is examined in a circular polariscope. Can you use the extrapolation technique to determine $K^*_{AP}/\sigma\sqrt{\pi a}$? How are the additional data points to be generated which are necessary for the success of the extrapolation method?

14.22. Write a computer program that will generate a photoelastic fringe pattern for a crack in a large plane body subjected to mode II loading.

14.23. Write a computer program that will generate an isopachic fringe pattern for a crack in a large plane body subjected to mode II loading.

14.24. Write a computer program that will generate a u-type moiré pattern for a crack in a large plane body subjected to mode II loading.

14.25. Write a computer program that will generate a v-type moiré pattern for a crack in a large plane body subjected to mode II loading.

14.26. Verify Eqs. (14.41) and (14.42).

14.27. Equation (14.43) is valid only if the fringe order N is determined along the line $\theta = \pi$. Derive similar relations for the cases where the fringe order N is determined along the lines $\theta = \pm\pi/2$. Show a procedure for determining K_I, K_{II}, and σ_{0x} from these relations.

14.28. Verify Eqs. (14.47) and (14.48).

14.29. Verify Eqs. (14.53), (14.54), and (14.56).

14.30. Verify Eq. (14.57).

14.31. Verify Eq. (14.63).

14.32. Prepare a scale for $N = 1$ and $N = 2$ corresponding to a direct-reading K_I gage exactly like the one described in Fig. 14.15 if $h = 1.5$ mm.

14.33. Repeat Exercise 14.32 but let $h = 0.5$ mm.

14.34. If you place a continuous birefringent coating on a through cracked panel as shown in Fig. 14.13 and observe a mode I isochromatic fringe loop of $N = 2$ with $r_{max} = 5$ mm and $\theta_{max} = 80°$, estimate $(K_I h/f_\sigma)$.

14.35. Plot several stress-strain curves using Eq. (14.67) to describe the functional relationship between σ and ϵ for steel. Let the strain-hardening coefficient vary between 1 and 30. Take $\sigma_0 = 50,000$ psi, $\epsilon_0 = 1666$ $\mu\epsilon$, and $\alpha = 1$. Vary n in the process.

14.36. Verify Eq. (14.73).

14.37. Write an engineering brief describing an experiment using birefringent coatings to determine the validity of the HRR theory for a new material which has not been classified as ductile or brittle.

REFERENCES

1. Toda, H.: *The Stress Analysis of Cracks Handbook*, Del Research Corp., Hallertown, 1973.
2. Post, D.: Photoelastic Stress Analysis for an Edge Crack in a Tensile Field, *Proc. SESA*, vol. XII, no. 1, pp. 99–116, 1954.
3. Wells, A., and D. Post: The Dynamic Stress Distribution Surrounding a Running Crack—A Photoelastic Analysis, *Proc. SESA*, vol. XVI, no. 1, pp. 69–92, 1958.
4. Irwin, G. R.: Discussion of Ref. 3, *Proc. SESA*, vol. XVI, no. 1, pp. 93–96, 1958.
5. Etheridge, J. M., and J. W. Dally: A Critical Review of Methods for Determining Stress Intensity Factors from Isochromatic Fringes, *Exp. Mech.*, vol. 17, pp. 248–254, 1977.

6. Bradley, W. B., and A. S. Kobayashi: An Investigation of Propagating Cracks by Dynamic Photoelasticity, *Exp. Mech.*, vol. 10, pp. 106–113, 1970.

7. Schroedl, M. A., and C. W. Smith: "Local Stress near Deep Surface Flaws under Cylindrical Bonding Fields," in *Progress in Flaw Growth and Fracture Toughness Testing*," ASTM publication STP 536, 1973, pp. 45–63.

8. Sanford, R. J., and J. W. Dally: A General Method for Determining Mixed Mode Stress Intensity Factors from Isochromatic Fringe Patterns, *Eng. Fract. Mech.*, vol. 11, pp. 621–633, 1979.

9. Sanford, R. J.: Determining Fracture Parameters with Full-Field Optical Methods, *Exp. Mech.*, vol. 29, pp. 241–247, 1989.

10. Dudderar, T. D., and H. J. Gorman: The Determination of Mode I Factors by Holographic Interferometry, *Exp. Mech.*, vol. 13, no. 4, pp. 145–149, 1973.

11. Dudderar, T. D., and R. O'Regan: Measurement of the Strain Field near a Crack Tip in Polymethylmethacrylate by Holographic Interferometry, *Exp. Mech.*, vol. 11, no. 2, pp. 49–56, 1971.

12. Barker, D. B., R. J. Sanford, and R. Chona: Determining K and Related Stress-Field Parameters from Displacement Fields, *Exp. Mech.*, vol. 25, no. 12, pp. 399–407, 1985.

13. Smith, C. W.: Use of 3D Photoelasticity in Fracture Mechanics, *Exp. Mech.*, vol. 13, pp. 539–544, 1973.

14. Smith, C. W., and A. S. Kobayashi: "Experimental Fracture Mechanics" in A. S. Kobayashi (ed.), *Handbook on Experimental Mechanics*, Prentice-Hall, Englewood Cliffs, N.J., 1987, pp. 891–956.

15. Hyde, T. H., and N. A. Warrior: "A Critical Assessment of Photoelastic Methods of Determining Stress Intensity Factors," in A. S. Tooth and J. Spence (eds.), *Applied Solid Mechanics—2*, Elsevier Applied Science, New York, 1987, pp. 23–40.

16. Dally, J. W., and R. J. Sanford: Classification of Stress-Intensity Factors from Isochromatic-Fringe Patterns, *Exp. Mech.*, vol. 18, no. 12, pp. 441–448, 1978.

17. Smith, D. G., and C. W. Smith: Photoelastic Determination of Mixed Mode Stress Intensity Factors, *Eng. Fract. Mech.*, vol. 4, no. 2, pp. 357–366, 1972.

18. Gdoutos, E. E., and P. G. Theocaris: A Photoelastic Determination of Mixed Mode Stress-Intensity Factors, *Exp. Mech.*, vol. 18, no. 3, pp. 87–97, 1978.

19. Der, V. K., D. B. Barker, and D. C. Holloway: *Mech. Res. Commun.*, vol. 5, no. 6, pp. 313–318, 1978.

20. Kobayashi, T., and J. W. Dally: Dynamic Photoelastic Determination of the a-K Relation for 4340 Alloy Steel, in G. T. Hahn and M. F. Kanninen (eds.), "Crack Methodology and Application," ASTM publication STP 711, 1980, pp. 189–210.

21. Dally, J. W., and R. J. Sanford: On Measuring the Instantaneous Stress Intensity Factor for Propagating Cracks, *Proc. 7th International Conference on Fracture*, ICF7, Houston, March 20–24, 1989, pp. 3223–3230.

22. Dally, J. W., J. R. Berger, and Y.-C. Ham: On the Use of Birefringent Coatings in Fracture Mechanics, *Proc. SEM Spring Conf.*, Cambridge, Mass., 1989, pp. 513–520.

23. Hutchinson, J. W.: Singular Behavior at the End of a Tensile Crack in a Hardening Material, *J. Mech. Phys. Sol.*, vol. 16, pp. 13–31, 1968.

24. Rice, J. R., and G. F. Rosengren: Plane Strain Deformation near a Crack Tip in a Power Law Hardening Material, *J. Mech. Phys. Sol.*, vol. 16, pp. 1–12, 1968.

25. Shih, C. F.: "Tables of Hutchinson-Rice-Rosengren Singular Field Quantities," Brown University Report MRL E-147, Material Research Laboratory, Brown University, June 1983.

PART

IV

COATING
METHODS

CHAPTER
15

PHOTOELASTIC COATINGS AND BRITTLE COATINGS

15.1 INTRODUCTION

In the application of coating methods, one applies a thin layer of a reactive material to the surface of the body that is to be analyzed. The thin coating is bonded to the surface and displacements at the coating-specimen interface are transmitted without amplification or attenuation. These displacements at the interface produce stresses and strains in the coating and the coating responds. The analyst observes the coating response and infers the stresses on the surface of the specimen based on the observed behavior of the coating.

The two most significant advantages of coating methods are first, the capability of applying the coating directly to the prototype. As with strain gages, it is not necessary to model the specimen. The prototype is used under operational conditions. The second advantage is the "whole"-field response of the coating. Stain gages respond over small regions of the field and give approximations to strain at a point. Coatings respond over the entire surface of the specimen and give field data rather than point data.

There are two markedly different coating methods that are used in stress analysis. The first is a birefringent coating that produces a photoelastic fringe pattern related to the coating stresses. The second is a brittle coating that fails by cracking when the coating stresses exceed some threshold value. Both of these coating methods will be covered in this chapter.

15.2 COATING STRESSES

Consider a thin coating that is bonded to a specimen as shown in Fig. 15.1. If the coating is thin relative to the thickness of the specimen, the strains developed at the surface of the specimen are transmitted without significant change to the coating and

$$\epsilon^c_{xx} = \epsilon^s_{xx} \quad \text{and} \quad \epsilon^c_{yy} = \epsilon^s_{yy} \tag{15.1}$$

Also it is reasonable to assume that the normal stresses vanish. Thus,

$$\sigma^c_{zz} = \sigma^s_{zz} = 0 \tag{15.2}$$

With these assumptions, it is clear that a state of plane stress exists in both the coating and on the surface of the specimen. Therefore, the strains may be expressed in terms of the stresses by employing Eqs. (2.19) to obtain

$$\epsilon^s_{xx} = \frac{1}{E^s}(\sigma^s_{xx} - v^s\sigma^s_{yy}) \qquad \epsilon^s_{yy} = \frac{1}{E^s}(\sigma^s_{yy} - v^s\sigma^s_{xx})$$

$$\epsilon^c_{xx} = \frac{1}{E^c}(\sigma^c_{xx} - v^c\sigma^c_{yy}) \qquad \epsilon^c_{yy} = \frac{1}{E^c}(\sigma^c_{yy} - v^c\sigma^c_{yy}) \tag{15.3}$$

Substituting Eqs. (15.3) into Eqs. (15.1) yields

$$\frac{1}{E^s}(\sigma^s_{xx} - v^s\sigma^s_{yy}) = \frac{1}{E^c}(\sigma^c_{xx} - v^c\sigma^c_{yy})$$

$$\frac{1}{E^s}(\sigma^s_{yy} - v^s\sigma^s_{xx}) = \frac{1}{E^c}(\sigma^c_{yy} - v^c\sigma^c_{xx}) \tag{a}$$

These equations are solved for σ^c_{xx} and σ^c_{yy} to give

$$\sigma^c_{xx} = \frac{E^c}{E^s(1 - v^{c2})}[(1 - v^c v^s)\sigma^s_{xx} + (v^c - v^s)\sigma^s_{yy}]$$

$$\sigma^c_{yy} = \frac{E^c}{E^s(1 - v^{c2})}[(1 - v^c v^s)\sigma^s_{yy} + (v^c - v^s)\sigma^s_{xx}] \tag{15.4}$$

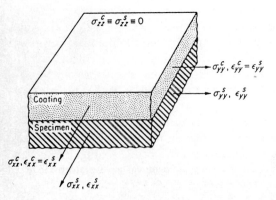

FIGURE 15.1
Elemental section of coating and specimen showing stresses and strains.

Equations (15.4) represent the stresses produced in the coating by the stresses on the surface of a specimen. Thus, they can be used to describe the response of both birefringent coatings and brittle coatings.

15.3 BIREFRINGENT COATINGS [1–5]

The method of birefringent coatings represents an extension of the procedures of photoelasticity to the determination of surface strains in opaque two- and three-dimensional bodies. The coating is a thin sheet of birefringent material, usually a polymer, which is bonded to the surface of the prototype being analyzed. The coating is mirrored at the interface to provide a reflecting surface for the light. When the prototype is loaded, the displacements on its surface are transmitted to the mirrored side of the coating to produce a strain field through the thickness of the coating. The distribution of the strain field over the surface of the prototype, in terms of principal-strain differences, is determined by employing a reflected-light polariscope to record the fringe orders, as illustrated in Fig. 15.2.

The birefringent-coating method has many advantages over other methods of experimental stress analysis. It provides full-field data that enable the investigator to visualize the complete distribution of surface strains. The method is nondestructive, and since the coatings can be applied directly to the prototype, the need for models is eliminated. Through proper selection of coating materials, the method can be made applicable over a very wide range of strain. The method is also very useful in converting complex nonlinear stress-analysis problems in the prototype into relatively simple linear elastic problems in the coating. For instance, plastic and viscoelastic response of a prototype can be measured in terms of the elastic response of the birefringent coating. Similarly, the anisotropic response of

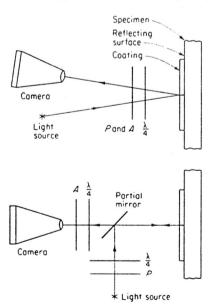

FIGURE 15.2
Reflection polariscopes commonly employed to record birefringent-coating data.

composite materials can be examined in terms of an isotropic response of the coating.

The concept of birefringent coatings was first introduced by Mesnager [1] in France in 1930, and later by Oppel [2] in Germany. These early efforts were not successful due to the lack of suitable adhesives, the low sensitivity of the available polymers, and the reinforcing effects occurring when glass was employed as a coating. When the epoxies became available, in the early 1950s, the bonding and sensitivity problems were alleviated. Significant developments of the method in the area of materials and technique were made by Fleury and Zandman [3] in France in 1954 and later by D'Agostino, Drucker, Liu, and Mylonas [4] in the United States and by Kawata [5] in Japan.

15.4 STRESS- AND STRAIN-OPTIC RELATIONS FOR COATINGS

When the specimen is loaded, the surface displacements of the specimen at the specimen-coating interface are transmitted to the birefringent coating if the bond is adequate. As the coating responds to these transmitted displacements, stresses and birefringence are induced. Observing the coating in a reflection polariscope gives a fringe pattern which is related to the surface strains of the specimen. If it is assumed that the coating is sufficiently thin, the strains occurring on the surface of the specimen are transmitted to the coating without distortion. Within the framework of this assumption, it is clear that

$$\sigma_3 = \sigma_{zz} = 0 \quad \text{in both coating and specimen}$$

$$\epsilon_1^c(x, y) = \epsilon_1^s(x, y) \qquad \epsilon_2^c(x, y) = \epsilon_2^s(x, y) \tag{15.5}$$

where the coordinate system is defined in Fig. 15.1.

Since the strains expressed in Eqs. (15.5) are identical to those described in Eqs. (15.1), it is possible to employ Eqs. (15.4) directly to give

$$\sigma_1^c - \sigma_2^c = \frac{E^c(1 + v^s)}{E^s(1 + v^c)} (\sigma_1^s - \sigma_2^s) \tag{15.6}$$

Inspection of Eq. (15.6) shows that the difference in the principal stresses acting in the coating $\sigma_1^c - \sigma_2^c$ is linearly related to the difference in the principal stresses acting on the surface of the specimen $\sigma_1^s - \sigma_2^s$. The elastic constants E^c, E^s, v^c, and v^s influence the magnitude of $\sigma_1^c - \sigma_2^c$. The photoelastic response of the coating is related to $\sigma_1^c - \sigma_2^c$ by employing Eq. (12.7) and noting that the optical path length is $2h^c$. Thus,

$$\sigma_1^c - \sigma_2^c = \frac{Nf_\sigma}{2h^c} = \frac{E^c(1 + v^s)}{E^s(1 + v^c)} (\sigma_1^s - \sigma_2^s) \tag{15.7a}$$

and the difference in the principal stresses for the specimen is given by

$$\sigma_1^s - \sigma_2^s = \frac{E^s(1 + v^c)}{E^c(1 + v^s)} \frac{Nf_\sigma}{2h^c} \tag{15.7b}$$

It is clear that $\sigma_1^s - \sigma_2^s$ can be determined from the isochromatic fringe order in the birefringent coating provided E^s, E^c, v^c, v^s, f_σ, and h^c are known. In certain instances it may be preferable to work in terms of strain instead of stress. This transformation is quite simple since it has been assumed that $\epsilon_1^c - \epsilon_2^c = \epsilon_1^s - \epsilon_2^s$; hence

$$\epsilon_1^s - \epsilon_2^s = \frac{N f_\epsilon}{2h^c} = \frac{1 + v^c}{E^c} \frac{N f_\sigma}{2h^c} \tag{15.8}$$

By using Eq. (15.8), the birefringent coating can be employed as a strain gage to give the difference in principal strains $\epsilon_1^s - \epsilon_2^s$.

The strain-optic relationship presented in Eq. (15.8) is often written in the form

$$\epsilon_1^s - \epsilon_2^s = \epsilon_1^c - \epsilon_2^c = \frac{N f_\epsilon}{2h^c} = \frac{N}{2h^c} \frac{\lambda}{K} \tag{15.9}$$

where K is the strain coefficient for the coating and λ is the wavelength in microinches (nanometers) of the light being used—that is, 21.5 μin, or 546.1 nm for mercury green light. This alternative form of the strain-optic law is more general since the same strain coefficient K for a coating (provided by the manufacturer) can be employed with light sources having different wavelengths. For a perfectly linear elastic photoelastic material, the constants f_σ, f_ϵ, and K are related by the expressions

$$f_\epsilon = \frac{\lambda}{K} = \frac{1 + v^c}{E^c} f_\sigma \tag{15.10a}$$

$$f_\sigma = \frac{E^c}{1 + v^c} \frac{\lambda}{K} = \frac{E^c}{1 + v^c} f_\epsilon \tag{15.10b}$$

The isochromatic and isoclinic data at a point in a coating are employed with Eq. (2.18) to establish the shearing strain γ_{xy}^s as

$$\gamma_{xy}^s = \gamma_{xy}^c = (\epsilon_1^c - \epsilon_2^c) \sin 2\phi_1 \tag{15.11}$$

where ϕ_1 is the isoclinic parameter defining the angle between σ_1 and the x axis. From Eqs. (15.8) and (15.11), it is clear that

$$\gamma_{xy}^s = \frac{N f_\epsilon}{2h^c} \sin 2\phi_1 \tag{15.12}$$

and from the stress-strain relationship expressed by Eq. (2.20)

$$\tau_{xy}^s = \frac{E^s}{2(1 + v^s)} \frac{N f_\epsilon}{2h^c} \sin 2\phi_1$$

$$= \frac{E^s(1 + v^c)}{E^c(1 + v^s)} \frac{N f_\sigma}{4h^c} \sin 2\phi_1 \tag{15.13}$$

The photoelastic data obtained from the isochromatic and isoclinic fringe patterns are not sufficient, in general, to determine the individual values of the principal stresses and strains. Auxiliary methods must be employed to determine σ_1, σ_2, and τ_{max}.

15.5 COATING SENSITIVITY

The response of a photoelastic coating to a stress field is controlled by a number of factors. The effects of these factors are evaluated by defining a stress sensitivity index S_σ^s as

$$S_\sigma^s = \frac{N}{\sigma_1^s - \sigma_2^s} \tag{15.14}$$

By substituting Eqs. (15.7b) and (15.10b) into Eq. (15.14), the stress sensitivity index S_σ^s is expressed in terms of coating properties and specimen properties as

$$S_\sigma^s = \frac{2h^c}{f_\epsilon} \frac{1 + v^s}{E^s} = C_c C_s \tag{15.15}$$

where $C_s = 2h^c/f_\epsilon$ is the coating coefficient of sensitivity
$C_s = (1 + v^s)/E^s$ is the specimen coefficient of sensitivity

The equation for the stress-sensitivity index indicates that optical response is increased, for a given coating and specimen material, only by increasing the coating thickness. Arbitrary increases in coating thickness are usually not possible, however, because of errors associated with thick coatings.

In applications of coatings to elastic-stress problems, the maximum response of the coating occurs when some point on the specimen yields. If the principal stresses in the specimen are of opposite sign, and if the material follows the Tresca yield criterion, the maximum stresses and strains at yielding are

$$(\sigma_1^s - \sigma_2^s)_{max} = S_y \tag{15.16a}$$

and

$$(\epsilon_1^s - \epsilon_2^s)_{max} = C_s S_y \tag{15.16b}$$

where S_y is the yield strength of the specimen material. The maximum fringe order observed as the specimen begins to yield follows from Eqs. (15.14), (15.15), and (15.16a) as

$$N_{max} = C_c C_s S_y = \frac{2h^c}{f_\epsilon} \frac{1 + v^s}{E^s} S_y \tag{15.17}$$

It is evident from Eq. (15.17) that N_{max} is a function of the three parameters S_y, C_s, and C_c. Typical values for N_{max} in a typical coating bonded to several engineering materials are listed in Table 15.1.

It is evident from Table 15.1 that the maximum optical response exhibited by a photoelastic coating depends strongly on the properties of the specimen

TABLE 15.1
Yield strength, specimen coefficient of sensitivity, strain difference, and maximum fringe order at yielding for a typical birefringent coating

Material	S_y		C_s		$(\epsilon_1^s - \epsilon_2^s)_{max}$ $\mu m/m$	N_{max},† fringes
	lb/in²	MPa	10^{-8} in²/lb	pm²/N		
Steel:						
HR 1020	35,000	240	4.3	6.2	1,500	1.86
CD 1020	45,000	310	4.3	6.2	1,940	2.40
HT 1040	80,000	550	4.3	6.2	3.450	4.26
HT 4140	130,000	900	4.3	6.2	5,600	6.92
HT 5210	180,000	1240	4.3	6.2	7,700	9.58
Maraging (18 Ni)	250,000	1720	4.9	7.1	12,200	15.20
Aluminum:						
1100 H16	20,000	140	13.3	19.3	2,660	3.30
3004 H34	29,000	200	13.3	19.3	3,860	4.78
2024 T3	50,000	345	13.3	19.3	6,670	8.25
7075 T6	73,000	500	13.3	19.3	9,700	12.00
Magnesium, AM 11	21,000	145	20.8	30.2	4,380	5.42
Cartridge brass	63,000	435	8.7	12.6	5,480	6.80
Phosphor bronze	75,000	515	8.7	12.6	6,520	8.10
Beryllium copper	70,000	480	7.7	11.2	5.400	6.70
Glass	3,000	21	12.5	18.1	375	0.46
Concrete (compression)	4,000	28	42.0	60.9	1,680	2.10
Plastic, nylon 6-6	12,000	83	467	677	56,000	69.00
Glass-reinforced plastic	120,000	830	31.8	46.1	38,200	47.40

† For $C_c = 1.24 \times 10^3$ fringes, which corresponds to $h^c = 0.10$ in $= 2.54$ mm, $f_c = 1.62 \times 10^{-4}$ in/fringe $= 4.10 \times \mu m/$fringe, $K = 0.14$, $\lambda = 22.7$ $\mu in = 578$ nm.

material. Maximum fringe orders range over two orders of magnitude, with very low responses on concrete and glass and relatively high responses on high-strength steel, aluminum, and composite materials. The coating thickness and the methods employed to determine fringe orders in any elastic-stress analysis will depend on the specific problem. Where the optical response is low, thick coatings or point-by-point compensation methods are required. With higher-strength materials, thin coatings can be employed, and the fringe orders can be photographed or observed directly on the model.

15.6 COATING MATERIALS [6–9]

One of the most important decisions in a photoelastic analysis is the selection of the proper material for the photoelastic model. Indeed, major advances in the application of two- and three-dimensional photoelasticity and birefringent coatings occurred only after the introduction of suitable materials. The physical properties which an ideal coating should exhibit include the following:

1. A high optical strain coefficient K to maximize coating response
2. A low modulus of elasticity E^c to minimize reinforcing effects
3. A high resistance to both optical and mechanical stress relaxation to ensure stability of the measurement with time
4. A linear strain-optical response to minimize data-reduction problems
5. A good adhesive bond to ensure perfect strain transmission between coating and specimen
6. A high proportional limit to increase the range of strain over which the coating can be utilized
7. Sufficient pliability to permit use on curved surfaces of three-dimensional components

In most instances, the selection of a coating material involves a compromise since no material exhibits all these characteristics. As an example, the first coating material employed by Mesnager was glass, which exhibited a relatively high sensitivity to strain. However, a glass coating reinforces the component significantly and it cannot be applied to curved surfaces. A number of polymers are available which can be employed as photoelastic coatings. Typical properties of these materials are listed in Table 15.2.

TABLE 15.2
Properties of different photoelastic materials for coating applications

Material	Modulus E		K	Sensitivity $1/f_\epsilon$		Strain limit, %	Ref.
	lb/in^2	MPa		fringes/in†	fringes/mm		
Polycarbonate	320,000	2,210	0.16	7,300	287		9, 10
PS-1	360,000	2,480	0.14	6,500	256	10.0	‡
Epoxy with anhydride	475,000	3,280	0.12	5,400	213	2.0	8
Epoxy with amines§	450,000	3,100	0.09	4,000	157		
PS-2	450,000	3,100	0.12	5,400	213	3.0	‡
Polyester¶	560,000	3,860	0.04	1,950	78	1.5	
Modified epoxy	2,800	19	0.02	1,000	39	15.0	6
PS-3	30,000	210	0.02	940	37	30.0	‡
Polyurethane††	500	3.5	0.008	380	15		
PS-4	1,000	7.0	0.005	230	9	150.0	‡
Glass	10,000,000	69,000	0.14	6,500	256	0.10	

† With mercury green light ($\lambda = 21.5\ \mu$in = 546 nm).
‡ Commercially available from Photoelastic, Inc., Malvern. PA.
§ 100 parts per hundred ERL 2774, 15 parts per hundred TETA.
¶ Homalite 100.
†† 100 parts per hundred Hysol 2085, 24 parts per hundred Hysol 3562.

An examination of Table 15.2 shows that polycarbonate exhibits a superior combination of properties. Unfortunately, it is available only in sheet form and techniques have not yet been established to allow contouring of the sheets to three-dimensional surfaces with compound curvatures. The epoxies also exhibit good sensitivity and are preferred when contouring is necessary. The casting and contouring process with epoxy materials will be discussed in Sec. 15.7. For large strains, which are often encountered in plastic analyses, modified epoxies (a blend of rigid and flexible resins of a copolymer of epoxy and polysulfide) are employed as coatings since they exhibit a linear response over a large range of strain (30 percent). Polyurethane rubber is used in applications involving very large strains where sensitivity is not an issue but where the strains encountered range from 30 to 100 percent. Glass is most commonly used in optical transducers, where its excellent stability with time and the environment is an important characteristic.

15.7 BONDING THE COATINGS [9]

The successful use of photoelastic coatings requires a perfect adhesive bond between the coating and the component. For this reason, careful attention must be given to component and coating surface preparation, the adhesive, and the bonding procedure. The surface of the component should be smooth and free of all foreign material. A satisfactory component surface can normally be obtained by sanding and cleaning with chemical agents and solvents. For plane surfaces, a precured sheet of coating can be machined to size, cleaned, and then bonded to the component. Proper matching of the geometry of the coating to that of the test specimen is required, during bonding, with this procedure. Alternatively, the sheet of coating can be bonded to the test specimen and, after the adhesive cures, the specimen can be used as a template to machine the coating to the proper shape.

The epoxy adhesive normally used with birefringent coatings is filled with aluminum powder. Once the adhesive is spread evenly over the surface of the component, the sheet of coating is applied. The coating should be positioned at one end and slowly rotated and pressed into position to work out the air bubbles. Pressure is not applied as the adhesive cures since residual stresses could be introduced in the coating by the clamping devices. The final layer of adhesive is between 0.003 and 0.010 in (0.08 and 0.25 mm) thick, depending primarily on the flatness of the specimen surface.

The application of photoelastic coatings to curved surfaces is best achieved by using the contoured-sheet method illustrated in Fig. 15.3. A sheet of epoxy is cast to the desired thickness on a plate coated with Teflon. The polymerization process is made to proceed slowly by selecting proper amine curing agents. During the first stage of polymerization, the coating material slowly transforms from a liquid (stage A) to a rubbery solid (stage B). In stage B, the sheet is soft and pliable. Also, the strain-optic coefficient is very low; therefore, large deformations can be imposed on the sheet without introducing photoelastic response. At this stage, the sheet should be stripped from the casting plate and contoured by hand to fit the curved surfaces of the test specimen.

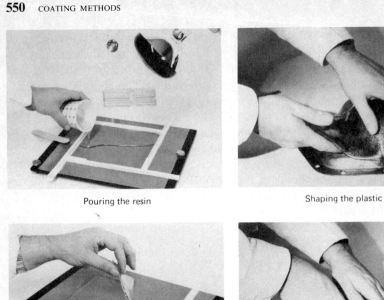

Pouring the resin

Shaping the plastic

Removing the sheet from the mold

Shaping the plastic

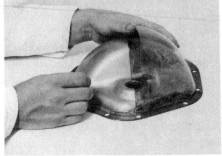

Cutting away the 1/4-in meniscus

Hardened contoured coating

FIGURE 15.3
Sequence of operations in applying contoured sheets.

During the next stage (C) of polymerization, as the coating becomes rigid and photoelastically responsive, it should be maintained in contact with the surface of the test specimen. When the polymerization is completed, the rigid contoured shell of coating can be stripped from the specimen, cleaned, and checked for thickness variations. The shell can then be trimmed to match the adjoining section of coating and cemented in place.

This contoured-sheet method enables the coating to be applied to specimens of almost any shape. It permits the coating to be formed into stress-free shells of reasonably uniform thickness which conform perfectly to the surface to be coated. In actual operation, some experience is required to develop the skills associated with forming the shells used to coat a complicated three-dimensional specimen.

15.8 EFFECTS OF COATING THICKNESS

When a photoelastic coating is bonded to a specimen, only in a few instances are the strains transmitted to the coating without some modification or distortion. More realistically, the coating is considered as a three-dimensional extension of the specimen which is loaded by means of shear and normal tractions at the interface. These tractions vary so that the displacements experienced by the coating and the specimen at the interface are identical (as dictated by perfect bonding). Thus, in the most general case:

1. The average strain in the coating does not equal the strain at the interface.
2. A strain gradient exists through the thickness of the coating.
3. The coating serves to reinforce the specimen.

It is evident that these effects of thickness tend to vanish as the coating thickness approaches zero. However, coatings with finite thickness (usually 0.50 to 3.00 mm, or 0.02 to 0.10 in) are required to obtain a high fringe count for accurate fringe-order determinations. As a result, the question naturally arises regarding the magnitude of the error associated with thickness effects in the application of the coating method.

The topic of thickness effects will be treated here by beginning with the simple model of the coating considered previously in Sec. 15.2. The model of the coating will be made progressively more complex as additional factors influencing the behavior of the coating are introduced. Where possible, experimental verifications will be used to justify assumptions and to minimize the complexity of the analysis.

15.8.1 Reinforcing Effects of Birefringent Coatings [10]

When a birefringent coating is applied to a specimen and subjected to loads, the coating carries a portion of the load and consequently the strain on the surface of the specimen is reduced. It is possible in many cases to calculate the reinforcing effect due to the birefringent coating and to establish correction factors which can be employed in a simple fashion to account for the reinforcement. In this section, the reinforcement due to the coating will be determined for plane-stress and flexural problems. For plane stress, an element from a coated specimen is isolated as shown in Fig. 15.4. A similar element from an uncoated specimen is also isolated,

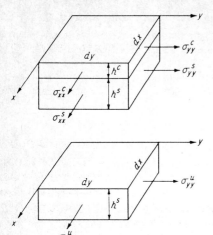

FIGURE 15.4
Comparison of stresses in coated and uncoated elements from a plane-stress specimen.

and the forces acting in the x direction on both elements can be equated to give

$$h^s \, dy \, \sigma_{xx}^u = h^s \, dy \, \sigma_{xx}^s + h^c \, dy \, \sigma_{xx}^c$$

from which

$$\sigma_{xx}^u = \sigma_{xx}^s + \frac{h^c}{h^s} \sigma_{xx}^c \tag{15.18a}$$

The corresponding expression for forces in the y direction is

$$\sigma_{yy}^u = \sigma_{yy}^s + \frac{h^c}{h^s} \sigma_{yy}^c \tag{15.18b}$$

If it is again assumed that

$$\epsilon_{xx}^c = \epsilon_{xx}^s \qquad \epsilon_{yy}^c = \epsilon_{yy}^s \qquad \sigma_{zz}^c = \sigma_{zz}^s = 0$$

both the coating and the specimen are in a state of plane stress, and from Eqs. (2.20) it is apparent that

$$\frac{E^s}{1 - v^{s2}} (\epsilon_{xx}^u + v^s \epsilon_{yy}^u) = \frac{E^s}{1 - v^{s2}} (\epsilon_{xx}^s + v^s \epsilon_{yy}^s) + \frac{h^c}{h^s} \frac{E^c}{1 - v^{c2}} (\epsilon_{xx}^c + v^c \epsilon_{yy}^c) \tag{a}$$

$$\frac{E^s}{1 - v^{s2}} (\epsilon_{yy}^u + v^s \epsilon_{xx}^u) = \frac{E^s}{1 - v^{s2}} (\epsilon_{yy}^s + v^s \epsilon_{xx}^s) + \frac{h^c}{h^s} \frac{E^c}{1 - v^{c2}} (\epsilon_{yy}^c + v^c \epsilon_{xx}^c) \tag{b}$$

Subtracting Eq. (b) from Eq. (a) and simplifying yields

$$\epsilon_{xx}^u - \epsilon_{yy}^u = \left(1 + \frac{h^c E^c (1 + v^s)}{h^s E^s (1 + v^c)}\right)(\epsilon_{xx}^c - \epsilon_{yy}^c) \tag{15.19}$$

This equation may be rewritten as

$$\epsilon_{xx}^u - \epsilon_{yy}^u = F_{CR}(\epsilon_{xx}^c - \epsilon_{yy}^c)$$

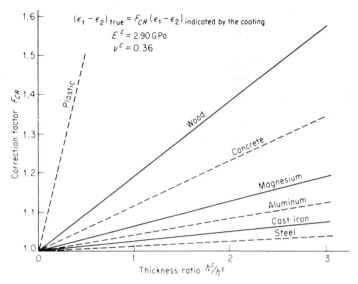

FIGURE 15.5
Correction factor F_{CR} for a number of different specimen materials.

where

$$F_{CR} = \left(1 + \frac{h^c E^c(1 + v^s)}{h^s E^s(1 + v^c)}\right) \tag{15.20}$$

The term F_{CR} represents a correction factor which must be applied to the strain difference $\epsilon^c_{xx} - \epsilon^c_{yy}$ obtained from the birefringent coating to establish the true value of the principal strain difference in the uncoated specimen. The correction factor F_{CR} accounts for the reinforcement due to the presence of the birefringent coating.

A graph showing F_{CR} as a function of the thickness ratio h^c/h^s is presented for a number of different materials in Fig. 15.5. These results are based on values of $E^c = 2.90$ GPa (420 ksi) and $v^c = 0.36$, which are representative of the rigid epoxy and polycarbonate coating materials. These results show that the correction factor is small for values of h^c/h^s less than 1 provided the specimen material is metallic. If, however, the specimen material has a low modulus like wood, concrete, or plastic, the correction factor becomes appreciable.

15.8.2 Strain Variations through the Coating Thickness [10]

A second example to illustrate the combined effect of strain variation through the thickness of the coating and the reinforcing effects is that of a plate subjected to a pure bending moment M. Consider an element from the central region of this plate, as indicated in Fig. 15.6. Since the strain distribution is linear and is

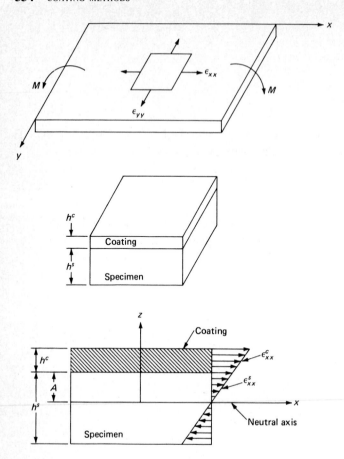

FIGURE 15.6
Element with a birefringent coating from the center of a wide plate in bending.

transmitted through the specimen-coating interface, then elementary plate theory gives the strain as a function of z in both the specimen and the coating as

$$\epsilon_{xx}^s = \frac{z}{\rho} \qquad \text{for } (h^s - A) \leq z \leq A$$

$$\epsilon_{xx}^c = \frac{z}{\rho} \qquad \text{for } A \leq z \leq (A + h^c) \qquad (15.21)$$

$$\epsilon_{yy}^s = \epsilon_{yy}^c = 0 \qquad \text{for all } z$$

where z is measured from the neutral axis, A is the distance from the neutral axis to the interface, and ρ is the radius of curvature. Since σ_{zz} is assumed to vanish for all values of z, Eqs. (2.20) are employed with Eqs. (15.21) to express the stress

σ_{xx} in terms of the strains:

$$\sigma_{xx}^s = \frac{E^s}{1 - v^{s2}} \frac{z}{\rho} \qquad \text{for } h^2 - A \leq z \leq A$$

$$\sigma_{xx}^c = \frac{E^c}{1 - v^{c2}} \frac{z}{\rho} \qquad \text{for } A \leq z \leq A + h^c$$

(15.22)

The position of the neutral axis, as described by the symbol A, can be obtained by considering equilibrium of the plate in the x direction and is

$$A = \frac{h^2(1 - BC^2)}{2(1 + BC)}$$

(15.23)

where

$$B = \frac{E^c(1 - v^{s2})}{E^s(1 - v^{c2})}$$

$$C = \frac{h^c}{h^s}$$

The radius of curvature ρ can be computed by considering equilibrium of the moments, to obtain

$$\frac{1}{\rho} = \frac{12M}{H} \frac{1 - v^{s2}}{E^s h^{s3}}$$

(15.24)

where

$$H = 4(1 + BC^3) - \frac{3(1 - BC^2)^2}{1 + BC}$$

If the coating is examined in a reflection polariscope, the fringe pattern is proportional to the average of the strain difference $\epsilon_{xx}^c - \epsilon_{yy}^c$ through the coating thickness. The average strain can be determined from Eqs. (15.21), (15.23), and (15.24) as

$$(\epsilon_{xx}^c - \epsilon_{yy}^c)_{\text{avg}} = \frac{12M}{H} \frac{1 - v^{s2}}{E^s h^{s3}} \frac{1}{h^c} \int_A^{A+h^c} z \, dz$$

which yields

$$(\epsilon_{xx}^c - \epsilon_{yy}^c)_{\text{avg}} = \frac{6M}{H} \frac{1 - v^{s2}}{E^s h^{s2}} \frac{1 + C}{1 + BC}$$

(15.25)

Since the true difference of strain on the surface of an uncoated plate is

$$(\epsilon_{xx}^s - \epsilon_{yy}^s)_{\text{true}} = 6M \frac{1 - v^{s2}}{E^s h^{s2}}$$

(15.26)

it is clear by comparison of Eqs. (15.25) and (15.26) that the coating does not indicate the true difference in the surface strains. It is possible to correct the resulting error by introducing a bending correction factor defined by

$$(\epsilon_{xx}^s - \epsilon_{yy}^s)_{true} = F_{CB}(\epsilon_{xx}^c - \epsilon_{yy}^c)_{avg}$$

where the bending correction factor F_{CB} is

$$F_{CB} = \frac{H(1 + BC)}{1 + C} = \frac{1 + BC}{1 + C}\left[4(1 + BC^3) - \frac{3(1 - BC^2)^2}{1 + BC}\right] \quad (15.27)$$

The bending correction factor, which is a function of the dimensionless ratios B and C, is shown in Fig. 15.7.

This correction factor accounts for the two thickness effects which occur in this example. The first is due to reinforcing, which reduces the strain at the interface between the coating and the specimen in comparison with a plate without the coating. The second effect is due to the gradient of strain through the coating. The optical response of the coating is related to the average strain, which in this instance represents the strain at the midpoint in the coating. Since the average coating strain is higher than the interface strain, the effects of the strain gradient tend to offset the reinforcing effect of the coating. In applications of coatings on beams and plates, the coating thickness h^c is usually less than the specimen thickness h^s, and the results of Fig. 15.7 indicate that the correction factor for $h^c/h^s < 1$ is appreciable and should not be neglected.

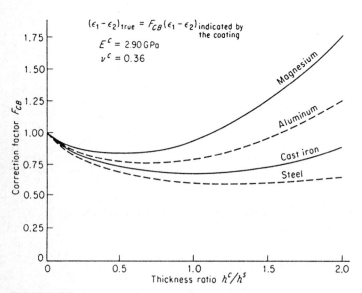

FIGURE 15.7
Correction factor F_{CB} for a number of different specimen materials.

15.8.3 Poisson's Ratio Mismatch [11]

In plane-stress problems, the errors arising from coating thickness effects, i.e., reinforcement, strain gradient, and curvature, are usually small. However, in almost all cases, a mismatch in Poisson's ratio occurs with v^c usually greater than v^s. This difference in Poisson's ratio produces a distortion of the displacement and strain fields through the thickness of the coating which is pronounced at the boundaries.

The significance of the Poisson's ratio mismatch effect can be examined by assuming that the strains ϵ_1 tangent to the boundary are transmitted without loss or amplification so that $\epsilon_1^c = \epsilon_1^s$. This assumption implies that the distortion of the strain field occurs in the transverse direction. At the interface, the transverse strain in the coating ϵ_2^c is controlled by the specimen and

$$\epsilon_2^c = \epsilon_2^s = -v^s \epsilon_1^s \qquad (a)$$

At the free surface of the coating, the transverse strain is controlled by the coating and

$$\epsilon_2^c = -v^c \epsilon_1^c = -v^c \epsilon_1^s \qquad (b)$$

The average value of ϵ_2^c through the thickness of the coating is bounded by these two limiting values; therefore, the fringe-order response of the coating at the boundary is bounded by

$$\frac{(1 + v^s)\epsilon_1^s}{F_\epsilon} < N < \frac{(1 + v^c)\epsilon_1^s}{F_\epsilon} \qquad (15.28)$$

where $F_\epsilon = f_\epsilon/2h^c$.

It is clear from this inequality that the magnitude of the distortion is controlled by the mismatch parameter $(1 + v^c)/(1 + v^s)$. For a constant value of $v^c = 0.36$ and variations in v^s between 0 and 0.5, the mismatch parameter ranges from 1.36 to 0.90. Experiments with tensile specimens, illustrated in Fig. 15.8, indicate that the fringe orders on the boundary and in interior regions of the specimen are given by

$$N = \frac{\epsilon_1^s(1 + v^c)}{F_\epsilon} \qquad \text{on the boundary}$$

$$\qquad\qquad\qquad\qquad\qquad\qquad (15.29)$$

$$N = \frac{\epsilon_1^s(1 + v^s)}{F_\epsilon} \qquad \text{in the interior}$$

A transition zone exists near the boundary, where the relation between fringe order and strain in a tension specimen can be expressed as

$$N = [1 + v^s + C_v(v^c - v^s)]\frac{\epsilon_1^s}{F_\epsilon} \qquad (15.30)$$

where C_v is a correction factor accounting for the mismatch. Experiments conducted with tension specimens with a large mismatch parameter (1.24) and with

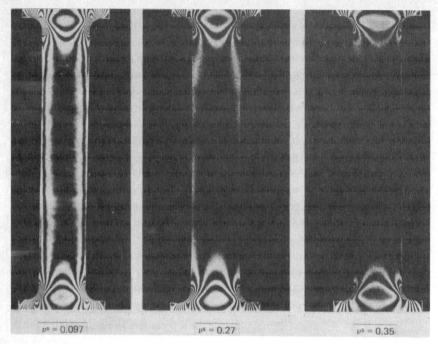

$\nu^s = 0.097$ $\nu^s = 0.27$ $\nu^s = 0.35$

FIGURE 15.8
Effect of Poisson's ratio mismatch on the response of birefringent coatings, $\nu^c = 0.36$.

coatings ranging in thickness from 0.55 to 3.25 mm (0.02 to 0.13 in) indicated that the width of the transition zone is about 4 times the thickness of the coating. The value of C_ν decreases from 1 at the boundary to zero at a position $4h^c$ from the boundary, as indicated in Fig. 15.9.

For metallic specimens, $\nu^c - \nu^s$ is usually less than 0.06, which is relatively small in comparison with $1 + \nu^s \approx 1.3$ in Eq. (15.30). As a result, the effect of

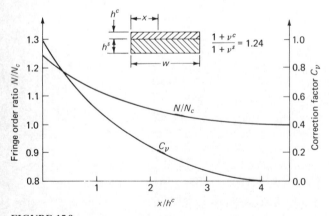

FIGURE 15.9
Correction factor C_ν as a function of x/h^c.

Poisson's ratio mismatch is often neglected in analyses of most metallic components.

15.9 FRINGE-ORDER DETERMINATIONS IN COATINGS

The methods employed to determine the order of fringes exhibited by the coating depend upon the response of the coating and the accuracy required in the analysis. If the response of the coating is large (four or more fringes), monochromatic light can be used to obtain photographs of the light- and dark-field isochromatic fringe patterns. These fringes can usually be interpolated or extrapolated to the nearest 0.2 fringe, giving an accuracy of about 5 percent based on a maximum of four fringes. As example of a high-response fringe pattern which can be analyzed as indicated above is shown in Fig. 15.10.

For fringe patterns exhibiting between two and four fringes, the use of colored patterns produced with white light is advantageous. The colored pattern is due to the attenuation and extinction of one or more colors from the white spectrum. The observed fringes represent the complementary color produced by the transmitted portion of the white-light spectrum. The sequence of colored fringes produced by an increasing stress field is listed in Table 15.3. The exact shade of color will be a function of the energy distribution in the white-light spectrum and the recording characteristics of the film being used; however, the sequence listed in the table is adequate for direct visual observations.

It is evident from Table 15.3 that the use of white light substantially increases the number of fringes that can be identified. For example, in the interval $0 \leq N \leq 2$, twelve different color bands exist, which can be used to establish fractional fringe orders. Moreover, the polariscope can also be used with light-field settings to yield a second family of colored fringes to effectively double the amount of data available

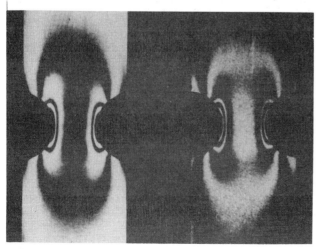

FIGURE 15.10
Fringe patterns from a thin coating on a glass-reinforced plastic specimen.

TABLE 15.3
Sequence of colors produced in a dark-field reflection polariscope with white light

Color	Retardation, nm	Fringe order
Black	0	0
Gray	160	0.28
White	260	0.45
Yellow	350	0.60
Orange	460	0.79
Red	520	0.90
Tint of passage no. 1†	577	1.00
Blue	620	1.06
Blue-green	700	1.20
Green-yellow	800	1.38
Orange	940	1.62
Red	1050	1.81
Tint of passage no. 2†	1150	2.0
Green	1350	2.33
Green-yellow	1450	2.50
Pink	1550	2.67
Tint of passage no. 3†	1730	3.00
Green	1800	3.10
Pink	2100	3.60
Tint of passage no 4†	2300	4.00
Green	2400	4.13

† The tint of passage is a sharp dividing zone between red and blue in the first-order fringe, red and green in the second-order fringe, and pink and green in the third-, fourth, and fifth-order fringes. Beyond five fringes, white-light analysis is not practical since the colors become very pale and difficult to distinguish.

for estimating fractional fringe orders. Thus, the fringe order can be established to within ± 0.1 fringe, giving an accuracy of about ± 5 percent based on a maximum of 2 fringes.

For precise fringe-order determinations, where the maximum fringe order is less than 2 and accuracies of 5 percent or less are required in the analysis, it is necessary to use compensation techniques. These techniques are point-by-point methods which significantly improve the accuracy of the fringe-order determination but require detailed involvement of the investigator. The compensation methods employed are similar to those described in Sec. 13.2.3.

15.10 BRITTLE COATINGS [12–16]

The first brittle coatings to be employed were those that naturally formed on structural members, such as mill scale on hot-rolled steel. These oxide coatings failed by flaking or cracking when the base material yielded under load and excessive strains were produced. To improve the visibility of the oxide coating response, the structural members were often coated with whitewash. The flaking of the coating produced dark lines which were visible against the white background. The first artificial brittle coating was a mixture of shellac and alcohol, which Sauerwald and Wieland employed in 1925 to indicate regions of plastic strain. This coating was an improvement over the whitewash technique, but the strains required for the coating to fail were still in the yield region for the base material and when the coating was used the prototype was destroyed. To reduce the strain sensitivity, other coatings were developed and introduced in Europe and the United States in the 1930s. A coating developed by Ellis, called *Stresscoat*, was widely used for about 30 years. The formulation of this coating was relatively simple, with three basic constituents: (1) a zinc resinate base, (2) carbon disulfide as a solvent, and (3) dibutyl phthalate as a plasticizer to vary the degree of brittleness of the coating. Stresscoat was modified later when the toxicity and flammability of carbon disulfide became objectionable. In the modified coating, the solvent carbon disulfide was replaced with methylene chloride to provide a nonflammable, low-toxicity product which is much safer to employ in typical industrial laboratories. The plasticizer in the modified coatings has also been changed in order to reduce the typical threshold strain to about 500 $\mu\varepsilon$. These commercial developments were preceded by the formulation of a new nonflammable, low-toxicity coating called *Straintec* by Racine, Taracks, and Killinger of the General Motors Truck and Coach Division [16].

Ceramic-based brittle coatings, developed by Singdale [15], are porcelain enamels mixed with lead borosilicate. The ceramic-based brittle coatings extend the applicability of the brittle-coating method to temperatures approaching 700°F (370°C). Unfortunately, the coatings must be fired onto the component at temperatures ranging from 950 to 1100°F (510 to 595°C), a procedure that severely limits their use.

The brittle coating is usually air-sprayed onto the component being studied until a layer of coating from 0.08 to 0.25 mm (0.003 to 0.010 in) thick has accumulated. After the coating has dried, the component is subjected to a system of loads which causes the coating to fail by cracking. The stresses in the coating which produce the crack pattern are related to the stresses in the specimen, as discussed previously in Sec. 15.2.

Since the coating is usually very thin relative to the thickness of the specimen, we assume that the surface strains occurring on the specimen are transmitted to the coating and that these coating strains are uniform through the thickness of the coating. Most of the investigators working with brittle coatings assume that the coating fails when it is strained to a certain critical value called the *threshold strain*. The value of the threshold strain ϵ_{t*} is determined by calibrating the coating in a manner which will be discussed in Sec. 15.10.1. The principal stress in the specimen, perpendicular to the crack in the coating, is computed by using the equation

$$\sigma_1^s = E^s \epsilon_{t*} \tag{15.31}$$

where σ_1^s = principal stress in specimen at location of coating crack

E^s = modulus of elasticity of specimen material

ϵ_{t*} = threshold strain required to crack coating when specimen is subjected to a uniaxial state of stress

This equation for computing the principal stresses is extremely simple and is widely accepted as being sufficiently accurate for brittle-coating work; however, this approach neglects the influence on the failure of the coating of the biaxial stress system which usually exists in the coating when it fails.

15.10.1 Calibration of Brittle Coatings

A brittle coating is usually calibrated by applying it to a calibration beam which is subjected to a known deflection, as shown in Fig. 15.11. The coating fails by cracking progressively from the fixed end of the beam toward the free (deflected) end. The position of the last crack from the free end is marked. Since this position is linearly proportional to strain, a scale can be used to measure the threshold strain ϵ_{t*} in calibration.

If the beam is considered as the specimen, the specimen stresses are given by

$$\sigma_{xx}^s = E^s \epsilon_{xx}^s \qquad \sigma_{yy}^s \approx 0 \qquad (15.32)$$

The procedure followed in calibration is to set

$$\epsilon_{xx}^s = \epsilon_{t*} \qquad (15.33)$$

to obtain the threshold strain for the coating. Substituting Eqs. (15.6) and (15.7)

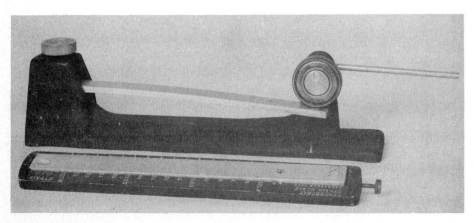

FIGURE 15.11
Calibration fixture, beams, and strain-measuring scale.

into Eqs. (15.4) gives the coating stresses for the calibration beam:

$$
\sigma_{xx}^c = \frac{E^c}{1 - v^{c2}}(1 - v^c v^*)\epsilon_{t^*} = \frac{E^c}{E^*(1 - v^{c2})}(1 - v^c v^*)\sigma_{xx}^*
$$

(15.34)

$$
\sigma_{yy}^c = \frac{E^c}{1 - v^{c2}}(v^c - v^*)\epsilon_{t^*} = \frac{E^c}{E^*(1 - v^{c2})}(v^c - v^*)\sigma_{xx}^*
$$

where v^* = Poisson's ratio for calibration beam
σ^* = uniaxial stress in calibration beam required to crack coating
E^* = modulus of elasticity of calibration beam

From Eqs. (15.34) it is apparent that a uniaxial state of stress in the calibration beam produces a biaxial state of stress in the coating. The biaxial stress system results from the mismatch in Poisson's ratio between the specimen and the coating. Since v^c of the resin-based coatings is about 0.42, this mismatch is appreciable when brittle coatings are employed on metals where v^s ranges between 0.29 and 0.33.

15.10.2 Failure Theories [17]

To predict stresses in the specimen from observations of the crack pattern in the coating, some law of failure for the coating must be utilized. Three laws of failure—maximum-normal-strain theory, maximum-normal-stress theory, and the Mohr theory—have been used in the past to represent the behavior of different coatings.

Unfortunately, basic experiments which define the law of failure for the two commercially available coatings (modified Stresscoat and All Temp) have not been conducted. The current practice in applying brittle coatings is to use Eqs. (15.5) and to assume that the coating does not respond to σ_2^s. This practice produces error in predicting specimen stresses based on coating failure. The amount of the error depends on the biaxiality of the specimen stresses and the law of failure of the coating. Exercises are provided at the end of this chapter to illustrate methods for determining the magnitude of the error when a law of failure is specified.

15.10.3 Brittle-Coating Crack Patterns [17]

The manner in which a brittle coating fails by cracking depends upon the state of stress in the specimen to which it adheres. The failure behavior of the coating is determined by the magnitudes of the principal stresses in the coatings. Consider the following special cases for a specimen subjected to direct loading.

CASE 1: $\sigma_1 > 0$, $\sigma_2 < 0$, $\sigma_3 = 0$. In this case, one set of cracks forms perpendicular to the principal stress σ_1 (see Fig. 15.12a). These cracks, in the direction of the principal stress σ_2, are stress trajectories or isostatics.

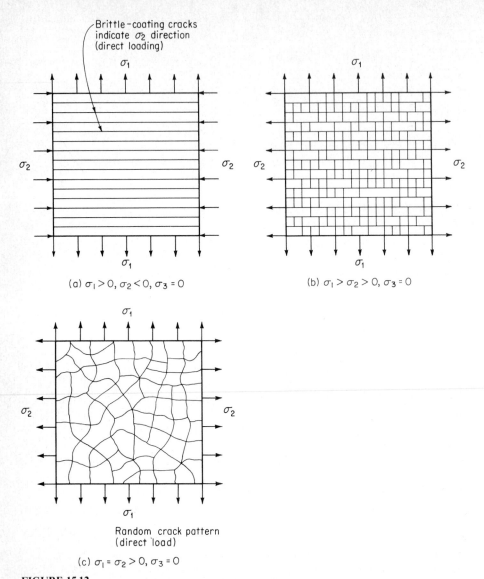

FIGURE 15.12
Brittle-coating crack patterns produced by different states of stress.

CASE 2: $\sigma_1 > \sigma_2 > 0$, $\sigma_3 = 0$. In this case, two families of cracks can form. The first set of cracks, due to σ_1, is perpendicular to σ_1 and parallel to σ_2 (see Fig. 15.12b). When the stress level of σ_2 becomes sufficiently high, a second family of cracks forms perpendicular to σ_2 and parallel to σ_1. When both families of cracks are produced, a complete set of isostatics, or stress trajectories, is obtained. This type of crack pattern is often encountered when testing cylindrical pressure vessels,

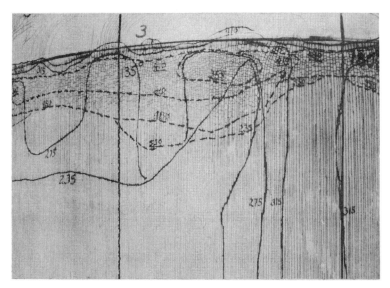

FIGURE 15.13
Enlarged view of the brittle-coating crack pattern over a portion of a cylindrical pressure vessel.

where σ_1 (the hoop stress) is twice as large as σ_2 (the axial stress). If the pressure is increased slowly, the first cracks due to the hoop stress will appear in the coating along the axis of the cylinder. Later, after the pressure applied to the cylindrical vessel has more than doubled, a second set of cracks will form in the hoop direction because of the axial stresses. An example of brittle-coating crack formation on a cylindrical pressure vessel is shown in Fig. 15.13. In the central portion of the cylinder, at the bottom of this figure, only the cracks due to σ_1 (the hoop stress) have formed. At the top of the photograph, near the weld which joins the cylinder to the hemispherical head, σ_2 (the axial stress) is large enough to produce the second set of cracks in the coating.

CASE 3 $\sigma_1 = \sigma_2 > 0$, $\sigma_3 = 0$. When $\sigma_1 = \sigma_2$ at any point on the body, the stress system is said to be *isotropic* and every direction is a principal direction. If the value of $\sigma_1 = \sigma_2$ is sufficiently high, the coating will fail; however, the crack pattern produced will be random in character. Crack patterns of this form are often referred to as *craze patterns* since the cracks have no preferential direction. A schematic illustration of this type of crack pattern is shown in Fig. 15.12c. Crack patterns of this type occur when pressure vessels with hemispherical head closures are tested. An illustration of a typical pattern obtained from the hemispherical head of a pressure vessel is shown in Fig. 15.14.

CASE 4: $\sigma_2 < \sigma_1 < 0$, $\sigma_3 = 0$. In this instance, the coating is subjected to a state of biaxial compressive stress and will not crack under direct load. However, if the compressive stresses are sufficiently large, the coating will fail by flaking from the

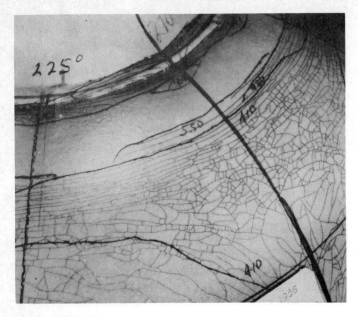

FIGURE 15.14
Brittle-coating crack pattern on a hemispherical head closure. (The random directional property which cracks exhibit indicates nearly equal stresses in the circumferential and meridional directions.)

surface of the specimen. Observations of regions where brittle coatings have failed by flaking are not common in elastic analyses, since loads are usually limited to maintain stresses in the specimen below the yield strength of the specimen material. However, when the loads are increased and regions of the specimen experience plastic flow, flaking is commonly observed, particularly near the points of application of concentrated loads.

In many practical problems where the brittle-coating method is applied, a particular region is often highly stressed and the remainder of the body is not stressed to a level sufficient to produce coating response. These low-stressed regions are often as important to an engineer seeking weight reduction as the highly stressed region, since material can be removed from the low-stressed regions without reducing the service life of the component. The only information provided by a coating which will not crack is that the stress is lower than σ_{xx}^*, the lowest specimen stress required to crack the coating. It is necessary to employ strain gages to determine the stresses in regions where the coating does not naturally respond. However, with some inducement, the brittle coating can provide information regarding the directions of the stresses in these regions. When the principal-stress directions are known, the number of strain gages required to determine the principal stresses is reduced to two.

It is possible to obtain coating cracks in low-stressed regions by employing refrigeration techniques with brittle coatings. First, the specimen is loaded until the stress in the critical region is just below the yield stress; then, the coating is subjected to a rapid temperature decrease (ΔT) while under load. The change in temperature ΔT produces a state of hydrostatic tension in the coating which is superimposed upon the existing stresses in the coating due to the load. The combined load-induced and thermal stresses are usually sufficient to produce a crack pattern. The direction of the cracks is coincident with one of the principal stresses due to the load since the thermal stresses are isotropic and have no preferential directions. This technique is simple to apply, and the crack patterns give an accurate representation of the isostatics.

In order to reduce the temperature of the coating, i.e., refrigeration, one of two techniques is commonly employed. In the first method ice water is sponged over the area of the coating which has not previously responded. This method is not always successful since it does not produce sufficiently high thermal stresses to give total coating response. A larger temperature change can be obtained by passing a stream of compressed air through a container of dry ice before it is directed onto the surface of the coated model. The stream of very cold air can be accurately directed, and the resulting crack patterns can be closely controlled.

In many applications of the brittle-coating method, situations are encountered where one or possibly both of the principal stresses are compressive. In these instances it is not possible to obtain the crack patterns associated with the compressive stresses because a brittle coating does not crack as a result of compressive stress. In order to circumvent this difficulty, a relaxation technique is employed which is exactly opposite to the direct-loading procedure. That is, a load is applied to the coated specimen before the coating has completely dried. The load is maintained on the coated specimen until drying is complete. Under these conditions the coating is stress-free (neglecting residual stresses, which are always present in the coating) while the specimen is stressed in compression. If a portion of the load is relieved, the specimen will stretch, since it was previously compressed, and tensile stresses will develop in the coating. The coating fails and indicates regions of compressive specimen stress. Thus, relaxation of the load on the specimen reverses the coating stresses and permits determination of the compressive stresses in the specimen.

15.10.4 Crack Detection

When a load is applied to a specimen treated with a brittle coating, cracks will form in the coating. If the coating has cured with a residual tensile stress, these cracks will remain open after the load is removed. To employ the coating to give either directional information or stress-level data, it is necessary to locate all cracks and to record the value of the load at which they occur.

The cracks produced in brittle coatings are very fine. They are V-shaped with a depth equal to the coating thickness and a width of about 5 to 10 percent of the coating thickness. To observe these fine cracks, a focused light source must be directed at oblique incidence to the surface and normal to the crack. Often the location, direction, and the load producing the first crack are not known.

Consequently, the time expended in a brittle-coating analysis to locate the regions of cracked coating is significant.

A method of electrified-particle inspection, available under the trade name Statiflux, reduces the experience needed in locating the crack patterns. Water containing wetting agents to reduce its surface tension is applied to the coating to be examined. If any cracks are present, the "wet water" flows into the cracks, fills them, and makes electrical contact with the metal specimen. The surface of the specimen is then dried with a facial tissue but the water is not removed from the cracks. Next, a talcum powder with negatively charged particles is sprayed on the specimen. The negatively charged particles of powder are electrostatically attracted to the grounded water contained in the cracks. The powder forms small, white mounds over the cracks and provides an excellent means of locating the crack pattern. A typical example of a Statiflux crack pattern is illustrated in Fig. 15.15. The method also can be used very successfully as an aid to the photographer in recording the crack patterns.

The prime disadvantage of the Statiflux method of crack detection is that the "wet water" must repeatedly be applied and removed from the coating. Any deviation from a uniform application of the water over the complete surface of

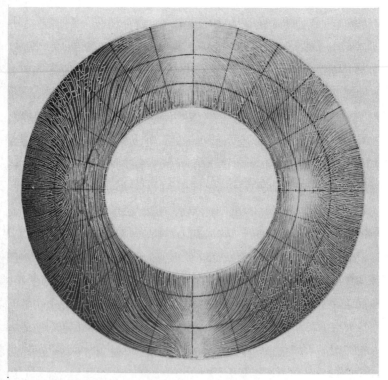

FIGURE 15.15
Statiflux method of crack detection. (*Courtesy of A. J. Durelli.*)

the coating will probably result in a temperature change due to differentials in evaporation rates. Since the sensitivity of the resin-based coatings changes markedly with temperature, a temperature variation over the surface of the coating is to be avoided in any quantitative analysis.

A red dye etchant is used with the resin-based coatings to increase the visibility of the crack patterns for photographic purposes. The dye etchant is a mixture containing solvents, oils, and red dye. The etchant is applied to the surface of a cracked brittle coating for approximately 10 s. During this time the etchant begins to attack the coating in the neighborhood of the cracks, thus making them wider. If the etchant is left in contact with the coating too long, it will attack the surface of the coating. After the etchant is wiped from the surface of the coating, the coating is cleaned with soap and water. The dye which has penetrated the cracks is not removed during the cleaning process and the cracks appear as fine red lines on a yellow background. The brittle-coating patterns shown in Figs. 15.13 and 15.14 were dye-etched before the photographs were taken.

It is not advisable to use dye etchant as a means of crack detection during the actual experiment since the solvent in the etchant affects the sensitivity of the coating and inconsistent results are obtained.

15.11 COMMERCIAL BRITTLE COATINGS [19–21]

15.11.1 Ceramic-Based Coatings

Two types of brittle coating are commercially available from Electrix Industries. The first is a ceramic-based coating marketed under the trade name All-Temp, and the second is a resin-based coating marketed under the trade name Stresscoat.† All-Temp consists of finely ground ceramic particles suspended in a volatile carrier. The coating is sprayed onto the specimen, air-dried to a soft powder, and fired at high temperature until the ceramic particles melt and coalesce to form a glaze.

Although use of the resin-based brittle coatings is much more common (because they are easier to apply and to cure), the ceramic-based coatings have several advantages. Since the coating is essentially a porcelain-glass, its strain sensitivity is not markedly affected by changes in humidity, test temperature, or loading time. The threshold strain of ceramic coatings is not sensitive to coating thickness. For these reasons, it appears that higher accuracies should be possible with ceramic coatings than with resin-based coatings.

Another major advantage of ceramic coatings is their ability to perform at elevated temperatures. The threshold strain of All-Temp coatings employed on a number of engineering materials is shown as a function of temperature in Fig.

---

† This is the modified Stresscoat, which differs significantly in composition from the original Stresscoat developed by Ellis in 1937.

15.16. These results indicate relatively small changes in the strain sensitivity of the coating with temperature over the temperature range from 24 to 288°C (75 to 550°F), and imply that small variations in temperature during the test period would not produce serious errors in the analysis. At about 288°C (550°F) the porcelain coating begins to soften, and the strain sensitivity increases rapidly with further increases in temperature. It appears then that these coatings can be used up to temperatures of approximately 300°C (575°F) before they become too insensitive to respond to elastic strains. This is a significant improvement in the upper limit of the temperature range since resin-based coatings can be used only at temperatures up to about 40°C (100°F).

Ceramic coatings have one serious disadvantage that greatly limits their use: They are difficult to apply by spraying, and they must be fired at temperatures approaching 600°C (1100°F). Unfortunately, this firing temperature is so high that components fabricated from aluminum, magnesium, and many heat-treated steels will be damaged by the firing process. Also, if the component is large, a furnace capable of 600°C (1100°F) may not be available. The firing temperature is also critical, as a 15°C (27°F) overtemperature will produce overfired coatings, while a 15°C (27°F) undertemperature will produce only a partially cured coating.

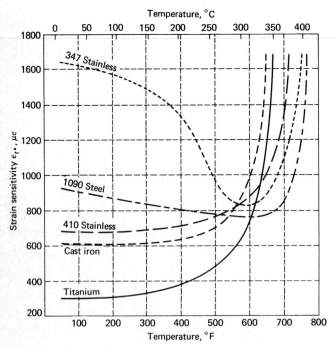

FIGURE 15.16
Effect of temperature on the strain sensitivity of All-Temp coatings. (*Electrix Industries, Inc.*)

Another disadvantage of ceramic coatings is associated with crack detection since cracks developed in the coating are usually not visible without enhancement. An electrified-particle crack-detecting system or a dye penetrant is commonly used to locate cracks between each loading increment during a test.

The All-Temp coatings are designed to match the thermal coefficient of expansion of the component. Since coatings with different coefficients of expansion are selected, the strain sensitivity of the coating can be controlled. If a coating with a coefficient of expansion greater than that of the component is selected, less strain is required to fail the coating because cooling from the firing temperature introduces residual tensile stresses in the coating. Conversely, coatings selected with a smaller coefficient of expansion than that of the specimen will require very large strains to produce cracking since the cooling process will introduce residual compressive stresses in the coating.

Eight different grades of All-Temp (AT-20 to AT-90) are available with a wide range of coefficients of expansion. Selection of a particular grade depends upon the coefficient of expansion of the specimen and the threshold strain specified for the test. The coating selection chart presented in Fig. 15.17 is used to select the particular grade of All-Temp to be employed. For example, with a component fabricated from 1095 carbon steel and a threshold strain of 600 $\mu\varepsilon$, the coating AT-40 would be selected.

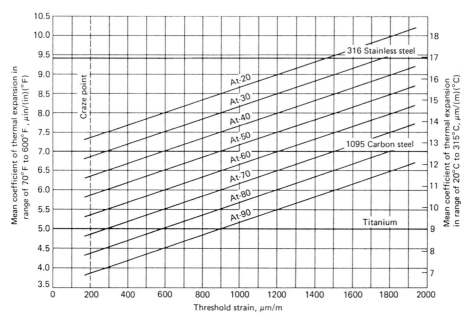

FIGURE 15.17
All-Temp coating selection chart. (*Electrix Industries, Inc.*)

15.11.2 Resin-Based Coatings [16, 22–23]

There is only one resin-based coating commercially available in the United States today. The exact composition of this coating is proprietary; however, the constituents [16] are probably similar to those found in Straintec, which consists of zinc and calcium resinate dissolved in the solvent methylene chloride with oleic acid and dibutyl phthalate used as the plasticizer.

The sensitivity of Stresscoat is influenced by atmospheric conditions. Since the thermal coefficient of expansion of the resin base is an order of magnitude greater than the expansion coefficients of typical metals, the coating is very sensitive to small changes in temperature. Relative humidity, in addition to temperature, also affects the behavior of these coatings since it influences drying time.

To account for varying but predictable temperature and humidity conditions, resin-based coatings are available in several different grades. A particular grade is selected to give a specified threshold strain (usually 500 $\mu\epsilon$) for some specified conditions of temperature and humidity. The selection chart for Stresscoat presented in Fig. 15.18 shows that 14 different grades are available. If the temperature and relative humidity for both the cure and testing periods are 21°C (70°F) and 20 percent, respectively, the coating identified as ST70F/21C should be selected to give a threshold strain of 500 $\mu\epsilon$. The different grades of these coatings are essentially the same except for the amount of plasticizer added to the resin-solvent base.

Variations in the test temperature during the experiment, when the brittle-coating crack patterns are being developed, are a major source of error in any attempt to predict the specimen stresses. The influence of temperature on a resin-based brittle coating is shown graphically in Fig. 15.19. Since it is unlikely that the strength of the coating varies appreciably with temperature over this range, the marked change in the strain required to crack the coating is due to the change in the residual stresses developed when the coating expands or contracts relative to the specimen during a temperature change.

A change of 2°C (3.5°F) in the testing temperature will produce a change of approximately 120 $\mu\epsilon$ in the threshold strain. If the coating sensitivity was about 500 $\mu\epsilon$ initially, the error produced by a temperature shift of 2°C (3.5°F) during the test would be about 24 percent.

It is obvious that, for consistent results, a laboratory must have at least a temperature-controlled atmosphere. If the component for some reason must be tested in the field or factory, where temperature control is impossible, there are measures which can be taken to minimize the errors due to temperature changes: (1) The weather bureau should be consulted to find when the temperature will be the most stable. Usually on a summer day the temperature increases rapidly during the morning and more slowly in the early afternoon, finally peaking later in the afternoon and then slowly decreasing until sunset. Temperature changes are minimized by conducting the brittle-coating analysis at this peak period in the late afternoon. (2) The test should be conducted as rapidly as possible to shorten

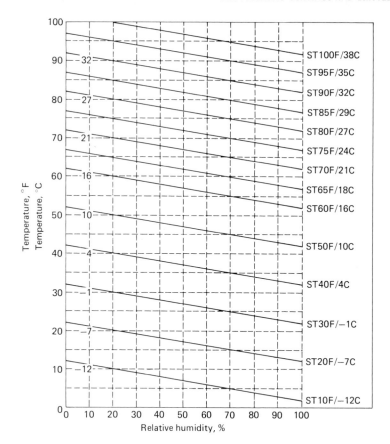

FIGURE 15.18
Stresscoat selection chart. (*Electrix Industries, Inc.*)

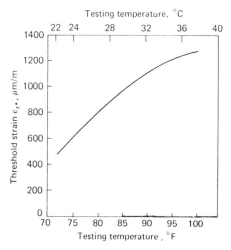

FIGURE 15.19
Influence of testing temperature on threshold strain
for a resin-based brittle coating.

the time interval over which the temperature has an opportunity to change. (3) The coating should be calibrated frequently so that the threshold strain is established as a function of time and temperature during the test.

The coatings are sprayed onto the components being analyzed by using aerosol cans or compressed-air spray guns. For resin-based coatings with methylene chloride as the solvent, the recommended range of thickness is only 0.05 to 0.10 mm (0.002 to 0.009 in). At the recommended thicknesses, the coatings require approximately 24 h to cure at room temperature.

In spite of the fact that resin-based coatings are called "brittle coatings," they are viscoelastic with properties that vary as a function of time. The recommended procedure for determining the threshold strain with a calibration beam involves deflecting the beam in about 1 s and holding the beam in the deflected position for approximately 15 s. The position of the last crack is marked before unloading the beam. Since the threshold strain is a function of this load-time relation, the time required to load a component under study is extremely important. If the loading of the component being tested cannot be accomplished in the same manner (a loading time of 1 s), the value of the threshold strain for the coating must be corrected.

Suppose, for example, that the load must be applied slowly to the component. Viscoelastic behavior of the coating during the loading interval results in a

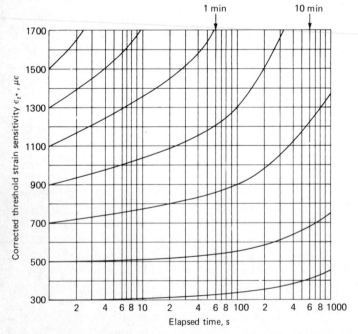

FIGURE 15.20
Stress-relaxation chart for Stresscoat.

relaxation of the stresses in the coating before it fails by cracking. The effect of this stress relaxation is to increase the strain required to crack the coating. The effects of loading times longer than 1 s can be taken into account by increasing the value of the threshold strain by the proper amount. Corrected values of threshold strain as a function of loading time are presented in Fig. 15.20 for several grades of Stresscoat with different sensitivities.

To illustrate the use of the correction chart shown in Fig. 15.20, consider a coating with a 1-s calibration value of threshold strain equal to 500 $\mu\epsilon$ which is to be used on a component that requires 240 s to load. In this instance, the corrected value of threshold strain is 600 $\mu\epsilon$, which is 20 percent higher than the original calibration.

Resin-based brittle coatings are inexpensive, easy to apply, and provide a simple but direct method for determining the magnitudes and directions of the principal stresses in actual prototypes. The amount of error involved in predicting the magnitudes of the principal stresses depends on the ability of the stress analyst to control the effects of temperature, relative humidity, the load-time function, and other factors which influence the coating. If control of these parameters can be exercised, the accuracy can approach 15 to 20 percent of the maximum stress levels, which is within the accuracy requirements for many industrial applications.

15.12 TEST PROCEDURES FOR RESIN-BASED COATINGS

Careful preparation of the surface of the component to be tested is essential for successful brittle-coating analysis. All paint or other coatings must be completely removed. The surface should then be lightly sanded or sand-blasted and degreased with one or more solvents such as acetone, Freon, or methylene chloride. Next, an aluminum undercoating should be sprayed on the component to provide a uniform reflecting background which increases the visibility of the crack patterns. It is often advantageous to draw a reference grid on the aluminum undercoating for convenience in referring to the crack patterns.

Selection of the proper coating to be sprayed is easily made with the aid of coating-selection charts (see Fig. 15.18), which are provided by the manufacturer of the coating. The usual practice is to select a coating with a threshold strain of approximately 500 $\mu\epsilon$, which is above the craze limit of the coatings (300 $\mu\epsilon$) but well below the yield or fracture strains associated with most engineering materials. The ability to select the proper coating depends on the accuracy achieved in predicting tomorrow's temperature and humidity if the test is conducted outdoors or on the ability to control the temperature in the laboratory if the test is conducted indoors.

In an air-conditioned laboratory (even one without humidity control), precise adjustment of the threshold strain can be achieved quite easily. After several experiments with brittle coatings, a logbook of threshold strains for various coating numbers and curing cycles is established and then used to select coatings for future tests. If for some reason the proper value of the threshold strain is not obtained, the temperature of the laboratory can be adjusted (a 1°C decrease in temperature reduces the threshold strain by 60 $\mu\epsilon$) to give the precise value of ϵ_{t*} specified for the experiment.

Curing the coating should be accomplished in an oven with a uniform temperature distribution if possible. A curing temperature of 30°C (86°F) for 16 h is recommended to minimize the influence of variations in coating thickness on the threshold strain. If the component is too large for an available oven, it is often possible to increase the temperature of the room in which the component is housed to 30°C (86°F) during the curing cycle.

Once the coating has been selected, the surface of the component prepared, and the coating sprayed to the proper thickness and cured with the recommended cycle, the actual brittle-coating test can be initiated. The component must be handled carefully, to avoid damage to the coating, during placement of the specimen in the loading device. Loads are commonly applied with universal testing machines, hydraulic cylinders, dead weights, or some form of pressure. It is important to control the loading increments and apply them relatively quickly. The first increment of load should be limited to a value which will produce cracking of the coating only over a small area in the highest stressed region. The time to apply the load should be recorded since this time is needed to correct the calibration value of threshold strain for time of loading effects. The load increment is maintained on the component for approximately 15 s to permit the coating crack pattern to develop fully, and then is removed. After unloading, the entire surface of the coating should be examined carefully for coating cracks. As indicated later, when analysis of data is discussed, the ability to identify the load level at which the first crack appears affects the maximum-stress determinations. After each examination for cracks, the component is loaded to a level approximately 20 percent higher than the previous level and the entire process repeated. Care is taken not to load more frequently than once every 5 min to allow the coating to recover from stresses produced by relaxation of the coating under the load. The crack patterns located after each loading cycle are encircled with a line (isoentatic line) and marked to identify the load which produced the pattern (see Fig. 15.21). The isoentatic lines which encircle the crack pattern separate the cracked and uncracked zones of the coating. The load increments can be continued until the highly stressed regions approach either the yield or fracture conditions.

The isoentatic lines constructed after a given increment of load are loci of points of approximately constant principal stress σ_1. In the example shown in Fig. 15.21, the coating cracked a second time under direct load, giving a second family of isostatics and a second isoentatic family. This second pattern of cracks, which is superimposed on the first pattern, has been encircled with a dashed line, representing a locus point along which the secondary principal stress σ_2 is approximately constant.

The isoentatics represent lines of constant stress and are analogous to contour lines on a topographic map. To determine the stress associated with each line, the coating is calibrated. This calibration procedure, described in Sec. 15.10.1, is repeated 10 to 15 times during the course of the experiment so that a number of values of ϵ_{t*} are obtained. These calibration results are then evaluated statistic-

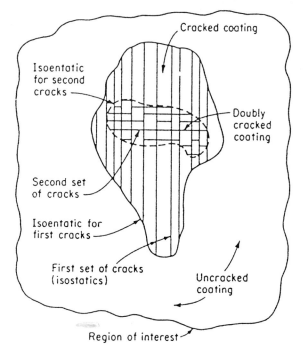

Cracked coating

Isoentatic for second cracks

Doubly cracked coating

Second set of cracks

Isoentatic for first cracks

First set of cracks (isostatics)

Uncracked coating

Region of interest

FIGURE 15.21
Method of constructing isoentatic lines on a field of brittle coating.

ally. The mean value of the threshold strain used to compute the stresses associated with each isoentatic is obtained from the relation

$$\bar{\epsilon}_{t*} = \frac{1}{N} \sum_{i=1}^{N} \epsilon_i \tag{15.35}$$

where $\bar{\epsilon}_{t*}$ = estimated mean value of threshold strain
ϵ_i = ith value of the threshold strain
N = total number of calibration values

If the estimated mean of ϵ_{t*} is employed in the analysis instead of a single calibration value, a much more accurate estimate of the true threshold strain is obtained.

By employing multiple values of ϵ_{t*}, an estimate of the standard deviation of the threshold strain can be computed and predictions on the accuracy of the brittle-coating determinations can be made.

An estimate of the standard deviation S_x of the threshold strain is computed from

$$S_x = \sqrt{\frac{1}{N-1} \sum_{i=1}^{N} (\epsilon_i - \bar{\epsilon})^2} = \sqrt{\frac{1}{N-1} \left[\sum_{i=1}^{N} \epsilon_i^2 - \frac{1}{N} \left(\sum_{i=1}^{N} \epsilon_i \right)^2 \right]} \tag{15.36}$$

The true threshold strain ϵ_{true} lies within the limits

$$\epsilon_{true} = \begin{cases} \bar{\epsilon} \pm S_x & 68\% \text{ of the time} \\ \bar{\epsilon} \pm 2S_x & 95\% \text{ of the time} \end{cases} \tag{15.37}$$

and the coefficient of variability is

$$\begin{aligned} &\pm \frac{S_x}{\bar{\epsilon}} & 68\% \text{ of the time} \\ &\pm \frac{2S_x}{\bar{\epsilon}} & 95\% \text{ of the time} \end{aligned} \tag{15.38}$$

Typical values of the standard deviation S_x are in the range from 30 to 60 $\mu\epsilon$ for a typical brittle-coating test conducted under laboratory conditions with temperature control to $\pm 0.5°C$ ($\pm 1°F$). For a coating with $\bar{\epsilon} = 500 \, \mu\epsilon$ for the threshold strain and $S_x = 50 \, \mu\epsilon$, the error associated with the brittle-coating determination will be within ± 10 percent 68 percent of the time and within ± 20 percent 95 percent of the time. This is the calibration error due to the inherent variation in the strength of the coating from point to point.

In calibrating the coating, it is often impossible to duplicate the load-time function applied to the component. Loading with manually operated testing machines often requires 10 s or longer, while deflecting the calibration beam requires only 1 s. Corrections for this difference in time to apply the load are made by using data similar to those presented in Fig. 15.20. It is often advantageous to use an independent method to check the accuracy of the brittle-coating calibration. Strain gages can be used for this independent check, as illustrated in Fig. 15.22. Two strain gages are mounted on an isoentatic line, perpendicular and parallel to the isostatics. The vessel is loaded to the pressure which formed the isoentatic, and the strain-gage readings are taken. These strains are employed with the

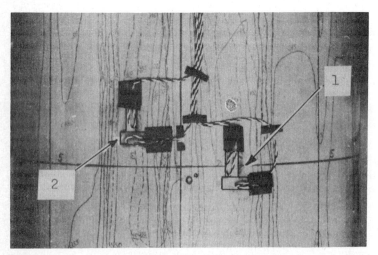

FIGURE 15.22
Strain gages mounted on the pressure vessel to calibrate the coating.

well-known stress-strain relations to compute the values for the two principal stresses:

$$\sigma_1^s = \frac{E^s}{1 - v^{s2}} \left(\epsilon_1^s + v^s \epsilon_2^s \right)$$

$$\sigma_2^s = \frac{E^s}{1 - v^{s2}} \left(\epsilon_2^s + v^s \epsilon_1^s \right)$$

(15.39)

where σ_1^s, σ_2^s = two principal stresses in specimen
E^s = modulus of elasticity of specimen
$\epsilon_1^s, \epsilon_2^s$ = two principal strains from the gages
v^s = Poisson's ratio for specimen material

Once the principal stresses σ_1^s and σ_2^s are determined, the threshold strains are computed from Eqs. (15.5) as

$$\epsilon_{t*1} = \frac{\sigma_1^2}{E^s} \quad \text{and} \quad \epsilon_{t*2} = \frac{\sigma_2^2}{E^s}$$

where ϵ_{t*1} and ϵ_{t*2} are the threshold strains for the first and second families of isoentatics, respectively.

Another advantage of utilizing strain gages on the actual component is that the calibration tends to account for the effect of the biaxial stress field on the threshold strain. When a double set of cracks is obtained (one before the other), it is essential that the value of ϵ_{t*2} be calibrated in this manner since it is always greater than ϵ_{t*1}.

15.13 ANALYSIS OF BRITTLE-COATING DATA [27]

If the stresses are linear with respect to the load, the principal stresses associated with a given isoentatic are determined from Eq. (15.31), which is modified to give the stresses for a specified reference load. Thus,

$$\sigma_1^s|_i = E^s \epsilon_{t*1} \frac{p_s}{p_i}$$

$$\sigma_2^s|_i = E^s \epsilon_{t*2} \frac{p_s}{p_i}$$

(15.40)

where $\sigma_1^s|_i, \sigma_2^s|_i$ = principal stresses in specimen associated with ith isoentatic
p_s = reference pressure or load to which all stresses are related; a value of 100 psi was arbitrarily selected as the reference pressure in the following example
p_i = pressure which caused crack pattern to extend to isoentatic numbered i
E_s = modulus of elasticity of specimen material
$\epsilon_{t*1}, \epsilon_{t*2}$ = threshold strains for isoentatic families 1 and 2, respectively.

As an example, consider the brittle-coating pattern for the thin-walled pressure vessel shown in Fig. 15.23. Values of σ_1^s and σ_2^s computed for each isoentatic by using Eqs. (15.40) are shown in Table 15.4.

By comparing these brittle-coating patterns with the values shown in Table 15.4 it is possible to obtain the magnitudes of the principal stresses over nearly the entire surface of the pressure vessel. The stresses are assumed to be linearly proportional to the pressure. This assumption is justified over most of the surface

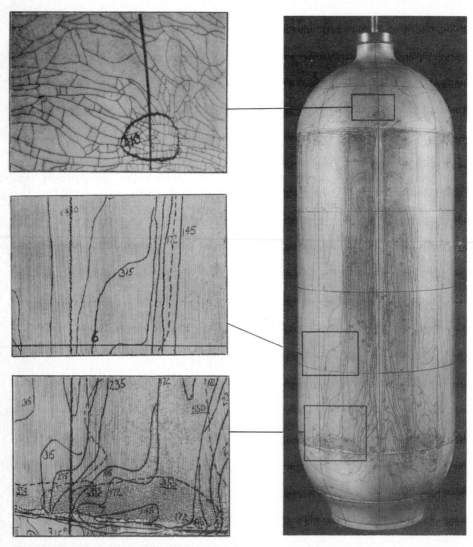

FIGURE 15.23
Brittle-coating crack patterns on the pressure vessel.

TABLE 15.4
Values of σ_1^s and σ_2^s associated with the isoentatics on the pressure vessel†

Isoentatic load p_i, psi	σ_1^s, psi	σ_2^s, psi
145	14,200	
172	11,900	
195	10,500	
235	8,750	
275	7,460	
315	6,520	
355	5,790	
410	5,010	5,520
430	4,280	4,720
550	3,740	4,120

† $\epsilon_{t \cdot 1} = 685 \ \mu\epsilon$ and $\epsilon_{t \cdot 2} = 775 \ \mu\epsilon$.

of the vessel; however, at the longitudinal weld, the cylinder is flattened on each side of the weld. At these imperfections the stresses are not linear with respect to the pressure, because the flattening of the shell becomes less pronounced with increasing pressure and the effects of the imperfections on the stress level decrease. The results of the brittle-coating analysis showed clearly that the stresses in the cylindrical pressure vessel were appreciably different from those predicted by using elementary shell theory. The local imperfections adjacent to the weld produced significant stress concentrations.

It should be noted that the brittle coating responded over nearly the entire outer surface of the vessel, giving information regarding the stresses near the longitudinal weld, the girth weld, and the forward and aft heads. Large amounts of data were obtained in a very short time. It was necessary to employ a few strain gages at selected points to complete the stress analysis. It should be emphasized that the strain gages were used to supplement the brittle-coating data, for often the reverse situation is true, namely, that the brittle-coating data are used to supplement the strain-gage data.

EXERCISES

15.1. Verify Eqs. (15.4).

15.2. Explain why a birefringent coating can be considered as a strain gage. Discuss the concept of the gage length associated with the coating.

15.3. A polycarbonate with $K = 0.14$ is used as a birefringent coating in a polariscope with a sodium light source. If the modulus of elasticity and Poisson's ratio of the polycarbonate are 2.48 GPa and 0.38, respectively, determine f_ϵ and f_σ for the coating.

15.4. Determine the specimen coefficient of sensitivity for the following materials:
- (a) Mild steel AISI 1010
- (b) High-strength steel AISI 4340
- (c) Aluminum 24S
- (d) Aluminum 2S
- (e) Hastelloy A
- (f) Inconel X
- (g) Magnesium M1
- (h) Red brass
- (i) Titanium
- (j) Plexiglas

15.5. Verify Eq. (15.15).

15.6. For a coating with $h^c = 1.50$ mm and $f_\varepsilon = 4.00$ μm/fringe, determine the maximum fringe order N which could be developed in the materials listed in Exercise 15.4.

15.7. Specify a coating (material and thickness) to use for the analysis of the following steel ($E^s = 207$ GPa and $v^s = 0.30$) structures:
- (a) A curved beam where $\sigma_{max} = 150$ MPa
- (b) A spherical shell with radius $R = 4.00$ m and wall thickness $t = 15$ mm subjected to a pressure $p = 1500$ kPa
- (c) A cylindrical shell with radius $R = 1500$ mm and wall thickness $t = 20$ mm subjected to a pressure $p = 5.00$ MPa

15.8. Determine the maximum fringe order developed in the coatings specified in Exercise 15.7.

15.9. Discuss the procedure to be used to install a coating on a panel with an elliptical hole.

15.10. Discuss the procedure to be used to install a coating at the intersection between two circular cylinders of the same diameter.

15.11. A coating is placed on an aluminum tension strip of known dimensions w and h. The tension strip is loaded with a known load P and the resulting fringe order N is measured with a reflection polariscope.
- (a) Outline the procedure for determining the material fringe value f_ε for the coating from these data.
- (b) If a sodium light source ($\lambda = 589.3$ nm) is used, determine the strain coefficient K for the coating.
- (c) If a helium-neon laser light source ($\lambda = 632.8$ nm) is used, determine the strain coefficient K for the coating.

15.12. At an interior point on a steel specimen with a polycarbonate coating ($h^c = 4.00$ mm and $f_\varepsilon = 5.10$ μm/fringe), a value of $N = 2.00$ and $\phi_1 = 30°$ is measured. Describe the state of stress in the specimen at this point.

15.13. If the coating and measurements of Exercise 15.12 are associated with a point on the free boundary of a panel specimen with in-plane loads, describe the state of stress at the point.

15.14. Write an engineering brief describing the development of a contoured coating to match the geometry of a complex surface.

15.15. Verify Eq. (15.19).

15.16. Determine the correction factor F_{CR} needed to account for reinforcing effects of the coating for a plane-stress specimen fabricated from the materials listed in Exercise 15.4 if

- (a) $\dfrac{h^c}{h^s} = 0.1$
- (b) $\dfrac{h^c}{h^s} = 0.2$
- (c) $\dfrac{h^c}{h^s} = 0.5$

- (d) $\dfrac{h^c}{h^s} = 1.0$
- (e) $\dfrac{h^c}{h^s} = 1.5$
- (f) $\dfrac{h^c}{h^s} = 2.0$

15.17. Verify Eq. (15.23).

15.18. Verify Eq. (15.24).

15.19. Verify Eq. (15.27).

15.20. For the materials listed in Exercise 15.4, determine the correction factor F_{CB} for wide plates in bending if

(a) $\dfrac{h^c}{h^s} = 0.1$ (b) $\dfrac{h^c}{h^s} = 0.2$ (c) $\dfrac{h^c}{h^s} = 0.5$

(d) $\dfrac{h^c}{h^s} = 1.0$ (e) $\dfrac{h^c}{h^s} = 1.5$ (f) $\dfrac{h^c}{h^s} = 2.0$

Prepare a graph showing F_{CB} as a function of h^c/h^s for each material.

15.21. A birefringent coating ($E^c = 2.50$ GPa and $v^c = 0.36$) is used on a plane-stress tensile specimen fabricated from glass-reinforced plastic ($E^s = 27.5$ GPa and $v^s = 0.20$) to measure the stress concentration factor resulting from a centrally located hole. If $N_{max} = 4.5$ on the boundary of the hole and $N_0 = 1.00$ at an interior point well removed from the hole, determine the stress concentration factor due to the hole.

15.22. A polycarbonate coating ($h^c = 2.0$ mm) is used on an aluminum 2024T3 plane-stress specimen. The specimen is loaded (in plane) until certain regions have yielded. If $\varepsilon_2 < 0$ in these regions, determine the color of the fringe delineating the elastic plastic boundary. Assume that the observation is made with white light in a dark-field polariscope.

15.23. Two fringe patterns are available from a plane-stress specimen with a birefringent coating. The first pattern gives $N_0(x, y)$ due to incident light. The second pattern gives $N_{\theta 1}(x, y)$ due to an oblique incidence observation made by rotating the direction of the light about the ε_1 direction by an angle θ_1. Derive the oblique-incidence relations shown below that permit separation of the strains from the data of the normal and oblique incidence fringe patterns.

$$\epsilon_1 = \frac{f_\epsilon}{2h^c(1 + v^c)\sin^2\theta_1}\left\{[(1 - v^c)\cos\theta_1]N_{\theta 1} - (\cos^2\theta_1 - v^c)N_0\right\}$$

$$\epsilon_2 = \frac{f_\epsilon}{2h^c(1 + v^c)\sin^2\theta_1}\left\{[(1 - v^c)\cos\theta_1]N_{\theta 1} - (1 - v^2\cos^2\theta_1)N_0\right\}$$

15.24. A specimen with a birefringent coating is subjected to a specified load P, and fringe orders $N_0(x, y)$ and isoclinics $\phi(x, y)$ are recorded over the field. Next, a slitting saw is used to cut a narrow channel through the thickness of the coating along an isostatic. A second fringe pattern is recorded to obtain the distribution of fringes N_1 along the edge of the saw cut. Derive a relation for the stresses σ_1^s and σ_2^s along the isostatic coinciding with the saw cut.

15.25. List the advantages of the brittle-coating method.

15.26. What are the primary disadvantages of the brittle-coating method?

15.27. Determine the stresses in a brittle coating applied to a component fabricated from steel ($E = 207$ GPa, $v = 0.30$) when the specimen stresses are $\sigma_1^s = 280$ MPa and $\sigma_2^s = -210$ MPa for

(a) A resin-based coating with $E^c = 1.40$ GPa and $v^c = 0.42$

(b) A ceramic-based coating with $E^c = 70$ GPa and $v^c = 0.25$.

15.28. If the threshold strain in a brittle coating is 500 $\mu\epsilon$, determine the corresponding state of stress in the coating during calibration for
(a) A resin-based coating with $E^c = 1.40$ GPa and $v^c = 0.42$
(b) A ceramic-based coating with $E^c = 70$ GPa and $v^c = 0.25$

15.29. Describe qualitatively how the state of stress in the coating on a calibration beam changes after the coating fails by cracking. Extend this description to include the dependence of crack density (number of cracks per unit length) on the magnitude of the stress applied to the coating after it first fails. Also, describe how coating thickness influences the crack density.

15.30. The maximum-normal-strain theory of failure for brittle coatings indicates that cracks will form whenever

$$\epsilon_1^c \geq \epsilon_{t*} \qquad \epsilon_2^c \geq \epsilon_{t*} \qquad \epsilon_3^c \geq \epsilon_{t*}$$

Show that the failure equations for the coating in terms of the stresses applied to the specimen are

$$\frac{\sigma_1^s}{\sigma_1^*} - \frac{v^s \sigma_2^s}{\sigma_1^*} \geq 1 \qquad\qquad (15.41a)$$

$$\frac{\sigma_2^s}{\sigma_1^*} - \frac{v^s \sigma_1^s}{\sigma_1^*} \geq 1 \qquad\qquad (15.41b)$$

$$-\frac{v^c(1 - v^s)}{1 - v^c}\left(\frac{\sigma_1^s}{\sigma_1^*} + \frac{\sigma_2^s}{\sigma_1^*}\right) \geq 1 \qquad\qquad (15.41c)$$

15.31. Using the results of Exercise 15.30, construct the failure diagram for the maximum-normal-strain theory of failure.

15.32. Failure of the coating according to Eqs. (15.41a) and (15.41b) results in an orthogonal set of coating cracks, as illustrated in Fig. 15.12b. Describe the failure exhibited by the coating when the conditions of Eq. (15.41c) are satisfied.

15.33. The maximum-normal-stress theory of failure for brittle coatings indicates that cracks will form whenever

$$\sigma_1^c \geq \sigma_{ut} \qquad \sigma_2^c \geq \sigma_{ut}$$

where σ_{ut} is the ultimate tensile strength of the coating. Show that the failure equations for the coating in terms of the stresses applied to the specimen are

$$\frac{\sigma_1^s}{\sigma_1^*} + \frac{(v^c - v^s)\sigma_2^s}{(1 - v^c v^s)\sigma_1^*} \geq 1 \qquad \frac{\sigma_2^s}{\sigma_1^*} + \frac{(v^c - v^s)\sigma_1^s}{(1 - v^c v^s)\sigma_1^*} \geq 1 \qquad (15.42)$$

15.34. Using the results of Eqs. (15.42), construct the failure diagram for the maximum-normal-stress theory of failure.

15.35. Describe why the condition $\sigma_3^c \geq \sigma_{ut}$ was not considered in Exercise 15.33.

15.36. Sketch the brittle-coating pattern which would be observed on the following components:
(a) A tensile specimen
(b) A circular shaft subjected to pure torsion

(c) A circular disk subjected to diametrical compression

(d) A circular ring subjected to diametrical compression

(e) A cylindrical thin-walled vessel subjected to internal pressure

(f) A spherical thin-walled vessel subjected to internal pressure

15.37. What is the significance of uncracked areas of the brittle-coating field at the conclusion of a brittle-coating test? Are these regions of any importance with respect to the economic aspects of a given design? What are some procedures which can be employed to gain further information regarding the stresses in these regions?

15.38. For a resin-based brittle coating, compute the maximum possible crack width with a coating thickness of 0.10 mm. Assume the cracks have a uniform spacing of 1.5 mm.

15.39. What are the consequences of being unable to locate a cracked region of coating during a brittle-coating test? Can dye etchant or the electrified-particle method of crack detection be employed during the test with (a) resin-based coatings and (b) ceramic-based coatings?

15.40. Why are ceramic-based brittle coatings less sensitive to temperature changes than resin-based coatings? What is the anticipated change in threshold strain of a resin-based coating for a 3°C change in temperature? What change in strain sensitivity would be encountered with All-Temp on 410 stainless steel as the temperature is increased from 40 to 200°C?

15.41. Select the proper grade of All-Temp for a strain sensitivity of $600 \, \mu\epsilon$ on the following materials:

(a) 1040 carbon steel   (b) Titanium

(c) Class 60 gray cast iron   (d) 301 stainless steel

(e) 316 stainless steel   (f) Monel

(g) Hastelloy A   (h) Inconel X

15.42. Select the proper grade of a resin-based coating for curing and testing under the following atmospheric conditions:

Temperature, °C	30	25	20	10
Relative humidity, %	90	60	40	10

15.43. What is the effect of poor thickness control during application of a resin-based brittle coating? What are normal variations in the thickness of a coating applied to a component with an irregular shape? How can the influence of these variations in the thickness of the coating be minimized?

15.44. Discuss temperature control to minimize errors in a test with a resin-based coating when the test is conducted in the field.

15.45. A resin-based brittle coating with a strain sensitivity of $700 \, \mu\epsilon$ is employed in a test of a structure loaded with sand bags. If 12 min is required to position all the bags, compute the corrected strain sensitivity and comment on the advisability of the test procedure. If the loading is accomplished with hydraulic rams, the loading time (with manual control) can be reduced to 10 s. Determine the corrected sensitivity in this instance.

15.46. Comment on the advisability of using servo-controlled loading methods for brittle-coating testing.

15.47. Prepare test specifications, which include surface preparation, coating selection, coating application procedures, coating curing cycle, loading schedule, and coating inspection and marking procedures for a brittle-coating test of the following components:
(a) An internal combustion engine head
(b) A large T joint on a gas transmission line
(c) A crane hook
(d) A passenger-car wheel
(e) A pillow block bearing housing
(f) A structure beam
(g) A pressure vessel with pipe supports

15.48. The isoentatic pattern shown in Fig. E15.48 was produced during a direct-loading test of an aluminum ($E = 70$ GPa, $v = 0.33$) component. Load levels of 400, 500, 750, 1000, and 1500 N were used during the test. Using only this isoentatic pattern and a corrected calibration value of threshold strain of 520 $\mu\epsilon$, estimate the distribution of σ_1^s along the lines
(a) $y = 0$ (b) $y = 25$ mm   (c) $y = -25$ mm
(d) $x = 75$ mm   (e) $x = 100$ mm   (f) $x = 125$ mm
(g) What is your estimate of the maximum value of σ_1^s?

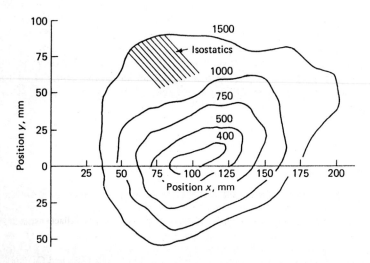

FIGURE E15.48

15.49. The same aluminum component described in Exercise 15.48 was retested using the relaxation loading method to obtain the isostatic and isoentatic patterns associated with σ_2^s shown in Fig. E15.49. The corrected calibration value of threshold strain was 550 $\mu\epsilon$. Estimate the distribution of σ_2^s along the lines

(a) $y = 0$ (b) $y = 25$ mm   (c) $y = -25$ mm
(d) $x = 75$ mm   (e) $x = 100$ mm   (f) $x = 125$ mm
(g) What is your estimate of the maximum value of σ_2^s?

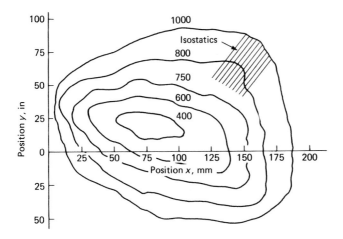

FIGURE E15.49.

15.50. A steel (E = 207 GPa and $v = 0.30$) machine component is being studied using brittle-coating methods. At a critical point on the specimen, cracks begin to appear when the load is raised to 10 kN. The corrected calibration threshold strain for the coating at this time is 400 $\mu\epsilon$. The load is raised in increments, and at a level of 40 kN cracks perpendicular to the first set begin to appear at the same point. The corrected calibration threshold strain for the second family of cracks is 500 $\mu\epsilon$. Determine the two principal stresses in the specimen at a load level of 10 kN by assuming that the coating cracks according to the maximum-normal-stress theory of failure. Assume that the coating is a resin-based coating with the following properties: $E^c = 1.40$ GPa and $v^c = 0.42$.

REFERENCES

1. Mesnager, M.: Sur la determination optique des tensions interieures dans les solides a trois dimensions, *C. R. (Paris)*, vol. 190, p. 1249, 1930.
2. Oppel, G.: Das polarisationsoptische Schichtverfahren zur Messung der Oberflachenspannung am beanspruchten Bauteil ohne Modell, *Z. Ver. Dtsch. Ing.*, vol. 81, pp. 803–804, 1937.
3. Fleury, R., and F. Zandman: Jauge d'efforts photoelastique, *C. R. (Paris)*, vol. 238, p. 1559, 1954.
4. D'Agostino, J., D. C. Drucker, C. K. Liu, and C. Mylonas: Epoxy Adhesives and Casting Resins as Photo-Elastic Plastics, *Proc. SESA*, vol. XII, no. 2, pp. 123–128, 1955.
5. Kawata, K.: Analysis of Elastoplastic Behavior of Metals by Means of Photoelastic Coating Method, *J. Sci. Res. Instrum.*, *Tokyo*, vol. 52, pp. 17-40, 1958.
6. Leven, M. M.: "Epoxy Resins for Photoelastic Use," in M. M. Frocht (ed.), *Photoelasticity 1963*, Pergamon Press, New York, 1963, pp. 145–168.
7. Ito, K.: New Model Materials for Photoelasticity and Photoplasticity, *Exp. Mech.*, vol. 2, no. 12, pp. 373–376, 1962.
8. Melver, R. W.: Structural Test Applications Utilizing Large Continuous Photoelastic Coatings, *Exp. Mech.*, vol. 5, no. 1, pp. 19A–25A, 1965; vol. 5, no. 2, pp. 19A–26A, 1965.
9. Zandman, F., S. Redner, and J. W. Dally: *Photoelastic Coatings*, Iowa State University Press, Ames, 1977.

10. Zandman, F., S. S. Redner, and E. I. Riegner: Reinforcing Effect of Birefringent Coatings, *Exp. Mech.*, vol. 2, no. 2, pp. 55–64, 1962.
11. Dally, J. W., and I. Alfirevich: Application of Birefringent Coatings to Glass-Fiber-Reinforced Plastics, *Exp. Mech.*, vol. 9, no. 3, pp. 97–102, 1969.
12. Sauerwald, F., and H. Wieland: Uber die Kerbschlagprobe nach Schule-Moser, *A. Metalkd.*, vol. 17, pp. 358–364, 392–399, 1925.
13. Durelli, A. J., J. Hall, and F. Stern: in chap. 11, "Brittle Coating," A. S. Kobayashi (ed.), *Handbook on Experimental Mechanics*, Prentice-Hall, Englewood Cliffs, N.J., 1987, pp. 516–554.
14. Ellis, G.: Practical Strain Analysis by Use of Brittle Coatings, *Proc. SESA*, vol. I, no. 1, pp. 46–53, 1943.
15. Singdale, F. N.: Improved Brittle Coatings for Use under Widely Varying Temperature Conditions, *Proc. SESA*, vol. XI, no. 2, pp. 173–178, 1954.
16. Racine, F. L., H. G. Taracks, and J. J. Killinger: Development of a Non-Flammable Resin System for Engineering Stress-Strain Applications, *Gen. Mot. Eng. J.*, 4th quarter issue, pp. 8–12, 1965.
17. Durelli, A. J., and T. N. DeWolf: Law of Failure of Stresscoat, *Proc. SESA*, vol. VI, no. 2, pp. 68–83, 1949.
18. Durelli, A. J.: Experimental Determination of Isostatic Lines, *J. Appl. Mech.*, vol. 64, pp. A155–A160, 1942.
19. Staats, H. N., and S. J. Baranowski: Calibrated Porcelain Enamel Coatings, *Am. Ceram. Soc. Bull.*, vol. 35, No. 4, pp. 143–145, 1956.
20. Rollins, C. T., and E. K. Lynn: Experience with Ceramic Coatings, *Exp. Mech.*, vol. 8, no. 3, pp. 19N–27N, 1968.
21. Singdale, F. N.: Method of Determining Strain Values in Rigid Articles, U.S. Patent 2,724,964, Nov. 29, 1955.
22. Ellis, G.: Resinous Composition for Determining the Strain Concentration in Rigid Articles, U.S. Patent 2,428,559, Oct. 7, 1947.
23. Murthy, P. N.: Theoretical Investigation of Creep and Crack Density Studies in Stresscoat, *Proc. SESA*, vol. XV, no. 1, pp. 57–64, 1957.
24. Dally, J. W., and A. J. Durelli: Stress Analysis of a Reactor Head Closure, *Proc. SESA*, vol. XVII, no. 2, pp. 71–87, 1959.
25. Durelli, A. J., J. W. Dally, and S. Morse: Experimental Study of Large-Diameter Thin-Walled Pressure Vessels, *Exp. Mech.*, vol. 1, no. 2, pp. 33–42, 1961.
26. Durelli, A. J., S. Okubo, and R. H. Jacobson: Study of Some Properties of Stresscoat, *Proc. SESA*, Vol. XII, no. 2, pp. 55–76, 1955.
27. Durelli, A. J., E. A. Phillips, and C. H. Tsao: *Introduction to the Theoretical and Experimental Analysis of Stress and Strain*, McGraw-Hill, New York, 1958.

PART
V

APPLICATION
OF STATISTICS

CHAPTER
16

STATISTICAL ANALYSIS OF EXPERIMENTAL DATA

16.1 INTRODUCTION

Experimental measurements of quantities such as pressure, temperature, length, force, stress, or strain will always exhibit some variation if the measurements are repeated a number of times with precise instruments. This variability, which is fundamental to all measuring systems, is due to two different causes. First, the quantity being measured may exhibit significant variation. For example, in a materials study to determine fatigue life at a specified stress level, large differences in the number of cycles to failure are noted when a number of specimens are tested. This variation is inherent in the fatigue process and is observed in all fatigue life measurements. Second, the measuring system, which includes the transducer, signal conditioning equipment, analog-to-digital (A/D) converter, recording instrument, and the operator, may introduce error in the measurement. This error may be systematic or random, depending upon its source. An instrument operated out of calibration produces a systematic error, whereas reading errors due to interpolation on a chart are random. The accumulation of random errors in a measuring system produces a variation that must be examined in relation to the magnitude of the quantity being measured.

The data obtained from repeated measurements represent an array of readings, not an exact result. Maximum information can be extracted from such an array of readings by employing statistical methods. The first step in the

591

statistical treatment of data is to establish the distribution. A graphical representation of the distribution is usually the most useful form for initial evaluation. Next, the statistical distribution is characterized with a measure of its central value, such as the mean, the median, or the mode. Finally, the spread or dispersion of the distribution is determined in terms of the variance or the standard deviation.

With elementary statistical methods, the experimentalist can reduce a large amount of data to a very compact and useful form by defining the type of distribution, establishing the single value that best represents the central value of the distribution (mean), and determining the variation from the mean value (standard deviation). Summarizing data in this manner is the most meaningful form of presentation for application to design problems or for communication to others who need the results of the experiments.

The treatment of statistical methods presented in this chapter is relatively brief; therefore, only the most commonly employed techniques for representing and interpreting data are presented. A formal course in statistics, which covers these techniques in much greater detail as well as many other useful techniques, should be included in the program of study of all engineering students.

16.2   CHARACTERIZING STATISTICAL DISTRIBUTIONS

For purposes of this discussion, consider that an experiment has been conducted n times to determine the ultimate tensile strength of a fully tempered beryllium copper alloy. The data obtained represent a sample of size n from an infinite population of all possible measurements that could have been made. The simplest way to present these data is to list the strength measurements in order of increasing magnitude, as shown in Table 16.1.

These data can be arranged into seven groups to give a frequency distribution as shown in Table 16.2. The advantage of representing data in a frequency distribution is that the central tendency is more clearly illustrated.

16.2.1   Graphical Representations of the Distribution

The shape of the distribution function representing the ultimate tensile strength of beryllium copper is indicated by the data groupings of Table 16.2. A graphical presentation of this group data, known as a *histogram*, is shown in Fig. 16.1. The histogram method of presentation shows the central tendency and variability of the distribution much more clearly than the tabular method of presentation of Table 16.2. Superimposed on the histogram is a curve showing the relative frequency of the occurrence of a group of measurements. Note that the points for the relative frequency are plotted at the midpoint of the group interval.

A cumulative-frequency diagram, shown in Fig. 16.2, is another way of representing the ultimate-strength data from the experiments. The cumulative

TABLE 16.1
Data on ultimate tensile strength of beryllium copper, listed in order of increasing magnitude

Sample number	Strength, ksi (MPa)	Sample number	Strength, ksi (MPa)
1	170.5 (1175)	21	176.2 (1215)
2	171.9 (1185)	22	176.2 (1215)
3	172.6 (1190)	23	176.4 (1217)
4	173.0 (1193)	24	176.6 (1218)
5	173.4 (1196)	25	176.7 (1219)
6	173.7 (1198)	26	176.9 (1220)
7	174.2 (1201)	27	176.9 (1220)
8	174.4 (1203)	28	177.2 (1222)
9	174.5 (1203)	29	177.3 (1223)
10	174.8 (1206)	30	177.4 (1223)
11	174.9 (1206)	31	177.7 (1226)
12	175.0 (1207)	32	177.8 (1226)
13	175.4 (1210)	33	178.0 (1228)
14	175.5 (1210)	34	178.1 (1228)
15	175.6 (1211)	35	178.3 (1230)
16	175.6 (1211)	36	178.4 (1230)
17	175.8 (1212)	37	179.0 (1236)
18	175.9 (1213)	38	179.7 (1239)
19	176.0 (1214)	39	180.1 (1242)
20	176.1 (1215)	40	181.6 (1252)

TABLE 16.2
Frequency distribution of ultimate tensile strength

Group intervals, ksi (MPa)	Observations in the group	Relative frequency	Cumulative frequency
169.0–170.9 (1166–1178)	1	0.025	0.025
171.0–172.9 (1179–1192)	2	0.050	0.075
173.0–174.9 (1193–1206)	8	0.200	0.275
175.0–176.9 (1207–1220)	16	0.400	0.675
177.0–178.9 (1221–1234)	9	0.225	0.900
179.0–180.9 (1235–1248)	3	0.075	0.975
181.0–182.9 (1249–1261)	1	0.025	1.000
Total	40		

frequency is the number of readings having a value less than a specified value of the quantity being measured (ultimate strength) divided by the total number of measurements. As indicated in Table 16.2, the cumulative frequency is the running sum of the relative frequencies. When the graph of cumulative frequency versus the quantity being measured is prepared, the end value for the group intervals is used to position the point along the abscissa.

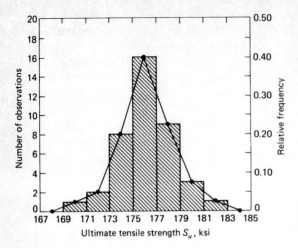

FIGURE 16.1
Histogram with a superimposed relative-frequency diagram.

16.2.2   Measures of Central Tendency

While histograms or frequency distributions are used to provide a visual representation of a distribution, numerical measures are used to define the characteristics of the distribution. One basic characteristic is the central tendency of the data. The most commonly employed measure of the central tendency of a distribution of data is the sample mean $\bar{x}$, which is defined as

$$\bar{x} = \sum_{i=1}^{n} \frac{x_i}{n} \tag{16.1}$$

where x_i = the ith value of the quantity being measured
n = the total number of measurements

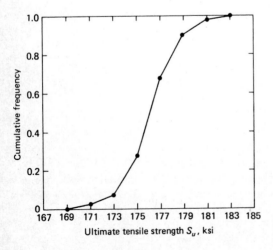

FIGURE 16.2
Cumulative-frequency diagram.

Because of time and costs involved in conducting tests, the number of measurements is usually limited; therefore, the sample mean $\bar{x}$ is only an estimate of the true arithmetic mean μ of the population. It is shown later that $\bar{x}$ approaches μ as the number of measurements increases. The mean value of the ultimate-strength data presented in Table 16.1 is $\bar{x} = 176.1$ ksi (1215 MPa).

The median and mode are also measures of central tendency. The median is the central value in a group of ordered data. For example, in an ordered set of 41 readings, the 21st reading represents the median value with 20 readings lower than the median and 20 readings higher than the median. In instances when an even number of readings is taken, the median is obtained by averaging the two middle values. For example, in an ordered set of 40 readings, the median is the average of the 20th and 21st readings. Thus, for the ultimate tensile strength data presented in Table 16.1, the median is $\frac{1}{2}(176.1 + 176.2) = 176.15$ ksi (1215 MPa).

The mode is the most frequent value of the data; therefore, it is the peak value on the relative-frequency curve. In Fig. 16.1, the peak of the relative probability curve occurs at an ultimate tensile strength $S_u = 176.0$ ksi (1214 MPa); therefore, this value is the mode of the data set presented in Table 16.1.

It is evident that a typical set of data may give different values for the three measures of central tendency. There are two reasons for this difference. First, the population from which the samples were drawn may not be gaussian where the three measures are expected to coincide. Second, even if the population is gaussian, the number of measurements n is usually small and deviations due to a small sample size are to be expected.

16.2.3   Measures of Dispersion

It is possible for two different distributions of data to have the same mean but different dispersions, as shown in the relative-frequency diagrams of Fig. 16.3. Different measures of dispersion are the range, the mean deviation, the variance, and the standard deviation. The standard deviation S_x is the most popular and

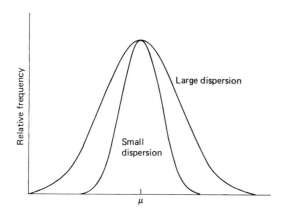

FIGURE 16.3
Relative-frequency diagrams with large and small dispersions.

is defined as

$$S_x = \left[\sum_{i=1}^{n} \frac{(x_i - \bar{x})^2}{n - 1} \right]^{1/2} \tag{16.2}$$

Since the sample size n is small, the standard deviation S_x of the sample represents an estimate of the true standard deviation σ of the population. Computation of S_x and $\bar{x}$ from a data sample is easily performed with most scientific-type calculators.

Expressions for the other measures of dispersion, namely, range R, mean deviation d_x, and variance S_x^2, are as follows:

$$R = x_L - x_S \tag{16.3}$$

where x_L = the largest value of the quantity in the distribution
x_S = the smallest value of the quantity in the distribution

$$d_x = \frac{\sum_{i=1}^{n} x_i - \bar{x}}{n} \tag{16.4}$$

Equation (16.4) indicates that the deviation of each reading from the mean is determined and summed. The average of the n deviations is the mean deviation. The absolute value of the difference $(x_i - \bar{x})$ must be used in the summing process to avoid cancelation of positive and negative deviations.

$$S_x^2 = \frac{\sum_{i=1}^{n} (x_i - \bar{x})^2}{n - 1} \tag{16.5}$$

The variance of the population σ^2 is estimated by S_x^2 where the denominator $n - 1$ in Eqs. (16.2) and (16.5) serves to reduce error introduced by approximating the true mean μ with the estimate $\bar{x}$. As the sample size n is increased, the estimates of $\bar{x}$, S_x, and S_x^2 improve as shown in the discussion of Sec. 16.4. Variance is an important measure of dispersion because it is used in defining the normal distribution function.

Finally, a measure known as the coefficient of variation C_v is used to express the standard deviation S_x as a percentage of the mean $\bar{x}$. Thus,

$$C_v = \frac{S_x}{\bar{x}} (100) \tag{16.6}$$

The coefficient of variation represents a normalized parameter that indicates the variability of the data in relation to its mean.

16.3 STATISTICAL DISTRIBUTION FUNCTIONS

As the sample size is increased, it is possible in tabulating the data to increase the number of group intervals and to decrease their width. The corresponding relative-frequency diagram, similar to the one illustrated in Fig. 16.1, will approach a smooth curve (a theoretical distribution curve) known as a *distribution function*.

A number of different distribution functions are used in statistical analyses. The best-known and most widely used distribution in experimental mechanics is the gaussian or normal distribution. This distribution is extremely important because it describes random errors in measurements and variations observed in strength determinations. Other useful distributions include binomial, exponential, hypergeometric, chi-square (χ^2), F, Gumbel, Poisson, Student's t, and Weibull distributions. The reader is referred to Refs. 1 through 5 for a complete description of these distributions. Emphasis here will be on gaussian and Weibull distribution functions because of their wide range of application in experimental mechanics.

16.3.1   Gaussian Distribution

The gaussian or normal-distribution function, as represented by a normalized relative-frequency diagram, is shown in Fig. 16.4. The gaussian distribution is completely defined by two parameters: the mean μ and the standard deviation σ. The equation for the relative frequency f in terms of these two parameters is

$$f(z) = \frac{1}{\sqrt{2\pi}} e^{-(z^2/2)} \tag{16.7}$$

where

$$z = \frac{x - \mu}{\sigma} \tag{16.8}$$

Experimental data (with finite sample sizes) can be analyzed to obtain $\bar{x}$ as

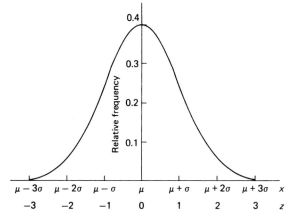

FIGURE 16.4
The normal or gaussian distribution function.

an estimate of μ and S_x as an estimate of σ. This procedure permits the experimentalist to use data drawn from small samples to represent the entire population.

The method for predicting population properties from a gaussian (normal) distribution function utilizes the normalized relative-frequency diagram shown in Fig. 16.4. The area A under the entire curve is given by Eq. (16.7) as

$$A = \frac{1}{\sqrt{2\pi}} \int_{-\infty}^{\infty} e^{-(z^2/2)} \, dz = 1 \tag{16.9}$$

Equation (16.9) implies that the population has a value z between $-\infty$ and $+\infty$ and that the probability of making a single observation from the population with a value $-\infty \leq z \leq +\infty$ is 100 percent. While the previous statement may appear trivial and obvious, it serves to illustrate the concept of using the area under the normalized relative-frequency curve to determine the probability P of observing a measurement within a specific interval. Figure 16.5 shows graphically (shaded area under the curve) the probability that a measurement will occur within the interval between z_1 and z_2. Thus, from Eq. (16.7),

$$P(z_1, z_2) = \int_{z_1}^{z_2} f(z) \, dz = \frac{1}{\sqrt{2\pi}} \int_{z_1}^{z_2} e^{-(z^2/2)} \, dz \tag{16.10}$$

Evaluation of Eq. (16.10) is most easily made by using tables that list the areas under the normalized relative-frequency curve as a function of z. Table 16.3 lists one-sided areas between limits of $z_1 = 0$ and z_2 for the normal-distribution function.

Since the distribution function is symmetric about $z = 0$, this one-sided table is sufficient for all evaluations. For example,

$$A(-1, 0) = A(0, +1)$$

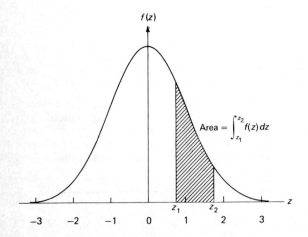

FIGURE 16.5
Probability of a measurement of x between limits of z_1 and z_2. The total area under the curve $f(z)$ is 1.

TABLE 16.3
Areas under the normal-distribution curve from $z_1 = 0$ to z_2 (one side)

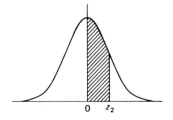

$z_2 = \dfrac{x - \bar{x}}{S_x}$	0.00	0.01	0.02	0.03	0.04	0.05	0.06	0.07	0.08	0.09
0.0	.0000	.0040	.0080	.0120	.0160	.0199	.0239	.0279	.0319	.0359
0.1	.0398	.0438	.0478	.0517	.0557	.0596	.0636	.0675	.0714	.0753
0.2	.0793	.0832	.0871	.0910	.0948	.0987	.1026	.1064	.1103	.1141
0.3	.1179	.1217	.1255	.1293	.1331	.1368	.1406	.1443	.1480	.1517
0.4	.1554	.1591	.1628	.1664	.1700	.1736	.1772	.1808	.1844	.1879
0.5	.1915	.1950	.1985	.2019	.2054	.2088	.2123	.2157	.2190	.2224
0.6	.2257	.2291	.2324	.2357	.2389	.2422	.2454	.2486	.2517	.2549
0.7	.2580	.2611	.2642	.2673	.2704	.2734	.2764	.2794	.2823	.2852
0.8	.2881	.2910	.2939	.2967	.2995	.3023	.3051	.3078	.3106	.3233
0.9	.3159	.3186	.3212	.3238	.3264	.3289	.3315	.3340	.3365	.3389
1.0	.3413	.3438	.3461	.3485	.3508	.3531	.3554	.3577	.3599	.3621
1.1	.3643	.3665	.3686	.3708	.3729	.3749	.3770	.3790	.3810	.3830
1.2	.3849	.3869	.3888	.3907	.3925	.3944	.3962	.3980	.3997	.4015
1.3	.4032	.4049	.4066	.4082	.4099	.4115	.4131	.4147	.4162	.4177
1.4	.4192	.4207	.4222	.4236	.4251	.4265	.4279	.4292	.4306	.4319
1.5	.4332	.4345	.4357	.4370	.4382	.4394	.4406	.4418	.4429	.4441
1.6	.4452	.4463	.4474	.4484	.4495	.4505	.4515	.4525	.4535	.4545
1.7	.4554	.4564	.4573	.4582	.4591	.4599	.4608	.4616	.4625	.4633
1.8	.4641	.4649	.4656	.4664	.4671	.4678	.4686	.4693	.4699	.4706
1.9	.4713	.4719	.4726	.4732	.4738	.4744	.4750	.4758	.4761	.4767
2.0	.4772	.4778	.4783	.4788	.4793	.4799	.4803	.4808	.4812	.4817
2.1	.4821	.4826	.4830	.4834	.4838	.4842	.4846	.4850	.4854	.4857
2.2	.4861	.4864	.4868	.4871	.4875	4878	.4881	.4884	.4887	.4890
2.3	.4893	.4896	.4898	.4901	.4904	.4906	.4909	.4911	.4913	.4916
2.4	.4918	.4920	.4922	.4925	.4927	.4929	.4931	.4932	.4934	.4936
2.5	.4938	.4940	.4941	.4943	.4945	.4946	.4948	.4949	.4951	.4952
2.6	.4953	.4955	.4956	.4957	.4959	.4960	.4961	.4962	.4963	.4964
2.7	.4965	.4966	.4967	.4968	.4969	.4970	.4971	.4972	.4973	.4974
2.8	.4974	.4975	.4976	.4977	.4977	.4978	.4979	.4979	.4980	.4981
2.9	.4981	.4982	.4982	.4983	.4984	.4984	.4985	.4985	.4986	.4986
3.0	.49865	.4987	.4987	.4988	.4988	.4988	.4989	.4989	.4989	.4990

Therefore

$$A(-1, +1) = p(-1, +1) = 0.3413 + 0.3413 \quad = 0.6826$$
$$A(-2, +2) = p(-2, +2) = 0.4772 + 0.4772 \quad = 0.9544$$
$$A(-3, +3) = p(-3, +3) = 0.49865 + 0.49865 = 0.9973$$
$$A(-1, +2) = p(-1, +2) = 0.3413 + 0.4772 \quad = 0.8185$$

Since the normal-distribution function has been well characterized, predictions can be made regarding the probability of a specific strength value or measurement error. For example, one may anticipate that 68.3 percent of the data will fall between limits of $\bar{x} \pm S_x$, 95.4 percent between limits of $\bar{x} \pm 2S_x$, and 99.7 percent between limits of $\bar{x} \pm 3S_x$. Also, 81.9 percent of the data should fall between limits of $\bar{x} - S_x$ and $\bar{x} + 2S_x$.

In many problems, the probability of a single sample exceeding a specified value z_2 must be determined. It is possible to determine this probability by using Table 16.3 together with the fact that the area under the entire curve is unity ($A = 1$); however, Table 16.4, which lists one-sided areas between limits of $z_1 = z$ and $z_2 \to \infty$, yields the results more directly.

Use of Tables 16.3 and 16.4 can be illustrated by considering the ultimate-tensile-strength data presented in Table 16.1. By using Eqs. (16.1) and (16.2), it is easy to establish estimates for the mean $\bar{x}$ and standard deviation S_x as $\bar{x} = 176.1$ ksi (1215 MPa) and $S_x = 2.25$ ksi (15.5 MPa). These values of $\bar{x}$ and S_x characterize the population from which the data of Table 16.1 were drawn. It is possible to establish the probability that the ultimate tensile strength of a single specimen drawn randomly from the population will be between specified limits (by using Table 16.3), or that the ultimate tensile strength of a single sample will not be above or below a specified value (by using Table 16.4). For example, one determines the probability that a single sample will exhibit an ultimate tensile strength between 175 and 178 ksi by computing z_1 and z_2 and using Table 16.3. Thus,

$$z_1 = \frac{175 - 176.1}{2.25} = -0.489 \qquad z_2 = \frac{178 - 176.1}{2.25} = 0.844$$

$$p(-0.489, 0.844) = A(-0.489, 0) + A(0, 0.844)$$
$$= 0.1875 + 0.3006 = 0.4981$$

This simple calculation shows that the probability of obtaining an ultimate tensile strength between 175 and 178 ksi from a single specimen is 49.8 percent. The probability of the ultimate tensile strength of a single specimen being less than 173 ksi is determined by computing z_1 and using Table 16.4. Thus,

$$z_1 = \frac{173 - 176.1}{2.25} = -1.37$$

$$p(-\infty, -1.37) = A(-\infty, -1.37) = A(1.37, \infty) = 0.0853$$

TABLE 16.4
Areas under the normal-distribution curve from z_1 to $z_2 \to \infty$ (one side)

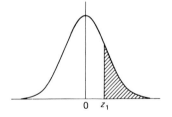

$z_1 = \dfrac{x - \bar{x}}{S_x}$	0.00	0.01	0.02	0.03	0.04	0.05	0.06	0.07	0.08	0.09
0.0	.5000	.4960	.4920	.4880	.4840	.4801	.4761	.4721	.4681	.4641
0.1	.4602	.4562	.4522	.4483	.4443	.4404	.4364	.4325	.4286	.4247
0.2	.4207	.4168	.4129	.4090	.4052	.4013	.3974	.3936	.3897	.3859
0.3	.3821	.3783	.3745	.3707	.3669	.3632	.3594	.3557	.3520	.3483
0.4	.3446	.3409	.3372	.3336	.3300	.3264	.3228	.3192	.3156	.3121
0.5	.3085	.3050	.3015	.2981	.2946	.2912	.2877	.2843	.2810	.2776
0.6	.2743	.2709	.2676	.2643	.2611	.2578	.2546	.2514	.2483	.2451
0.7	.2420	.2389	.2358	.2327	.2296	.2266	.2236	.2206	.2177	.2148
0.8	.2119	.2090	.2061	.2033	.2005	.1977	.1949	.1922	.1984	.1867
0.9	.1841	.1814	.1788	.1762	.1736	.1711	.1685	.1660	.1635	.1611
1.0	.1587	.1562	.1539	.1515	.1492	.1469	.1446	.1423	.1401	.1379
1.1	.1357	.1335	.1314	.1292	.1271	.1251	.1230	.1210	.1190	.1170
1.2	.1151	.1131	.1112	.1093	.1075	.1056	.1038	.1020	.1003	.0985
1.3	.0968	.0951	.0934	.0918	.0901	.0885	.0869	.0853	.0838	.0823
1.4	.0808	.0793	.0778	.0764	.0749	.0735	.0721	.0708	.0694	.0681
1.5	.0668	.0655	.0643	.0630	.0618	.0606	.0594	.0582	.0571	.0559
1.6	.0548	.0537	.0526	.0516	.0505	.0495	.0485	.0475	.0465	.0455
1.7	.0446	.0436	.0427	.0418	.0409	.0401	.0392	.0384	.0375	.0367
1.8	.0359	.0351	.0344	.0336	.0329	.0322	.0314	.0307	.0301	.0294
1.9	.0287	.0281	.0274	.0268	0.262	0.256	.0250	.0244	.0239	.0233
2.0	.0228	.0222	.0217	.0212	.0207	.0202	.0197	.0192	.0188	.0183
2.1	.0179	.0174	.0170	.0166	.0162	.0158	.0154	.0150	.0146	.0143
2.2	.0139	.0136	.0132	.0129	.0125	.0122	.0119	.0116	.0113	.0110
2.3	.0107	.0104	.0102	.00990	.00964	.00939	.00914	.00889	.00866	.0084
2.4	.00820	.00798	.00776	.00755	.00734	.00714	.00695	.00676	.00657	.00639
2.5	.00621	.00604	.00587	.00570	.00554	.00539	.00523	.00508	.00494	.00480
2.6	.00466	.00453	.00440	.00427	.00415	.00402	.00391	.00379	.00368	.00357
2.7	.00347	.00336	.00326	.00317	.00307	.00298	.00288	.00280	.00272	.00264
2.8	.00256	.00248	.00240	.00233	.00226	.00219	.00212	.00205	.00199	.00193
2.9	.00187	.00181	.00175	.00169	.00164	.00159	.00154	.00149	.00144	.00139

Thus, the probability of drawing a single sample with an ultimate tensile strength less than 173 ksi is 8.5 percent.

16.3.2 Weibull Distribution

In investigations of the strength of materials due to brittle fracture, of crack-initiation toughness, or fatigue life, researchers often find that the Weibull distribution provides a more suitable approach to the statistical analysis of the available data. The Weibull distribution function $P(x)$ is defined as

$$P(x) = 1 - e^{-[(x-x_0)/b]^m} \qquad \text{for } x > x_0$$
$$P(x) = 0 \qquad \qquad \text{for } x < x_0$$

(16.11)

where x_0, b, and m are the three parameters which define this distribution function. In studies of strength, $P(x)$ is taken as the probability of failure when a stress x is placed on the specimen. The parameter x_0 is the zero strength since $P(x) = 0$ for $x < x_0$. The constants b and m are known as the *scale parameter* and the *Weibull slope parameter (modulus)*, respectively.

Four Weibull distribution curves are presented in Fig. 16.6 for the case where $x_0 = 3$, $b = 10$, and $m = 2, 5, 10,$ and 20. These curves illustrate two important features of the Weibull distribution. First, there is a threshold strength x_0 and if the applied stress is less than x_0, the probability of failure is zero. Second, the Weibull distribution curves are not symmetric, and the distortion in the S-shaped curves is controlled by the Weibull slope parameter m. Application of the Weibull distribution to predict failure rates of 1 percent or less of the population is particularly important in engineering projects where reliabilities of 99 percent or greater are required.

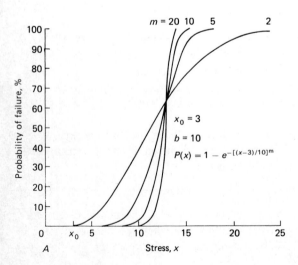

FIGURE 16.6
The Weibull distribution function.

To utilize the Weibull distribution requires knowledge of the Weibull parameters. In experimental investigations, it is necessary to conduct experiments and obtain a relatively large data set for the determination of x_0, b, and m. Consider as an illustration Weibull's own work in statistically characterizing the fiber strength of Indian cotton. In this example, an unusually large sample ($n = 3000$) was studied by measuring the load to fracture (in grams) for each fiber. The strength data obtained was placed in sequential order with the lowest value corresponding to $k = 1$ first and the largest value corresponding to $k = 3000$ last. The probability of failure $P(x)$ at a load x is then determined from

$$P = \frac{k}{n + 1} \tag{16.12}$$

where k = the order number of the sequenced data
n = the total sample size

At this stage it is possible to prepare a graph of probability of failure $P(x)$ as a function of strength x to obtain a curve similar to that shown in Fig. 16.6. However, to determine the Weibull parameters x_0, b, and m requires additional conditioning of the data. From Eqs. (16.11), it is evident that

$$e^{[(x - x_0)/b]^m} = [1 - P(x)]^{-1} \tag{16.13}$$

Taking the natural log of both sides of Eq. (16.13) yields

$$\left[\frac{(x - x_0)}{b}\right]^m = \ln\,[1 - P(x)]^{-1} \tag{16.14}$$

Taking $\log_{10}$ of both sides of Eq. (16.14) gives a relation for the slope parameter m. Thus,

$$m = \frac{\log_{10} \ln\,[1 - P(x)]^{-1}}{\log_{10}(x - x_0) - \log_{10} b} \tag{16.15}$$

The numerator of Eq. (16.15) is the reduced variate $y = \log_{10}\ln\,[1 - P(x)]^{-1}$ used for the ordinate in preparing a graph of the conditioned data, as indicated in Fig. 16.7. Note that y is a function of P alone and for this reason both the P and y scales can be displayed on the ordinates (see Fig. 16.7). The lead term in the denominator of Eq. (16.15) is the reduced variate $x = \log_{10}(x - x_0)$ used for the abscissa in Fig. 16.7.

In the Weibull example, x_0 was adjusted to 0.46 g so that the data would fall on a straight line when plotted against the reduced x and y variates. The constant b is determined from the condition that

$$\log_{10} b = \log_{10}\,(x - x_0) \qquad \text{when } y = 0 \tag{16.16}$$

Note from Fig. 16.7 that $y = 0$ when $\log_{10}(x - x_0) = 0.54$, which gives $b = 0.54$. Finally, m is given by the slope of the straight line when the data is plotted in terms of the reduced variates x and y. In this example problem, $m = 1.48$.

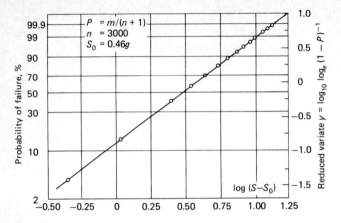

FIGURE 16.7
Fiber strength of Indian cotton shown as a graph with Weibull's reduced variate (from data by Weibull).

16.4 CONFIDENCE INTERVALS FOR PREDICTIONS

Once experimental data are represented with a normal distribution by using estimates of the mean $\bar{x}$ and standard deviation S_x and predictions are made about the occurrence of measurements, questions arise concerning the confidence that can be placed on either the estimates or the predictions. One cannot be totally confident in the predictions or estimates because of the effects of sampling error.

Sampling error can be illustrated by drawing a series of samples (each containing n measurements) from the same population and determining several estimates of the mean $\bar{x}_1, \bar{x}_2, \bar{x}_3, \ldots$. A variation in $\bar{x}$ will occur, but fortunately, this variation can also be characterized by a normal distribution function, as shown in Fig. 16.8. The mean of the x and $\bar{x}$ distributions is the same; however, the standard deviation of the $\bar{x}$ distribution $S_{\bar{x}}$ (sometimes referred to as the *standard*

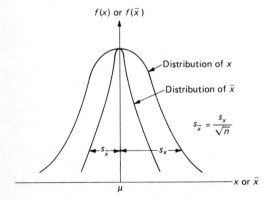

FIGURE 16.8

Normal distribution of individual measurements of the quantity x and of measurements of the mean $\bar{x}$ from samples of size n.

error) is less than S_x since

$$S_{\bar{x}} = \frac{S_x}{\sqrt{n}} \tag{16.17}$$

Once the standard deviation of the population of $\bar{x}$'s is known, it is possible to place confidence limits on the determination of the true population mean μ from a sample of size n, provided n is large ($n > 25$). The confidence interval within which the true population mean μ is located is given by the expression

$$(\bar{x} - zS_{\bar{x}}) < \mu < (\bar{x} + zS_{\bar{x}}) \tag{16.18}$$

where $\bar{x} - zS_{\bar{x}}$ = the lower confidence limit
$\qquad \bar{x} + zS_{\bar{x}}$ = the upper confidence limit

The width of the confidence interval depends upon the confidence level required. For instance, if $z = 3$ in Eq. (16.18), a relatively wide confidence interval exists; therefore, the probability that the population mean μ will be located within the confidence interval is high (99.7 percent). As the width of the confidence interval decreases, the probability that the population mean μ will fall within the interval decreases. Commonly used confidence levels and their associated intervals are shown in Table 16.5.

When the sample size is very small ($n < 20$), the standard deviation S_x does not provide a reliable estimate of the standard deviation σ of the population and Eq. (16.18) should not be employed. The bias introduced by small sample size can be removed by modifying Eq. (16.18) to read as

$$[\bar{x} - t(\alpha)S_{\bar{x}}] < \mu < [\bar{x} + t(\alpha)S_{\bar{x}}] \tag{16.19}$$

where $t(\alpha)$ = the statistic known as Student's t
$\qquad \alpha$ = the level of significance (the probability of exceeding a given value of t)

The Student's t distribution is defined by a relative frequency equation $f(t)$,

TABLE 16.5
Confidence interval variation with confidence level interval = $\bar{x} + zS_{\bar{x}}$

Confidence level, percent	z	Confidence level, percent	z
99.9	3.30	90.0	1.65
99.7	3.00	80.0	1.28
99.0	2.57	68.3	1.00
95.0	1.96	60.0	0.84

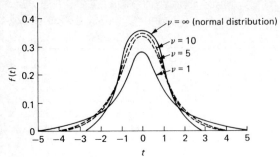

FIGURE 16.9
Student's t distribution for several degrees of freedom v.

which can be expressed as

$$f(t) = F_0\left(1 + \frac{t^2}{v}\right)^{(v+1)/2}$$
(16.20)

where F_0 = the relative frequency at $t = 0$ required to make the total area
under the $f(t)$ curve equal to unity
v = the number of degrees of freedom

The distribution function $f(t)$ is shown in Fig. 16.9 for several different degrees of freedom v. The degrees of freedom equal the number of independent measurements employed in the determination. It is evident that as v becomes large, Student's t distribution approaches the normal distribution. One-sided areas for the t distribution are listed in Table 16.6 and illustrated in Fig. 16.10.

The term $t(\alpha)S_{\bar{x}}$ in Eq. (16.19) represents the measure from the estimated mean $\bar{x}$ to one or the other of the confidence limits. This term may be used to estimate the sample size required to produce an estimate of the mean $\bar{x}$ with a specified reliability. Noting that one-half the bandwidth of the confidence interval is $\delta = t(\alpha)S_{\bar{x}}$ and using Eq. (16.17), it is apparent that the sample size is given by

$$n = \left[\frac{t(\alpha)S_x}{\delta}\right]^2$$
(16.21)

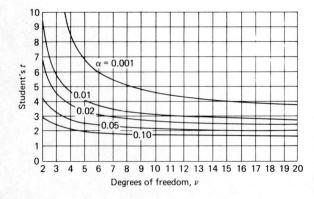

FIGURE 16.10
Student's t statistic as a function of degrees of freedom v with α the probability of exceeding t as a parameter.

TABLE 16.6
Student's *t* distribution for v degrees of freedom showing $t(\alpha)$ as a function of area A (one side)

v	Confidence level α									
	0.995	0.99	0.975	0.95	0.90	0.80	0.75	0.70	0.60	0.55
1	63.66	31.82	12.71	6.31	3.08	1.376	1.000	.727	.325	.158
2	9.92	6.96	4.30	2.92	1.89	1.061	.816	.617	.289	.142
3	5.84	4.54	3.18	2.35	1.64	.978	.765	.584	.277	.137
4	4.60	3.75	2.78	2.13	1.53	.941	.741	.569	.271	.134
5	4.03	3.36	2.57	2.02	1.48	.920	.727	.559	.267	.132
6	3.71	3.14	2.45	1.94	1.44	.906	.718	.553	.265	.131
7	3.50	3.00	2.36	1.90	1.42	.896	.711	.549	.263	.130
8	3.36	2.90	2.31	1.86	1.40	.889	.706	.546	.262	.130
9	3.25	2.82	2.26	1.83	1.38	.883	.703	.543	.261	.129
10	3.17	2.76	2.23	1.81	1.37	.879	.700	.542	.260	.129
11	3.11	2.72	2.20	1.80	1.36	.876	.697	.540	.260	.129
12	3.06	2.68	2.18	1.78	1.36	.873	.695	.539	.259	.128
13	3.01	2.65	2.16	1.77	1.35	.870	.694	.538	.259	.128
14	2.98	2.62	2.14	1.76	1.34	.868	.692	.537	.258	.128
15	2.95	2.60	2.13	1.75	1.34	.866	.691	.536	.258	.128
16	2.92	2.58	2.12	1.75	1.34	.865	.690	.535	.258	.128
17	2.90	2.57	2.11	1.74	1.33	.863	.689	.534	.257	.128
18	2.88	2.55	2.10	1.73	1.33	.862	.688	.534	.257	.127
19	2.86	3.54	2.09	1.73	1.33	.861	.688	.533	.257	.127
20	2.84	2.53	2.09	1.72	1.32	.860	.687	.533	.257	.127
21	2.83	2.52	2.08	1.72	1.32	.859	.686	.532	.257	.127
22	2.82	2.51	2.07	1.72	1.32	.858	.686	.532	.256	.127
23	2.81	2.50	2.07	1.71	1.32	.858	.685	.532	.256	.127
24	2.80	2.49	2.06	1.71	1.32	.857	.685	.531	.256	.127
25	2.79	2.48	2.06	1.71	1.32	.856	.684	.531	.256	.127
26	2.78	2.48	2.06	1.71	1.32	.856	.684	.531	.256	.127
27	2.77	2.47	2.05	1.70	1.31	.855	.684	.531	.256	.127
28	2.76	2.47	2.05	1.70	1.31	.855	.683	.530	.256	.127
29	2.76	2.46	2.04	1.70	1.31	.854	.683	.530	.256	.127
30	2.75	2.46	2.04	1.70	1.31	.854	.683	.530	.256	.127
40	2.70	2.42	2.02	1.68	1.30	.851	.681	.529	.255	.126
60	2.66	2.39	2.00	1.67	1.30	.848	.679	.527	.254	.126
120	2.62	2.36	1.98	1.66	1.29	.845	.677	.526	.254	.126
∞	2.58	2.33	1.96	1.65	1.28	.842	.674	.524	.253	.126

Use of Eq. (16.21) can be illustrated by considering the data in Table 16.1, where $S_x = 2.25$ ksi and $\bar{x} = 176.1$ ksi. If this estimate of μ is to be accurate to ± 1 percent with a reliability of 99 percent, then

$$\delta = (0.01)(176.1) = 1.76 \text{ ksi}$$

Since $t(\alpha)$ depends on n, a trial-and-error solution is needed to establish sample size n needed to satisfy the specifications. For the data of Table 16.1, $n = 40$; therefore $v = 39$ and $t(\alpha) = t(0.995) = 2.71$ from Table 16.6. The value $t(\alpha) = t(0.995)$ is used since 0.5 percent of the distribution must be excluded on each end of the curve to give a two-sided area corresponding to a reliability of 99 percent. Substituting into Eq. (16.21) yields

$$n = \left[\frac{2.71\,(2.25)}{1.76}\right]^2 = 12.00$$

This result indicates that a much smaller sample than 40 will be sufficient.
Now try $n = 12$, $v = 11$, and $t(\alpha) = 3.11$:

$$n = \left[\frac{3.11\,(2.25)}{1.76}\right]^2 = 15.80$$

Finally, with $n = 15$, $v = 14$, and $t(\alpha) = 2.98$:

$$n = \left[\frac{2.98\,(2.25)}{1.76}\right]^2 = 14.50$$

Thus, a sample size of 15 would be sufficient to ensure an accuracy of ± 1 percent with a confidence level of 99 percent. The sample size of 40 listed in Table 16.1 is too large for the degree of accuracy and confidence level specified. This simple example illustrates how sample size can be reduced and cost savings effected by using statistical methods to determine the required sample size.

16.5   COMPARISON OF MEANS

Since Student's t distribution compensates for the effect of small-sample bias and converges to the normal distribution in large samples, it is a very useful statistic in engineering applications. A second important application utilizes the t distribution as the basis for a test to determine if the difference between two means is significant or due to random variation. For example, consider the yield strength of a steel determined with a sample size of $n_1 = 20$, which gives $\bar{x}_1 = 78.4$ ksi and $S_{x_1} = 6.04$ ksi. Suppose now that a second sample from another supplier is tested to determine the yield strength and the results are $n_2 = 25$, $\bar{x}_2 = 81.6$ ksi, and $S_{x_2} = 5.56$ ksi. Is the steel from the second supplier superior in terms of yield strength? The standard deviation of the difference in means $S_{(\bar{x}_2 - \bar{x}_1)}$ can be expressed as

$$S^2_{(\bar{x}_2 - \bar{x}_1)} = S^2_p\left(\frac{1}{n_1} + \frac{1}{n_2}\right) = S^2_p\frac{n_1 + n_2}{n_1 n_2} \qquad (16.22)$$

where S_p^2 is the pooled variance that can be expressed as

$$S_p^2 = \frac{(n_1 - 1)S_{x_1}^2 + (n_2 - 1)S_{x_2}^2}{n_1 + n_2 - 2} \tag{16.23}$$

The statistic t can be computed from the expression

$$t = \frac{|\bar{x}_2 - \bar{x}_1|}{S_{(\bar{x}_2 - \bar{x}_1)}} \tag{16.24}$$

A comparison of the value of t determined from Eq. (16.24) with a value of $t(\alpha)$ obtained from Table 16.6 provides a statistical basis for deciding whether the difference in means is real or due to random variations. The value of $t(\alpha)$ to be used depends upon the degrees of freedom $v = n_1 + n_2 - 2$ and the level of significance required. Levels of significance commonly employed are 5 and 1 percent. The 5 percent level of significance means that the probability of a random variation being taken for a real difference is only 5 percent. Comparisons at the 1 percent level of significance are 99 percent certain; however, in such a strong test, real differences can often be attributed to random error.

In the example being considered, Eq. (16.23) yields $S_p^2 = 33.37$ ksi, Eq. (16.22) yields $S_{(\bar{x}_2 - \bar{x}_1)}^2 = 3.00$ ksi, and Eq. (10.24) yields $t = 1.848$. For a 5 percent level of significance test with $d = 43$ and $\alpha = 0.05$ (the comparison is one-sided, since the t test is for superiority), Table 16.6 indicates that $t(\alpha) = 1.68$. Since $t > t(\alpha)$, it can be concluded with a 95 percent level of confidence that the yield strength of the steel from the second supplier was higher than the yield strength of steel from the first supplier.

16.6 STATISTICAL SAFETY FACTOR

In experimental mechanics, it is often necessary to determine the stresses acting on a component and its strength in order to predict whether failure will occur or if the component is safe. This can be a difficult prediction if both the stress σ_{ij} and the strength S_y are variables since failure will occur only in the region of overlap of the two distribution functions, as shown in Fig. 16.11.

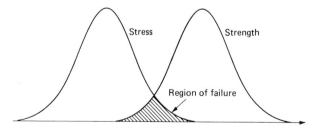

FIGURE 16.11
Superimposed normal distribution curves for strength and stress showing region of failure.

TABLE 16.7
Reliability R as a function of the statistic z_R

R, percent	z_R	R, percent	z_R
50	0	99.9	3.091
90	1.288	99.99	3.719
95	1.645	99.999	4.265
99	2.326	99.9999	4.753

To determine the probability of failure, or conversely, the reliability, the statistic z_R is computed by using the equation

$$z_R = \frac{\bar{x}_S - \bar{x}_\sigma}{S_{S-\sigma}} \tag{16.25}$$

where

$$S_{S-\sigma} = \sqrt{S_S^2 + S_\sigma^2} \tag{16.26}$$

and the subscripts S and σ refer to strength and stress, respectively.

The reliability associated with the value of z_R determined from Eq. (16.25) may be determined from a table showing the area $A(z_R)$ under a standard normal distribution curve by using

$$R = 0.5 + A(z_R) \tag{16.27}$$

Typical values of R as a function of the statistic z_R are given in Table 16.7.

The reliability determined in this manner incorporates a safety factor of 1. If a safety factor of N is to be specified together with a reliability, then Eqs. (16.25) and (16.26) are rewritten to give a modified relation for z_R as

$$z_R = \frac{\bar{x}_S - N\bar{x}_\sigma}{\sqrt{S_S^2 + S_\sigma^2}} \tag{16.28}$$

which can be rearranged to give the safety factor N as

$$N = \frac{1}{\bar{x}_\sigma}\left(\bar{x}_S - z_R\sqrt{S_s^2 + S_\sigma^2}\right) \tag{16.29}$$

16.7 STATISTICAL CONDITIONING OF DATA

Previously it was indicated that measurement error can be characterized by a normal distribution function and that the standard deviation of the estimated mean $S_{\bar{x}}$ can be reduced by increasing the number of measurements. In most situations, cost places an upper limit on the number of measurements to be made. Also, it must be remembered that systematic error is not a random variable; therefore, statistical procedures cannot serve as a substitute for precise, accurately calibrated, and properly zeroed measuring instruments.

One area where statistical procedures can be used very effectively to condition experimental data is with the erroneous data point resulting from a measuring or recording mistake. Often, this data point appears questionable when compared with the other data collected, and the experimentalist must decide whether the deviation of the data point is due to a mistake (hence to be rejected) or due to some unusual but real condition (hence to be retained). A statistical procedure known as *Chauvenet's criterion* provides a consistent basis for making the decision to reject or retain such a point from a sample containing several readings.

Application of Chauvenet's criterion requires computation of a deviation ratio DR for each data point, followed by comparison with a standard deviation ratio DR_0. The standard deviation ratio DR_0 is a statistic that depends on the number of measurements, while the deviation ratio DR for a point is defined as

$$DR = \frac{x_i - \bar{x}}{S_x} \tag{16.30}$$

The data point is rejected when $DR > DR_0$ and retained when $DR \leq DR_0$. Values for the standard deviation ratio DR_0 are listed in Table 16.8.

If the statistical test of Eq. (16.30) indicates that a single data point in a sequence of n data points should be rejected, then the data point should be removed from the sequence and the mean $\bar{x}$ and the standard deviation S_x should be recalculated. Chauvenet's method can be applied only once to reject a data point that is questionable from a sequence of points. If several data points indicate that $DR > DR_0$, then it is likely that the instrumentation system is inadequate or that the process being investigated is extremely variable.

TABLE 16.8
Deviation ratio DR_0 used for statistical conditioning of data

Number of measurements n	Deviation ratio DR_0	Number of measurements n	Deviation ratio DR_0
4	1.54	25	2.33
5	1.65	50	2.57
7	1.80	100	2.81
10	1.96	300	3.14
15	2.13	500	3.29

16.8   REGRESSION ANALYSIS

Many experiments involve the measurement of one dependent variable, say, y, which may depend upon one or more independent variables, $x_1, x_2, \ldots, x_k$. Regression analysis provides a statistical approach for conditioning the data obtained from experiments where two or more related quantities are measured.

16.8.1   Linear Regression Analysis

Suppose measurements are made of two quantities that describe the behavior of a process exhibiting variation. Let y be the dependent variable and x the independent variable. Since the process exhibits variation, there is not a unique relationship between x and y, and the data, when plotted, exhibit scatter, as illustrated in Fig. 16.12. Frequently, the relation between x and y that most closely represents the data, even with the scatter, is a linear function. Thus,

$$Y_i = mx_i + b \tag{16.31}$$

where Y_i is the predicted value of the dependent variable y_i for a given value of the independent variable x_i.

A statistical procedure used to fit a straight line through scattered data points is called the *least-squares method*. With the least-squares method, the slope m and the intercept b in Eq. (16.31) are selected to minimize the sum of the squared deviations of the data points from the straight line shown in Fig. 16.12. In utilizing the least-squares method, it is assumed that the independent variable x is free of measurement error and the quantity

$$\Delta^2 = \sum (y_i - Y_i)^2 \tag{16.32}$$

is minimized at fixed values of x. After substituting Eq. (16.31) into Eq. (16.32) this implies that

$$\frac{\partial \Delta^2}{\partial b} = \frac{\partial}{\partial b} \sum (y_i - mx - b)^2 = 0$$

$$\frac{\partial \Delta^2}{\partial m} = \frac{\partial}{\partial m} \sum (y_i - mx - b)^2 = 0 \tag{a}$$

Differentiating yields

$$2 \sum (y_i - mx - b)(-x) = 0$$

$$2 \sum (y_i - mx - b)(-1) = 0 \tag{b}$$

Solving Eqs. (b) for m and b yields

$$m = \frac{\sum x \sum y - n \sum xy}{(\sum x)^2 - n \sum x^2}$$

$$b = \frac{\sum y - m \sum x}{n} \tag{16.33}$$

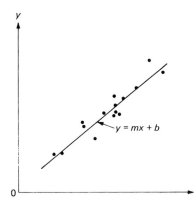

FIGURE 16.12
Linear-regression analysis is used to fit a least-squares line through scattered data points.

where n is the number of data points. The slope m and intercept b define a straight line through the scattered data points such that Eq. (16.32) is minimized.

In any regression analysis it is important to establish the correlation between x and y. Equation (16.31) does not predict the values that were measured exactly, because of the variation in the process. To illustrate, assume that the independent quantity x is fixed at a value x_1 and that a sequence of measurements is made of the dependent quantity y. The data obtained would give a distribution of y, as illustrated in Fig. 16.13. The dispersion of the distribution of y is a measure of the correlation. When the dispersion is small, the correlation is good and the regression analysis is effective in describing the variation in y. If the dispersion is large, the correlation is poor and the regression analysis may not be adequate to describe the variation in y.

The adequacy of regression analysis can be evaluated by determining a correlation coefficient R^2 that is given by the following expression:

$$R^2 = 1 - \frac{n-1}{n-2}\left[\frac{\{y^2\} - m\{xy\}}{\{y^2\}}\right] \qquad (16.34)$$

where $\{y^2\} = \sum y^2 - (\sum y)^2/n$
$\{xy\} = \sum xy - (\sum x)(\sum y)/n$

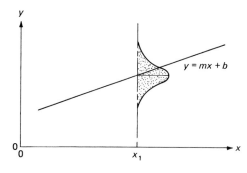

FIGURE 16.13
Distribution of y at a fixed value of x superimposed on the linear-regression display.

When the value of the correlation coefficient $R^2 = 1$, perfect correlation exists between y and x. If $R^2 = 0$, no correlation exists and the variations observed in y are due to random fluctuations and not changes in x. Because random variations in y exist, a value of $R^2 = 1$ is not obtained even if $y(x)$ is linear. To interpret correlation coefficients $0 < R^2 < 1$, the data in Table 16.9 are used to establish the probability of obtaining a given R^2 due to random variations in y.

As an example, consider a regression analysis with $n = 15$, which gives $R^2 = 0.65$ as determined by Eq. (16.34). Reference to Table 16.9 indicates that the probability of obtaining $R^2 = 0.65$ due to random variations is slightly less than 1 percent. Thus one can be 99 percent certain that the regression analysis represents a true correlation between y and x.

16.8.2 Multivariate Regression

Many experiments involve measurements of a dependent variable y, which depends upon several independent variables $x_1, x_2, x_3, \ldots$, etc. It is possible to represent y as a function of $x_1, x_2, x_3, \ldots$ by employing the multivariate regression equation

$$Y_i = a + b_1 x_1 + b_2 x_2 + \cdots + b_k x_k \qquad (16.35)$$

where $a, b_1, b_2, \ldots, b_k$ are regression coefficients.

The regression coefficients $a, b_1, b_2, \ldots, b_k$ are determined by using the method of least squares in a manner similar to that employed for linear regression analysis where the quantity $\Delta^2 = \sum (y_i - Y_i)^2$ is minimized. Substituting Eq. (16.32) into Eq. (16.35) yields

$$\Delta^2 = \sum (y_i - a - b_1 x_1 - b_2 x_2 - \cdots - b_k x_k)^2 \qquad (16.36)$$

TABLE 16.9
Probability of obtaining a correlation coefficient R^2 due to random variations in y

	Probability			
n	0.10	0.05	0.02	0.01
5	0.805	0.878	0.934	0.959
6	0.729	0.811	0.882	0.917
7	0.669	0.754	0.833	0.874
8	0.621	0.707	0.789	0.834
10	0.549	0.632	0.716	0.765
15	0.441	0.514	0.592	0.641
20	0.378	0.444	0.516	0.561
30	0.307	0.362	0.423	0.464
40	0.264	0.312	0.367	0.403
60	0.219	0.259	0.306	0.337
80	0.188	0.223	0.263	0.291
100	0.168	0.199	0.235	0.259

Differentiating yields

$$\frac{\partial \Delta^2}{\partial a} = 2[\sum (y_i - a - b_1 x_1 - b_2 x_2 - \cdots - b_k x_k)(-1)] = 0$$

$$\frac{\partial \Delta^2}{\partial b_1} = 2[\sum (y_i - a - b_1 x_1 - b_2 x_2 - \cdots - b_k x_k)(-x_1)] = 0$$

$$\frac{\partial \Delta^2}{\partial b_2} = 2[\sum (y_i - a - b_1 x_1 - b_2 x_2 - \cdots - b_k x_k)(-x_2)] = 0 \qquad (16.37)$$

$$\cdots\cdots\cdots\cdots\cdots\cdots\cdots\cdots\cdots\cdots\cdots\cdots\cdots\cdots$$

$$\frac{\partial \Delta^2}{\partial b_k} = 2[\sum (y_i - a - b_1 x_1 - b_2 x_2 - \cdots - b_k x_k)(-x_k)] = 0$$

Equations (16.37) lead to the following set of $k + 1$ equations, which can be solved for the unknown regression coefficients $a, b_1, b_2, \ldots, b_k$:

$$an + b_1 \sum x_1 + b_2 \sum x_2 + \cdots + b_k \sum x_k = \sum y_i$$
$$a \sum x_1 + b_1 \sum x_1^2 + b_2 \sum x_1 x_2 + \cdots + b_k \sum x_1 x_k = \sum y_i x_1$$
$$a \sum x_2 + b_1 \sum x_1 x_2 + b_2 \sum x_2^2 + \cdots + b_k \sum x_1 x_k = \sum y_i x_2 \qquad (16.38)$$
$$\cdots\cdots\cdots\cdots\cdots\cdots\cdots\cdots\cdots\cdots\cdots\cdots\cdots$$
$$a \sum x_k + b_1 \sum x_1 x_k + b_2 \sum x_2 x_k + \cdots + b_k \sum x_k^2 = \sum y_i x_k$$

The correlation coefficient R^2 is again used to determine the degree of association between the dependent and independent variables. For multiple regression equations, the correlation coefficient R^2 is given as

$$R^2 = 1 - \frac{n-1}{n-k}\left[\frac{\{y^2\} - b_1\{yx_1\} - b_2\{yx_2\} - \cdots - b_k\{yx_k\}}{\{y^2\}}\right] \qquad (16.39)$$

where

$$\{yx_k\} = \sum yx_k - \frac{(\sum y)(\sum x_k)}{n}$$

$$\{y^2\} = \sum y^2 - \frac{(\sum y)^2}{n}$$

This analysis is for linear, noninteracting, independent variables; however, the analysis can be extended to include cases where the regression equations would have higher-order and cross-product terms. The nonlinear terms can enter the regression equation in an additive manner and are treated as extra variables. With well-established computer routines for regression analysis, the set of $k + 1$ simultaneous equations given by Eqs. (16.38) can be solved quickly and inexpensively and no difficulties are encountered in adding extra terms to account for nonlinearities and interactions.

16.8.3 Field Applications of Least-Squares Methods

The least-squares method is an important mathematical process used in regression analysis to obtain regression coefficients. Sanford showed that the least-squares method could be extended to field analysis of data obtained with optical techniques (photoelasticity, moiré, holography, etc.). With these optical methods, a fringe order N, related to a field quantity such as stress, strain, or displacement, can be measured at a large number of points over a field (x, y). The applications require an analytical representation of the field quantities as a function of position (x, y) over the field. Several important problems including calibration, fracture mechanics, and contact stresses have analytical solutions where select terms require experimental data for complete evaluation. Two examples will be described which introduce both the linear and nonlinear least-squares method applied over a field (x, y).

LINEAR LEAST-SQUARES METHOD. Consider a calibration model in photoelasticity as the first example and write the equation for the fringe order $N(x, y)$ as

$$N(x, y) = \frac{h}{f_\sigma} G(x, y) + E(x, y) \tag{a}$$

where $G(x, y) =$ the analytical representation of the difference of the principal
stresses $\sigma_1 - \sigma_2$ in the calibration model
$h =$ the model thickness
$f_\sigma =$ the material fringe value
$E(x, y) =$ the residual birefringence

Assume a linear distribution for $E(x, y)$ which can be expressed as

$$E(x, y) = Ax + By + C \tag{b}$$

For any selected point (x_k, y_k) in the field where N_k is determined,

$$N_k(x_k, y_k) = \frac{h}{f_\sigma} G_k(x_k, y_k) + Ax_k + By_k + C \tag{16.40}$$

Note that Eq. (16.40) is linear in terms of the unknowns h/f_σ, A, B, and C. For m selected data points with $m > 4$, an overdeterministic system of linear equations results from Eq. (16.40). This system of equations can be expressed in matrix form as

$$[N] = [a][w]$$

where

$$[\mathbf{N}] = \begin{bmatrix} N_1 \\ N_2 \\ \vdots \\ N_m \end{bmatrix} \qquad [\mathbf{a}] = \begin{bmatrix} G_1 & x_1 & y_1 & 1 \\ G_2 & x_2 & y_2 & 1 \\ \vdots & \vdots & \vdots & \vdots \\ G_m & x_m & x_m & 1 \end{bmatrix} \qquad [\mathbf{w}] = \begin{bmatrix} h/f_\sigma \\ A \\ B \\ C \end{bmatrix}$$

The solution of the set of m equations for the unknowns h/f_σ, A, B, and C can be achieved in a least-squares sense through the use of matrix methods. Note that

$$[\mathbf{a}]^T[\mathbf{N}] = [\mathbf{c}][\mathbf{w}]$$

where

$$[\mathbf{c}] = [\mathbf{a}]^T[\mathbf{a}]$$

and that

$$[\mathbf{w}] = [\mathbf{c}]^{-1}[\mathbf{a}]^T[\mathbf{N}]$$

where $[\mathbf{c}]^{-1}$ is the inverse of $[\mathbf{c}]$. Solution of the matrix $[\mathbf{w}]$ gives the column elements which are the unknowns. This form of solution is easy to accomplish on a small computer which can be programmed in BASIC to perform the matrix manipulations.

The matrix algebra outlined above is equivalent to minimizing the cumulative error $\mathscr{E}$, which is

$$\mathscr{E} = \sum_{k=1}^{m} \left[\frac{h}{f_\sigma} G(x_k, y_k) + Ax_k + By_k + C - N_k \right]^2 \tag{16.41}$$

The matrix operations apply the least-squares criteria which require that

$$\frac{\partial \mathscr{E}}{\partial (h/f_\sigma)} = \frac{\partial \mathscr{E}}{\partial A} = \frac{\partial \mathscr{E}}{\partial B} = \frac{\partial \mathscr{E}}{\partial C} = 0 \tag{16.42}$$

The advantage of this statistical approach to calibration of model materials in optical arrangements is the use of full-field data to reduce errors due to discrepancies in either the model materials or the optical systems.

NONLINEAR LEAST-SQUARES METHOD. In the preceding section a linear least-squares method provided a direct approach to improving the accuracy of calibration with a single-step computation of an overdeterministic set of linear equations. In other experiments involving either the determination of unknowns arising in stresses near a crack tip or contact stresses near a concentrated load, the governing equations are nonlinear in terms of the unknown quantities. In these cases, the procedure to be followed involves linearizing the governing equations, applying the least-squares criteria to the linearized equations, and finally iterating to converge to an accurate solution for the unknowns.

To illustrate this statistical approach, consider a photoelastic experiment that yields an isochromatic fringe pattern near the tip of a crack in a specimen subjected to mixed-mode loading. In this example, there are three unknowns K_I, K_{II}, and σ_{0x} which are related to the experimentally determined fringe orders N_k at positions (r_k, θ_k). The governing equation for this mixed-mode fracture problem is

$$\left(\frac{Nf_\sigma}{h}\right)^2 = \frac{1}{2\pi r} [(K_I \sin \theta + 2K_{II} \cos \theta)^2 + (K_{II} \sin \theta)^2]$$

$$+ \frac{2\sigma_{0x}}{\sqrt{2\pi r}} \sin \frac{\theta}{2} [K_I \sin \theta(1 + 2 \cos \theta)$$

$$+ K_{II}(1 + 2 \cos^2 \theta + \cos \theta)] + \sigma_{0x}^2 \qquad (16.43)$$

Equation (16.43) can be solved in an overdeterministic sense by forming the function $f(K_I, K_{II}, \sigma_{0x})$ as

$$f_k(K_I, K_{II}, \sigma_{0x}) = \frac{1}{2\pi r_k} [(K_I \sin \theta_k + 2K_{II} \cos \theta_k)^2 + (K_{II} \sin \theta_k)^2]$$

$$+ \frac{2\sigma_{0x}}{\sqrt{2\pi r_k}} \sin \frac{\theta_k}{2} [K_I \sin \theta_k(1 + 2 \cos \theta_k)$$

$$+ K_{II}(1 + 2 \cos^2 \theta_k + \cos \theta_k)] + \sigma_{0x}^2 - \left(\frac{N_k f_\sigma}{h}\right)^2 = 0 \quad (16.44)$$

where $k = 1, 2, 3, \ldots, m$ and (r_k, θ_k) are coordinates defining a point on an isochromatic fringe of order N_k. A Taylor series expansion of Eq. (16.44) yields

$$(f_k)_{i+1} = (f_k)^i + \left(\frac{\partial f_k}{\partial K_I}\right)_i \Delta K_I + \left(\frac{\partial f_k}{\partial K_{II}}\right)_i \Delta K_{II} + \left(\frac{\partial f_k}{\partial \sigma_{0x}}\right)_i \Delta \sigma_{0x} \qquad (16.45)$$

where i refers to the ith iteration step and ΔK_I, ΔK_{II}, and $\Delta \sigma_{0x}$ are corrections to the previous estimate of K_I, K_{II}, and σ_{0x}. It is evident from Eq. (16.45) that corrections should be made given $f_k(K_I, K_{II}, \sigma_{0x}) = 0$. This fact leads to the iterative equation

$$\left(\frac{\partial f_k}{\partial K_I}\right)_i \Delta K_I + \left(\frac{\partial f_k}{\partial K_{II}}\right)_i \Delta K_{II} + \left(\frac{\partial f_k}{\partial \sigma_{0x}}\right)_i \Delta \sigma_{0x} = -(f_k)_i \qquad (16.46)$$

In matrix form the set of m equations represented by Eq. (16.46) can be written as

$$[f] = [a][\Delta K] \qquad (16.47)$$

where the matrices are defined as

$$[f] = \begin{bmatrix} f_1 \\ f_2 \\ \vdots \\ f_m \end{bmatrix} \quad [a] = \begin{bmatrix} \partial f_1/\partial K_I & \partial f_1/\partial K_{II} & \partial f_1/\partial \sigma_{0x} \\ \partial f_2/\partial K_I & \partial f_2/\partial K_{II} & \partial f_2/\partial \sigma_{0x} \\ \vdots & \vdots & \vdots \\ \partial f_m/\partial K_I & \partial f_m/\partial K_{II} & \partial f_m/\partial \sigma_{0x} \end{bmatrix} \quad [\Delta K] = \begin{bmatrix} \Delta K_I \\ \Delta K_{II} \\ \Delta \sigma_{0x} \end{bmatrix}$$

The least-squares minimization process is accomplished by multiplying, from the left, both sides of Eq. (16.47) by the transpose of matrix $[\mathbf{a}]$, to give

$$[\mathbf{a}]^T[\mathbf{f}] = [\mathbf{a}]^T[\mathbf{a}][\mathbf{\Delta K}]$$

or

$$[\mathbf{d}] = [\mathbf{c}][\mathbf{\Delta K}]$$

where

$$[\mathbf{d}] = [\mathbf{a}]^T[\mathbf{f}]$$
$$[\mathbf{c}] = [\mathbf{a}]^T[\mathbf{a}]$$

Finally, the correction terms are given by

$$[\mathbf{\Delta K}] = [\mathbf{c}]^{-1}[\mathbf{d}] \tag{16.48}$$

The solution of Eq. (16.48) gives ΔK_{I}, K_{II}, and $\Delta \sigma_{0x}$, which are used to correct initial estimates of K_{I}, K_{II}, and σ_{0x} and obtain a better fit of the function f to m data points.

The procedure is executed on a small computer programmed in BASIC. One starts by assuming initial values for K_{I}, K_{II}, and σ_{0x}. Then, the elements of the matrices $[\mathbf{f}]$ and $[\mathbf{a}]$ are computed for each of the m data points. The correction matrix $[\mathbf{\Delta K}]$ is then computed from Eq. (16.48), and finally the estimates of the unknowns are corrected by noting that

$$(K_{\mathrm{I}})_{i+1} = (K_{\mathrm{I}})_i + \Delta K_{\mathrm{I}}$$
$$(K_{\mathrm{II}})_{i+1} = (K_{\mathrm{II}})_i + \Delta K_{\mathrm{II}} \tag{16.49}$$
$$(\sigma_{0x})_{i+1} = (\sigma_{0x})_i + \Delta \sigma_{0x}$$

The procedure is repeated until each element in the correction matrix $[\mathbf{\Delta K}]$ becomes acceptably small. As convergence is quite rapid, the number of iterations required for accurate estimates of the unknowns is usually small.

16.9   CHI-SQUARE TESTING

The χ^2 test is used in statistics to verify the use of a specific distribution function to represent the population from which a set of data has been obtained. The χ^2 statistic is defined as

$$\chi^2 = \sum_{i=1}^{k} \left[\frac{(n_o - n_e)^2}{n_e} \right]_i \tag{16.50}$$

where n_o = the actual number of observations in the ith group interval
n_e = the expected number of observations in the ith group interval based on the specified distribution
k = the total number of group intervals

The value of χ^2 is computed to determine how well the data fit the assumed statistical distribution. If $\chi^2 = 0$, the match is perfect. Values of $\chi^2 > 0$ indicate the possibility that the data are not represented by the specified distribution. The probability P that the value of χ^2 is due to random variation is illustrated in Fig. 16.14. The degree of freedom is defined as

$$v = n - k \tag{16.51}$$

where n = the number of observations
  k = the number of conditions imposed on the distribution

As an example of the χ^2 test, consider the ultimate tensile strength data presented in Table 16.1 and judge the adequacy of representing the ultimate tensile strength with a normal probability distribution described with $\bar{x} = 176.1$ ksi and $S_x = 2.25$ ksi. By using the properties of a normal distribution function, the number of specimens expected to fall in any strength group can be computed. The observed number of specimens in Table 16.1 exhibiting ultimate tensile strengths within each of seven group intervals, together with the computed number of specimens in a normal distribution in the same group intervals, are listed in Table 16.10. The computation of the χ^2 value ($\chi^2 = 12.785$) is also illustrated in the table. The number of groups is 7 ($n = 7$), and since the two distribution parameters $\bar{x}$ and S_x were determined by using these data ($k = 2$), the number of degrees of freedom is $v = n - k = 7 - 2 = 5$.

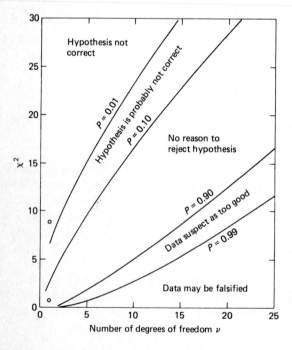

FIGURE 16.14
Probability of χ^2 values exceeding those shown as a function of the number of degrees of freedom.

TABLE 16.10
χ^2 **Computation for grouped ultimate tensile strength data**

Group interval	Number observed	Number expected	$(n_o - n_e)^2/n_e$
0–170.9	1	0.468	0.604
171–172.9	2	2.904	0.281
173–174.9	8	9.096	0.132
175–176.9	16	13.748	0.369
177–178.9	9	9.836	0.071
179–180.9	3	3.360	0.039
180 – ∞	1	0.588	0.289
			$1.785 = \chi^2$

Plotting these results ($v = 5$ and $\chi^2 = 1.785$) in Fig. 16.14 shows that the point falls in the region where there is no reason to expect that the hypothesis is not correct. The hypothesis is to represent the tensile strength with a normal probability function. The χ^2 test does not prove the validity of this hypothesis, but instead fails to disprove it.

The lines dividing the $\chi^2 - v$ graph of Fig. 16.14 into five different regions are based on probabilities of obtaining χ^2 values greater than the values shown by the curves. For example, the line dividing the regions "no reason to reject hypothesis" and "hypothesis probably is not correct" has been selected at a probability level of 10 percent. Thus, there is only 1 chance in 10 that data drawn from a population correctly represented by the hypothesis would give a χ^2 value exceeding that specified by the $P > 0.10$ curve. The hypothesis rejected region is defined with the $P > 0.01$ curve, indicating only 1 chance in 100 of obtaining a χ^2 value exceeding those shown by this curve.

The χ^2 function can also be used to question if the data have been adjusted. Probability levels of 0.90 and 0.99 have been used to define regions where "data suspect as too good" and "data may be falsified." For the latter classification there are 99 chances out of 100 that the χ^2 value will exceed that determined by a χ^2 analysis of the data.

The χ^2 statistic can also be used in contingency testing where the sample is classified under one of two categories—say pass or fail. Consider, for example, an inspection procedure with a particular type of strain gage where 10 percent of the gages are rejected due to etching imperfections in the grid. In an effort to reduce this rejection rate, the manufacturer has introduced new clean-room techniques that are expected to improve the quality of the grids. On the first lot of 2000 gages, the failure rate was reduced to 8 percent. Is this reduced failure rate due to chance variation, or have the new clean-room techniques improved the manufacturing process? A χ^2 test can establish the probability of the improvement being the result of random variation. The computation of χ^2 for this example is illustrated in Table 16.11.

TABLE 16.11
Observed and expected inspection results

Group interval	Number observed	Number expected	$(n_o - n_e)^2/n_e$
Passed	1840	1800	0.89
Failed	160	200	8.00
			$\overline{8.89} = \chi^2$

Plotting the results from Table 16.11 on Fig. 16.14 after noting that $v = 1$ shows that $\chi^2 = 8.89$ falls into the region "hypothesis is not correct." In this case the hypothesis is that there has been no improvement. The χ^2 test has shown that there is less than 1 chance in 100 of the improvement in rejection rate (8 percent instead of 10 percent) being due to random variables. Thus, one can with confidence conclude that the new clean-room techniques were effective in improving yield.

16.10   ERROR PROPAGATION

Previous discussions of error have been limited to error arising in the measurement of a single quantity: however, in many engineering applications, several quantities are measured (each with its associated error) and another quantity is predicted on the basis of these measurements. For example, the volume V of a cylinder could be predicted on the basis of measurements of two quantities (diameter D and length L). Thus, errors in the measurements of diameter and length will propagate through the governing mathematical formula $V = \pi D^2 L/4$ to the quantity (volume, in this case) being predicted. Since the propagation of error depends upon the form of the mathematical expression being used to predict the reported quantity, standard deviations for several different mathematical operations are listed below.

For addition and/or subtraction of quantities ($y = x_1 \pm x_2 \pm \cdots \pm x_n$), the standard deviation $S_{\bar{y}}$ of the mean $\bar{y}$ of the projected quantity y is

$$S_{\bar{y}} = \sqrt{S_{\bar{x}_1}^2 + S_{\bar{x}_2}^2 + \cdots + S_{\bar{x}_n}^2} \qquad (16.52)$$

For multiplication of quantities ($y = x_1 x_2 \cdots x_n$), the standard deviation $S_{\bar{y}}$ is

$$S_{\bar{y}} = (\bar{x}_1 \bar{x}_2 \cdots \bar{x}_n) \sqrt{\frac{S_{\bar{x}_1}^2}{\bar{x}_1^2} + \frac{S_{\bar{x}_2}^2}{\bar{x}_2^2} + \cdots + \frac{S_{\bar{x}_n}^2}{\bar{x}_n^2}} \qquad (16.53)$$

For division of quantities ($y = x_1/x_2$), the standard deviation $S_{\bar{y}}$ is

$$S_{\bar{y}} = \frac{\bar{x}_1}{\bar{x}_2} \sqrt{\frac{S_{\bar{x}_1}^2}{\bar{x}_1^2} + \frac{S_{\bar{x}_2}^2}{\bar{x}_2^2}} \qquad (16.54)$$

For calculations of the form $y = x_1^k$, the standard deviation $S_{\bar{y}}$ is

$$S_{\bar{y}} = k\bar{x}_1^{k-1}S_{\bar{x}_1} \tag{16.55}$$

For calculations of the form $y = x_1^{1/k}$, the standard deviation $S_{\bar{y}}$ is

$$S_{\bar{y}} = \frac{\bar{x}_1^{1/k}}{k\bar{x}_1}S_{\bar{x}_1} \tag{16.56}$$

Consider, for example, a rectangular rod where independent measurements of its width, thickness, and length have yielded $\bar{x}_1 = 2.0$ with $S_{\bar{x}_1} = 0.005$, $\bar{x}_2 = 0.5$ with $S_{\bar{x}_2} = 0.002$, and $\bar{x}_3 = 16.5$ with $S_{\bar{x}_3} = 0.040$ where all dimensions are in inches. Since the volume of the bar is

$$V = \bar{x}_1\bar{x}_2\bar{x}_3$$

the standard error of the volume can be determined by using Eq. (16.53). Thus,

$$S_{\bar{y}} = (\bar{x}_1\bar{x}_2\bar{x}_3)\sqrt{\frac{S_{\bar{x}_1}^2}{\bar{x}_1^2} + \frac{S_{\bar{x}_2}^2}{\bar{x}_2^2} + \frac{S_{\bar{x}_3}^2}{\bar{x}_3^2}}$$

$$= (2.0)(0.5)(16.5)\sqrt{\frac{(0.005)^2}{(2.0)^2} + \frac{(0.002)^2}{(0.5)^2} + \frac{(0.040)^2}{(16.5)^2}}$$

$$= 0.0875 \text{ in}^3$$

This determination of $S_{\bar{y}}$ for the volume of the bar can be used together with the properties of a normal probability distribution to predict the number of bars with volumes within specific limits.

The method of computing the standard error of a quantity $S_{\bar{y}}$ as given by Eqs. (16.52) to (16.56), which are based on the properties of the normal probability-distribution function, should be used where possible. However, in many engineering applications, the number of measurements that can be made is small; therefore, the data $\bar{x}_1, \bar{x}_2, \ldots, \bar{x}_n$ and $S_{\bar{x}_1}, S_{\bar{x}_2}, \ldots, S_{\bar{x}_n}$ needed for statistically based estimates of the error are not available. In these instances, error estimates can still be made but the results are less reliable.

A second method of estimating error when data are limited is based on the chain rule of differential calculus. For example, consider a quantity y that is a function of several variables:

$$y = f(x_1, x_2, \ldots, x_n) \tag{16.57}$$

Differentiating yields

$$dy = \frac{\partial y}{\partial x_1}dx_1 + \frac{\partial y}{\partial x_2}dx_2 + \cdots + \frac{\partial y}{\partial x_n}dx_n \tag{16.58}$$

In Eq. (16.58), dy is the error in y and $dx_1, dx_2, \ldots, dx_n$ are errors involved in the measurements of $x_1, x_2, \ldots, x_n$. The partial derivatives $\delta y/\delta x_1, \delta y/\delta x_2, \ldots, \delta y/\delta x_n$ can be determined exactly from Eq. (16.57). Frequently, the errors $dx_1, dx_2, \ldots, dx_n$ are estimates based on the experience and judgment of the experimentalist. An estimate of the maximum possible error can be obtained by summing the individual error terms in Eq. (16.58). Thus

$$dy\Big|_{max} = \left|\frac{\partial y}{\partial x_1} dx_1\right| + \left|\frac{\partial y}{\partial x_2} dx_2\right| + \cdots + \left|\frac{\partial y}{\partial x_n} dx_n\right| \tag{16.59}$$

Use of Eq. (16.59) gives a worst-case estimate of error, since the maximum errors $dx_1, dx_2, \ldots, dx_n$ are assumed to occur simultaneously and with the same sign.

A more realistic equation for estimating error is obtained by squaring both sides of Eq. (16.58) to give

$$(dy)^2 = \sum_{i=1}^{n} \left(\frac{\partial y}{\partial x_i}\right)^2 (dx_i)^2 + \sum_{i=1, j=1}^{n} \left(\frac{\partial y}{\partial x_i}\right)\left(\frac{\partial y}{\partial x_j}\right) dx_i\, dx_j \tag{16.60}$$

where $i \neq j$.

If the errors dx_i are independent and symmetrical with regard to positive and negative values, then the cross-product terms will tend to cancel and Eq. (16.60) reduces to

$$dy = \sqrt{\left(\frac{\partial y}{\partial x_1} dx_1\right)^2 + \left(\frac{\partial y}{\partial x_2} dx_2\right)^2 + \cdots + \left(\frac{\partial y}{\partial x_n} dx_n\right)^2} \tag{16.61}$$

16.11 SUMMARY

Statistical methods are extremely important in engineering, since they provide a means for representing large amounts of data in a concise form that is easily interpreted and understood. Usually, the data are represented with a statistical distribution function that can be characterized by a measure of central tendency (the mean $\bar{x}$) and a measure of dispersion (the standard deviation S_x). A normal or gaussian probability distribution is by far the most commonly employed; however, in some cases, other distribution functions may have to be employed to adequately represent the data.

The most significant advantage resulting from the use of a probability distribution function in engineering applications is the ability to predict the occurrence of an event based on a relatively small sample. The effects of sampling error are accounted for by placing confidence limits on the predictions and establishing the associated confidence levels. Sampling error can be controlled if the sample size is adequate. Use of Student's t distribution function, which characterizes sampling error, provides a basis for determining sample size consistent with specified levels of confidence. The Student's t distribution also permits a comparison to be made of two means to determine whether the observed difference is significant or whether it is due to random variation.

Statistical methods can also be employed to condition data and to eliminate an erroneous data point (one) from a series of measurements. This is a useful technique that improves the data base by providing strong evidence when something unanticipated is affecting an experiment.

Regression analysis can be used effectively to interpret data when the behavior of one quantity (say, y) depends upon variations in one or more independent quantities (say, $x_1, x_2, \ldots, x_n$). Even though the functional relationship between quantities exhibiting variation remains unknown, it can be characterized statistically. Regression analysis provides a method to fit a straight line or a curve through a series of scattered data points on a graph. The adequacy of the regression analysis can be evaluated by determining a correlation coefficient. Methods for extending regression analysis to multivariate functions exist. In principle, these methods are identical to linear regression analysis; however, the analysis becomes much more complex. The increase in complexity is not a concern, however, since computer subroutines are available that solve the tedious equations and provide the results in a convenient format.

Many probability functions are used in statistical analyses to represent data and predict population properties. Once a probability function has been selected to represent a population, any series of measurements can be subjected to a χ^2 test to check the validity of the assumed function. Accurate predictions can be made only if the proper probability function has been selected.

Finally, statistical methods for accessing error propagation were discussed. These methods provide a means for determining error in a quantity of interest y based on measurements of related quantities $x_1, x_2, \ldots, x_n$ and the functional relationship $y = f(x_1, x_2, \ldots, x_n)$ between quantities.

EXERCISES

16.1. Ten measurements of the fracture strength (ksi) of an aluminum alloy are:

25.0	25.2	24.9	25.5	24.6
24.8	25.2	25.0	24.8	25.0

Determine the mean, median, and mode which represent the central tendency of this data.

16.2. Verify the mean value $\bar{x}$ of the data listed in Table 16.1.

16.3. Determine the range R, mean deviation d_x, and variance S_x^2 for the data given in Exercise 16.1.

16.4. Determine the range R, mean deviation d_x, and variance S_x^2 for the data listed in Table 16.1.

16.5. Find the coefficient of variation for the data given in Exercise 16.1.

16.6. Find the coefficient of variation for the data listed in Table 16.1.

16.7. Consider a gaussian population with a mean $\mu = 100$ and a standard deviation $\sigma = 10$. Determine the probability of selecting a single sample with a value in the interval between

(a) 75–80   (b) 98–102   (c) 92–97
(d) 115–123   (e) greater than 125

16.8. Determine the percent of data which will probably be within limits of

(a) $\bar{x} \pm 2.5S_x$
(b) $\bar{x} - S_x$ and $\bar{x} + 1.5S_x$
(c) $\bar{x} - 1.5S_x$ and $\bar{x} + S_x$
(d) $\bar{x} \pm 0.5S_x$

16.9. A manufacturing process yields aluminum rods with a mean yield strength of 35,000 psi and a standard deviation of 1000 psi. A customer places a very large order for rods with a minimum yield strength of 32,000 psi. Prepare a letter for submission to the customer that describes the yield strength to be expected and outline your firm's procedures for assuring that this quality level will be achieved and maintained.

16.10. Determine the Weibull distribution function corresponding to the parameters $x_0 = 5$, $b = 5$, and $m = 10$. Plot $P(x)$ for $0 < x < 50$.

16.11. For the Weibull distribution of Exercise 16.10 predict the probability of failure if $x = 6.5$.

16.12. A Weibull distribution function describing the strength of a brittle ceramic uses $x_0 = 3$ ksi, $b = 2$ ksi, and $n = 3$. Compute the expected values of the 10 lowest strengths measured if a total of 400 specimens were tested.

16.13. Repeat 16.12 but let $b = 3$ ksi.

16.14. In calibrating a brittle coating 12 calibration beams are tested to determine that $\bar{x} = 470 \mu\epsilon$ and $S_x = 60 \mu\epsilon$. Determine the standard distribution of the mean $S_{\bar{x}}$. Give the confidence level associated with the statement that the true mean μ of the calibration was between 460 and 480 $\mu\epsilon$.

16.15. In the calibration test of the brittle coating in Exercise 16.14, the number of beams is increased from 12 to 40. The new results give $\bar{x} = 465 \mu\epsilon$ and $S_x = 70 \mu\epsilon$. Determine $S_{\bar{x}}$ and give confidence limits on the statement that the true threshold strain is between 455 and 475 $\mu\epsilon$.

16.16. Use Eq. (16.20) with 2 degrees of freedom and evaluate $f(t)$ as a function of t. Plot your results on Fig. 16.9. Select F_0 so that

$$\int_{-\infty}^{+\infty} f(t)\, dt = 1$$

16.17. Compare the results of Exercise 16.16 with the distribution function from a normal distribution.

16.18. Determine the sample size necessary to ensure that the average strength of a material exceeds 70 MPa if the sample is drawn from a population with an estimated mean of 65 MPa and an estimated standard deviation of 2 MPa. Use a confidence level of 5 percent.

16.19. An inspection laboratory samples two large shipments of dowel pins by measuring both length and diameter. For shipment A, the sample size was 40, the mean diameter was 6.12 mm, the mean length was 25.3 mm, the estimated standard deviation on diameter was 0.022 mm, and the estimated standard deviation on length was 0.140 mm. For shipment B, the sample size was 60, the mean diameter was 6.04 mm, the mean length was 25.05 mm, the estimated standard deviation on diameter was 0.034 mm, and the estimated standard deviation on length was 0.203 mm.

(*a*) Are the two shipments of dowel pins the same?

(*b*) What is the level of confidence in your predictions?

(*c*) Would it be safe to mix the two shipments of pins? Explain.

16.20. Repeat Exercise 16.19 for the following two shipments of dowel pins:

	Shipment A	Shipment B
Number	20	10
Diameter	$\bar{x} = 6.05$ mm, $S_x = 0.03$ mm	$\bar{x} = 5.98$ mm, $S_x = 0.04$ mm
Length	$\bar{x} = 24.9$ mm, $S_x = 0.22$ mm	$\bar{x} = 25.4$ mm, $S_x = 0.18$ mm

16.21. Fatigue tests of a component under simulated load conditions indicated that the mean failure load for the specified life was $\bar{F} = 4115$ N with an estimated standard deviation of 340 N. In service this component will be subjected to an average load of 2800 N but this load may vary and the estimated standard deviation is 700 N. Determine the reliability of the component over the specified life.

16.22. For the data in Exercise 16.21, determine the safety factor for a reliability of (*a*) 90, (*b*) 95, and (*c*) 99 percent.

16.23. The following sequence of measurements of atmospheric pressure (millimeters of mercury) was obtained with a barometer:

764.3	764.6	764.4	765.2	764.5
764.5	765.7	765.4	764.8	765.3
765.2	764.9	764.6	765.1	764.6

Determine the mean and the standard deviation after employing Chauvenet's criterion to statistically condition the data.

16.24. Determine the slope m and the intercept b for a linear regression equation $y = mx + b$ and the correlation coefficient R^2 for the data listed below:

x	1	2	3	4	5	6	7	8	9	10
y	2.1	3.9	6.4	8.2	9.5	12.0	13.6	16.7	17.9	19.4

16.25. The drying of a coating containing a solvent is a diffusion-controlled process which can be described by the equation $C = ae^{-mt}$ where C is the concentration (% solvent), t is the time (seconds), and a and m are diffusion constants. Use linear regression methods to determine a and m for the following data, which were obtained by weighing a sample of coating during drying:

t	100	200	300	400	500	700	1000
C	3.40	1.40	0.561	0.195	0.067	0.0097	0.0004

16.26. Determine the regression coefficients a, b_1, b_2, and b_3 and the correlation coefficient R^2 for the following data set:

y	x_1	x_2	x_3
6.8	1.0	2.0	1.0
7.8	1.5	2.0	1.5
7.9	2.0	2.0	2.0
8.0	2.5	3.0	1.0
8.3	3.0	3.0	1.5
8.4	3.5	3.0	2.0
8.5	4.0	4.0	1.0
8.6	4.5	4.0	1.5
8.9	5.0	4.0	2.0
9.1	5.5	5.0	1.0
9.3	6.0	5.0	1.5
9.6	6.5	5.0	2.0

16.27. Write a computer program which utilizes field data taken from a calibration disk in photoelasticity. The program should accept 20 data points $N(x, y)$, the model thickness h, and analytic functions $G(x, y)$ and $E(x, y)$. The output [see Eq. (16.40)] should give f_σ, A, B, and C.

16.28. Write a computer program to implement the nonlinear least-squares method. Use the mixed-mode relations given in Eq. (16.43) as the example. The program input is field data $N(r, \theta)$ and calibration data f_σ/h. The output is K_{I}, K_{II}, and σ_{0x}.

16.29. A die-casting operation produces bearing housings with a rejection rate of 4 percent when the machine is operated over an 8-h shift to produce a total output of 3200 housings. The method of die cooling was changed in an attempt to reduce the rejection rate. After 2 h of operation under the new cooling conditions, 775 acceptable castings and 25 rejects had been produced.
(a) Did the change in the process improve the output?
(b) How certain are you of your answer?

16.30. The stress σ_x at a point on the free surface of a structure or machine component can be expressed in terms of the normal strains ϵ_x and ϵ_y measured with electrical resistance strain gages as

$$\sigma_x = \frac{E}{1 - v^2}(\epsilon_x + v\epsilon_y)$$

If ϵ_x and ϵ_y are measured within ± 2 percent and E and v are measured within ± 5 percent, estimate the error in σ_x.

16.31. A gear-shaft assembly consists of a shaft with a shoulder, a bearing, a sleeve, a gear, a second sleeve, and a nut. Dimensional tolerances for each of these components are listed below:

Component	Tolerance, mm
Shoulder	0.050
Bearing	0.025
First sleeve	0.100
Gear	0.050
Second sleeve	0.100
Nut	0.100

    (*a*) Determine the anticipated tolerance of the series assembly.

    (*b*) What will be the frequency of occurrence of the tolerance of part *a*?

16.32. Estimate the error in determining the weight of a cylindrical rod if the dimensional measurements are accurate to 0.5 percent and the specific weight is accurate to 0.1 percent.

16.33. Estimate the error in determining strain from two displacement measurements if the displacements are measured with an accuracy of 2 percent and the positions x_1 and x_2 each are measured with an accuracy of 1 percent.

REFERENCES

Introductory Statistics Books

1. Blackwell, D.: *Basic Statistics*, McGraw-Hill, New York, 1969.
2. Snedecor, G. W., and W. G. Cochran: *Statistical Methods*, 6th ed., Iowa State University Press, Ames, 1967.
3. Zehna, P. W.: *Introductory Statistics*, Prindle, Weber & Schmidt, Boston, 1974.
4. Chou, Y.: *Probability and Statistics for Decision Making*, Holt, Rinehart & Winston, New York, 1972.

Statistical Books for Engineering, Production, and Research

5. Bethea, R. M., B. S. Duran, and T. L. Boullion: *Statistical Methods for Engineers and Scientists*, Dekker, New York, 1975.
6. Davies, O. L., and P. L. Goldsmith: *Statistical Methods in Research and Production*, Hafner, New York, 1972.
7. McCall, C. H., Jr.: *Sampling and Statistics Handbook for Research*, Iowa State University Press, Ames, 1982.

Books with Application of Statistics to Experimental Mechanics

8. Young, H. D.: *Statistical Treatment of Experimental Data*, McGraw-Hill, New York, 1962.
9. Bragg, G. M.: *Principles of Experimentation and Measurement*, Prentice-Hall, Englewood Cliffs, N.J., 1974.
10. Weibull, W.: *Fatigue Testing and Analysis of Results*, Pergamon Press, New York, 1961.
11. Holman, J. P.: *Experimental Methods for Engineers*, 2d ed., McGraw-Hill, New York, 1966.
12. Durelli, A. J., E. A. Phillips, and C. H. Tsao: *Theoretical and Experimental Analysis of Stress and Strain*, McGraw-Hill, New York, 1958.
13. "A Tentative Guide for Fatigue Testing and the Statistical Analysis of Fatigue Data," ASTM publication STP-91-A, 1958.

Selected Articles

14. Landes, J. D., and D. H. Shaffer: "Statistical Characterization of Fracture in the Transition Region," Scientific Paper 79-ID3-JINTF-P4, Westinghouse Research Center, 1979.
15. Andrews, W. R.: "Small Specimen Brittle Fracture Toughness Testing," Report No. DF 79MPL250, Materials and Processing Laboratory, General Electric Co., 1979.
16. Sanford, R. J.: Application of the Least Squares Method to Photoelastic Analysis, *Exp. Mech.*, vol. 20, pp. 192–197, 1980.
17. Sanford, R. J., and J. W. Dally: A General Method for Determining Mixed-Mode Stress Intensity Factors from Isochromatic Fringe Patterns, *Eng. Fract. Mech.*, vol. 11, pp. 621–633, 1979.

INDEX